Fuels and Petrochemicals Division

Core Programming Topic at the 2011 AIChE Annual Meeting

Minneapolis, Minnesota, USA
16-21 October 2011

ISBN: 978-1-61839-736-2

Printed from e-media with permission by:

Curran Associates, Inc.
57 Morehouse Lane
Red Hook, NY 12571

Some format issues inherent in the e-media version may also appear in this print version.

Printed by Curran Associates, Inc. (2012)

AIChE
3 Park Avenue
New York, NY 10016-5991

Phone: (203) 702-7660
Fax: (203) 775-5177

www.aiche.org

Additional copies of this publication are available from:

Curran Associates, Inc.
57 Morehouse Lane
Red Hook, NY 12571 USA
Phone: 845-758-0400
Fax: 845-758-2634
Email: curran@proceedings.com
Web: www.proceedings.com

TABLE OF CONTENTS

(38a) Designing Nanostructures for High Performance Batteries and Supercapacitors

Yi Cui, Materials Science and Engineering, Stanford University, Stanford, CA

Storing energy electrochemically involves electronic and ionic processes and chemical transformation inside and at the interface of materials. The ability to engineer nanostructures and their interfaces afford the great opportunities for controlling these fundamental processes, which can ultimately lead to high performance devices such as batteries and supercapacitors. Here I will present several examples on designing nanostructures and their interfaces to realize high performance energy storage devices. One example is on designing nanowires and heterostructured nanowires for ultrahigh capacity storage of lithium ions in silicon battery anodes. The challenges of large volume expansion and solid-electrolyte-interphase (SEI) have been overcome. In another example, we have designed nanoporous carbon to encapsulate sulfur for lithium ion battery cathodes, which can addresses many issues related to sulfur. In the third example, paper and textile batteries and supercapacitors have been developed.

See more of this Session: Nanomaterials for Energy Storage I
See more of this Group/Topical: Topical 5: Nanomaterials for Energy Applications

Materials with Hierarchical Porosity for Energy Storage and Conversion

Andreas Stein, Jane Davidson,
Nicholas Petkovich, Yuqiang Qian, Stephen Rudisill, Luke Venstrom, Anh Vu
University of Minnesota, Minneapolis, MN 55455, U.S.A.

Introduction

Materials containing pores in size ranges from micropores to macropores offer interesting features for a number applications involving energy storage and conversion, such as batteries, fuel cells, hydrogen storage, and sunlight–to–fuel conversion. Depending on the specific application, they can provide large surface areas for reaction, interfacial transport, or dispersion of active sites; they can provide nanostructured features which enhance reactivity, alter materials properties, or shorten diffusion paths; they can act as host materials to stabilize other active components; or, in the case of porous carbons, they can provide electrically conductive phases as well as intercalation sites. However, the higher reactivity brought about by nanostructured features can also lead to decreased stability, particularly in applications where phase changes may alter the structure of the material or where high temperatures are employed. This talk will focus on two redox systems (lithium-ion batteries and sunlight-to-fuel conversion), in which templated porous materials provide a platform for either electrical energy storage or light-to-chemical energy conversion. In the first system the role of pore architecture in carbon-based electrodes will be discussed. In the second system, we investigate the role of porosity in ceria-based materials of interest for solar thermal splitting of water or carbon dioxide to produce hydrogen or carbon monoxide fuels, respectively.

Experimental Section

Three-dimensionally ordered macroporous (3DOM) carbon monoliths and 3DOM carbon with mesoporous walls (3DOM/m C) were synthesized by colloidal crystal templating and nonionic surfactant templating as decribed previously.[1] Tin, tin oxide, sulfur or $LiFePO_4$ were incorporated in the porous carbon materials by solution impregnation, vapor phase impregnation or direct triconstituent syntheses.[2,3] 3DOM CeO_2 and 3DOM $Ce_{0.8}Zr_{0.2}O_2$ powders were synthesized by colloidal crystal templating from $CeCl_3{\cdot}7H_2O$ and $ZrOCl_2{\cdot}8H_2O$ precursors. These samples were compared to a mesoporous CeO_2 sample prepared by thermal decomposition of $Ce(NO_3)_3{\cdot}6H_2O$ in air at 723 K (DS CeO_2) and a commercial polycrystalline ceria sample (Alfa-Aesar, 99.9%). In-situ ceria reduction with hydrogen and oxidation with water or CO_2 were carried out in a fixed bed, vertical tubular flow reactor. Additional details about syntheses, characterization and test reactions are provided in references 1–4.

Results and Discussion

Improved rate capabilities for lithiation/delithiation are observed for hierarchically porous carbon electrodes. In composites with tin or tin oxide for anodes of lithium ion batteries, these structures maintain electrical contact between tin-based particles, even when those particles undergo significant volume changes during cycling, and hence the composite anode maintains good capacities over multiple cycles (Figure 1). Composites of hierarchically structured carbon with poorly conducting but otherwise desirable electrode materials (like sulfur or $LiFePO_4$) can be used to overcome limitations in electrical conductivity of those materials, increasing the choice of useful electrode materials.

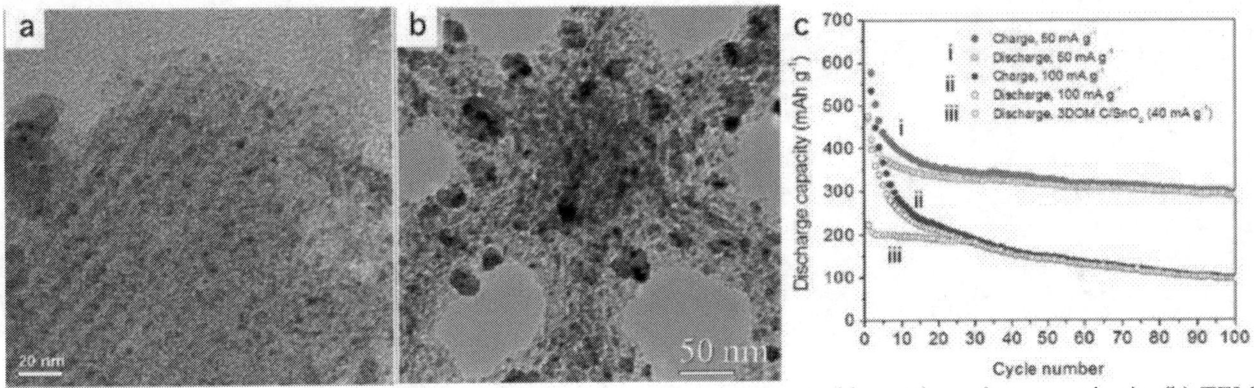

Figure 1. (a) TEM image of a mesostructured carbon-tin composite prepared by a triconstituent synthesis. (b) TEM image of a carbon-tin oxide composite structure with hierarchical porosity. (c) Lithiation charge and discharge curves of 3DOM/m C/SnO_2 (i), 3DOM C coated with SnO_2 (ii) and 3DOM carbon (iii) as function of cycle number.[2]

In reactions targeting the production of chemical fuels (H_2, CO) from water and carbon dioxide, the heterogeneous conversion processes benefit from added surface area in 3DOM CeO_2 samples that facilitate the oxidation cyle. While overall productivity does not depend strongly on sample morphology (given sufficient reaction time), the addition of porosity improves peak fuel production rates and reactivities (Figure 2). However, a mesoporous ceria sample displaying an even higher surface area (DS CeO_2) quickly lost activity as the material sintered during the reaction. The loss in surface area due to sintering of 3DOM CeO_2 was less and could be improved further by doping of ceria with zirconia, which stabilized the porous structures at the high reaction temperatures. The interplay between composition and morphology of these materials, thermal stability, and conversion efficiencies will be discussed.

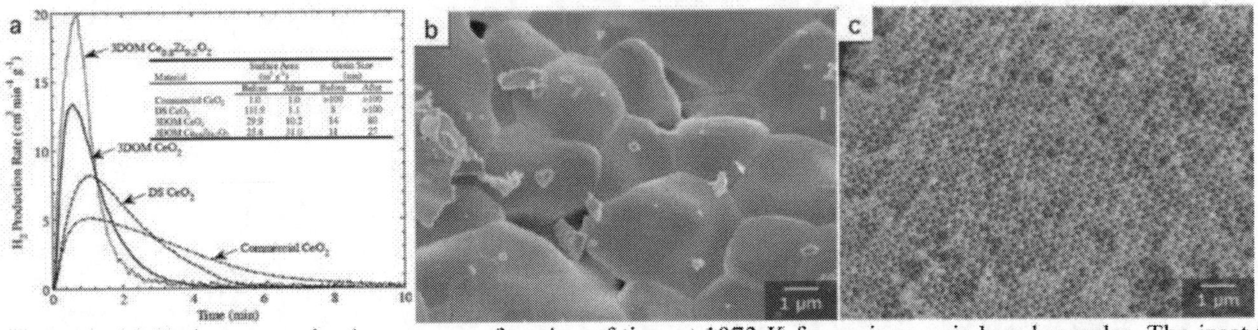

Figure 2. (a) Hydrogen production rate as a function of time at 1073 K for various ceria-based samples. The inset tabulates specific surface areas and grain sizes of the various ceria-based samples before redox cycling and after redox cycling at 1073 K. The SEM images show (b) a chemically cycled commercial CeO_2 sample and (c) a cycled 3DOM $Ce_{0.8}Zr0._{2}O_2$ sample.[4]

References

1. Wang, Z.; Stein, A. *Chem. Mater.* **2008**, *20*, 1029-1040
2. Wang, Z.; Fierke, M. A.; Stein, A. *J. Electrochem. Soc.*, **2008**, *155*, A658-A663.
3. Vu, A.; Stein, A. *Chem. Mater.* **2011**, *23*, 3237-3245.
4. Venstrom, L.; Petkovich, N.; Rudisill, S.; Stein, A.; Davidson, J. H. *J. Sol. Energ.-T. ASME* **2011**, submitted.

(609c) Solar Cells Based On Sensitized Nanowire Arrays

Jason B. Baxter, Department of Chemical and Biological Engineering, Drexel University, Philadelphia, PA

Solar cells can provide an abundant, clean, and sustainable source of electricity, but high costs have limited their implementation. The use of sensitized nanostructured architectures may enable both low-cost processing and high efficiency by decoupling the functions of light harvesting and charge transport into different materials. Sensitized solar cells consist of a bicontinuous interpenetrating network of electron- and hole-transporting materials with an interfacial absorber layer. In this architecture, interfacial recombination is the dominant loss process, so controlling the interfacial chemistry during deposition is critical. I will present our group's work on the use of ZnO nanowire arrays sensitized by thin CdSe coatings to produce extremely thin absorber (ETA) solar cells. Nanowire arrays can provide direct pathways for electron transport as well as sufficient surface area for sensitization. Progress in sensitized nanowire-based solar cells will be presented, as well as some of the challenges for improving their efficiencies. Overcoming these challenges will require detailed fundamental understanding and control of charge transfer in the materials and interfaces within the cells.

See more of this Session: Semiconductor Nanowires and Nanotubes for Next Generation Photovoltaics - From Fundamentals to Devices
See more of this Group/Topical: Topical 5: Nanomaterials for Energy Applications

(470b) Advanced Nanostructures for Thermoelectric Energy Harvesting

Yue Wu, School of Chemical Engineering, Purdue University, West Lafayette, IN

Substantial efforts have been devoted to use nanostructured materials for thermoelectric energy harvesting and solid-state cooling in the past decade. However, the majority of the research is still limited in lab scale due to the incapability to mass-produce well-defined nanostructured materials with low yet industrial-compatible process. In addition, a lot of widely used thermoelectric materials contain toxic and expensive elements that prevent the large-scale deployment of the thermoelectric devices. In this presentation, we will discuss our research on the development of mass production of molecular scale nanostructures of chalcogenides and metal oxides, as well as their heterostructures, for the manufacture of thermoelectric generators operating at different temperature ranges. Particularly, we will talk about our approach to discover and investigate the non-toxic and abundant nanostructured materials to achieve an environmentally friendly process. Our preliminary research indicated that thermoelectric figure of merit (ZT) close to 2 can be achieved in the molecular scale nanowires of certain chalcogenide due to significantly enhanced power factor and reduced thermal conductivity as a result of quantum confinement.

See more of this Session: Nanomaterials for Thermal-to-Electric Conversion I
See more of this Group/Topical: Topical 5: Nanomaterials for Energy Applications

DEEPWATER REGULATIONS IN THE GULF OF MEXICO

Ian Sutton
AMEC Paragon
Houston, TX 77041
ian.sutton@amec.com

Summary

The Deepwater Horizon incident (DWH) of April 2010 led to many changes in deepwater operations in the Gulf of Mexico. Not the least of those changes were to do with regulations. First, the agency that had the primary responsibility for offshore safety, the Minerals Management Service (MMS), was renamed the Bureau of Environmental Management and Regulatory Enforcement (BOEMRE). The revenue collection and safety enforcement functions of the old MMS were split away from the safety management role so as to avoid perceived conflict of interest problems. Associated with this change was the installation of new senior management at the agency.

The second short-term consequence of DWH was that BOEMRE put in place a drilling moratorium and issued various specific instructions to do with blowout preventers. At the time of writing (April 2011) the moratorium has been lifted, but very few licenses for drilling have yet been issued. This restriction continues to be a continuing source of tension between industry and the agency, although, at the time of writing, it does appear as if a path forward is developing.

Not only did DWH generate short-term regulatory initiatives, it also led to the creation of longer term rule changes to do with process safety. In particular, BOEMRE now requires all companies operating on the Outer Continental Shelf (OCS) to prepare and implement a Safety and Environmental Management Standard (SEMS). This standard covers all types of offshore activity, including drilling, production and certain pipelines.

As can be seen from Table 1 SEMS is very similar in structure to the Process Safety Management (PSM) used in onshore processing facilities such as refineries and chemical plants.

Table 1
SEMS and PSM

SEMP	PSM
1. Safety and Environmental Information	1. Employee Participation
2. Hazards Analysis	2. Process Safety Information
3. Operating Procedures	3. Process Hazards Analysis
4. Training	4. Operating Procedures
5. Pre-Startup Review	5. Training
6. Assurance of Quality and Mechanical Integrity of Equipment	6. Contractors
7. Safe Work Practices	7. Prestartup Safety Review
8. Management of Change	8. Mechanical Integrity
9. Investigation of Incidents	9. Hot Work
10. Emergency Response and Control	10. Management of Change
11. Audit of Safety and Environmental Management Program Elements	11. Incident Investigation
12. Records and Documentation	12. Emergency Planning and Response
	13. Compliance Audits
	14. Trade Secrets

Although there are differences in wording between the two systems (for example *Pre-Startup Review* and *Prestartup Safety* Review) the similarities between the two lists are striking. And some of the differences are industry-specific. For example, the offshore oil and gas industries do not generally use proprietary chemical processes, so there is no need for the *Trade Secrets* element in SEMS.

At different times different process industries have shown leadership in the development of Safety Management Systems. For example, the tragedy at Bhopal in India in the year 1984, followed by a series of accidents in the United States, generated sufficient public concern that the chemical and refining industries made great strides in the ensuing decade in the development of PSM programs — particularly OSHA's PSM standard (29 CFR 1910.119), introduced in the year 1992.

It is suggested in this paper that currently the offshore oil and gas industry finds itself in the role an industry leader with respect to Safety Management Systems for the following reasons.

- As drill rigs and production platforms move to deeper and deeper water, the technical challenges that they face are increasingly formidable — and the industry's Safety Management Systems have to keep up with those challenges. This includes making a clear distinction between Occupational, Process and Technical Safety. It also means understanding the role of safety cases as distinct from management systems such as those shown in Table 1.
- There is an increasing need to understand risk not just in terms of safety, but also financial loss.
- The Deepwater Horizon incident shook up the industry. One of the many consequences of that event was the introduction of a new SMS rule called the *Safety and Environmental Management Systems (SEMS)* that goes into force November 15th 2011.

Historical Background

It is an unfortunate truth that Safety Management Systems are generally introduced following major accidents. The closing of stable doors invariably follows the bolting of horses. In the offshore oil and gas industry the two most important events were Piper Alpha (1988) and Deepwater Horizon (2010). The first demonstrated the need for Safety Management Systems, the second showed that the systems that had been developed and put in place over the last twenty years were not sufficient; further improvements were and are needed.

In order to provide a context for the safety management systems used offshore, a very brief overview of the industry is provided below.

Development of the Offshore Oil and Gas Industry

The first offshore drilling rigs and production platforms were installed in the 1890s in fresh water lakes in the State of Pennsylvania. The first marine production facilities were probably off the coast of California at the close of the 19th century.

Gulf of Mexico

Development of offshore oil and gas in the Gulf of Mexico (GoM) started off the coast of Louisiana in the 1920s and 30s. These early GoM platforms were in very shallow water — indeed they are often described as marsh platforms, being offshore only when the tide is in.

But it was in the year 1947 that the first true offshore facility was developed. Kerr-McGee (now part of Anadarko) working with Phillips Petroleum (now part of ConocoPhillips) and Stanolind (now part of BP) developed the first drilling and production facility out of the sight of land (although the water depth was only 20 feet).

In the years that followed that first platform the oil and gas industry moved gradually into deeper and deeper waters, with many thousands of platforms being built.

Virtually all the facilities in this time period were fixed, *i.e.,* they had legs that rested on the sea bottom.

The management culture that was followed in the GoM was largely that of the onshore oil and gas industry (sometimes referred to as the "oil patch"). This culture affected the development and management of safety programs.

North Sea

In the 1960s gas, and then oil, were discovered in the North Sea. The first discoveries were in Dutch waters, but the largest developments were in the UK (mostly Scottish) and then Norwegian sectors. Unlike the Gulf of Mexico, the development of prospects in the North Sea was not gradual. Nor was there an existing onshore safety culture to draw from.

From the beginning conditions in the North Sea were challenging. Work was carried out in what was then considered to be deep water, often in very severe weather conditions. There were relatively few platforms compared with the GoM, but those platforms were generally bigger and had more on-board processing equipment.

Safety was managed through the use of Safety Cases. In other words, conditions on each rig and platform were analyzed separately and, based on those analyses, design and operating conditions were determined.

In both the Gulf of Mexico and the North Sea the industry developed without many major incidents.

And then came Piper Alpha.

Piper Alpha (1988)

Piper Alpha was a production platform located northeast of Aberdeen, Scotland. The incident, in which 167 men died and the platform itself went to the bottom of the sea, showed unequivocally that the manner in which safety was being managed offshore needed drastic improvement.

The official investigation into the Piper Alpha catastrophe was headed by Lord Cullen, a Scottish high court judge. The report of his committee was published in 1990, and was highly critical of the safety cases that had been developed and implemented up until that time. The report strongly recommended that the quality of offshore safety cases be radically upgraded; indeed, it is probably not an exaggeration to say that offshore safety cases as they are known now got their start in the early 1990s.

The Piper Alpha accident had an equally dramatic impact on the management of safety in United States waters. However, because the Gulf of Mexico has so many rigs and platforms, and because there are so many similarities between those facilities, it was felt that a more prescriptive approach was appropriate. Therefore, over a period of years a series of industry standards, mostly from the American Petroleum Institute (API), were published. Although some of these standards have been incorporated into law they are generally applied voluntarily.

From the point of view of Safety Management Systems, the most important of the standards was API's Recommended Practice (RP) 75, Development of a Safety and Environmental Management Program for Offshore Operations and Facilities. RP 75 recommended that companies develop a Safety and Environmental Management Program, or SEMP. Table 1 showed that there is a high degree of similarity between SEMP and OSHA's PSM.

So, by the mid 1990s the two safety cultures had been developed, and were following their different paths. And both approaches were successful. For example, Figure 1 shows safety trends for the U.S. Outer Continental Shelf. (It should be noted that Figure 1 reports mostly *occupational* safety as distinct from *process* safety data.)

Figure 1
Safety Trends — United States Outer Continental Shelf

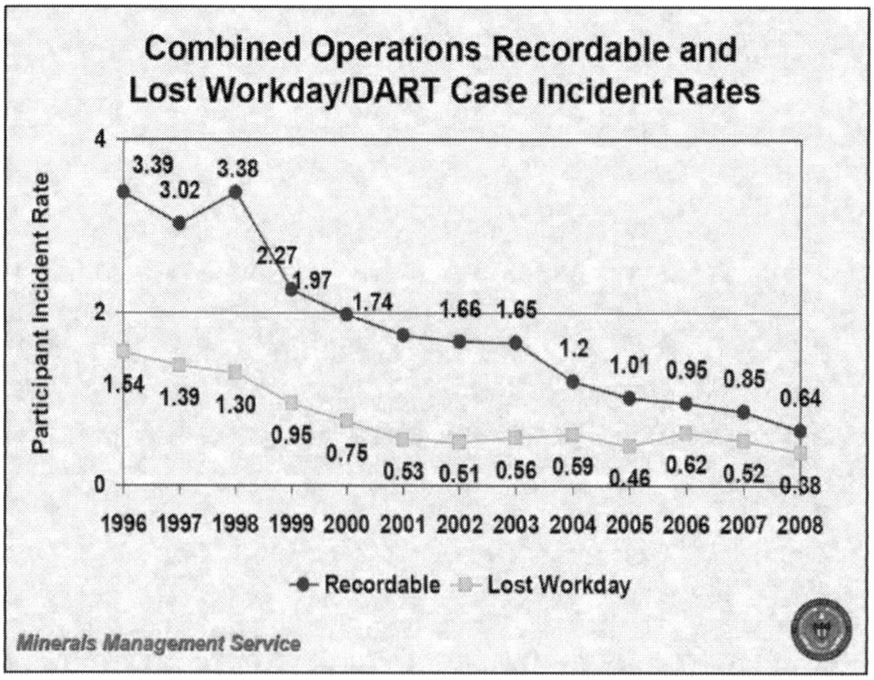

The above chart was published by the United States Mineral Management Service (MMS) — the government agency responsible for offshore safety regulation and enforcement at that time. It provides data to do with the number of offshore safety incidents for the period 1996-2008. The trend is impressive: in just a twelve year period the recordable injury rate declined from 3.39 to 0.64: a drop of around 80%. The number of lost workdays dropped by a similar percentage. Moreover the trend is quite smooth and steady, showing that the results are not a fluke or one-time event. And these advances were made as the industry has worked in ever-more challenging conditions — particularly as it has moved into very deep water operations.

And then came Deepwater Horizon

Deepwater Horizon

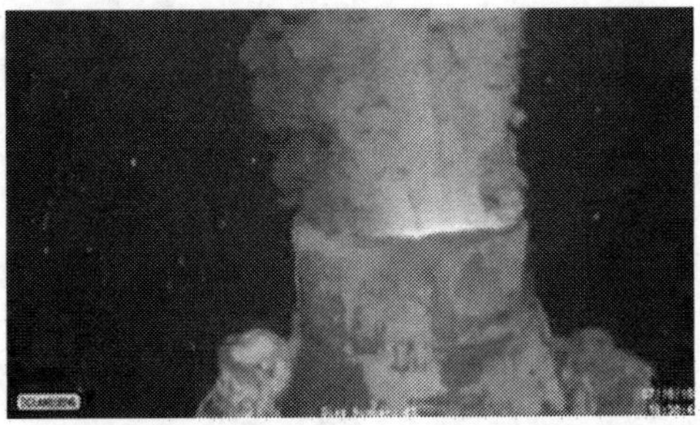

The Deepwater Horizon incident that occurred in April 2010 is still fresh in the minds of most process safety professionals. A world-scale drilling rig suffered a blowout in which eleven men died, an enormous amount of oil was spilled into the ocean, and close to $1 billion of investment wound up on the bottom of the sea.

From the point of view of this paper, the key lesson that came out of this event is that it showed that industry's safety programs had not "arrived" — major changes and improvements needed to be made.

TECHNOLOGY

As the world's sources of low cost, easily extracted oil are used up the industry is forced to find oil and gas in locations of ever increasing difficulty and complexity. These include the Canadian tar sands, arctic waters and deepwater exploration in the Gulf of Mexico, and other areas such as offshore Brazil. This means that those responsible for developing and running Safety Management Systems need to make sure that those systems match the challenges that the new technology poses.

Subsea Challenges

For those working with production platforms topsides these changes are not so dramatic since the manner in which oil, gas and water are processed — most of the major changes are occurring subsea, as illustrated by the focus that BOEMRE and others placed on ensuring reliable Blowout Preventer (BOP) operations in the weeks following the Deepwater Horizon disaster.

In a different vein, almost all the post-accident regulatory and political discussions were to do with drilling, particularly the "moratorium" on new drilling activities. This would have been an opportunity for the safety management community to have helped with a formal analysis of the usefulness of the moratorium.

Types of Safety

A small, but useful, lesson for the process industries overall comes from the clear distinction that is made offshore between occupational, process and technical safety — particularly in the preparation of safety cases.

The distinction between occupational and process safety was highlighted in the Baker report (2007), written after the 2005 explosion at a Texas City refinery.

> *"BP's executive management tracked the trends in BP's personal safety metrics, and they understood that BP's performance in this regard was both better than industry averages and consistently improving. Based upon these trends, BP's executive management believed that the focus on metrics such as OSHA recordables . . . were largely successful. With respect to personal safety, that focus evidently was effective. BP's executive management, however, mistakenly believed that injury rates, such as days away from work case frequency and recordable injury frequency, were indicators of acceptable process safety performance . . . it was not until after the Texas City accident that management understood that those metrics do not correlate with the state of process safety."*

However, a further distinction can be made between process and technical safety. Most of the elements of a process safety program such as OSHA's PSM or BOEMRE's SEMS are to do with operational management. These elements include topics such as operating procedures, management of change and pre-startup reviews. However, the best way of achieving safety is to design it in in the first place (Sutton 2010). This is done through the use of technical safety work such as fire and explosion modeling, vapor dispersion analysis and the design of escape routes.

This distinction between process and technical safety is very clearly defined in safety case work, where technical issues are grouped under the title *Formal Safety Analysis.*

Financial Risk

Offshore oil and gas production in the Gulf of Mexico is very big business indeed. Figure 2 (Oil Drum 2011) shows the production of oil from various States, along with the production from various offshore platforms.

Figure 2
Oil Production

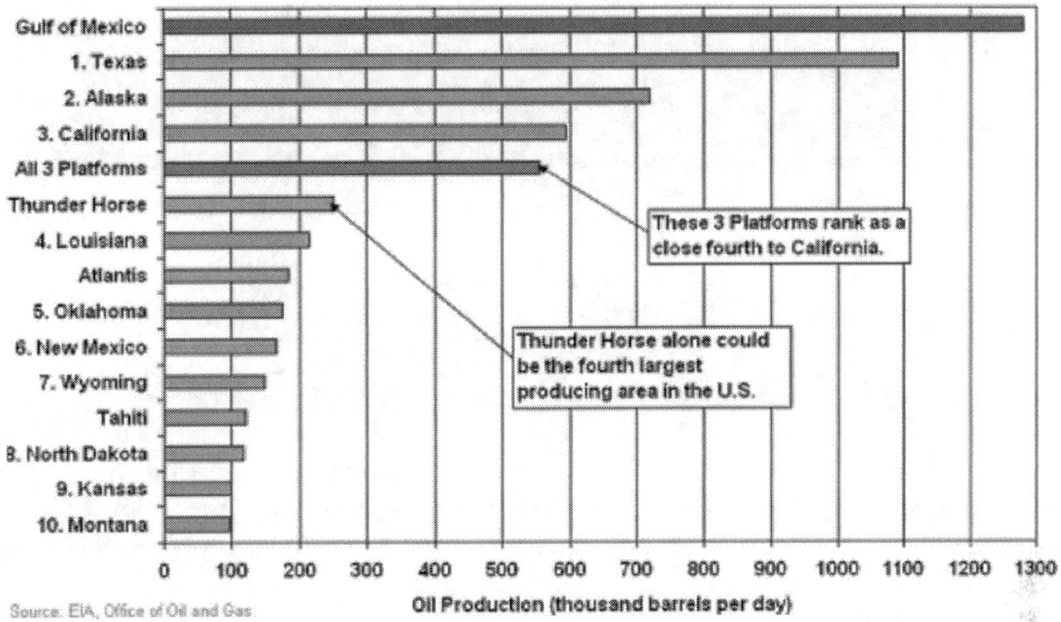

Some interesting, even startling, conclusions can be drawn from Figure 3.

- The Thunder Horse platform alone produces more oil than the entire state of Louisiana.
- The production of oil from just three platforms (Thunder Horse, Atlantis and Tahiti) nearly equals that of the state of California.
- Atlantis produces more oil than the state of Oklahoma.

Using Atlantis as an example and assuming a crude oil price of $100 per barrel, the production of 180,000 bbl/d from that platform is worth $18 million per day of revenue. That's a lot of money, even for a major oil company.

Yet there is a risk associated with these very large platforms. Figure 3 shows the platform Thunder Horse when she nearly sank in the year 2005. A hurricane had been approaching the facility and, as is standard policy, the crew was evacuated. Ironically the hurricane changed direction and did not hit the platform. While the crew was absent a series of mechanical and instrumentation errors led to water being pumped onboard into two of the legs. Fortunately, on its return the crew was able to correct the problem and save the platform. Still, the damage to onboard equipment and instrumentation was very serious and it took a long time before Thunder Horse was brought back into production.

Figure 3
Thunder Horse

Had the platform sunk its replacement cost would probably have been in excess of $2 billion, and production would have been curtailed for years. Even so, the operators must have lost hundreds of millions of dollars in revenue as a result of the facility being out of service for so long.

SEMS

One of the many consequences of the *Deepwater Horizon* catastrophe was that the Minerals Management Service (MMS) was renamed the Bureau of Ocean Energy Management, Regulation and Enforcement, or BOEMRE. As already noted, most of the energies of the new agency were focused on improving drilling standards, but they also decided to implement a Safety and Environmental Management Systems (SEMS) standard. (They had started work on this in 2006, but were not moving expeditiously until Deepwater Horizon occurred.)

Final Rule

Essentially the SEMS rule converts the existing Recommended Practice 75 to a requirement. (Its full title is Oil and Gas and Sulphur Operations in the Outer Continental Shelf — Safety and Environmental Management Systems (SEMS). It is Subpart S of 30 CFR Part 250.) The key sentence in the new rule is,

> *BOEMRE is incorporating by reference, and making mandatory, the American Petroleum Institute's Recommended Practice for Development of a Safety and Environmental Management Program for Offshore Operations and Facilities (API RP 75), Third Edition, May 2004, reaffirmed May 2008. This recommended practice, including its appendices, constitutes a complete Safety and Environmental Management System (SEMS) program.*

The updated rule was published by the BOEMRE in October 15th 2010 with an effective date of November 15th 2010. Covered facilities were given one year in which to comply, *i.e.,* their programs should be finalized by November 15th 2011. It should be noted that nothing actually happens on that date — it is not like a tax document that must be mailed in on or before the deadline. However, once November 15th 2011 has passed, owner/operators are subject to full SEMS audits and incident investigations.

As the final SEMS rule was being developed during mid 2010 there was considerable discussion as to whether it would include a Safety Case requirement. In the end it did not do so.

Content of the Rule

Table 1 shows that the SEMS rule is technically very similar to the OSHA Process Safety Management standard. This similarity is likely to increase; BOEMRE has stated that they wish to align the rule more closely with the corresponding standards from OSHA and the EPA.

Operators / Contractors

The area of greatest discussion within the industry regarding SEMS concerns the relationship between operators and contractors. The rule places the full responsibility for compliance with the operators (lessees) of the offshore facilities. In most cases the operators is a large company that has its own safety systems and that can ensure that its contractor conform to the company and SEMS standard. However, the operator of a platform can also be a small organization, sometimes little more than a holding company. In such cases it is unlikely that the operator has the systems or depth of resources to manage a large number of organizations — many of whom are large organizations in their own right.

What is needed are bridging documents between the operators and contractors. However, given the very large number of companies involved it is suggested that all parties bridge through SEMS, as shown in Figure 4 (Sutton 2011).

Figure 4
Bridging through SEMS

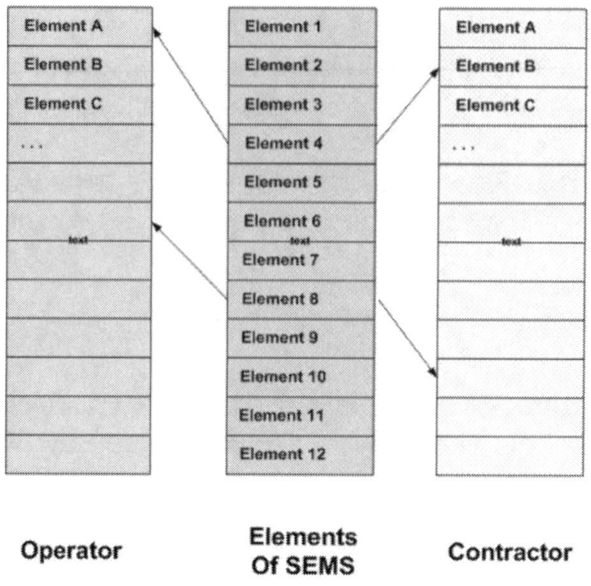

MANAGEMENT AND LEADERSHIP

This paper started with an overview of the two events that had the greatest impact on offshore safety standards and regulations: Piper Alpha and Deepwater Horizon. These two events led to the creation of safety cases, SEMP/SEMS and many other types of Safety Management System. Yet achieving safety goes beyond just management. With regard to Piper Alpha, for example, there is no doubt that the management systems such as the use of work permits needed major improvement. But the lessons to be learned from that event go beyond mere management, and should address questions such as:

1. It is thought that 7 men died in the initial explosion, yet the final death toll was 167. What led to those additional 160 deaths?
2. The inventory of oil and gas on a production platform such as Piper Alpha is quite low — typically an on-board fire should burn itself out within 20 to 30 minutes. Why did Piper burn for hours before eventually sinking?
3. Many of the deaths were caused by smoke inhalation. Workers escaped from the fire to the living quarters which filled with smoke. It is likely that some of those lives could have been saved had someone on board led the workers to locations where they could have escaped to the sea. Why were the men left to die?

The answer to the first two questions is that Piper Alpha was a hub platform, and that other platforms pumped their oil and (hydrocarbon) condensate to Piper where the combined materials were sent to the beach. One of those satellite platforms continued to pump oil and condensate into the raging fire, even though they could see the fires and explosions on board Piper. The managers on these other platforms failed to show leadership.

With regard to the third question, no one showed the necessary leadership to help evacuate the men from the smoke-filled quarters.

Similarly with Deepwater Horizon. The event did not happen instantaneously; there had been warning signs for as much as two days that the well was out of control. A strong leader would have taken the decision to idle the $1 million/day rig, and find out what was going on, and then take the corrective action.

There is one incident that does, in fact, show how leadership can be applied offshore. That incident is Blackbeard West. This non-incident occurred in the year 2005 in the Gulf of Mexico about 28 miles off the coast of Louisiana, and about 70 miles from the site of the future Deepwater Horizon event. The field was a large one, and promised very attractive financial returns. However, the drill crew reported that well pressures and temperatures were very high and they were not sure that they could control the situation. The decision as to what to do wound up on the desk of the CEO of the operating company (ExxonMobil). They decided to declare a dry hole, and they shut in the prospect. The senior managers were willing to, "take a risk for safety".

Conclusions

This paper has explored some of the similarities and differences between the management of safety offshore and onshore. Both disciplines have much to contribute. The offshore industry provides lessons to do with:

- Technical issues — particularly the drilling and production of oil and gas from extremely deepwater wells.
- A distinction between safety cases and more traditional approaches to process safety, with a particular focus on the use of Formal Safety Analysis.
- The need to manage the financial as well as the safety aspects of risk.
- The issues arising from the new SEMS rule.
- The distinction between leadership and management.

Given the pace at which this industry is developing new technology and new management systems and new regulations, it is likely that the offshore oil and gas industry will continue to be a "Happening Place".

Citations

Baker, J. January 2007. "The Report of the U.S. Refineries Independent Safety Review Panel".

Oil Drum. August 7, 2011. http://www.theoildrum.com/

Sutton, I. Offshore Safety Management. Elsevier. 2011.

Who Are You Leading? How Do You Know?

Jack Hipple

Innovation-TRIZ
Tampa, FL

Successful leadership requires a clear vision, committed followers, and resources. Without all three the leadership goal may not be accomplished. Very frequently, a clear vision is established and resources are provided but the followers do not follow due to a disconnect between the style of the leader and the styles of the followers and team members. This situation can be resolved by changing the followers, but this can be costly and time consuming. It is not always possible for a leader to choose or change the followers that are needed. Command and control is one approach and the success of this can be a function of the supply and demand of the job market at the time. A far more reliable and robust approach is for the leader to take the time to understand his or her own style and behavior preference as well as that of their followers. Understanding these can help a leader be successful under any circumstance with any group. This presentation will overview some key behavior preference and style measurements that accurately assess social and problem solving styles and how they can be used to assist leaders and followers in achieving this goal.

Leadership and People

We use this term all the time without giving sufficient thought to what it really means. It may mean getting a group of people to see something that is not obvious to them, but is important to an organization and its senior management. It may be responsibility for leading a multi-national and multi-cultural team in building a multi-billion dollar capital project involving thousands of engineers in many locations. It may mean trying to figure out what the next new business, product, or service an organization should invest in when there is no clear indication of what it should be. However, all these responsibilities, and many other leadership roles involve one thing in common—the influencing and direction of people. A leader frequently does not get to choose who is on their team and occasionally needs the support of people who do not directly report to the leader. People are different. No two individuals are the same. They have technical backgrounds, different professional experiences, different personal preferences, and different biases. In order to lead a small team or a large multi-national group, it is imperative that a leader understand these differences for two primary reasons:

1. Task assignment. Everyone is not good at everything. Everyone has different preferences for what they would prefer to do. Assigning people responsibilities to people who do not enjoy that type of work when not necessary will only produce a drop in productivity and team morale.
2. Team relations. The time that a team spends together is valuable and expensive. If people on a team and its leader do not understand the differences in styles and

preferences, a certain percentage of the time and energy naturally present within each individual and the team will be wasted in arguments and debates relating to perspective, objectives, resources, etc. There will never be an answer that everyone will agree to, but an understanding of why a particular individual feels the way they do can go a long way to making everyone feel that their contribution and perspective is appreciated.

Team Member Preferences and Styles

Over the years, many different individual assessment instruments have been developed which are useful for assessing differences in social and problem solving styles. Both are important in a typical engineering team as the leader would like the team to "get along" but also solve a problem or design a process, product, or facility. These are different things. Social style relates to how individuals relate to each other in a social context, not necessarily one that includes technical problems. Problem solving style relates to how individuals analyze problems and how they solve them, independent of who they may be working with who is also involved in the problem solving or design challenge. It is possible for individuals to have identical social styles as measured by an assessment such as the Myers Briggs assessment and to have very different problem solving styles as measured by an assessment instrument such as the Kirton KAI™

Social Styles

The Myers Briggs MBTI™ was the original social style assessment instrument and has been given to millions of individuals. It has successors such as Insights™ and 16Types™ which provide feedback in the same summary form. Regardless of which instrument is used, a person's social style can be characterized along the following continuum:

Extroversion/Introversion (E vs. I) This describes the outward vs. inward focus of an individual. An extrovert enjoys socialization, verbal communication, cold calling, possibly parties, and other occasions and events where outward communication is enjoyed and expected. A strong introvert, on the other hand, may prefer to read a book, surf the web for information and become very nervous in social situations where direct oral communication seems to be required.

Sensor/iNtuitor (S vs. "N").Note: An "N" is used here, rather than an "I" to avoid confusion with the "I" used in the previous category. A Sensor is strongly concerned with data, facts, and measurable aspects of a problem or situation whereas an Intuitor is equally concerned with aspects of a situation that may not be directly measurable such as implications, indirect effects, gut feelings, emotions, and undefined possibilities that cannot be immediately quantified. This characteristic primarily applies to how information and data is taken into a mental analysis as opposed to what comes out afterwards. A simple question to ask someone to find out which one of these they may be is to ask them to describe a leaf. A sensor will say things like "green" or "scaly", terms that are physically descriptive. An intuitor will describe it as "light" or "reflective".

Thinking/Feeling (T vs. F) This attribute characterizes how someone treats the analysis of the information received. A "T" person will be much more concerned with the impact of a decision on economics, structure, and costs while an "F" person will be just as strongly concerned about changes might affect the personal lives of individuals and their careers. In a downsizing situation, the "T" will want to know how much money will be saved and what the outplacement services will cost, while the "F" will want to know about the impact on peoples' families and their psychological well being.

Judging/Perceiving (J vs. P) This characteristic describes the focus of an individual and their approach to implementation and planning. A "J" will be very concerned with schedules, commitments, deadlines, and critical path diagrams. A "P" will be much more open to loose schedules, leaving the door open for changes and more comfortable with uncertainty. Deadlines are critical to the "J" and far less important to the "P" who is more worried about missing possibilities that cannot be seen as opposed to meeting commitments to do what is known.

You can imagine, as a leader, the challenges in managing a team where these characteristics were not uniform. It's easy to work with and manage people that are like us, but that does not necessarily make for the best team structure. If everyone thinks alike, there is no one to ask the difficult questions or propose alternatives and options

The social style assessment will come back, independent of the type of instrument used, as four letters, such as ESTJ or INTP. Though the descriptions above emphasize the extremes, it is possible for an individual to be borderline or in the middle. There can be small changes in an individual's preference over time, but not large ones. This type of assessment also provides an insight as to how a person will respond under stressful situations. For example, a strong introvert will become very extroverted when fundamental beliefs or values are questioned. Obviously, this type of situation can occur within a project team and if social style preferences are known and understood, these types of situations can be handled more effectively.

Problem Solving Style

An individual's problem solving style can be measured very accurately with an assessment instrument known as the Kirton KAI™. This assessment, consisting of 32 simple questions, analyzes a person's approach to problem solving. The assessment comes back as a number from 32-160 with an average of around 90 and a 2 sigma deviation between 70-110. A lower KAI™ number would describe someone who problem solves with known parameters, likes schedules and agendas, and generates "filtered" ideas whose probability of being practical is high. They are described as an "adapter". At the other extreme would be someone who tends to generate a larger quantity of ideas without necessarily worrying about their practicality, is less concerned about agendas and schedules, and will "bi-associate" ideas and make connections between disparate concepts not made by others. In KAI™ jargon, this person is described as an innovator. A simple question to ask someone to get a sense of where they are on the KAI™ scale is what they would do as a tea bag manufacturing plant manager who has just lost a major customer amounting to over 2/3 of the plant's business. An adaptor will suggest ideas relating to cost reduction and layoffs and encouraging sales and marketing to find some more tea bad customers. An "innovator" will ask what new product ideas might come from packaging something else in the bad or what else the bag fabric could be used for.

As mentioned earlier, people with the same social style profile can have differing problem solving style profiles. This can set up a situation where a team leader thinks that everyone on a team "likes" each other and their job will be easy, but if the team's challenges include serious problem solving, things may not be as easy as anticipated.

<u>Leadership Situations</u>

Let's consider some situations and the implications of not understanding the nature of the team being "led". First, an international new product introduction. If the leader's future income is dependent upon how fast and how many countries a new product can be introduced into, there is a high level of risk, especially if manufacturing and distribution are under contract in some sense. If the team members do not have the same stomach for risk and "fuzzy" situations, they may throw up artificial roadblocks and record "remember I told you" notes to be vindicated later if the introduction does not go as planned. If the leader knows about this style preference ahead of time, then a rational discussion can be had regarding how the risks can be minimized.

Consider a multi-country engineering team whose meetings may be as often over the web as in person. These various social and problem solving types are not uniformly distributed and can be affected by cultural biases. It is critical, especially when deadlines and commitments are involved, to make sure that a leader ensures that the meaning of deadlines is clear. A "J" will understand easily while a "P" may interpret a deadline as simply a target or a goal. A high scoring KAI person may take risks that seem acceptable to him/her but may not be acceptable to the team leader. Interpretation of boundaries will vary and the language needs to be understood at the beginning of a project, not in the middle of a crisis.

To summarize, it is critical that a team leader understand both social and problem solving styles of the people on their teams. Without such an understanding, there is great risk of misunderstood goals, decision making based on misunderstood criteria, and problem solving done with either too little or too large a boundary condition.

(496h) Fractionating Fatty Acid Methyl Esters by Urea Inclusion Process and Its Impact On the Cloud Point

Junli Liu, Agricultural and Biological Engineering, Purdue University, West Lafayette, IN and Bernie Tao, ABE, Purdue University, West Lafayette, IN

See more of this Session: Alternative Fuels and Enabling Technologies II

See more of this Group/Topical: Fuels and Petrochemicals Division

Reducing Process Complexity for Cellulosic Ethanol

Lonnie O. Ingram, Department of Microbiology and Cell Science
University of Florida, Box 110700, Gainesville, FL 32611 Ingram@UFL.edu

Lignocellulosic biomass represents a potential source of carbohydrates (cellulose and hemicellulose) for microbial fermentation into fuels and chemicals. Steam pretreatment with dilute mineral acids serves as an efficient approach to depolymerize hemicellulose and enhance cellulose digestion with enzymes. However, side products (furfural, 5-hydroxymethyl furfural, acetate, phenolics,and others) formed during pretreatment inhibit microbial growth and retard fermentation. Potential engineering solutions to this problem (separation of hemicellulose hydrolysate, counter-current-washing of fibers, and mitigation of toxins) add complexity and increase cost. The level of inhibitors produced during acid pretreatment is affected by the severity of conditions and the choice of acid. Weaker acids such as phosphoric acid produce lower levels of toxic side products and reduce the need for exotic metal alloys. Since mineral acids are not consumed by the pretreatment process, the higher cost of phosphoric acid as compared to sulfuric acid could be offset in part by recovery and reuse as a dilute fertilizer when crops are nearby. Recent progress has been made in the development of more robust *E. coli*-based biocatalyst with increased resistance to dilute acid hydrolysates. Using this biocatalyst and phosphoric acid pretreatment, we have been able to co-ferment sugarcane bagasse hemicellulose hydrolysate and enzyme-hydrolyzed cellulose in a single vessel using a variation of the simultaneous saccharification and co-fermentation process (SScF; Fig. 1). Hexose and pentose sugars from phosphoric acid pretreated sugarcane bagasse were fermented in a single vessel, eliminating process steps for solid–liquid separation and sugar cleanup. After pretreatment, a liquefaction step (L) with cellulase was included to improve mixing and saccharification (L + SScF), analogous to a corn ethanol process. Bagasse slurries containing 10% and 14% dry weight (fiber plus solubles) were tested using pretreatment temperatures of 160–190 C (1% phosphoric acid, 10 min). Enzymatic saccharification and inhibitor production both increased with pretreatment temperature. The highest titer (30 g/L ethanol) was obtained after incubation for 122 h using 14% dry weight slurries of pretreated bagasse (180 C). The highest yield (0.27 g ethanol/g dry bagasse) was 82 gal ethanol per dry ton of bagasse using this L+SScF process and 10% dry weight slurries.

Comparison of conversion processes

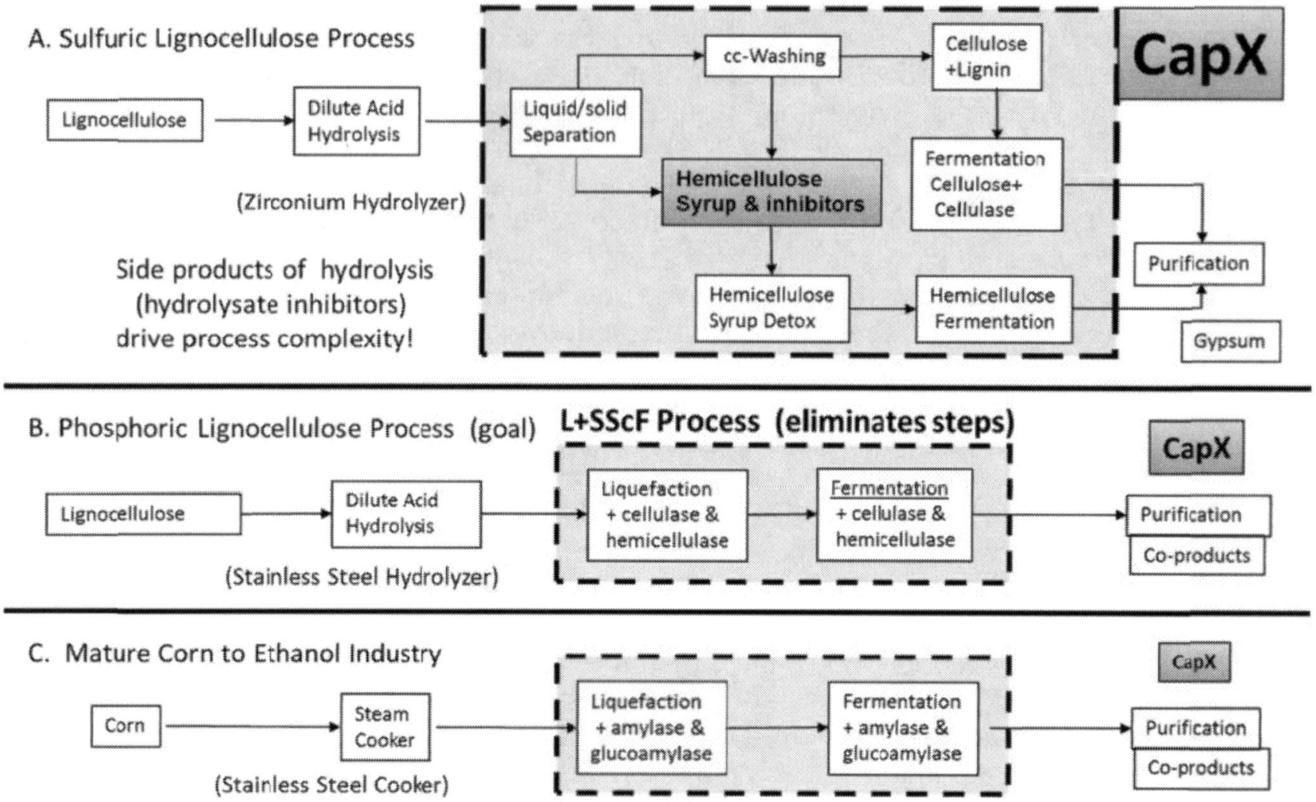

(364f) Progress Toward Biological Pretreatment - Understanding Plant Cell Wall Degradation by Termite and White-Rot Fungi

Shulin Chen[1], Jing Ke[2], Dhrubojyoti D. Laskar[2] and Jijiao Zeng[2], (1)Department of Biological Systems Engineering, Washington State University, Pullman, WA, (2)Washington State University, Pullman, WA

In an effort developing more efficient and environmentally benign lignocellulosic biomass processing technologies, the authors' laboratory has been studying plant cell wall degradation process in termite and white-rot fungi. This presentation summarizes the results so far obtained from various experimental investigations of the lab. First, termite guts provide both aerobic and anaerobic environments, the former is required for lignin degradation. The termite guts also have close to neutral pH range, which suggest existence of lignin degrading enzymes that are different from that of fungi. The organization of the gut system allows the conversion of biomass in termite to occur as a sequential process in which the pretreatment of the biomass occurs in the foregut and midgut, whereas the hydrolysis of the cellulose occurs mainly in the hindgut. There are synergies between termite and its symbiont in deconstructing plant cell walls. The chewing action of the termite increases the surface area for enzymatic actions, the enzymes produced by the termite tissues modifies the bonds that forms the barriers for cellulase access and symbionts produce the enzymes that are mainly responsible for cutting the cellulose polymer to sugar units. The results shows the actual changes of lignin structure as the biomass material pass through the termite guts. The lignin structure changes are more modification than degradation in nature. Equally import is the deconstruction of the association of lignin with other components of plant cell walls such as hemicelluloses that occurs in the foregut and midgut. It becomes apparent by comparing with other biological systems that the enzyme systems and their actions in biological plant cell wall degradation systems are more complex than what was previously known. A generality has also been observed in that most of the biological systems studied tended to reduce G unit from the original lignin structure. Additionally, termite is a more effective plant cell wall degradation system than white rot-fungus because of the chewing action of the termite and the better-controlled environment in the termite guts. Termite is also more effective than the cow stomach as the latter lacks capability of lignin modification. Developing an artificial system that can surpass the natural biological system in major performance criteria may offer the ultimate solutions to lignocellulosic biomass processing and utilizations.

See more of this Session: Developments In the Pretreatment of Lignocellulosics for Bioconversion I
See more of this Group/Topical: Sustainable Engineering Forum

(470a) Silicon-Based Bulk Nanostructured Thermoelectric Generators

Akram Boukai, Materials Science and Engineering, University of Michigan, Ann Arbor, MI

State-of-the-art thermoelectric generators are typically based on exotic alloys and have low device efficiencies. Therefore, research targeting new discoveries in earth abundant thermoelectric materials that increase device efficiency are needed. This talk will present recent progress on bulk nanostructured thermoelectric materials that are silicon-based. These include functional silicon nanowire devices and silicide/silicon eutectics. These bulk nanostructured thermoelectric devices exploit dramatic reductions of thermal conductivity due to phonon scattering at disordered interfaces, which is expected to increase ZT. Thermoelectric properties, structural characterization, and device efficiencies will be presented.

See more of this Session: Nanomaterials for Thermal-to-Electric Conversion I
See more of this Group/Topical: Topical 5: Nanomaterials for Energy Applications

(206a) Depolymerization and Catalytic Hydrodeoxygenation of Switchgrass Lignin with Formic Acid As a Hydrogen Source

Weiyin Xu[1], Stephen J. Miller[2], Pradeep K. Agrawal[1] and Christopher W. Jones[1], (1)School of Chemical & Biomolecular Engineering, Georgia Institute of Technology, Atlanta, GA, (2)Chevron Energy Technology Company, Richmond, CA

Second generation biofuels derived from environmentally friendly lignocellulosic crops that do not compete with food crops can potentially be processed into biofuels that can be used to supplement fossil fuels. Approximately a third of lignocellulosic biomass consists of lignin, a high molecular weight polyaromatic macromolecule that consists of units derived primarily from p-hydroxyphenyl (H), guaiacyl (G) and syringyl (S) units. In this study, organosolv lignin isolated from switchgrass was depolymerized and hydrodeoxygenated at 350 °C using a 20 wt% Pt/C catalyst with ethanol solvent in a batch reactor. Formic acid was also utilized in these reactions as a hydrogen source for hydrodeoxygenation. The impact of the Pt catalyst, formic acid and reaction time on liquid products formation was systematically evaluated. The switchgrass lignin was found to produce a wide variety of aromatic compounds, of which 9 compounds were identified and quantified. With increasing reaction time, the H/C ratios were shown to increase while the O/C ratios decreased, thus confirming the progression from raw lignin to"lignin-to-liquid" fuel blends. Additionally, average mixture or bulk properties such as functional group content and thermal stability were studied via NMR and thermogravimetric analysis of the depolymerized mixture, respectively.

See more of this Session: Catalytic Biofuels Refining I
See more of this Group/Topical: Fuels and Petrochemicals Division

(13f) Reforming of Residual Tars and Oils From Biomass Gasification

Lyman Frost[1], Elango Elangovan[1] and J. Hartvigsen[2], (1)Ceramatec, Inc, Salt Lake City, UT, (2) Electrochemical, Ceramatec, Inc, Salt Lake City, UT

One of the barriers to wider use of many kinds of biomass is the fact that they generate refractory tars and oils when gasified. These compounds cause difficulty when they are passed to downstream components in the system by causing plugging and drop in performance. The compounds can be condensed and removed but this involves extensive heat exchange, increased cost, and removal to a hazardous waste landfill.

Working as a subcontractor to Emery Energy, a Salt Lake City company with expertise in biomass gasification, Ceramatec will be applying its non-thermal plasma reformer technology to eliminate any refractory tars and oils that exit the Emery gasifier. Preliminary laboratory work indicates that it is possible for the plasma reformer to operate on the gas exiting the gasifier and convert these tars and oils to additional hydrogen and carbon monoxide. This work is being done at Western Research Institute in Laramie, Wyoming under a US Department of Energy award to Emery Energy.

This paper will present an overall process structure of the gasification process and the non-thermal plasma reformer in particular. Operational characteristics and laboratory results related to the reforming of the refractory tars and oils will be presented. The larger pilot sized facility at Laramie will also be described. Data is expected to be available from the larger pilot sized facility in Laramie by the time for the conference. The resultant gas stream is of a quality that could be used by a high temperature fuel cell.

See more of this Session: Catalytic Conversion of Renewable Resources to Synthesis Gases and Pyrolysis Oils
See more of this Group/Topical: Sustainable Engineering Forum

(564a) Development of a Compact Fischer Tropsch Reactor

Lyman Frost[1], Joseph Hartvigsen[2] and Elango Elangovan[2], (1)Ceramatec, Inc, Salt Lake City, UT, (2) Electrochemical, Ceramatec, Inc, Salt Lake City, UT

Fischer Tropsch (FT) synthesis is a polymerization process where a CH_X species is added to a growing aliphatic chain. There are several models of the catalytic chain propagation rate and the termination of the chain growth (e.g. Anderson – Schulz – Flory model). The exact methodology for the surface catalyzed formation and termination of the hydrocarbon chain is still a matter for study. Most FT facilities built today (e.g. the Oryx gas-to-liquids plant built in Qatar) are very large, expensive facilities that entail large capital expenditures and associated risk.

Ceramatec has been developing an alternative approach under funding from the Office of Naval Research and the State of Wyoming. The reactor design is a compact fixed bed reactor with integral cooling and unique components to insure the conduct of heat from the catalyst bed to maintain even heat through the bed. The basis of the reactor design was the reaction cooling duty requirement calculated using the process modeling tool VMGSim. The compact design will enable the transportation of the reactor and the use with any source of synthesis gas (e.g. coal gasification, biomass gasification, co-electrolysis of CO_2 and steam, etc).

This paper will discuss the modeling effort and the resultant patented design for the compact FT reactor. The presentation will also discuss the results obtained from the laboratory prototype of this reactor using a FeCuK catalyst and a Co based catalyst that were manufactured at Ceramatec. The initial results obtained from a larger scale unit (i.e. ~ 4 bpd with high activity Co catalyst) in place at Western Research Institute in Laramie, Wyoming will also be discussed.

See more of this Session: Alternative Fuels and Enabling Technologies III
See more of this Group/Topical: Fuels and Petrochemicals Division

(39b) Colloidal Nanomaterials As the Building Blocks for Next-Generation Solar Cells

Jeffrey M. Pietryga, Chemistry Division, Los Alamos National Laboratory, Los Alamos, NM

The use of colloidally synthesized nanomaterials in photovoltaic devices is attractive because of the low-cost and scalability of solution-based fabrication methods. Moreover, the facile control over optoelectronic properties, via the fine control over structure these methods afford, can potentially be exploited in highly efficient next-generation solar cells. Design, synthesis and characterization of novel nanomaterials to enable advanced-concept solar cells are an important part of the ongoing effort within the Center for Advanced Solar Photophysics, a DOE Energy Frontier Research Center. I will discuss a number of specific examples from this work that explore the range of complexities achievable through inexpensive wet-chemical methods. Themes to be examined include the straightforward rendering of established bulk materials into quantum-confined nanocrystal form, band-gap engineering in semiconductor nanoheterostructures, and finally the controlled coupling of plasmonic metal nanoparticles with semiconductor nanocrystals in a diverse series of hybrid superstructures.

See more of this Session: Nanomaterials for Photovoltaics I
See more of this Group/Topical: Topical 5: Nanomaterials for Energy Applications

(647a) Process Optimization of Bioethanol Production Via Hydrolysis of Switchgrass

Mariano Martn, Chemical Engineering, University of Salamanca, Salamanca, Spain and Ignacio E. Grossmann, Department of Chemical Engineering, Carnegie Mellon University, Pittsburgh, PA

Processoptimization of bioethanol production via Hydrolysis of Switchgrass

Mariano Martn[a,b],Ignacio E. Grossmann[b][1]

mariano.m3@usal.es; grossmann@cmu.edu

[a] Departamentode Ingeniera qumica y textil. Universidad de Salamanca. Plz. Caidos1-5 37008, Salamanca (Spain)

[b]Department of ChemicalEngineering. Carnegie Mellon University 5000 Forbes Avd. Pittsburgh PA 15213

TheEnergy act in 2007 established the increase in the production of bioethanolusing lignocellulosic raw materials to help reduce the dependence on crude oil.This was based on the fact that among the different possibilities only biomassprovides an alternative fuel that can be implemented in the short-term for the transportationsector due to its compatibility with currentautomobile engines (Cole, 2007) and to the fact that it can take advantage ofthe existing supply chain of liquid fuels that is already well established. Currentlythe production of ethanol from lignocellulosic raw materials still faces technical,economic and commercial barriers (Huang, 2008).

Twotypes of process technologies can be used to transform lignocellulosic rawmaterials into ethanol. The first one is based onthe gasification of the raw material into syngas, which is used to obtainethanol either via Fischer-Tropsch based catalytic reaction or via fermentationof the syngas (Phillips et al, 2007; Huhnke, 2008;Piccolo and Bezzo, 2009; Zhuet al, 2009). The second oneis based on the hydrolysis of the raw material to break down the physical andchemical structure of the crops to expose the sugars that are fermented toethanol. Due to its similarity with the current production of ethanol and theexpected lower capital cost, this technology has received the attention of manyresearchers, e.g. Hamelinck et al. (2005); Cardona& Snchez (2006); Zhang et al. (2009); Keshwani& Cheng (2009).

In thispaper we develop a conceptual design for the production of ethanol fromlignocellulosic raw materials with hydrolytic pretreatment of the biomass usingmathematical programming techniques

(Daichendt & Grossmann, 1997, Grossmannet al., 1999). The process consist of three steps, raw material pretreatment toexpose the sugars, sugar fermentation to ethanol and ethanol dehydration tofuel grade. Wepropose a superstructure optimization approach where we first construct aflowsheet embedding the various process units involved in ethanol productionconsidering a number of alternatives for switcgrass pretreatment such as ammoniafiber explosion or dilute acid pretreatment (Sung and Cheng, 2002; Keshwani & Cheng,2009) and for differenttechnologies that can operated in parallel and/ or in sequence for dehydratingthe ethanol like rectification, adsorption in corn grits, molecular sieves andpervaporation. These units are interconnected to each other through networkflows and other utility streams. The goal is to optimize the structure by minimizingthe energy input in the ethanol production process. The optimization of thesystem is formulated as a mixed-integer nonlinear programming (MINLP) problem,where the model involves a set of constraints representing mass and energybalances for all the units in the system. We then substitute the distillation columnsby multieffect column to reduce the consumption on cooling water and steam and wedesign the optimal heat exchanger network using SYNHEAT. The heat recoverynetwork, together with a modified distillation column design, further reducesthe energy consumption in the plant and in turn decreases the unit productioncost of ethanol. Finally, we perform an economic evaluation. For the optimalflowsheet we design the optimal water network based on the work by Ahmetovic,et al (2010). First, we identify the sources of water within the process(distillation columns, utilities units), the sinks (fermentor, pretreatment)and the process units (unitities units), and optimize the correspondingsuperstructure for reuse and recycle of water and treatment units so as tominimize the fresh water consumption.

Theresults of the optimized conceptual design indicates that under the currentyields reported for the pretreatment processes (Sung and Cheng, 2002) the optimalprocess for the production of ethanol from switchgrass involves the use ofdilute acid pretreatment, followed by enzymatic hydrolysis and sugarfermentation and the use of molecular sieves for the final dehydration ofethanol to fuel grade. The predicted production cost of this process is $0.8/galwith an investment cost of $161MM, while the consumption of water is1.6gal/gal. These results are compared with the themal based processes thatrequires a higher investment of $335MM, but has a reduced production cost of$0.41/gal and water consumption of 1 gal/gal (Martin & Grossmann, 2011,Martin et al ,2010)

References

Ahmetovic, E.; Martn, M.; Grossmann,I.E. (2010) "Optimization of Water Consumption in Process industry: Corn -based ethanol case study" Ind. Eng. Chem Res. 49 (17) 7972- 7982

Cardona, C.A., Snchez. O.J. (2006)Energy consumption analysis of integrated flowsheets for production of fuelethanol from lignocellulosic biomass. Enrgy , 31, 2447- 2459

Cole, D. E. (2007) Issues facing theAuto Industry: Alternative Fuels, Technologies, and Policies ACP Meeting EagleCrest Conference Center June 20, 2007

Daichendt, M. M., and I. E. Grossmann,(1997) Integration of Hierarchical Decomposition and Mathematical Programmingfor the Synthesis of Process Flowsheets, Comp. Chem. Eng., 22, 147.

Grossmann, I. E.; Caballero, J. A.;Yeomans, H. (1999) Mathematical Programming Approaches to the Synthesis ofChemical Process Systems, Korean J. Chem. Eng., 16, 407-426.

Keshwani, D. R., Cheng, J.J. (2009)Switchgrass for bioethanol and other value-added applications: A reviewBioresource Technology 100, 1515-1523

Hamelinck, C. N., Geertje vanHooijdonk, G., Faaij, A.P.C. (2005) Ethanol from lignocellulosic biomass: techno-economic performance in short-, middle- and long-termBiomass andBioenergy 28, 384-410

Huang, J. Qiu,H. and Scott Rozelle, S., (2008), More pain ahead for China's food prices, FarEastern Economic Review, 171, 5, 8-13.

Huhnke, R. L. (2008) Cellulosicethanol using gasification-fermentation. Resource:Engineering & Technology for a Sustainable World

http://www.articlearchives.com/energy-utilities/renewable-energy-biomass/896186-1.html

Martn, M.; Ahmetovic, E.; Grossmann,I.E. (2010) "Optimization of Water Consumption in Second Generation bio-EthanolPlants " I&ECR doi: 10.1021/ie101175p

Martn, M., Grossmann, I.E. (2011)"Energy optimization of lignocellulosic bioethanol production via gasification"accepted AIChE J. | DOI: 10.1002/aic.12544

Phillips, S., Aden, A., Jechura, J.and Dayton, D., Eggeman, T (2007) Thermochemical Ethanol via IndirectGasification and Mixed Alcohol Synthesis of Lignocellulosic Biomass TechnicalReport, NREL/ TP-510-41168, April 2007

Piccolo, C., Bezzo, F., (2009) Atechno-economic comparison between two technologies for bioethanol productionfrom lignocelluloses. Biomass and Bioenergy 33 (2009) 478 - 491

Sun, Y., Cheng, J., (2002) Hydrolysisof lignocellulosic materials for ethanol production: a review. Bioresour.Technol., 83, 1-11

Zhu, Y., Gerber, M.A., Jones, S.B.,Stevens, D.J (2009) Analysis of the effects of compositional andconfigurational assumptions on product costs for the thermochemical cornversionof lignocellulosic biomass to mixed alcohols FY 2007 Progress report. DOEPNNL.17949 Revision 1.

[1]Corresponding author. Tel.:+1-412-268-3642; Fax: +1-412-268-7139.

Email address:grossmann@cmu.edu (I.E. Grossmann)

See more of this Session: Energy Systems Design and Alternative Energy Sources I
See more of this Group/Topical: Computing and Systems Technology Division

Basic model of secondary zinc air batteries to quantify environmental impacts

Daniel Schröder[a], Ulrike Krewer[a,b]

[a]Portable Energy Systems, Chair for Process Systems Engineering, Otto-von-Guericke University Magdeburg, Universitätsplatz 2, 39106 Magdeburg, Germany

[b]Otto Hahn Group/Portable Energy Systems, Max Planck Institute for Dynamics of Complex Technical Systems, Magdeburg, Sandtorstrasse 1, 39106 Magdeburg, Germany

---

Zinc air batteries (ZABs) offer a high theoretical specific energy density (1353 Wh/kg[1]) and therefore appeal for application in portable and mobile devices. The availability of bi-functional air electrode catalysts, such as non precious perovskite structured metal, makes ZABs electrically rechargeable and even more attractive.

Figure 1 illustrates the reaction and transport processes occurring in a ZAB during discharge mode. In general, the battery consists of the zinc electrode, the air electrode and the separator. Equations (1)–(3) state the occurring reactions in caustic electrolyte.

$$\mathrm{Zn} + 4\,\mathrm{OH}^- \underset{}{\overset{r_1}{\rightleftharpoons}} \mathrm{Zn(OH)}_4^{2-} + 2\,\mathrm{e}^- \tag{1}$$

$$\mathrm{Zn(OH)}_4^{2-} \overset{r_2}{\rightleftharpoons} \mathrm{ZnO} + 2\,\mathrm{OH}^- + \mathrm{H_2O} \tag{2}$$

$$\mathrm{H_2O} + \frac{1}{2}\mathrm{O_2} + 2\,\mathrm{e}^- \overset{r_3}{\rightleftharpoons} 2\,\mathrm{OH}^- \tag{3}$$

---

[1]Linden, E., 2004. Handbook of Batteries, 3rd ed., McGraw-Hill, p.1.13

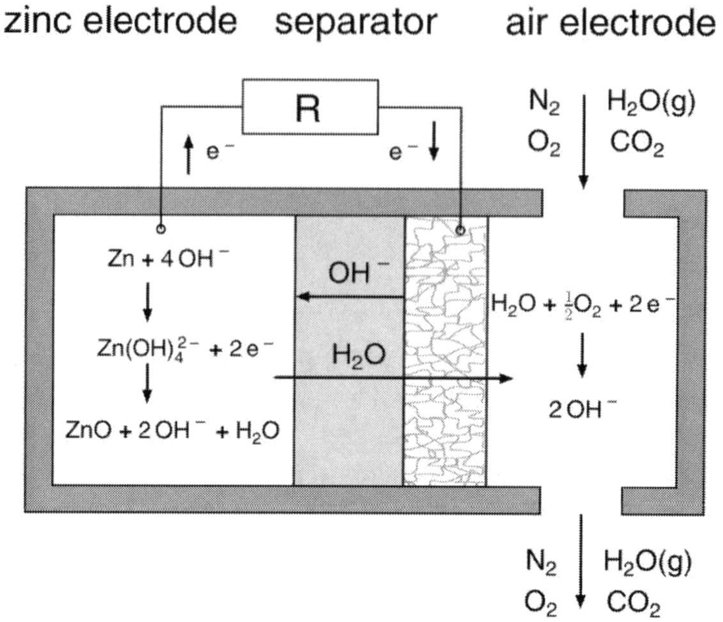

Figure 1: Schematic view of a zinc air battery showing the occurring reaction and transport processes during discharge. The charge process is then vice versa.

It is essential that OH^- and H_2O are transported through the separator during ZAB operation. As well, the ZAB is open to the environment which implies that uptake or loss of gaseous water, O_2 and even CO_2, might influence the operation behavior of the ZAB. For example, carbonates can be formed due to the reaction of the caustic electrolyte with CO_2.
Research on ZABs is almost exclusively experimental and focuses on the characterization of materials for zinc electrode, air electrode or separator, as well as on investigating the performance of entire ZABs.
It is suggested that ZABs are affected by the surrounding humidity; this holds especially for secondary batteries due to the longer operation time[2]. Models can help to analyze the influence of humidity on ZAB performance and get insight into the water balance of a ZAB.
Relatively few models exist to describe ZABs. They are mainly based on the macroscopic approach for porous battery electrodes by Newman and Tobias[3]. These models apply a set of coupled discretized non-linear dif-

[2]Haas, O., Müller, S. and Wiesener, K., 1996. Chem. Ing. Tech., 68(5), pp.524-542.
[3]Newman, J. and Tobias, C.W., 1962. J. Electrochem. Soc., 109, pp.1183-1191.

ferential algebraic equations (DAEs) to describe the transport and reaction phenomena occurring in the ZAB. They are used to analyze and optimize the ZABs design dimensions such as separator thickness and zinc electrode thickness. So far the existing models are not used to investigate the environmental impact on ZAB operation.

Our aim is to investigate the impact of the environmental gaseous water amount, O_2 amount and CO_2 amount on ZAB operation. For this, a flexible and expandable model is designed. In a first step, we introduce an isothermal basic model to describe the reaction and transport processes in secondary ZABs.

Our approach is as follows: The zinc electrode and the air electrode are described as an ideal continuously stirred tank reactor (CSTR), respectively. Each CSTR electrode consists of a fixed metal phase and a liquid phase containing the electrolyte and the dissolved and ionic species. The metal phase at the zinc electrode consists of Zn and ZnO. The metal phase at the air electrode is of inert catalyst structure. Both metal phases do not change their structure with time, however the zinc electrode metal phase ideally replaces its elements Zn and ZnO. The liquid volume in each CSTR electrode is varying with time and can be increased for example by uptake of gaseous water into the air electrode.

The here considered basic model consists of five molar balances for the ionic and solid species in the zinc electrode, three molar balances for the ionic and dissolved species at the air electrode, two charge balances for each electrode and two differential equations for the varying liquid electrolyte volume in the electrodes. These equations are a system of ordinary differential equations (ODEs). In our work, we solve the resulting ODE system with appropriate initial conditions using Matlab®.

The respective electrode over-potentials at the zinc electrode and the air electrode, originating from charge balances, and the ohmic drop through the separator, depending on the conductivity of the electrolyte, enable to calculate the total battery voltage.

The electrochemical reactions (1) and (3), are described by a Butler-Volmer approach. The chemical reaction (2) is applied as precipitation/saturation mechanism in the liquid electrolyte KOH.

Diffusion of OH^- and H_2O and migration of OH^- via the separator are included into the molar balances as exchange terms between the two electrodes. Both electrodes are apart by a certain distance, here the separator thickness in the range of μm. Due to the presence of a counter-ion (K^+) in the liquid electrolyte, electro-neutrality holds. Migration of $Zn(OH)_4^{2-}$ through the separator is omitted.

At this point we investigate three modifications of the basic model:

- Modification (a) is a reference, where gaseous water is neither entering nor leaving the ZAB. Here the O_2 amount in the air electrode is set constant and does not change with time.
- Modification (b) accounts for varying O_2 amount in the air electrode, whereas uptake or loss of water is neglected.
- Modification (c) allows gaseous water to enter or leave while the ZAB is operated with constant O_2 amount in the air electrode.

From transient simulations with galvanostatic charge and discharge cycles we obtain three results for the respective modification:

- Modification (a) shows that OH^- and H_2O transport is sufficient to maintain battery operation.
- Modification (b) shows how cycling may affect the reaction rates at the air electrode and the related ZAB performance
- Modification (c) shows how cycling may affect the water content and the related ZAB performance.

Further investigations target to include the carbonation at the air electrode.

PILOT TEST FOR FISCHER-TROPSCH SYNTHESIS USING A FIXED BED COMPACT HEAT EXCHANGE REACTOR

Zhijun Jia[1], Hani Gadalla[1], Steven Vallee[1], Burt Davis[2], and Dennis Sparks[2]*
[1]Chart Energy and Chemicals, La Crosse, WI
[2]University of Kentucky Center for Applied Energy Research, Lexington, KY

Extended Abstract, 2011 AIChE Annual Meeting, October 16-21 2011, Minneapolis, MN

Introduction

The Fischer-Tropsch reaction is a highly exothermic reaction (157 kJ/mol CO converted) used to convert syngas (a mixture of CO and H_2) to liquid hydrocarbons. Conventional fixed bed reactor technology suffers from heat transfer limitations. Operating conditions must be limited to prevent reactor runaway, resulting in lower productivity per unit reactor volume. Non-isothermal operating conditions make the process more difficult to start up and control and can lead to increased catalyst deactivation and unwanted methane side-product formation.

A Compact Heat Exchange Reactor (CHER) combines a reactor and compact heat exchanger into a single device, and provides a solution to the heat transfer limitations commonly observed for conventional fixed bed reactors. Multiple fixed bed compact heat exchange reactors were designed and constructed by Chart Energy & Chemicals in La Crosse WI, for the purpose of synthesis of wax and/or diesel fuel from syngas via the Fischer-Tropsch (FT) reaction. These reactors consisted of a plate-fin structure with alternating coolant stream layers and process stream layers. The heights of the channels in the process stream were ~3-6 mm. Testing of the reactors was performed by Dr. Davis's group at the Center for Applied Energy Research at the University of Kentucky (UKY CAER).

The design structure of the reactors allowed for nearly isothermal operation of the FT reaction. The temperature difference between the measured peak temperature in the catalyst bed and the coolant stream was within about 1-4°C of. Catalyst deactivation as a function of time online was much less in these reactors than in typical packed bed FT reactors. Methane selectivity was generally 7-10% with the catalyst used. Relatively high conversion and production was achieved using reasonable space velocities.

Experimental

The CHERs in these experiments were of a "core-in-kettle" type design, where the brazed aluminum plate-fin "core block" was placed inside a stainless steel "kettle" for added safety and manufacturing code compliance. Pictures of the reactor are shown in Figure 1. These CHERs utilized a core block that consisted of a plate-fin structure with alternating coolant stream layers and process stream layers. The core blocks were manufactured using vacuum brazing technology. The process stream layers contained channels that were packed with catalyst to a density of about 860 kg_{cat}/m^3. A commercial type cobalt catalyst of 90-120 µm particle size was used. Thermocouples were placed inside the process channels with the catalyst and were located axially at 2", 7", and 11" along the reactor length to measure the axial temperature profile.

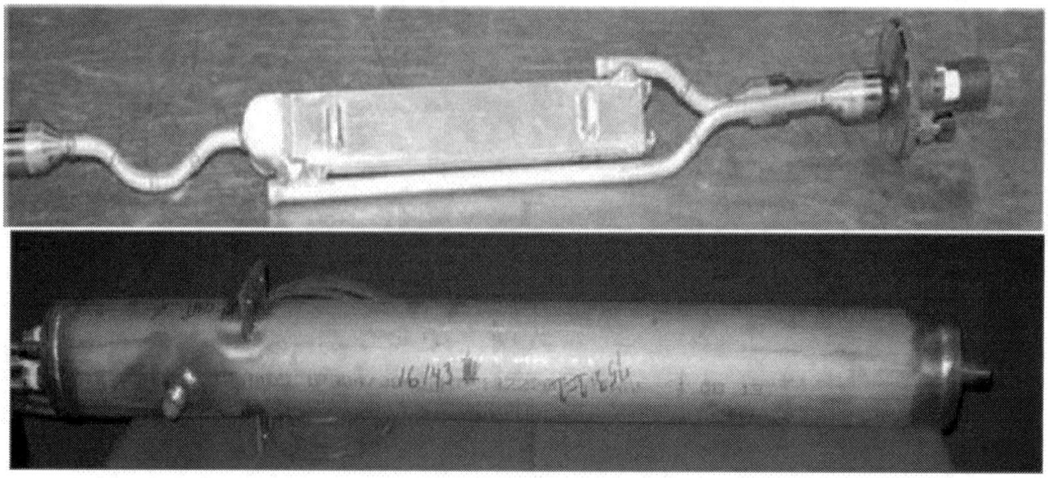

Figure 1a (top) - CHER core block and Figure 1b (bottom) - CHER "core-in-kettle" assembly.

These reactor core blocks were constructed of aluminum, which has superior heat transfer performance compared to stainless steel devices due to its higher thermal conductivity. These aluminum reactors are able to operate at commercially viable operating conditions for FT processes with cobalt catalysts, typically 210-240°C and up to 400 psig. They are also capable of catalyst activation conditions at 350°C at low pressure. An alternative construction to the CHER core blocks is using stainless steel for other chemical applications at higher temperatures and pressures, such as FT processes utilizing an iron catalyst.

Testing of the reactors was performed by Dr. Burt Davis's group at UKY CAER. The catalyst was activated in-situ at 350°C. Reaction conditions tested included temperatures from 190-231°C. In these devices at typical operating temperatures, a minimum pressure differential must be maintained between the process and coolant streams. Due to an equipment limitation from existing test facility, the process stream was limited to < 250 psig for these experiments. The pressure drop through the catalyst bed was 2-5 psi.

Results

The design structure of the reactors allowed for nearly isothermal operation of the FT reaction. In these reactors, the temperature difference between the measured peak catalyst bed temperature and the coolant stream was within 1-4°C. The axial temperature profiles shown in Figure 2 exemplify the near isothermal operation of the reactor. The reactors with a shorter process channel height showed better thermal performance, with a peak temperature rise of 0.6°C for a reactor with a ~5mm process channel height versus 1.4°C for the reactor with a ~6mm process channel height at comparable operating conditions at 215C. The conversion and product selectivity of the two reactors were similar. Although the shorter channel showed an improvement in heat dissipation, as a practical consideration in reactor design, the taller process channel allows for a larger amount of catalyst per total reactor volume, which reduces the total size and the lowers the cost for the reactor. In designing a reactor for a given catalyst, an optimum channel height would be one that is as high as possible while still meeting a specified process temperature difference tolerance.

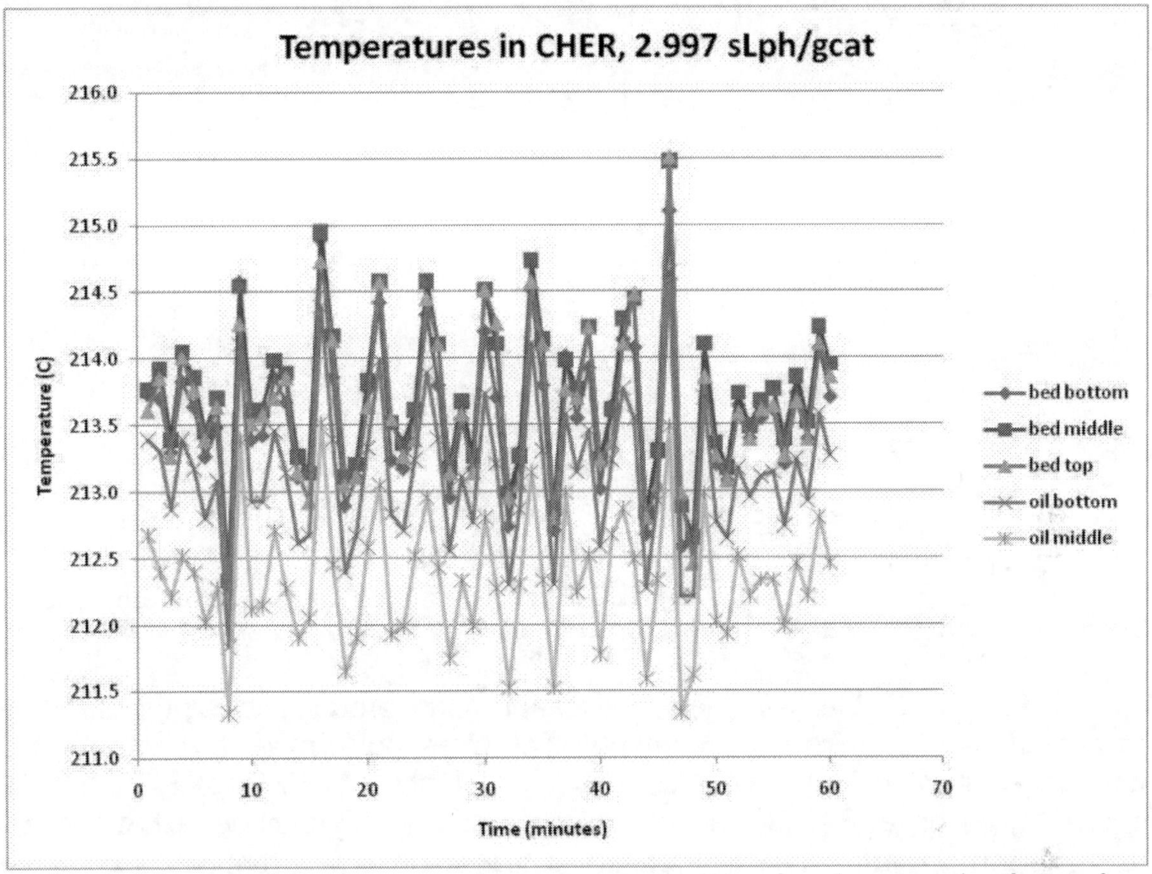

Figure 2 - Axial catalyst bed temperature measurements showing near isothermal reactor operation

The reactors were able to operate at high CO conversion with low methane selectivity at reasonable space velocities. An example of this is shown in Figure 3.

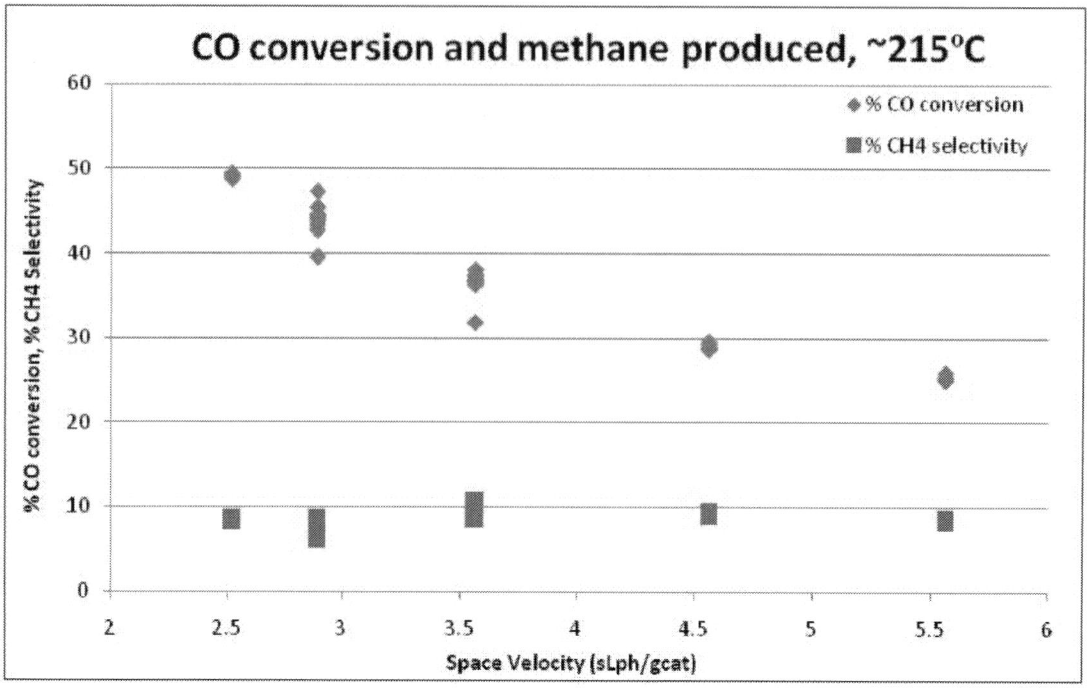

Figure 3 – % CO conversion and methane selectivity as a function of space velocity at ~215°C

The reactors were able to achieve reasonably high productivity at typical commercial operating conditions. During the experiments that were performed, the total hydrocarbon production ranged from 0.24 [g hydrocarbons / g_{cat} / hr] at 213°C to as high as 0.60 [g hydrocarbons / g_{cat} / hr] at 231°C. The productivity was a weak function of the space velocity over most of the experimental range; an increase in flow rate is accompanied by approximately a corresponding decrease in conversion as is shown in Figure 3. The product distribution followed the Anderson-Shultz-Flory equation relatively well (based on C8+ hydrocarbons) with alpha between 0.86 to 0.91, and an example is shown below in Table 1.

	CH4	C2 - C4	C5-C11	C12-C18	C19+			
TOS (hrs)	**Light Gas**	**Gas**	**Gasoline**	**Diesel**	**Wax**	**C5+**	**C12+**	**Alpha**
160.000	9.482	8.615	16.312	25.936	39.654	81.903	65.591	**0.8959**

Table 1 - Product distribution at 213°C and SV 2.997 sLph/g_{cat}, units in wt%.

Methane selectivity was generally 7-10% with the catalyst used, which was much lower than other experiments when using a catalyst with a larger particle diameter. The methane selectivity generally increased with increasing temperature.

Catalyst deactivation as a function of time online was much less in the CHERs than typical packed bed FT reactors. A plot of the catalyst activity is shown in Figure 4, showing that the deactivation is minimal during the operation period, and the profile is different from many other conventional catalyst which commonly show significant degradation during the first 100 hours of operation before achieving a near constant deactivation rate.

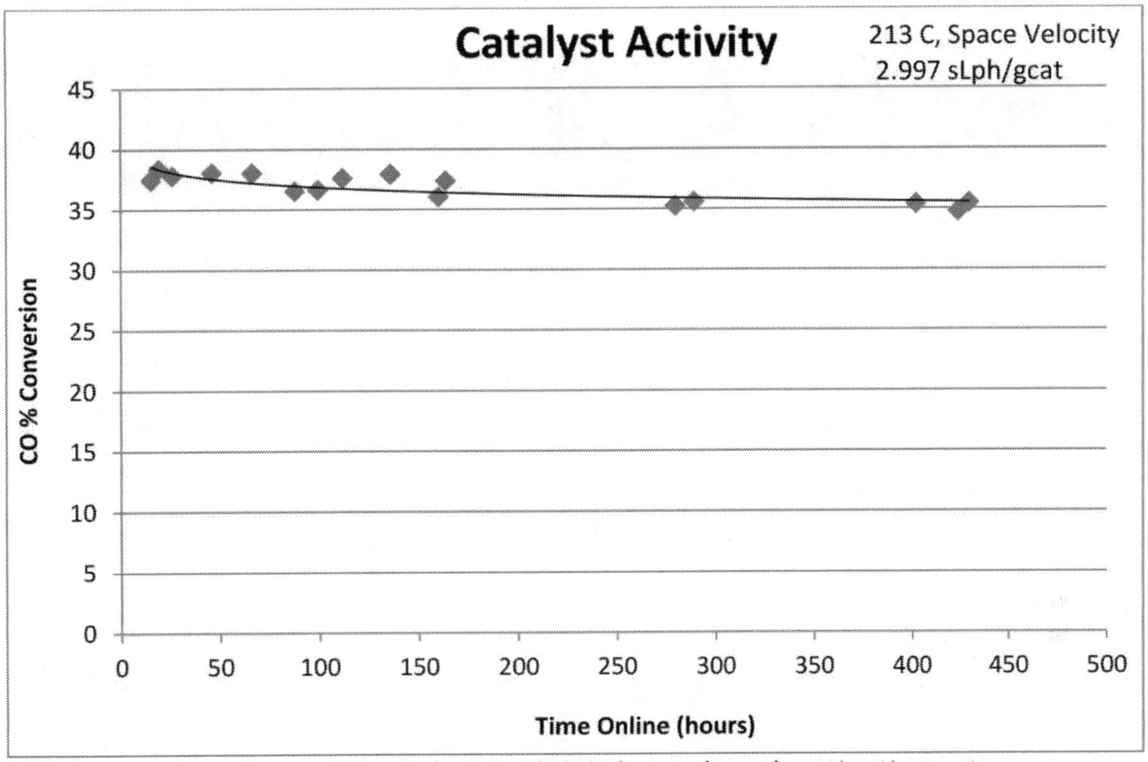

Figure 4 – FT catalyst in CHER has a low deactivation rate

Conclusions

Multiple fixed bed CHERs were designed and constructed by Chart Energy & Chemicals in La Crosse WI, for the purpose of synthesis of diesel fuel from syngas via the Fischer-Tropsch reaction in order to overcome the heat transfer limitations commonly observed with conventional fixed bed reactors. The design structure of the reactors allowed for nearly isothermal operation of the FT reaction. High productivity and low methane selectivity was obtained at commercially relevant operating conditions. The catalyst deactivation rate was much lower than typical packed bed FT reactors reported in the literature.

References

1. Bartholomew, C.H. and R.J. Farrauto (2006), Fundamentals of Industrial Catalytic Processes, John Wiley & Sons, Inc., Hoboken, NJ.
2. Das, Tapan K. and Burtron H. Davis et al. (2007), *Fischer-Tropsch Synthesis: Kinetics and Effect of Water for a Co/Al2O3 Catalyst*, Fischer-Tropsch Synthesis, Catalysts and Catalysis, Elsevier B.V.
3. Davis, Burtron H. (2002) *Overview of reactors for liquid phase Fischer-Tropsch synthesis*, Catalysis Today 71 249-300.
4. Davis, Burtron H. and Mario L. Occelli (2010), Advances in Fischer-Tropsch Synthesis, Catalysts, and Catalysis, Taylor and Francis Group, LLC, Boca Raton, FL.
5. Froment, Bischoff (1979), Chemical Reactor Analysis and Design, John Wiley & Sons.

6. Rytter, E. and E. Sigrid, et al. (2007), *Catalyst particle size of cobalt/rhenium on porous alumina and the effect on FT catalytic performance*, Ind Eng Chem Res, 46, 9032.
7. Syntroleum (2011), http://www.b2icc/Document/2029/110913.pdf, 4-12-2011.

Hydrogen-Bromine Flow Battery for Large-Scale Electrical Energy Storage

Trung Van Nguyen and Haley Kreutzer
Department of Chemical & Petroleum Engineering
University of Kansas
Lawrence, KS 66045, USA
Voice: 785-864-3938 – Fax: 785-864-4967 - Email: cptvn@ku.edu

Introduction

Renewable energy, including wind and solar, could supply a significant amount of electrical energy in the United States and around the world. However, despite the benefits of wind and solar, integration of these intermittent sources into the electrical grid system poses major challenges to the utility industry. Their unpredictability introduces more uncertainty to the power grid especially their unavailability when electricity is needed. Consequently, these intermittent energy sources can be fully exploited only if efficient, safe and reliable electrical energy storage (EES) systems are provided. Energy storage systems when used in a distributed manner, like at the point of use, also allow us to reduce the load on our existing electrical generation and transmission infrastructures during peak power demands.

Electrochemical energy storage devices such as batteries, fuel cells and electrochemical capacitors are among the leading EES technologies. Their power and energy density characteristics are shown in Figure 1.[1] Capacitors with their very high power densities, low energy densities, and sub-second response times are more suitable for power quality management. Only batteries and fuel cells/flow batteries have the energy densities needed for large-scale electrical energy storage.

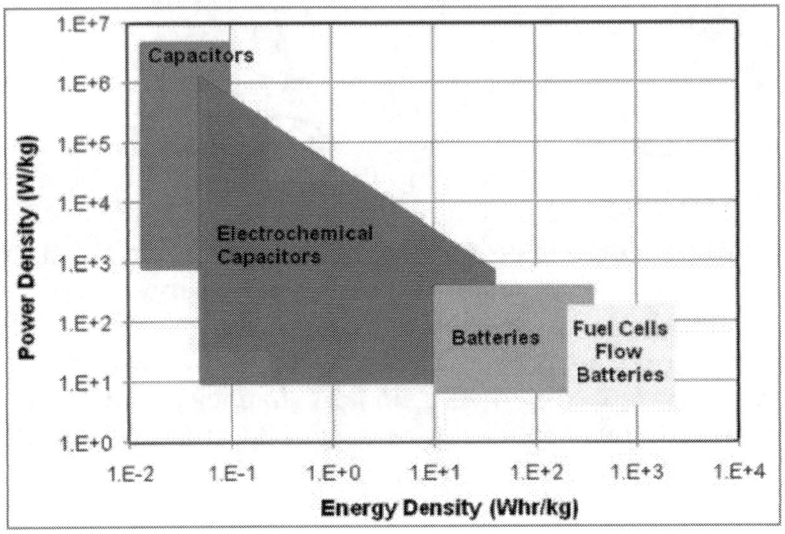

Figure 1
Power and energy densities of various electrical energy storage systems

Batteries and fuel cells/flow batteries differ in two main aspects. In a battery, the energy is stored internally, and the electrodes at which the energy conversion reactions occur are themselves the electrochemical fuel. The characteristics of the electrodes determine both the power density and the energy density of the battery. As a battery converts its chemical energy to electrical energy, its energy conversion components, the electrodes, are consumed and undergo significant physical and chemical changes which affect its electrical performance. Furthermore, because of the dual functions of the electrodes described above, a battery has minimal or no scale-up advantages. Instead, it can only be scaled-out. That is, if more energy is needed, then more battery modules with identical components are added. Consequently, a battery can never approach its theoretical energy density.

A flow battery or fuel cell, on the other hand, is an energy conversion device in which the active materials are stored externally. The electrodes of a fuel cell can be designed to have optimal energy conversion properties without the need to also maximize energy storage density. Furthermore, the electrodes do not undergo physical and chemical changes during operation, thus leading to more stable and durable performance. The energy capacity requirement of a fuel cell is addressed by the size of the external fuel and oxidant storage devices. Consequently, a fuel cell system could theoretically approach its theoretical energy density as the system is scaled up to a point where the weight or volume of the fuel cell is small relative to that of the stored fuel and oxidant. A conventional analogous system is the internal combustion engine system in which the power is determined by the size of the engine and the energy density is determined by the size of the fuel tank. A review of flow batteries characteristics is recently provided.[2] Table 1 lists the characteristics of some flow battery/fuel cell systems. Gasoline and lithium ion batteries are included for comparison.

Table I. Characteristics of some energy storage systems

	Theoretical Energy Density (based on active materials only)	**Electric-to-Electric Efficiency**
System	(Whr/kg)	(%)
H_2/O_2	32640 (fuel only)	40
HBr (acid)	365	80
HBr (alkaline)	425	—
Vanadium	90	80
Lithium ion	600	85
Gasoline	12200 (fuel only)	—

As shown in Table I, the H_2/O_2 (air) fuel cell has the highest gravimetric energy density. However, due to its low electric-to-electric energy conversion efficiency, it is less attractive than the HBr system. The low efficiency of the H_2/O_2 system is attributed to the slow kinetics of the oxygen reactions. Even with noble metal catalysts such as platinum, the rate of oxygen reduction is still four orders of magnitude slower than that of the hydrogen oxidation reaction. The vanadium ion flow battery system has received attention; however, there are concerns over toxicity and its low energy density. The HBr system is selected for this study because of its fast reaction kinetics which translates to lower voltage losses and higher electric-to-electric energy conversion efficiency, very high power density,[3] very high energy storage capacity, potentially low cost electrode materials, simplicity, and reliability. Note that hydrogen storage is not a major issue in this application because it does not have the same volume and weight constraints as in the transportation applications. Even though bromine gas is toxic, its containment in a closed-cycle process can be accomplished, as demonstrated by the containment and handling of chlorine gas in the chlor-alkaline processes.

The charge and discharge reactions occurring in an acid-based HBr flow battery are as follows. At the bromine electrode, during the charge cycle, bromide ions in an HBr solution are oxidized to form bromine and two electrons,

$$2(H^+Br^-) \xleftarrow{\textit{discharge}} \xrightarrow{\textit{charge}} Br_2 + 2H^+ + 2e\text{-},\ E^o = -1.09V.$$

A high concentration of bromine remains in the electrolyte due to its high solubility. Br_2 solubility in the electrolyte is further enhanced by the formation of a tri-bromide complex [4]

$$Br_2 + Br^- \qquad\qquad Br_3^-$$

where K_{eq} is the equilibrium constant for the reaction at 25°C. The hydronium ions migrate to the hydrogen side of the fuel cell and are reduced to H_2,

$$2H^+ + 2e- \qquad\qquad H_2, E^o = 0.0V.$$

The overall reaction is

$$3(H^+Br^-) \qquad\qquad (H^+Br_3^-) + H_2, E^o_{cell} = -1.09V.$$

During discharge the reverse reactions occur. Figure 2 shows a schematic of an HBr flow battery system with the chemical and physical processes occurring within an HBr cell. The main functions of the cation-exchange membrane, beyond acting as a physical separator to prevent an electrical short-circuit of the two electrodes, is to allow the hydronium ions to migrate to and from the hydrogen electrode during charge and discharge while minimizing the crossover of hydrogen to the bromine side and keeping bromide ions and bromine from contacting the hydrogen electrode.

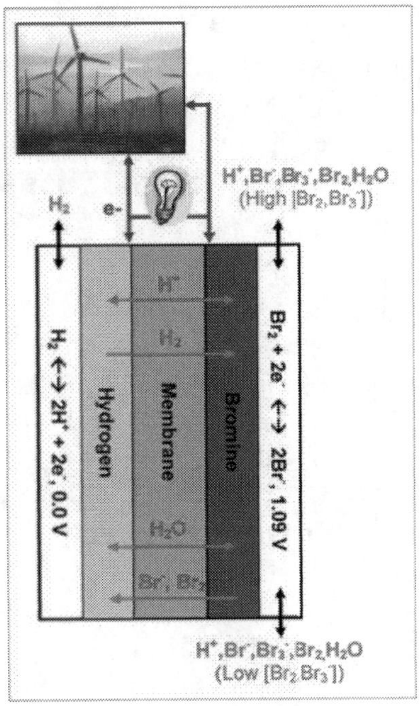

Figure 2
Schematic of an HBr flow battery storage system

Since bromide ions are known to poison hydrogen electro-catalysts, their crossover rate must be kept to a minimum. Similarly, if bromine is allowed to reach the hydrogen electrode, it will react electrochemically to form Br^-, leading to a chemical short-circuit, lower fuel utilization, and poisoning of the hydrogen catalyst.

Current research activities in this area by our group and our partners' group include design and synthesis of low cost and durable non-noble electrocatalysts with high reactivity and selectivity for the hydrogen and bromine reactions, development of durable and highly permselective membranes, and electrode microstructures and cell designs that minimize transport effects and maximize conversion efficiency. Preliminary test results on the HBr system using existing components from the hydrogen/air PEM fuel cell system are presented next to demonstrate the capability of this system.

Experimental

A 5 cm^2 HBr flow battery consisting of PTFE endplates, stainless steel current collectors, graphite flow fields and a seven-layer (cathode-GDL/MPL/CL/Membrane/CL/MPL/GDL-anode) membrane electrode assembly (MEA) made by TVN Systems, Inc., was used. Electrodes with 0.75 mg Pt/cm^2 loading and Nafion 112 membranes were used. Interdigitated flow fields were used to improve mass transport to the electrodes.[5] The cell was tested at room temperature (22°C). Hydrogen at 2-3 psig was recirculated through the anode (H_2), and a solution of 2M HBr/2M Br_2 was pumped through the cathode (Br_2) at a flow rate of 1.0 A/cm^2 equivalent using a peristaltic pump. A potentiostat/gavalnostat by Arbin Instruments was used to control the cell potential. The HBr flow cell was tested in potentiostatic mode. The data reported represent average values collected over 5 minutes.

Results

Figure 3 shows the discharge and charge performance of an HBr flow cell tested under the conditions described earlier. For comparison, the discharge performance of a H_2/Air PEM fuel cell under similar operating conditions is included. The first contrast observed from Figure 3 is the absence of a vertical voltage drop at low current density in the HBr discharge curves that is normally observed in H_2/O_2 (or air) systems as a result of the slow kinetics or high overpotential associated with the oxygen reduction reaction. The discharge peak power density obtained from the HBr flow cell using similar materials and under similar conditions is more than 100% higher than that from the H_2/Air fuel cell. If the ionic and mass transport resistances are reduced, which is equivalent to reducing the slope of the polarization curve, the HBr system is expected to yield much better performance. Achieving this goal is one the objectives of our research activities.

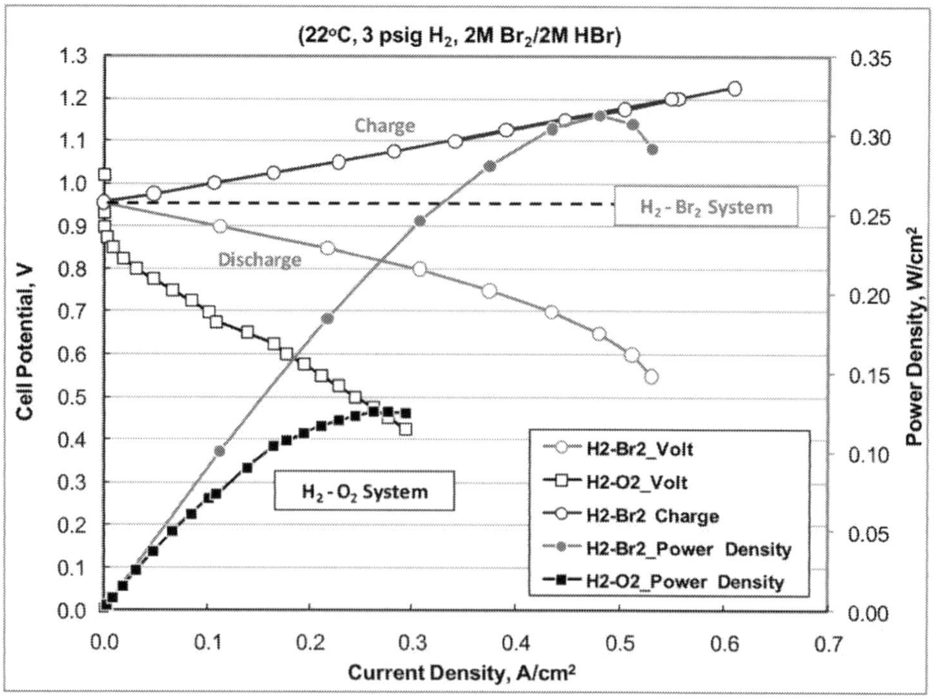

***Figure 3**. Performance of an HBr flow battery in comparison to a H_2/O_2 fuel cell*

Summary

A suitable electrical energy storage system enables us to exploit the full capacity of renewable energy sources such as wind and solar and to utilize our existing electrical energy generation and transmission infrastructures. The Hydrogen-Bromine (HBr) flow battery has been identified as a great candidate for this application because of its extremely fast reaction kinetics, high energy storage capacity, and high reliability. Areas that need to be overcome to make the commercialization of this system possible include low-cost, durable and highly active electrocatalysts; low-cost, durable and highly permselective membranes; and electrodes and cell designs that minimize transport resistance.

References

1. J. Goodenough, H. Abruna, and M. Buchanan, "Basic Research Needs for Electrical Energy Storage," Report of the DOE Basic Energy Sciences Workshop on Electrical Energy Storage, April 2-4, 2007.
2. T. Nguyen and R. Savinell, "Flow Batteries," *The Electrochemical Society's Interface*, Vol. 19, No. 3, 54-56 (2010).
3. V. Livshits, A. Ulus, and E. Peled, "High-power H2/Br2 fuel cell," *Electrochem. Comm.*, Vol. 8, 1358-1362 (2006).
4. R. Baldwin, "Electrochemical Performance and Transport Properties of a Nafion Membrane in a Hydrogen-Bromine Cell Environment," NASA Report TM-89862 (1987).
5. T. Nguyen, "A Gas Distributor Design for Proton Exchange Membrane Fuel Cells," *J. Electrochem. Society*, **143**, L103-L105 (1996).

Acknowledgements

The author wishes to acknowledge the financial support of this work by the National Science Foundation through grant no. EFRI-1038234.

(564g) High Throughput Experimentation As Efficient Tool for the Testing of Hydroprocessing Catalysts

Sascha Vukojevic and Josef Find, hte AG, Heidelberg, Germany

HighThroughput Experimentation as Efficient Tool for the Testing of HydroprocessingCatalysts

Increasing globaldemands for fuels, the tightening of environmental regulations and theoptimization of refineries has led to an increasing demand for catalyststesting capacity. High throughput catalyst testing is a time-efficient approachfor meeting this demand, however in the past the increase in testing throughputwas accompanied by a loss of data quality.

Over the last coupleof years hte has developed parallel reactor technology for testing ofHydroprocessing catalysts under a wide range of process conditions in order tomeet this demand. The latest development is capable of processing VGO, heavyVGO and bio-oil under industrially relevant conditions. Temperatures of up to450°C can be achieved routinely, as well as pressures from 20 up to 170 bar.

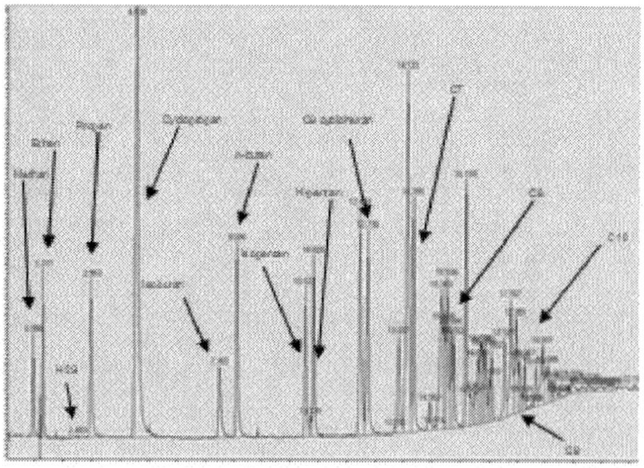

Figure 1: Hydrocracking gas phase products by online GC.

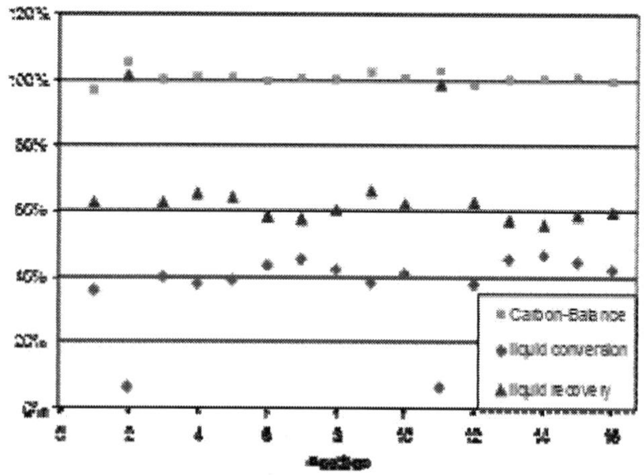

Figure 2: Carbon balance, liquid conversion and liquid recovery in a reactor-to-reactor comparison of HVGO hydrocracking in one of hte's parallel trickle-bed reactor systems.

A demanding examplefor the parallel testing of hydroprocessing catalysts is hydrocracking. The productsof this process depend strongly on the reaction conditions (temperature,pressure, catalyst activity). These products range from ethane, LPG to heavierhydrocarbons comprising mostly isoparaffins. Closing the total carbon massbalance covering gas and liquid products, under demanding reaction conditionsis one of the latest major achievements for high throughput reactor systems.Figure 1 reveals an example for the complex gas composition of hydrocrackingproducts.

The results that willbe presented will show the effectiveness of hte's parallel testing technologyand the data quality standard in a reactor-to-reactor and run-to-run comparisonfor hydrodesulphurization and hydrocracking processes over several monthsfeeding diesel, VGO, heavy VGO and bio-oil, as it is shown in Figure 2 for thehydrocracking of HVGO in a 16-fold parallel reactor system. To prove the valueof the obtained data, the experimental results will be compared to pilot plantdata. The latest developments in the field of high throughput reactortechnology for refining applications will be shown, including parallel catalysttesting technology in a sub-pilot scale with higher catalyst volumes.

See more of this Session: Alternative Fuels and Enabling Technologies III
See more of this Group/Topical: Fuels and Petrochemicals Division

(349a) Thermochemical Conversion of Coal and Biomass to Liquid and Gaseous Fuels

Venkata Abhijit Bhagavatula, Naresh Shah and Gerald Huffman, Consortium for Fossil Fuel Science, Department of Chemical and Materials Engineering, University of Kentucky, Lexington, KY

ThermochemicalConversion of Coal and Biomass to

Liquid andGaseous Fuels

Abstract

Venkata AbhijitBhagavatula[1], Naresh Shah[2]and Gerald Huffman[2]

Consortiumfor Fossil Fuel Science, Department of Chemical and Materials Engineering,

Universityof Kentucky, Lexington, KY - 40508

Introduction:

Theuse of biomass as a renewable source of energy has been increasingly gainingimportance over the past few years. The growing concerns over green house gasemissions, along with the need for energy independence have accentuated theneed for using alternative sources of fuel and chemical feedstocks. A flowchartillustrating the process of converting solid coal/biomass into gaseous andliquid fuel is shown in Figure 1.

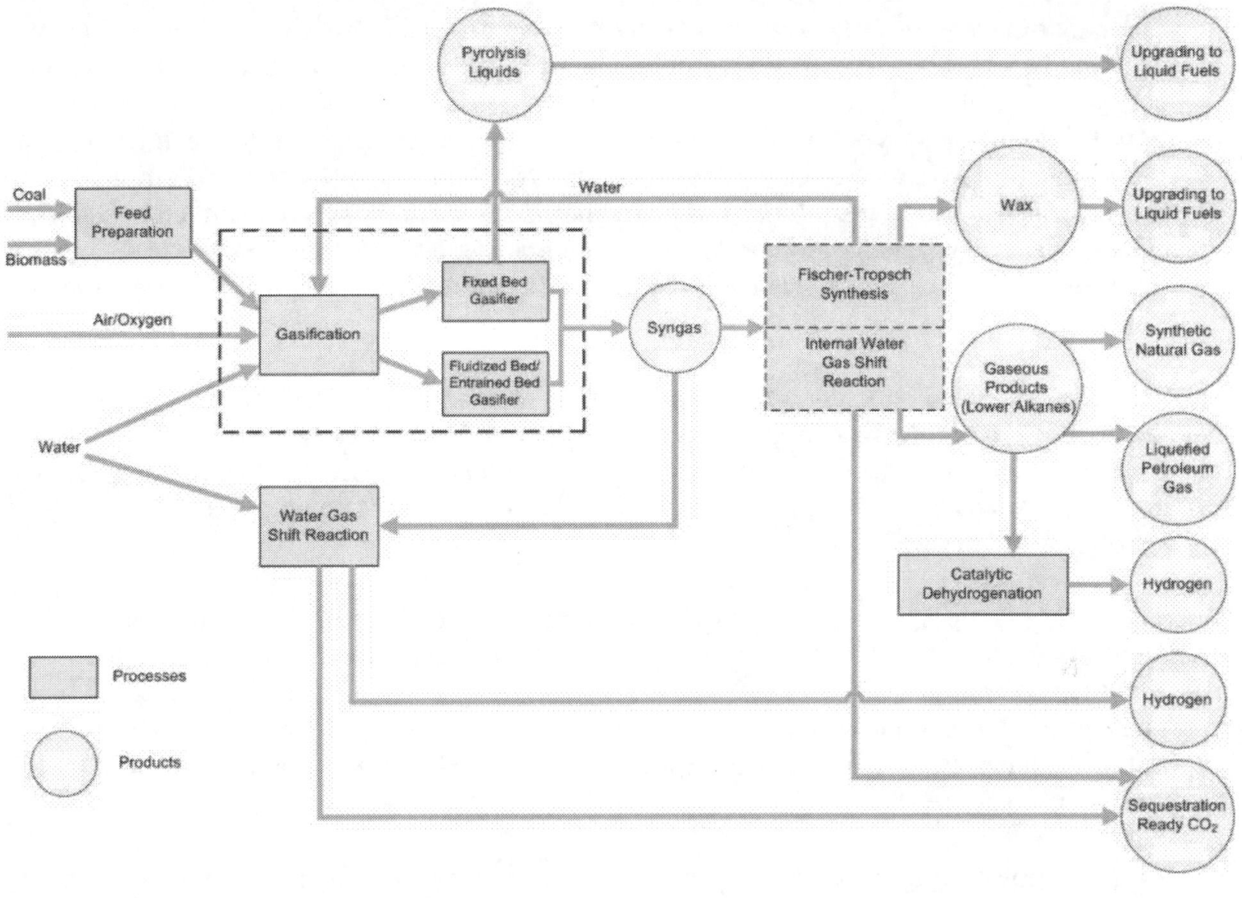

Figure 1. Flowchartillustrating the process of converting solid coal/biomass into gaseous andliquid fuel.

Gasificationis the thermochemical process of converting a solid carbonaceous source such ascoal or biomass into synthesis gas ($CO + H_2$) which is used as afeedstock for producing fuels and chemicals. Gasification, generally done inthe presence of mixtures of air/pure oxygen and steam at temperatures rangingbetween 600 °C and 900 °C, produces a gaseous mixture of hydrogen, carbonmonoxide, steam, methane and light hydrocarbons along with some undesirableeffluents such as inorganic particulates and condensable organic vapors ortars, as they are commonly known.

Using biomass alone asthe gasification feed greatly reduces the green house gas emissions but theprocess not only involves higher operating costs but also produces higheramount of tar when compared to that of coal. The high tar content of productgases from biomass gasification is a major and widely recognized problem. Thesehigh tar contents arise mainly from the lower temperatures and shorterresidence times in gasifiers constructed for biomass processing compared tothose designed for coal gasification. Tar yields from ligno-cellulosic biomassmaterials tend to be considerably higher than tar yields from coals. Co-gasificationof blends of coal and biomass to produce syngas ($CO + H_2$), which inturn can be used as feedstock for processes such as Fischer-Tropsch Synthesisused for producing liquid fuels, is a major area of research.

Themain objective of this research is to perform the co-gasification of differentblends of coal and biomass in a fixed bed reactor and analyze the products inorder to achieve high concentration of synthesis gas. Two different types ofcoal namely, sub-bituminous and lignite, and different types of biomass suchas, pine and switchgrass, have been used for the co-gasification process. Usingthermogravimetric analysis, the partial oxidation characteristics of mixturesof coal and different biomass samples have been analyzed. The effects of variouscoal and biomass blends, temperature, pressure and oxygen to steam ratio, inaddition to the particle size effects, on the gasification products have beenstudied thoroughly.

The main advantage ofusing a fixed bed reactor is because of the fact that in such a reactor highcarbon conversion, longer solid residence times and low ash carry over can beobtained. In a fixed-bed gasifier, the gasificationprocesses occur in four different zones known as combustion or oxidation,gasification or reduction, de-volatilization or pyrolysis, and drying.

The energy content ofthe gas produced through gasification depends on numerous factors, such as theoxidizing agent, reactor type, fuel type and form, etc. The oxidizing agent canbe chosen as air, oxygen, steam, or a mixture of these. When air is used, theresulting gas has a low calorific value. This can be increased by using oxygenor steam but in the latter case sufficient heat should be provided becausesteam gasification is an endothermic process.

Also, as shown inFigure 1, the conversion of synthesis gas, obtained from the gasification ofcoal and biomass, to hydrocarbons using Fischer-Tropsch Synthesis (FTS) hasbeen studied using both Cobalt and Iron based catalysts in a fixed bed reactor.The synthesis gas has to be cleaned of all impurities and adjusted to the needsof the subsequent Fischer-Tropsch synthesis step using Water Gas Shiftreaction, either internally using an iron based catalyst or externally using acobalt based catalyst. The detailed product distributions were measured usingonline gas chromatography. In addition to this, the carbon number distributionof the Fischer-Tropsch Synthesis products, known as the chain growthprobability factor, has been determined using the Anderson-Schultz-Flory (ASF)theory. The effect of various catalyst loadings and different heat transferagents on the product distribution would be presented in detail.

See more of this Session: Biomass Pyrolysis I
See more of this Group/Topical: Fuels and Petrochemicals Division

(218b) High Energy Density Lithium Flow Battery Design

Galen Suppes[1], Michael J. Gordon[2], Michael G. Heidlage[1] and David Guinta[1], (1)Chemical Engineering, University of Missouri - Columbia, Columbia, MO, (2)University of Missouri - Columbia, Columbia, MO

Traditional sandwich cell battery chemistries are used in a flow cell where a pump circulates fluid between electrodes through a permeable membrane. The energy densities of these batteries are many fold higher than flow batteries where one or more reagent is stored in the electrolyte. Side-by-side comparisons with sandwich (diffusion) batteries show considerable increases in delivered energy at high loads as well as higher power densities.

See more of this Session: Flow Batteries for Energy Storage (In Memory of Bernard Lee)
See more of this Group/Topical: Fuels and Petrochemicals Division

(436g) Pyrolysis of Marine Biomass in a Fixed-Bed Reactor

Kyung A. Jung, School of Environmental Science and Engineering, Advanced Environmental Biotechnology Research Center, POSTECH, Pohang, South Korea, Seong-Rin Lim, School of Environmental Science and Engineering, POSTECH, Pohang, South Korea and Jong Moon Park, Division of Advanced Nuclear Engineering, Department of Chemical Engineering, POSTECH, Pohang, South Korea

A thermochemical platform including a pyrolysis reaction is an attractive promising biomass conversion process since utilization of biomass is not limited comparing with other biochemical processes. The quantitative and qualitative efficiency of the pyrolysis process under anaerobic conditions relies on biomass types as well as process parameters such as heating rates, reactor types and mass loading. Therefore, a study on thermal characteristics of diverse biomass under different conditions is the first stage to study feasibility for application of pyrolysis systems and their products. In this study, we investigated the pyrolytic characteristics of the green (*Ulva* sp.), brown (*Laminaria* sp. and *Sargassum* sp.) and red (*Gelidium* sp.) seaweed and obtaind py-products from the seaweed biomass using a fixed-bed pyrolysis reactor. The thermal decomposition characteristics of the seaweed biomass were explored through thermogravimetric analysis (TGA) with differential thermal analysis (DTA). The chemical characteristics of py-products in three phases: oil, char and gas obtained under different process conditions were analyzed using GC/MS and FT-IR.

See more of this Session: Biomass Pyrolysis II
See more of this Group/Topical: Fuels and Petrochemicals Division

(13a) Mechanistic Modeling of Fast Pyrolysis of Cellulose to Predict Bio-Oil Composition

Vinu Ravikrishnan and Linda J. Broadbelt, Chemical and Biological Engineering, Northwestern University, Evanston, IL

Fast pyrolysis, a potential strategy for the production of transportation fuels from biomass, involves a complex network of competing reactions, which result in the formation of bio-oil, non-condensable gaseous species, and solid char. Bio-oil is a mixture of anhydro sugars, furan derivatives, and oxygenated aromatic and low molecular weight (LMW) compounds. Currently, the successful modeling of fast pyrolysis reactors for biomass conversion is hampered by lumped kinetic models, which fail to predict the bio-oil composition. Hence, a fundamental understanding of the chemistry and kinetics of biomass pyrolysis is important to evaluate the effects of process parameters like temperature, residence time and pressure on the composition of bio-oil.

We have recently developed a comprehensive mechanistic model to characterize the primary products of fast pyrolysis of pure cellulose. The model incorporates the following condensed phase reactions with ionic and electrocyclic fragmentation mechanisms for the formation of various LMW products in bio-oil: glucosidic bond scission, retro-aldol, retro-Diels-Alder, 1,2-dehydration, 1,3-dehydration, hydrolysis, cyclic Grob fragmentation and enol-keto transformation reactions. The kinetic rate coefficients for most of the above reaction steps were obtained from the literature, based on either experimentally determined Arrhenius parameters for cellulose and model compounds, or theoretically computed values of activation energies using quantum chemical calculations. A computational framework based on continuous distribution kinetics was constructed to solve the kinetic model. The model tracks 96 distinct polymer species and 25 LMW products, and incorporates over 350 reactions. The model predictions compare well with the experimental data (Patwardhan *et al.*, *J. Anal. Appl. Pyrolysis* **2009**, 86, 323-330) for the formation of important products like levoglucosan (60 wt.%), glycolaldehyde (6 wt.%), 5-hydroxymethyl furfural (3 wt.%), furfural (1 wt.%) and formic acid (6 wt. %). Levoglucosan was found to be formed by the initiation of ionic chain ends by mid-chain glucosidic bond scission, followed by end-chain ionic unzipping. Some of our efforts in extending the kinetic model of pyrolysis of pure cellulose to cellulose with mineral matter and hemicellulose will also be discussed in this presentation.

See more of this Session: Catalytic Conversion of Renewable Resources to Synthesis Gases and Pyrolysis Oils
See more of this Group/Topical: Sustainable Engineering Forum

Production of Synthetic Paraffinic Kerosene by Hydrocracking Fischer-Tropsch Wax

Integrity ★ Service ★ Excellence

Presented at the 2011 AIChE Annual Meeting, Minneapolis, MN

19 October 2011

HN Conkle, GM Marcum, EW Griesenbrock, and WA Jones, Battelle;
RW Morris, Jr. , AFRL/RZPF;
HJ Robota and DK Thomas, UDRI

Battelle
The Business of Innovation

Public Release Case No. : 88ABW-2011-4512

 AFRL 1

Abstract

- The US Air Force has committed to an alternative aviation-fuel initiative to implement a roadmap for testing and certifying alternative fuel components
- The impact of these new fuel blends on performance, the environment, safety and occupational health, and the fuel storage and supply infrastructure are being examined
- The characteristics of petroleum-based jet fuels as well as coal and natural gas based synthetic jet fuels are well known. In contrast, the fuel properties of biobased components are less well defined. A way to generate specialized fuel surrogates is needed to determine the effect of these new fuel components on combustion efficiency, emissions, and fit-for-purpose performance
- To generate these surrogates, the Air Force has erected a Sample Preparation Unit (SPU) at the Assured Aerospace Fuels Research Facility (AAFRF) located at Wright-Patterson Air Force Base
- The SPU was operated 24/7 to hydrocrack Fischer Tropsch (F-T) wax to produce approximately 10 gallons/day of on-spec synthetic paraffinic kerosene (SPK)
- This paper will describe the AAFRF SPU, operational testing results, and future fuel-surrogate production plans

 AFRL 2

Presentation Outline

- Acknowledgement
- Background
- Sample Preparation Unit (SPU)
- Fischer-Tropsch (F-T) Wax Hydrocracking Results
- Next Test
- Future Plans

 AFRL 3

Background

Air Force alternative aviation-fuel initiative

Early 2011 Certification Goal

Certify entire AF Fleet to use a 50/50 SynFuel blend

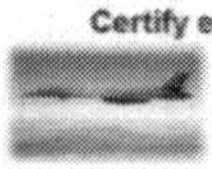
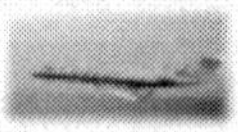

2016 Acquisition Goal

Cost competitively acquire 50% of Air Force's domestic aviation fuel requirement via alternative fuel blends in which the alternative component is derived from domestic sources produced in a manner that is greener than fuels produced from conventional petroleum

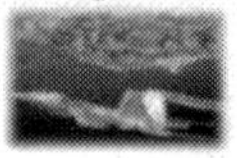

 AFRL 4

AAFRF Supports AFRL Mission

- The AAFRF provides a research test bed that will allow fuels researchers to answer practical questions about fuels from alternative sources and how these fuels might impact or relate to Air Force weapons systems
- This information will allow researchers to explore the full range of variables that can be manipulated in the formulation of aviation turbine fuels as they prepare for synthetic-rich fuels in the future

US Air Force C-17A with SJ-8 fuel, a 50/50 blend of JP-8 and synthetic fuel.

http://www.wpafb.af.mil/news/story.asp?id=123040425

Approved for public release; distribution unlimited

AFRL 5

SPU Features

- Fixed-bed reactors can be operated in series, parallel, or independently
- Three distillation columns for intermediate and product separation
- Ability to distill final product from 80 psig to ~2 psia
- Mass flow measurement on all streams
- Recycle of H_2 and vacuum column underflow
- iFIX® SCADA control of the system
- Safety PLC

Approved for public release; distribution unlimited

AFRL 6

SPU Process Diagram (Configuration Used for Commissioning)

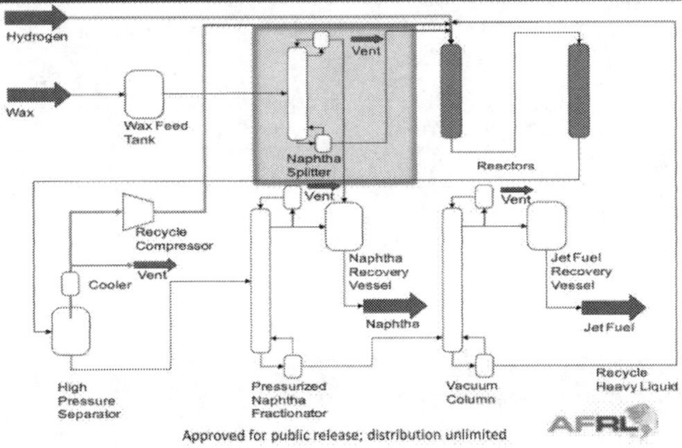

Approved for public release; distribution unlimited

AFRL 7

SPU Parameters

SPU Parameters	
System Pressure Design Limits	
Reaction System	600-2000 psig
Naphtha Splitter	10-25 psig
Naphtha Fractionator	20-85 psig
Vacuum Column	2-10 psia
H_2 Booster Pump	1200-2500 psig
System Temperature Design Limits	
Reaction System	550 – 850 °F
Naphtha Splitter	Reboiler : 750 °F Condenser : 100 °F (end)
Naphtha Fractionator	Reboiler : 750 °F Condenser : 100 °F (end)
Vacuum Column	Reboiler : 650 °F Condenser : 200 °F (end)
Wax Containing Equip	150-375 °F
H_2/Light Gas Recycle	70-150 °F

Approved for public release; distribution unlimited

AFRL 8

SPU
Process Safety Features

- Class I Div 2 Group B design
  - Deviation approved for reactor furnaces
- Dual PLC controls (Honeywell)
  - Safety shut-down separated from process controls
  - Both systems report to IFIX® Supervisory Control and Data Acquisition system
- H_2 detection systems
  - 6 detectors on equipment over most likely leak points
  - Integrated with ventilation and facility alarms
- Continuous N_2 purge for every vessel/process element not containing hydrogen
- Pressure relief on every vessel
- Dual Header System
  - Process vent header
  - Separate pressure relief header
  - Flame arrestor on each header
- Remote H_2 storage to minimize quantity in cell
- O_2 sensor on vacuum column
- Safety Permit required before operations

F-T Wax Hydrocracking
Goals and Objectives

- **Goal: Establish a baseline set of operating conditions required to meet product distributions and throughput of synthetic paraffinic kerosene (SPK: C_9-C_{16})**
- **Feed Conversion Objective:**
  - Reactor output C_9-C_{16} level: > 40 wt% SPK
- **Naphtha Separation Objectives:**
  - Overhead C_9-C_{16} retention: < 5 lb SPK in overhead/100 lb SPK fed
  - Bottoms C_3-C_8 level: < 1 wt% naphtha
- **SPK Recovery and Properties Objectives:**
  - Overhead C_9-C_{16} recovery: > 70 lb SPK in overhead/100 lb SPK fed
  - Overhead C_{17+} level: < 1.3 wt% total in product

Test Results
For 3 Campaigns

Run 5 Campaign No.			1	2	3
Duration			10 days	10 days	6 days
Objectives				Initial Results	
Reactor output C_9-C_{16} level, %	>	40	37	39	40
Fractionator overhead C_9-C_{16} retention	<	0.05	0.01	0.03	0.13
Fractionator bottoms C_3-C_8 level, %	<	1	4	0	0
Vacuum column overhead C_9-C_{16} yield	>	0.7	1.0	0.8	0.7
Vacuum column overhead C_{17+} level, %	<	1.3	5	0.9	0.3
SPK production rate, gal/day	>	10	5	6	10

Test met Run goals and objectives

Reactor Operation

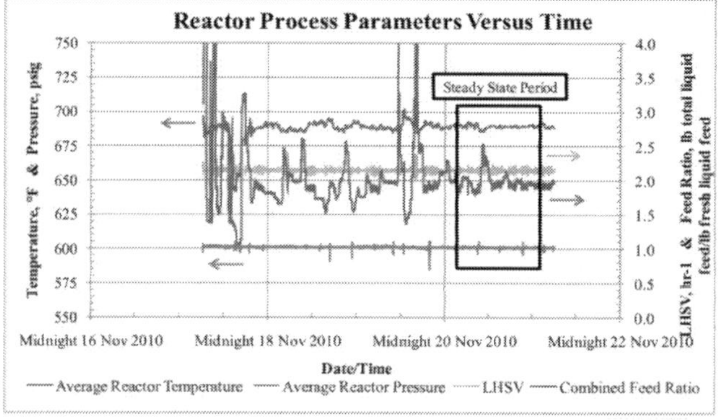

Distillation

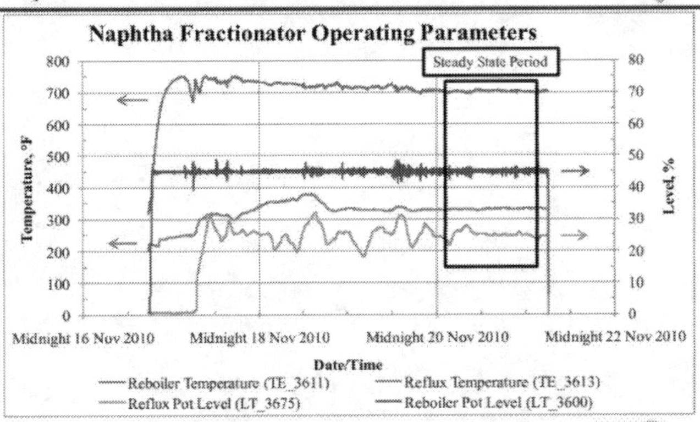

Approved for public release; distribution unlimited

AFRL 13

Distillation

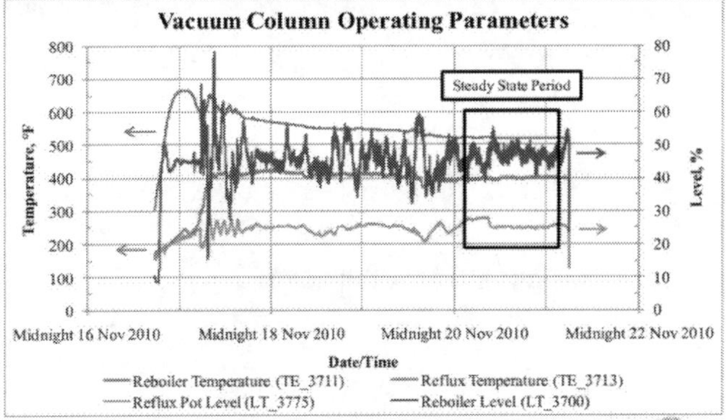

Approved for public release; distribution unlimited

AFRL 14

AAFRF SPK Produced from F-T Wax

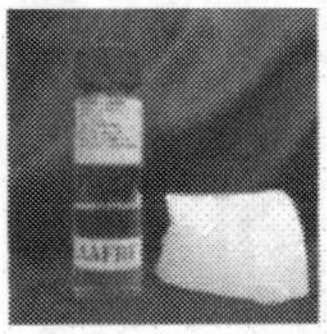

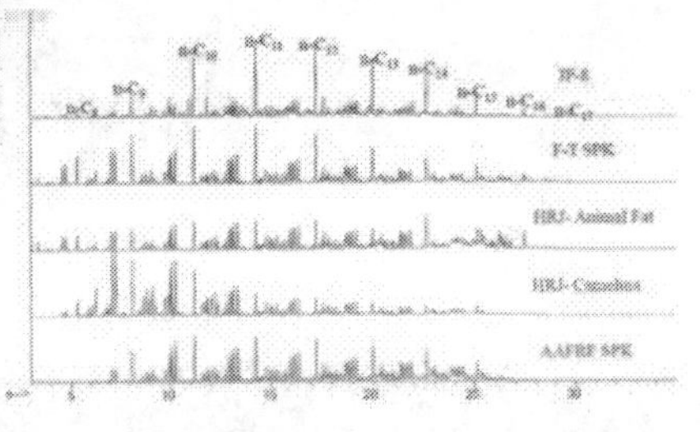

AAFRF SPK properties are nearly identical to other non JP-8 jet fuels

Approved for public release; distribution unlimited

AFRL 15

Conclusions from Initial F-T Wax Testing

- Hydrocracked F-T wax to SPK
- Distillation columns allowed separation and recovery of SPK
- Hydrocracker operated at design capacity
- Sample met SPK (MIL-DTL-83133G) specifications

Approved for public release; distribution unlimited

AFRL 16

Transition
Move SPU from Bldg 490 to 496

- Completed commissioning and initial F-T wax hydrocracking studies in Test Cell 154 in Building 490
- Move SPU to Building 496; a new dedicated building
- Start up planned for fall 2011

Building 496
Artists Conception

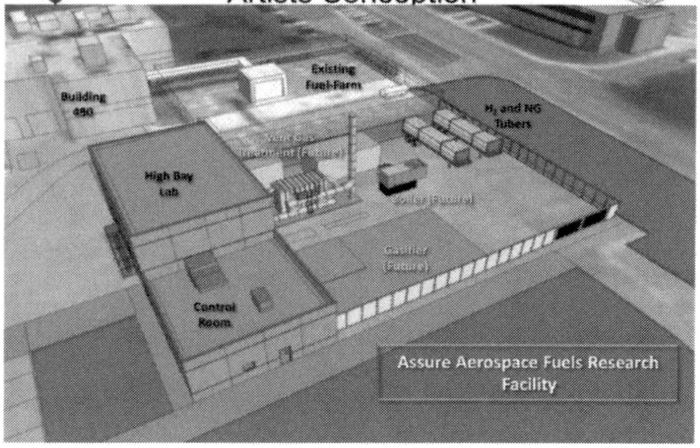

Ground Clearing

Frame Erection

Concrete Walls Attached

Approved for public release; distribution unlimited AFRL 21

Building 496 Today

Approved for public release; distribution unlimited AFRL 22

SPU Brought To New Building

Approved for public release; distribution unlimited AFRL 23

Advantages of 496

- A dedicated permanent home for AAFRF-related alternative fuels research
- Facility includes better H_2 supply systems
- Reduced distractions
- Operating space to include
  - Room for other skids for collaborative research programs
  - Room for 'Green' research programs
- Other collaborative efforts and programs

Approved for public release; distribution unlimited AFRL 24

Control Room

- 40 ft by 40 ft by 18 ft high
  - Gas sampling
  - Data analysis
  - "Up and Ready"
  - SPU
  - Video
  - RCS (facility support)
- Can support SPU as well as a separate operating unit via the "Up and Ready" station

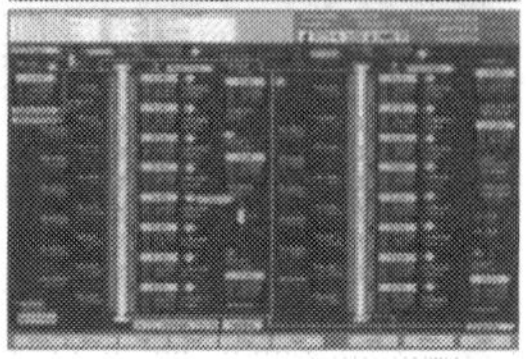

Approved for public release; distribution unlimited AFRL 25

SPU Being Installed in High Bay

- High Bay: 40 ft by 60 ft by 32 ft high

Approved for public release; distribution unlimited AFRL 26

2011 Transition Schedule

- Complete installation – September
- Complete commissioning – October
- Initiate operation – November

Approved for public release; distribution unlimited AFRL 27

First SPU Test in 496 Will Support Alternative Fuels

- JP-8 spec includes physical and chemical requirements based on experience with petroleum-derived fuels.
- However, with fuels derived from alternative sources, the experience base is not so broad, and it is difficult to obtain large, representative samples
- The AAFRF can convert raw feedstocks into fuel surrogates with properties similar to biobased fuels for Air Force testing and evaluation

Approved for public release; distribution unlimited AFRL 28

Biobased Path to Single Carbon-Number Alkanes

Jet fuel consists of distribution of C_9 through C_{16}

Biobased process can produce smaller building (C_4) that can be combined (C_8, C_{12}, C_{16}...). The carbon number of biobased C_4 and C_8 are too low and C_{16} is too high

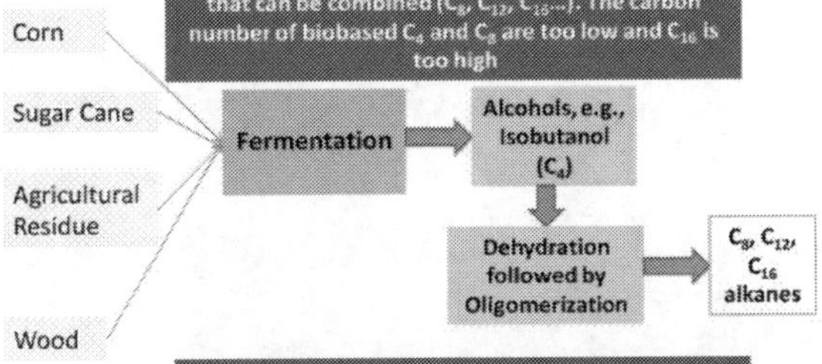

Tetradecane (C_{14}) as selected for testing

Production of an Isomerized Tetradecane (C_{14})

- **Reactor 1: Hydrogenation** (1-Tetradecene → n-Tetradecane)
- **Reactor 2: Isomerization** (n-Tetradecane → mix of isomers and normal C_{14})
  - Mono- and multi-methyl tetradecane
  - Cracked hydrocarbon byproducts (C_4 through C_{11})
- **Distillation Columns: Separation**
  - Tetradecane isomers
  - Light (C_4-C_{10}) and intermediate (C_8-C_{11}) hydrocarbons
- **New Unit: Purification**
  - Separate and recover highly isomerized C_{14} from n-C_{14}

Future Options

- Addition of batch/high-efficiency distillation
- Addition of an oligomerization unit
- Algae/other bio-oil upgrading
- Evaluation of specialty catalysts for upgrading and oligomerization / isomerization

Industrial Partner Participation Opportunities

- Feedstock upgrading evaluation
- New reactor designs
- Oligomerization / isomerization evaluation
- Optimized recovery of products
- Catalyst evaluation
- Production of unique fuels

Acknowledgements

This research was supported, in part, by the U. S. Air Force Cooperative Grant Numbers F33615-03-2-2347 and FA8650-10-2-2934 with Mr. Robert W. Morris Jr. serving as the Air Force Grant Monitor. The research was also sponsored by the State of Ohio Subrecipient Award No. COEUS # 005909 to the University of Dayton (Dr. Dilip Ballal as the Grant Monitor) under the "Center for Intelligent Propulsion and Advanced Life Management," program with the University of Cincinnati (Prime Award NO. TECH 09-022). The authors gratefully acknowledge this grant support.

Thank you to UDRI personnel: Steve Zabarnick, Matthew de Witt, Rich Striebich, Linda Shafer, Ryan Adams, Zachary West, Dave Thomas, Gordon Dieterle, James Shardo, Jerry Grieselhuber, Jeff Coleman, Jeff Unroe, Alan Wendel, Dennis Davis, Ted Williams, David Gasper, Scott Breitfield, James Higgins, Rhonda Cook, Zachary Sander, Jhoanna Alger

Thank you to Battelle personnel: Satya Chauhan, Russ Smith, Mike O'Brian, George Wrenn

Thank you to Air Force Personnel: Lt. Mark Roosz, Lt. Adam Parks, Milissa Flake

Thank you to UTC Personnel: Jennifer Stafford, Steve Procuniar

Approved for public release; distribution unlimited

AFRL 33

Battelle
The Business of Innovation

DISTRIBUTION STATEMENT A – Unclassified, Unlimited Distribution

AFRL 34

Benjamin Levie, Dwight Anderson, Johnway Gao, Alan Lovas

Catchlight Energy LLC (CLE)

A 50/50 Chevron and Weyerhaeuser joint venture.

Our Mission: To commercialize the large-scale production of liquid transportation fuels from sustainable forest-based resources

Weyerhaeuser | Catchlight Energy | Chevron

A Unique End-to-End Value Chain Solution
Leveraging the strengths of two natural resource leaders

Catchlight Energy (CLE) brings strengths not found elsewhere - feedstocks development and procurement, fuels testing and refining, operations, conversion engineering.

CLE drivers aligned with national interests to develop renewable energy platform, increase energy independence & security and promote rural development.

October 15, 2011

Softwood – High Availability/High Recalcitrance

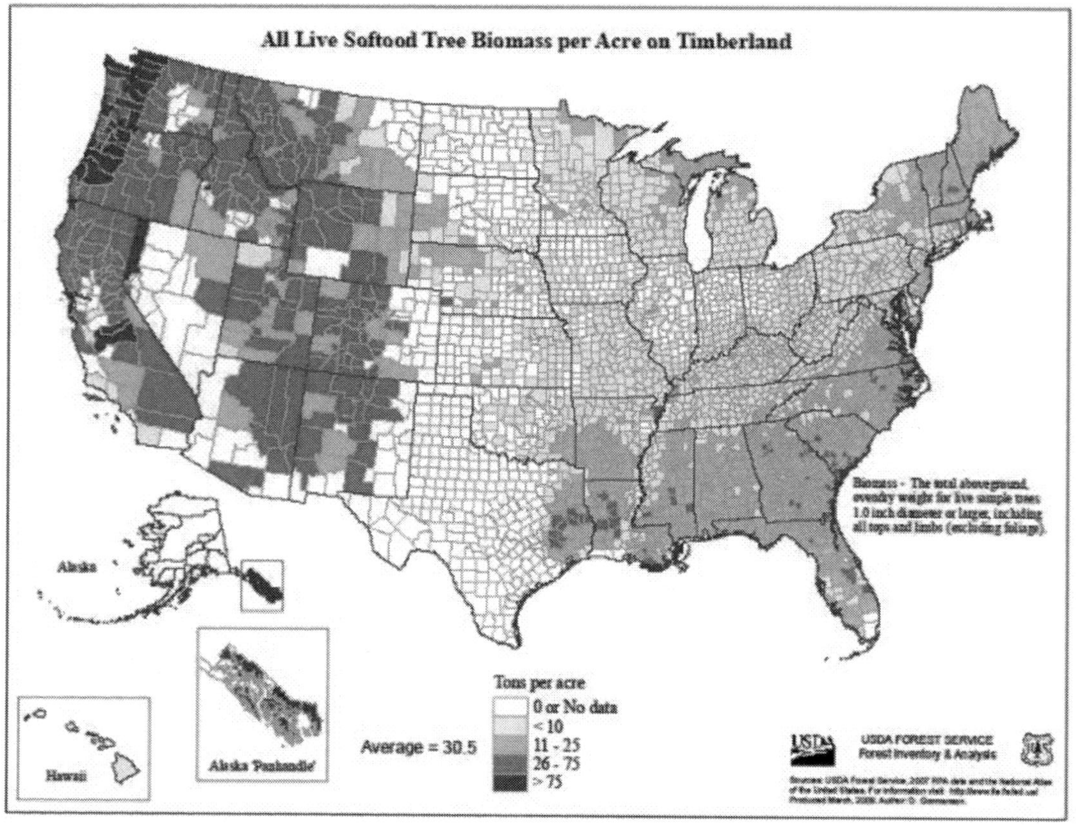

The use of softwood is of high critical value to the biofuels industry, but it remains a difficult feedstock for most pure bio-conversion processes.

October 15, 2011

Pretreatment Issues for Softwood and other Recalcitrant Biomass

Catchlight Energy
A Chevron | Weyerhaeuser Joint Venture

- **Dilute acid and base systems ineffective at reasonable levels.**
- **Known and proposed "Fractionation" pretreatments are excessively expensive, and not commercial scale.**
- **Pulping processes effective because of removal of lignin, but result in high and unsupportable costs.**
  - ➢ High capex (unless repurposed mill that is obtained at very low cost).
  - ➢ High opex (chemical costs or chemical recovery costs, wastewater treatment).
  - ➢ Fuels value can't compete with pulp value.

October 15, 2011

CLE Pretreatment

Catchlight Energy
A Chevron | Weyerhaeuser Joint Venture

- **Pulping, especially sulfite pulping, is an excellent but expensive pretreatment**
- **CLE Pretreatment has**
  - Higher throughput
  - Lower chemical consumption
  - Lower inhibitors
  - Excellent enzymatic hydrolysis results
  - Higher sugar yield

October 15, 2011

CLE Pretreatment Characteristics

Catchlight Energy
A Chevron | Weyerhaeuser Joint Venture

- Adaptable to existing infrastructure
- Scalable from pilot operations
- Feedstock flexible

October 15, 2011

CLE Pretreatment and Process to Sugar

Catchlight Energy

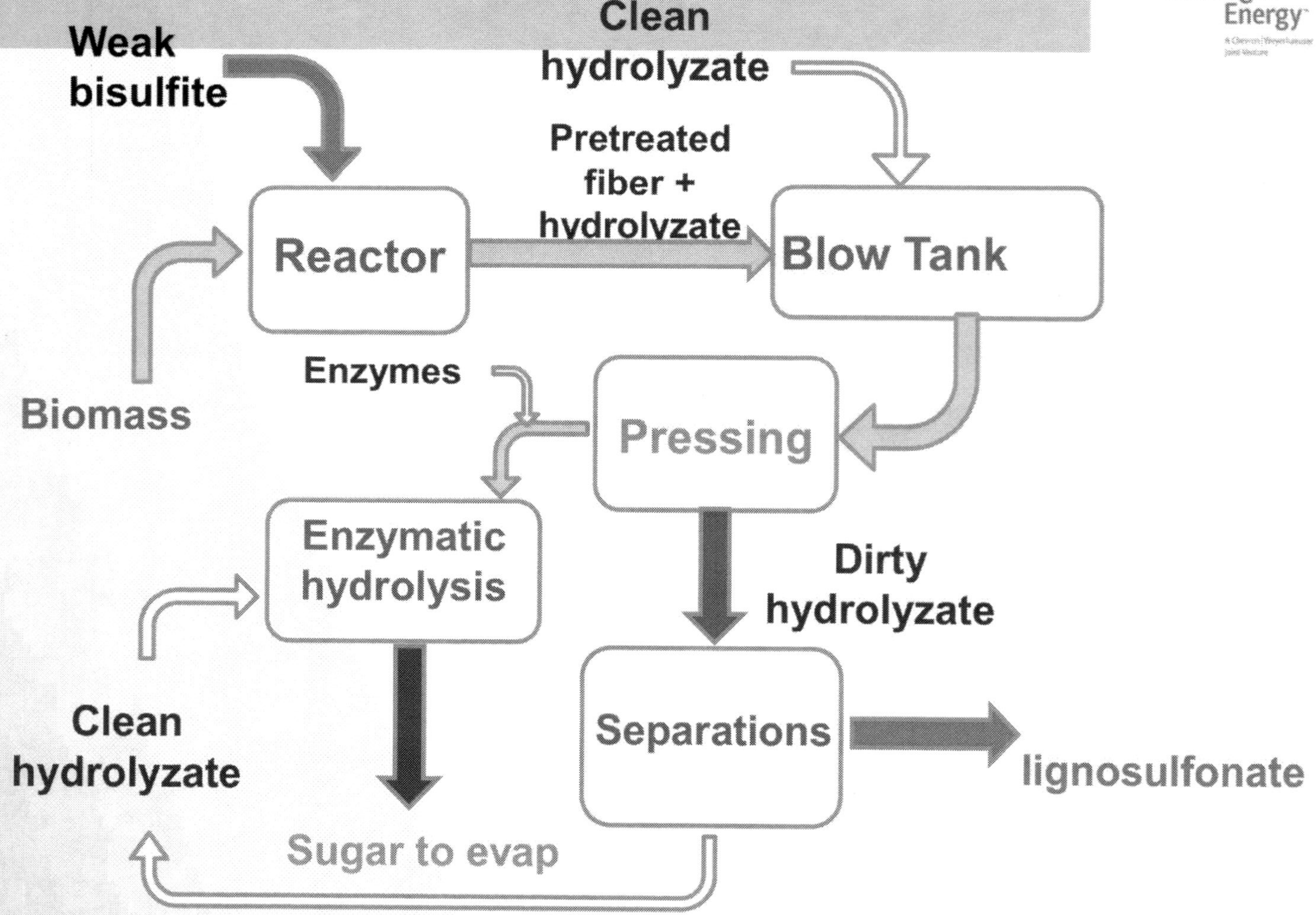

October 15, 2011

Softwood Sugar

Pretreated Softwood–Mill Residuals

Catchlight Energy

Softwood Residuals

Calcium Bisulfite Pretreatment

Pretreated Refined Pulp

Pretreated Residual Sugar Composition:

Arabinan	0.21%
Galactan	0.55%
Glucan	49.9%
Xylan	3.42%
Mannan	3.63%
Total	57.7%

Softwood Residual Sugar Composition:

Arabinan	1.01%
Galactan	2.09%
Glucan	39.5%
Xylan	6.05%
Mannan	10.5%
Total	59.2%

Pretreatment Liquor

Liquor Sugar Composition:

Arabinose	0.07%
Galactose	0.17%
Glucose	0.36%
Xylose	0.48%
Mannose	1.10%
Total	2.18%

October 15, 2011

Softwood Mill Residual Chips

Pilot Pretreatment Results

Pilot Pretreatment Size: 3 kg OD chips/1 cu ft digester	
Overall Pretreatment Sugar Yield on Wood Sugar	88-95%
Furan Formation on Wood Sugar	<1.5%
Acetic Acid on Wood	~1.0%
Initial Cook Liquor pH	2.0-4.0
Pulp Yield on Wood	65%-80%

CLE pretreatment produces high yield of pulp and ultimately sugar.

October 15, 2011

Enzymatic Hydrolysis

Catchlight Energy

- Near 80% conversion of glucan up to 17% solids in the hydrolysis step using low dose of enzymes.
- Overall sugar yields of 87% attained on clean softwood with economic enzyme loading.
- Overall sugar yields of 80% attained on softwood forest residuals.
- Switchgrass and hardwood behave similarly although have higher C5 sugar in the product.

October 15, 2011

CLE Pretreated Hydrolysis

Glucose Yield vs. CTec2 Enzyme Dosage

Catchlight Energy

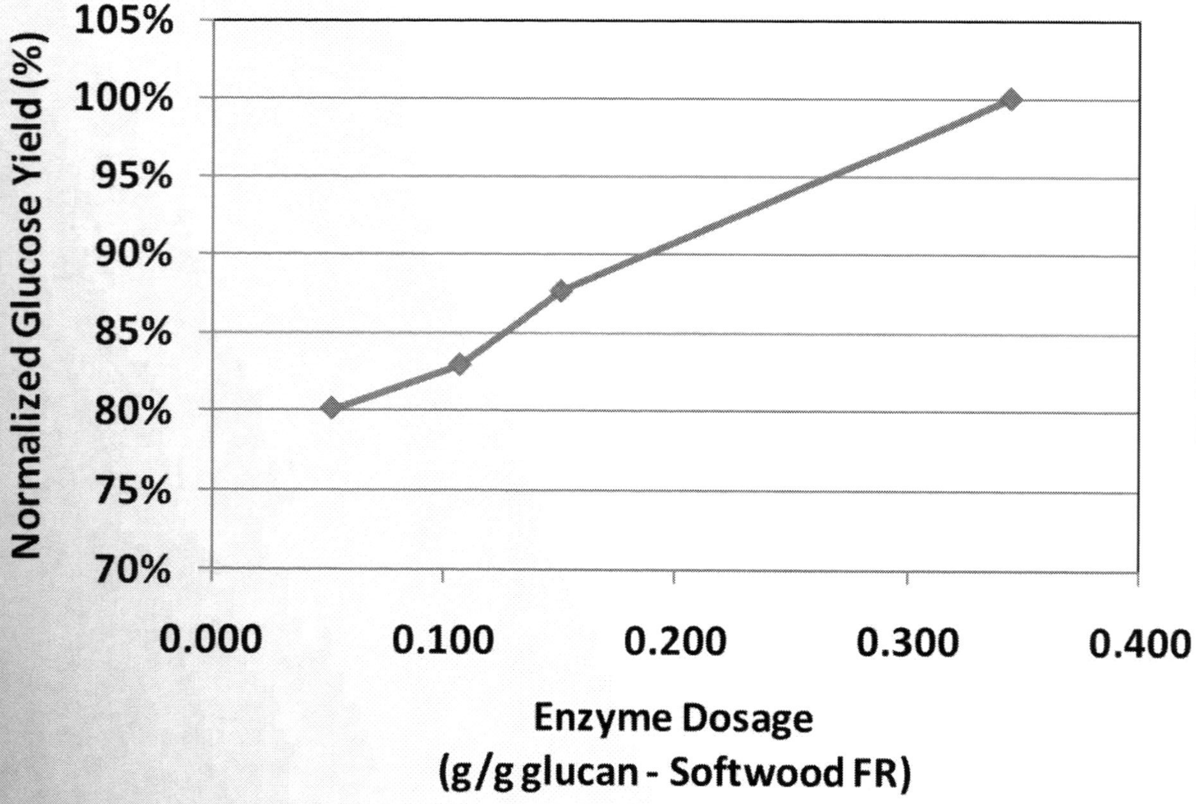

Hydrolysis Conditions:
Solid: 17%, unwashed
Temperature: 50°C
pH: 4.8-5.3
Shaking Speed: 200 rpm
Hydrolysis Time: 96 hrs

October 15, 2011

CLE Pretreatment Works on Multiple Feedstocks

Hydrolysis results at 4 days

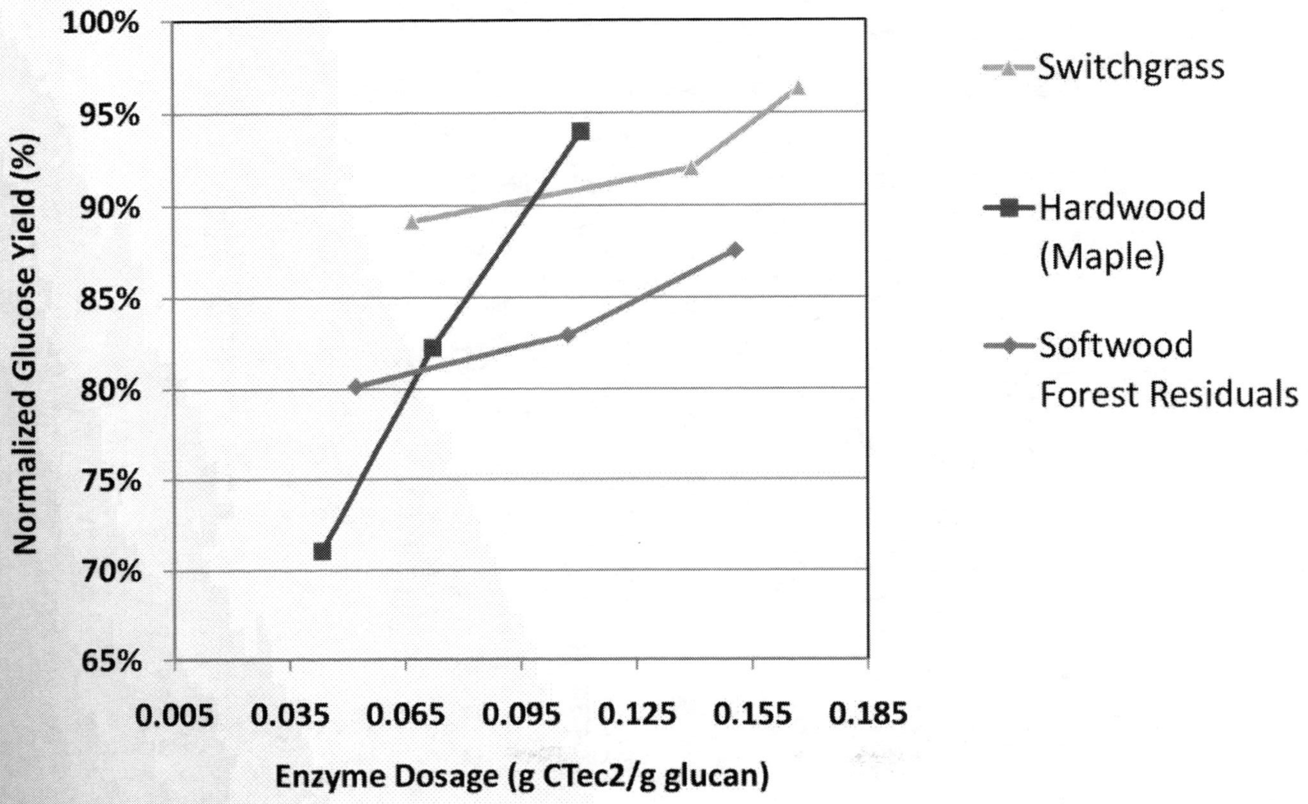

October 15, 2011

Commercial Data

CTec2 Dosage on Pretreated Corn Stover (Dilute Acid)

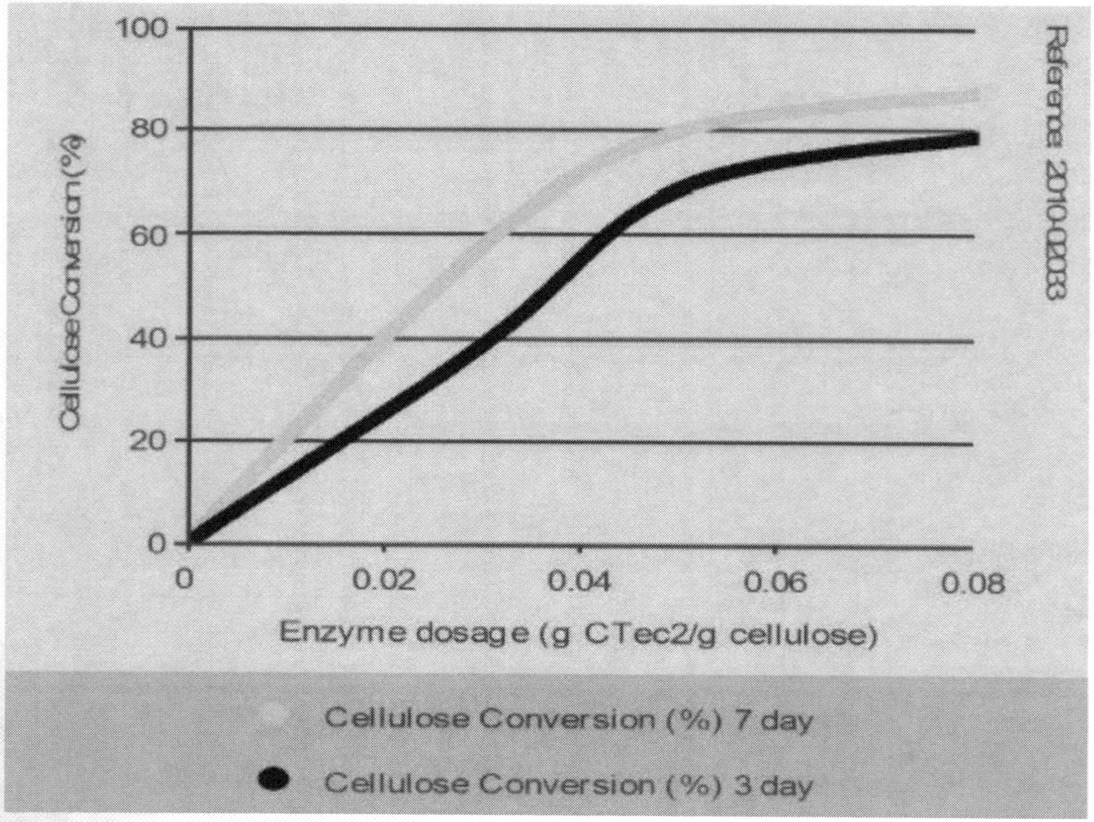

Reference: Corn stover data from CTec2 and HTec2 Application Sheet , www.novozymes.com
Used with permission

October 15, 2011

CLE Pretreatment Makes Softwood Hydrolyze Like Acid-Pretreated Corn Stover

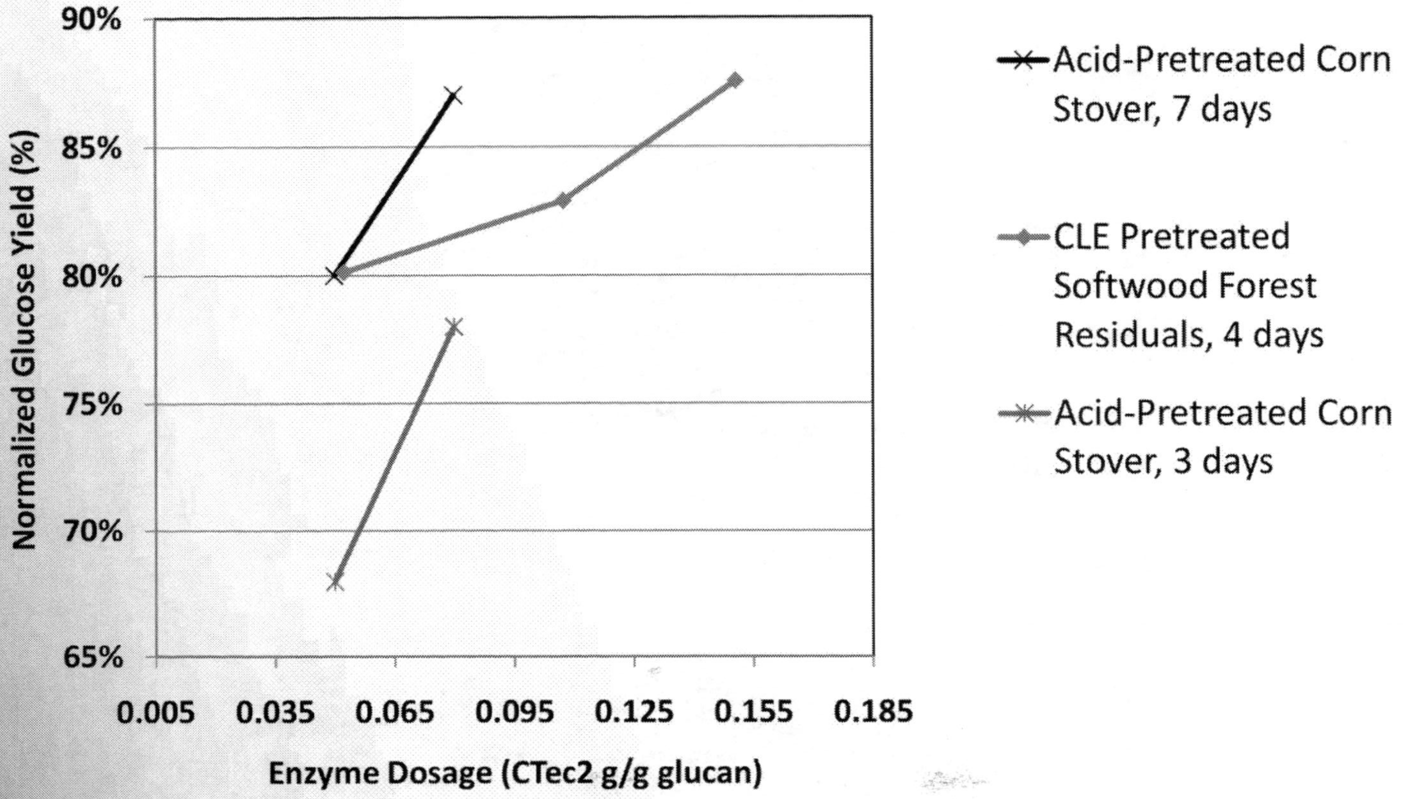

October 15, 2011

Catchlight Energy

CLE Sugar: Relative Cost Components

Catchlight Energy

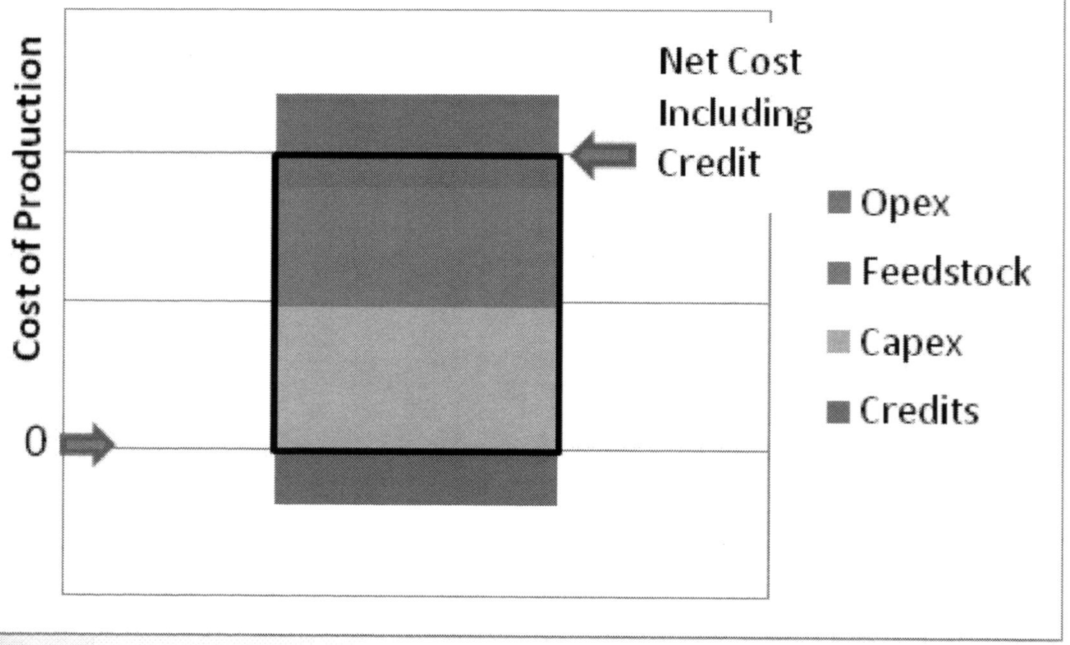

October 15, 2011

Conclusions

Catchlight Energy

- **CLE's pretreatment technology is advantageous for difficult to hydrolyze feedstocks, such as softwood residuals which are available today at commercial scale.**
- **The technology is proven, scalable, and cost effective.**
  - ➢ An engineering design has been developed for a large scale greenfield biofuels with downstream processing and separations for sugar, lignosulfonate, and energy production.
- **CLE is actively working with select partners and welcomes interest in the production of sugar, ethanol, and drop-in biofuels.**

October 15, 2011

Study of Culture Strategies of Microalgae In a Continuous Photobioreactor System for Biofuel Production

Haiying Tang, Meng Chen, Nadia Abunasser, Simon K. Y. Ng, Steven Salley

Abstract

New biomass sources for alternative fuels has become a subject of increasing importance as the nation strives to resolve the economic and strategic impacts of limited fossil fuel resources on our national security, environment, and global climate. Algae are among the most promising non-food-crop-based biomass feedstocks. However, there are currently no commercially viable microalgae-based production systems for biofuel production that have been developed, as limitations include less-than optimal oil content, growth rates, and cultivation techniques. While batch studies are critical for determining basic growth phases and characteristics of the algal species, steady-state studies are necessary to better understand and measure the specific growth parameters. This study evaluated the effects of dilution rate on microalgal biomass productivity, lipid content, and fatty acid profile under a steady-state condition with continuous illumination and carbon dioxide supplemented conditions for two types of algae. Continuous cultures were conducted for more that 3 months. Our results show that the productivity of *Chlorella minutissima* varied from 39 to 137 mg/L/day (dry mass) when the dilution rate varied from 0.08 to 0.64 day^{-1}. The biomass productivity of *Chlorella minutissima* reached the maximum value (137 mg/L/day) at a dilution rate of 0.33 day^{-1}; while the productivity of *Dunaliella tertiolecta* varied from 46 to 91 mg/L/day at a dilution rate of 0.17 to 0.74 day^{-1}. The biomass productivity of *Dunaliella tertiolecta* reached the maximum value, 91 mg/L/day at a dilution rate of 0.42 day^{-1}. Moreover, the lipid content had no significant change with various dilution rates.

Optimal Charge Rates for a Lithium Ion Cell

Saeed Khaleghi Rahimian, Sean Rayman, Ralph E. White

Department of Chemical Engineering, University of South Carolina, Columbia, SC, USA

Extended Abstract

The optimum charge rate for a lithium ion cell at each cycle is determined to maximize the useful life of the cell that is defined as the number of cycles before the end of discharge voltage (EODV) drops below 3.0 V or the cell discharge capacity becomes less than 20% of the original discharge capacity. First, we showed that by applying a dynamic optimization routine the number of cycles can be increased by approximately 29.4 % with respect to the case with one optimal charge current. The dynamic optimization results indicated that the optimum charge rates are the minimum currents at which the constraints for the useful life are satisfied. This is due to the minimum charge rate resulting in minimum side reaction rate and capacity fade. Then, a new method has been developed to find the optimal charge without using optimization algorithms. The cell capacity and the EODV are both a monotonic ascending function of the charge current; as a result the optimal charge current can be obtained by some iterations of the charge –discharge processes for each cycle. The new approach improves the useful cell life by approximately 41.6 % with respect to using only one optimal charge current in a few minutes while the previous optimization algorithm takes at least one day. Thus, the new technique enables us to find the best charge profile in online application in reasonable computation time by using any rigorous model for the cell simulation.

Introduction

The capacity of the Li-ion batteries decreases with the number of charge-discharge cycles due to undesirable side reactions. Since these batteries are used for tens to thousands cycles, depending on application, there is a demand to seek the best operating conditions to improve the cycling performance to minimize the capacity fade, and thereby increasing cell life. Only a few studies have attempted to optimize the charge rates for battery longevity and energy efficiency using optimization algorithms. S. Bashash et. al. applied multiobjective genetic algorithm to obtain the charge pattern of a plug-in hybrid electric vehicle to simultaneously minimize the total cost of fuel and electricity and the battery life degradation [1]. In previous work [2] we showed that the lithium ion cell life can be improved by ~30% with respect to the case with single optimal charge rate used by optimization of charge currents, as a function of cycle number during cycling. Choosing twenty rates, the optimization needs to evaluate the objective function (number of cycles) more than 80000 times and if the objective function evaluation takes only 1 sec, at the best condition, it will take about one day to find the optimal profile. In this work, a new approach is proposed to find the best charge rate at each cycle that improves the useful life 10% more than the dynamic optimization in much less computation time.

Lithium-ion cell Simulation

A single particle (SP) physics based model, which includes capacity fade, was applied to simulate the cell performance under low-earth-orbit (LEO) cycling conditions [2]. The main assumptions associated with the SP model are as follows:

- The concentration of the electrolyte and the potential in the solution phase are constant and uniform for all time across the cell sandwich (cathode, separator, anode).

- Positive and negative electrode potentials depend on time only.
- The model considers the capacity fade by a continuous occurrence of a side reaction (reduction of ethylene carbonate) only during the charge near the surface of the negative electrode. Moreover, the state of the charge (SOC) of the positive electrode is diminished during cycling.
- The concentration of Li ions in the spherical particle is estimated using a two term polynomial approximation.

A volume average technique is applied to predict the diffusion of the lithium ions for the cathode ($LiCoO_2$ with no Ni) and the carbon anode [4]:

$$\frac{dx_{p,avg}}{dt} = \frac{-3J_p}{FR_p c_{p,\max}} \qquad x_{p,surf} - x_{p,avg} = \frac{-J_p R_p}{5FD_{s,p} c_{p,\max}} \qquad J_p = \frac{I_{app}}{S_p}$$

$$\frac{dx_{n,avg}}{dt} = \frac{-3J_n}{FR_n c_{n,\max}} \qquad x_{n,surf} - x_{n,avg} = \frac{-J_n R_n}{5FD_{s,n} c_{n,\max}} \qquad J_n = \frac{-I_{app}}{S_n} - J_s$$

where $x_{i,avg}$ is the ratio of the solid average concentration to the maximum solid concentration for each electrode ($c_{i,max}$) , $x_{i,surf}$ is the ratio of the solid surface concentration to the maximum solid. J_s, the side reaction rate expression, is calculated by using Tafel kinetics:

$$\begin{cases} J_s = -i_{0f} \exp\left(\dfrac{-\alpha_{c,f} F}{RT} \eta_s\right) [charge] \\ J_s = 0 \; [discharge] \end{cases}$$

$$\eta_s = \phi_n - U_{ref,f} - \frac{I_{app}}{S_n} R_{film}$$

R_{film} is defined as:

$$R_{film} = R_{SEI} + \frac{\delta_{film}}{k_f}$$

and the rate at which the film thickness increases is calculated by:

$$\frac{d\delta_{film}}{dt} = \frac{-J_s M_f}{\rho_f F}$$

The SOC of the positive electrode updated at the end of each charging process as follows [4]:

$$\theta_{p,N} = \theta_{p,N-1} - \theta_{1,N-1}$$

where θ_l is the loss of SOC obtained by dividing the capacity loss to the maximum capacity of the cell (subscript N-1 indicates the previous cycle):

$$\theta_1 = \frac{Q_S}{Q_{\max}}$$

$$Q_s = -\int_0^{t=chsrge\ time} (J_s S_n)dt$$

$$Q_{\max} = \int_0^{t=chsrge\ time} I_{app} dt \qquad (first\ cycle)$$

LEO cycling [3], which contains the following steps, was applied for the cell simulation after the cell is completely charged initially:

1) Constant current discharge (0.6857 C-rate (40 % DOD)) for 35 min discharge time. Unless the voltage drops below 3.0 V or the cell capacity reaches the 20% of the total discharge capacity, go to step 2.
2) Constant current charging (up to 1 C rate) for 61 min charge time. If the voltage reaches the EOCV (4.05 V), go to step 3, if not go to step 1.
3) Constant voltage (4.05 V) charging for the remaining charge time, go to step 1.

Dynamic Optimization Results

To compare the results of the dynamic optimization and the new method, the best profile obtained by the optimization using single charge rate and twenty charge rates are presented in this work as well as [2]. Figures 1 and 2 show the variation of the cell capacity and the end of

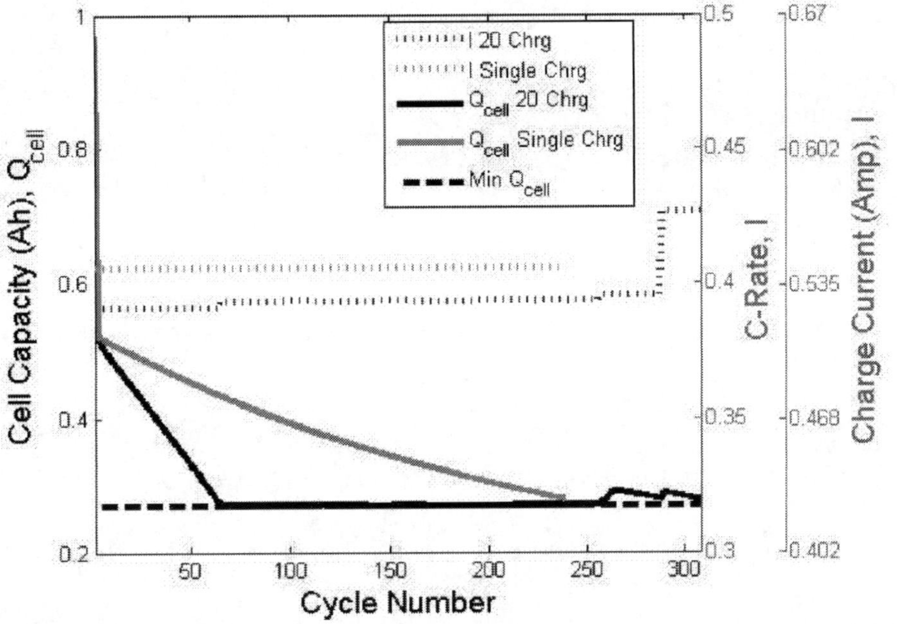

Figure 1: Remaining cell capacity at the end of discharge versus cycle number for single and 20 charge currents [2]

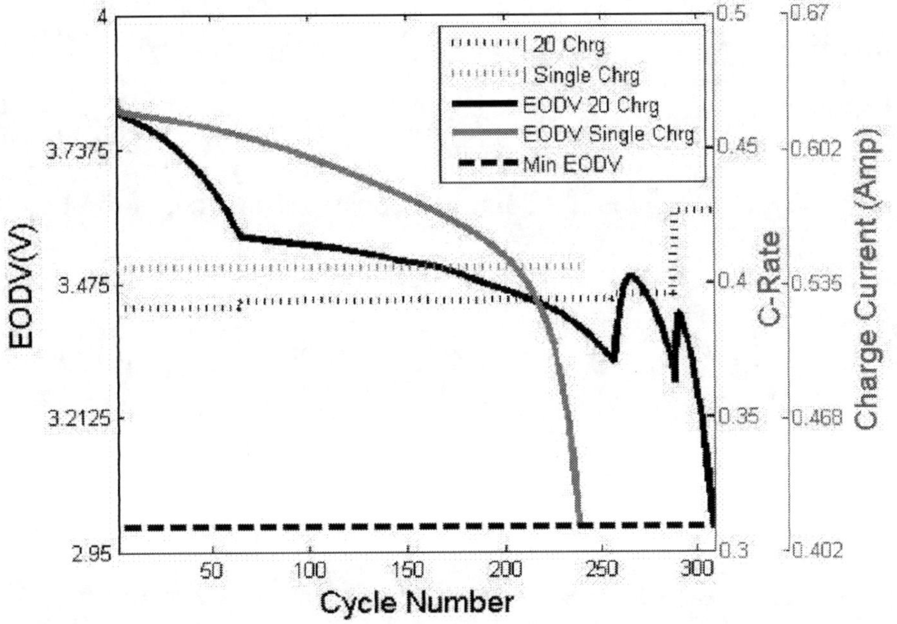

Figure 2: EODV versus cycle number for single charge and 20 charge currents [2]

voltage (EODV) respectively as a function of cycle number for single and twenty charge rates. Using twenty charge currents increases the number of cycles to 309, 29.4 % increase with respect to one optimum current.

New Approach

The optimization results indicate that the best charge current is the minimum rate at which, the cell capacity constraint and the EODV constraint are not violated. As a result we tried to develop a new method to find the optimal rate at each cycle as shown in Figure 3. By using this algorithm, any other constraints (e.g. end of charge voltage (EOCV) at 4.0 V) can be included without any difficulties.

Results and Discussion

The optimal charge rate and the cell capacity as a function of the cycle number obtained by the new approach that improves the number of cycles to 337, 41.6 % increase with respect to one optimal charge rate, shown in Figure 4. Variation of the EODV and the EOCV with the cycle number is presented in Figure 5. Adding the constraint that keeps the EOCV greater than 4.0 V (*constraint in Figure 3), decreases the number of cycles to 317. This is due to increase in charge rate in some cycles to not allow the EOCV dropping below the constraint. The computation time for the new approach is 228 sec on a Dell Precision T7500, with 2 Quad Core 2.53 GHz Zenon Processors CPUs and 12.285 GB of RAM.

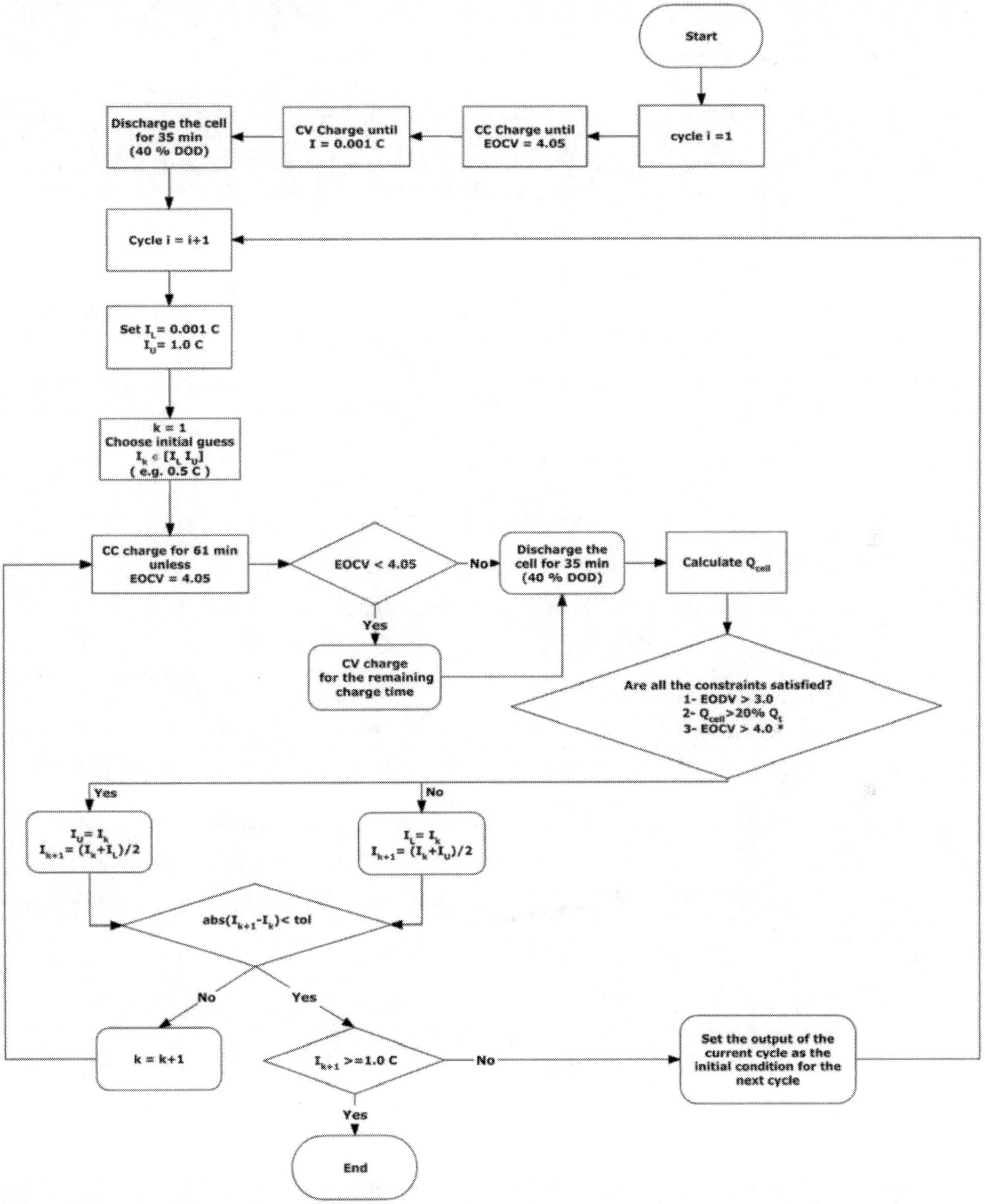

Figure 3: The new approach flowchart

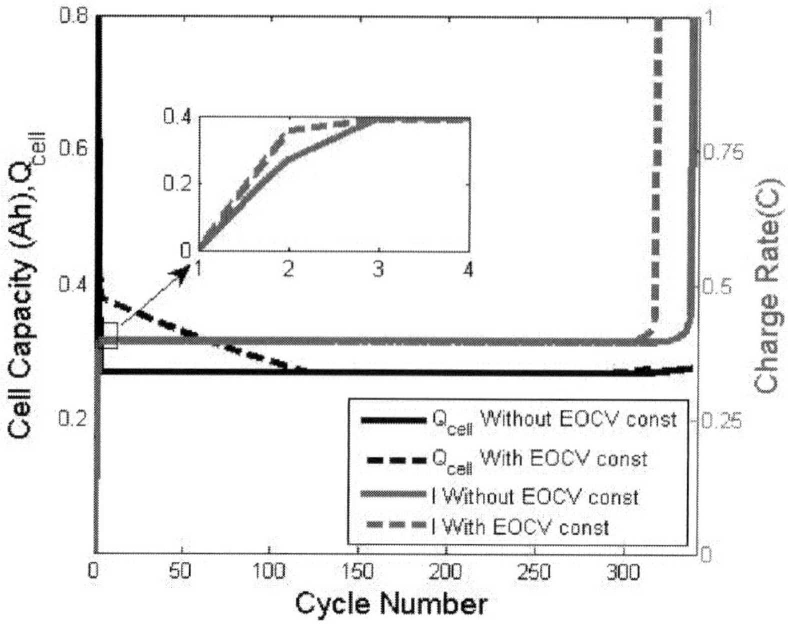

Figure 4: Remaining cell capacity at the end of discharge and Charge rate versus cycle number obtained by new approach with and without EOCV constraint

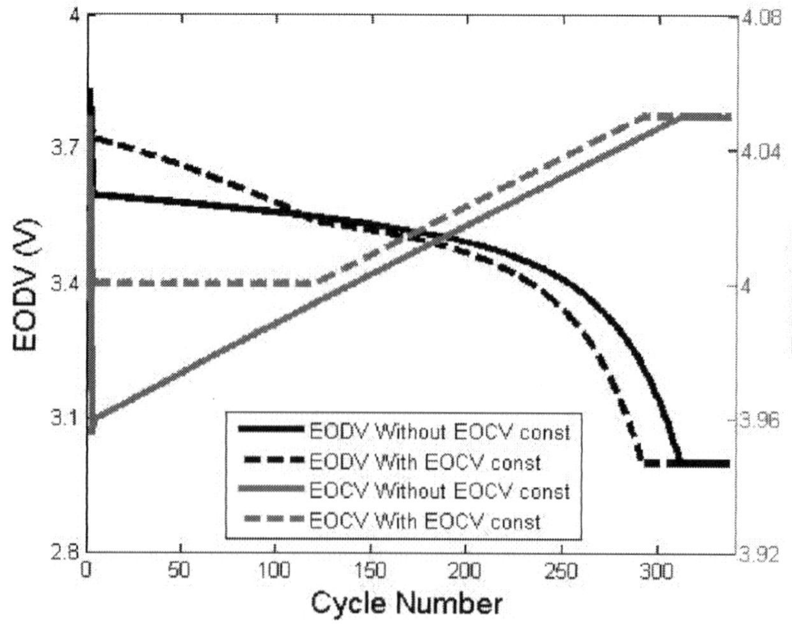

Figure 5: EODV (black) and EOCV (gray) versus cycle number obtained by new approach with and without EOCV constraint

Conclusion

A new approach was developed to provide the optimal charge rates at each cycle to maximize the useful life of a Lithium-ion cell. The computation time is greatly reduced by this approach with respect to the optimization method. For online application, the new method only requires the prediction of the next cycle by using prior cycling data while the entire cell life needs to be predicted for the optimization. Moreover, the new approach allows one to use a more rigorous model (i.e. P2D) to find the best charge rates without prohibitively increasing the computational time.

References

1. S. Bashash, S. J. Moura, J. C. Forman, H. K. Fathy, *J. Power Sources*, **196**, 541 (2011).
2. S. Khaleghi Rahimian, S.C. Rayman, and R.E. White, Maximizing the life of a lithium-ion cell by optimization of charging rates, *J. Electrochem. Soc.* **152** (12), A1302–A1308 (2010).
3. J. Lee, Y. K. Anguchamy, B. N. Popov, J. Power Sources 162 (2006) 1395.
4. P. Ramadass, B. Haran, P. M. Gomadam, R. E. White, B. N. Popov, J. Electrochem. Soc. 151(2) (2004) A196.

(647f) Decomposition Strategy for the Global Optimization of Energy Polygeneration Systems

Yang Chen[1], Xiang Li[1], Thomas A. Adams II[2] and Paul Barton[1], (1)Chemical Engineering, Massachusetts Institute of Technology, Cambridge, MA, (2)Chemical Engineering, McMaster University, Hamilton, ON, Canada

Coal- and biomass-based polygeneration with multiple energy products is a promising technology to supplement the current oil-based energy processes. A certain degree of operational flexibility can be introduced into a polygeneration process by over-sizing equipment so that one can alter the production rates of individual products in response to changing market conditions. The resulting flexible polygeneration plant achieves potentially higher profit and lower economic risk than the static plant [1].

In the design of flexible polygeneration systems, the optimal trade-off between operational flexibility and equipment capacity (or capital cost) needs to be determined. This problem can be formulated as a multi-period optimization problem, which is a potentially large-scale nonconvex mixed-integer nonlinear program (MINLP) and cannot be solved to global optimality by state-of-the-art global optimization solvers, such as BARON [2], within a reasonable time. A duality-based decomposition method, recently developed for the design and operation of natural gas production networks [3,4], can exploit the special structure of this mathematical programming problem and enable faster solution. In this method, the nonconvex MINLP is relaxed into a convex lower-bounding MINLP which can be further reformulated into a relaxed master problem according to the principles of projection, dualization and relaxation. The relaxed master problem yields an increasing sequence of lower bounds for the original problem, but these lower bounds may converge to a global optimum slowly due to the high nonconvexity of the problem.

In this work, an enhanced decomposition method with an improved relaxed master problem is developed by introducing additional dual information into the relaxed master problem, which requires solving additional nonconvex NLPs. The solutions of these nonconvex NLPs lead to stronger dual cuts that can significantly improve the convergence rate of the algorithm. The enhanced decomposition algorithm guarantees to find an ε-optimal solution in a finite number of iterations.

The computational advantages of the enhanced decomposition method are demonstrated via several case studies on the optimal design and operation of a flexible energy polygeneration system using coal and biomass as feedstocks and co-producing electricity, liquid fuels (naphtha and diesel) and chemicals (methanol). The computational results show that the enhanced decomposition algorithm achieves much faster convergence than both BARON and the original decomposition algorithm, and it solved the large-scale nonconvex MINLPs to ε-optimality in practical times (e.g., a problem with 70 binary variables and 14712 continuous variables was solved within 5 hours).

Reference

[1] Chen Y, Adams TA, Barton PI. Optimal design and operation of flexible energy polygeneration systems. *Industrial & Engineering Chemistry Research*. 2011;50(8):4553-4566.

[2] Tawarmalani M, Sahinidis NV. Global optimization of mixed-integer nonlinear programs: A theoretical and computational study. *Mathematical Programming*. 2004;99(3):563-591.

[3] Li X, Armagan E, Tomasgard A, Barton PI. Stochastic pooling problem for natural gas production network design and operation under uncertainty. *AIChE Journal*. DOI: 10.1002/aic.12419.

[4] Li X, Tomasgard A, Barton PI. Decomposition strategy for natural gas production network design under uncertainty. *Proceedings of the 49th IEEE Conference on Decision and Control*, 188-193 (2010).

See more of this Session: Energy Systems Design and Alternative Energy Sources I
See more of this Group/Topical: Computing and Systems Technology Division

(388d) Enhanced Xylitol Production Through Simultaneous Co-Consumption of Cellobiose and Xylose by An Engineereed Saccharomyces Cerevisiae Strain

Eun Joong Oh[1], Suk-Jin Ha[1], Soo Rin Kim[1], Jonathan M. Galazka[2], Jamie H.D. Cate[2] and Yong-Su Jin[1], (1)Department of Food Science and Human Nutrition, University of Illinois at Urbana-Champaign, Urbana, IL, (2)Department of Molecular and Cell Biology,, University of California at Berkeley, Berkeley, CA

Xylitol is a five-carbon sugar alcohol and has been used as a sugar substitute in the food industry because of its low caloric and anti-carcinogenic characteristics. In addition, xylitol is a building block for a variety of commodity chemicals. Biological routes for producing xylitol from cellulosic hydrolysates have been developed through metabolic engineering of *Saccharomyces cerevisiae*. *S. cerevisiae* cannot utilize xylose as a carbon source, but the expression of a xylose reductase (XR) gene from *Pichia stipitis* facilitates the production of xylitol from xylose with high yields. However, insufficient supply of NAD(P)H for XR, and inhibition of xylose transport by glucose, are major constraints for high productivity production of xylitol.

To overcome these problems, we engineered *S. cerevisiae* to convert xylose into xylitol through the simultaneous consumption of cellobiose and xylose. Specifically, xylose reductase from *P. stipitis* was chromosomally integrated into the *S. cerevisiae* genome, while a cellobiose transporter (*cdt-1*) and an intracellular β-glucosidase (*ghl-1*) from *Neurospora crassa* were introduced on a multi-copy plasmid. The resulting transformant was able to produce xylitol from a mixture of cellobiose and xylose. In contrast to glucose, cellobiose did not repress the transport of xylose, allowing the co-consumption of xylose and cellobiose. The co-consumption and metabolism of cellobiose regenerated NADPH for XR. As a result, an engineered *S. cerevisiae* strain (D-10-BT) exhibited 40% higher xylitol productivity when cellobiose was used as a co-substrate as compared to glucose. This result suggests that co-consumption of cellobiose and xylose allows efficient xylose uptake and cofactor regeneration without catabolic repression for the production of xylitol.

See more of this Session: Microbial Engineering for the Synthesis of Bulk and High-Value Products
See more of this Group/Topical: Sustainable Engineering Forum

(583f) Biomimicry of the Fungal Consolidation of Biomass Pretreatment with Saccharification

Justin T. Kaffenberger and Jonathan S Schilling, Bioproducts and Biosystems Engineering, University of Minnesota, St Paul, MN

Efficient breakdown of ligno-cellulosic material prior to fermentation for bio-fuel production is a key step in optimizing bio-fuel production. Current industrial practice involves chemical pretreatment and acid or enzymatic saccharification to prepare ligno-cellulosic biomass for fermentation. The commercial viability of bio-fuels hinges in large part on production cost. One approach to reducing production cost is process consolidation. Consolidation not only eliminates the capital costs associated with the additional processing steps, but can also trim operational costs by reducing total reaction time and down time between steps.

In nature, the brown-rot fungus *Postia placenta* successfully performs both chemical pretreatment and enzymatic saccharification of biomass in a consolidated manner. As evidenced by its recently sequenced genome, *P. placenta* produces a limited suite of cellulases, lacking genes for exoglucanases. Despite this limitation, it is still capable of rapidly degrading wood. Brown-rot fungi quickly reduce the degree of polymerization of wood with little weight loss through a reduction-oxidation pathway that yields highly reactive hydroxyl radicals through a chelator-mediated fenton reaction. While hydroxyl radicals would readily damage its cellulases, the fungus manages to conduct both reaction types, enzymatic and oxidative, in the same general location. Examination of the spatial and temporal relationship of these two reaction systems as brown-rot fungal degradation of wood progresses will be presented.

See more of this Session: Developments In the Pretreatment of Lignocellulosics for Bioconversion III
See more of this Group/Topical: Sustainable Engineering Forum

Investigation on electrode reaction mechanism

for vanadium redox flow battery

XIE Xiaofeng[1], LI Xiaoshan[1], YANG Chun[1], WANG Jinhai[1], WANG Shubo[1], Shang Yuming[1], MATHUR V. K[2]

1. Institute ofNuclear andNew EnergyTechnology, Tsinghua University, Beijing100084, China

2.Department of Chemical Engineering, University of New Hampshire, NH 03824, USA

Abstract:

The effect of glycerine addition to the anodic electrolyte, on the anodic reaction, was investigated. The results show that 1% glycerine added to the electrolyte had positive effect on the performance of electrolyte reaction on the surface electrode. The solution resistance, capacitance value and polarization resistance also improved. The reaction of VO_2^+ to VO^{2+} was helped as a result of lower polarization resistance.

Key words: Vanadium Redox Flow Battery ; Glycerine ; Electrochemical impedance spectroscopy; Electrolyte, Cyclic voltammetry; Potentiodynamic

1. Intruduction

Vanadium redox flow battery (VRB) employs vanadium ions of different valences as liquid electrolytes. It is independent of storage capacity, output power, normal atmospheric temperature operation, fast response, and long operating life. The VRB system can widely be used for load leveling, distributing power, standby power supply, and municipal transport. Cyclic voltammetry and Electrochemical impedance spectroscopy (EIS) methods are useful in characterizing the polarization resistance corrosion rate and electrochemical mechanism.

When the positive electrolyte vanadium concentration is up to 2M or the temperature is elevated above 310K, the fully charged electrolyte has poor stability. The precipitate of V(V) would block the electrode and prevent the electrolyte from flowing through the pipe. In order to keep the electrolyte from precipitating, and improve the energy storage density of electrolyte used in VRB, one method is to increase the concentration of vanadium ions. Many workers have chosen inorganic and organic additives to be added to the electrolyte for solving this problem. Some additives added to electrolyte can broaden the electrochemical window and increase electrochemical activity. The dissolved V(V) in concentrated sulfuric acid solution becomes more stable and effective at high concentration of vanadium. Glycerine has been used to enhance the conductivity and stability of anodic electrolyte for improving the performance of VRB. In this work, the effect of glycerine addition, to the anodic electrolyte, on the mechanism has been studied.

2 EXPERIMENTS

2.1 Preparation of anodic electrolyte

$VOSO_4 \cdot 3H_2O$ was dissolved in the deionized water, and H_2SO_4 was added to form $VOSO_4$+3 M solution. The solution was added to anodic electrolytic cell, and H2SO4 of the same concentration was added to cathodic electrolytic cell. The electrolytic cell was separated by an ion exchange membrane (Nafion 117, DuPont), and the lead electrodes were used for anode and cathode electrolysis. The prepared electrolyte was 1.8 mol/L, with the active area of $16cm^2$ (4cm × 4cm) and electrolytic current density of 20 ~ $200mA/cm^2$.

2.2 Electrochemical tests

A three-electrode electrochemical cell was used for electrochemical tests. The graphite electrode (with 5mm diameter) was a working electrode. The reference electrode was SCE, and a platinum mesh was used as auxiliary electrode. Different concentration (v/v) of glycerine was added in the anodic electrolyte.

The electrode characterization included the determination of the polarization resistance and electrochemical mechanism. A Zanher IM6ex instrument was used to conduct the potentiodynamic, cyclic voltammetry and EIS measurements. Potentiodynamic tests were used to measure the corrosion resistance. The initial potentials were -0.15 V, and the final potentials were 2.0 V, the scan rate was 1 mVs-1. Cyclic voltammetry was used to measure the electrochemical activities. The scan range was -0.8 ~ 1.6 V. The EIS measurements were performed at different potential and the perturbation amplitude was 5 mV. The applied frequencies ranged from 100K to 10^{-1} Hz.

3. RESULTS AND DISCUSSION

A potentiodynamic test is commonly employed to measure a material's resistance to corrosion. Potentiodynamic tests were conducted to estimate the corrosion resistances. The "corrosion resistances" here was considered as the electrolyte's ionization after the glycerine was added. Fig. 1 presents the potentiodynamic polarization scans at different glycerine contents. From the Tafel curves in Fig. 1, the related polarization kinetic parameters were analyzed by potentiodynamic polarization model in Zanher IM6ex electrochemical workstation. From the potentiodynamic polarization results and using Eq. (1), polarization resistance, Rp was estimated. Table 1 presents the polarization parameters of anodic electrolyte. The polarization resistances for 0.1% and 10% glycerine were 44744 Ω·cm2 and 13027 Ω·cm2, respectively.

$$R_P = \frac{\beta_a \beta_c}{2.3 i_{corr}(\beta_a + \beta_c)} \tag{1}$$

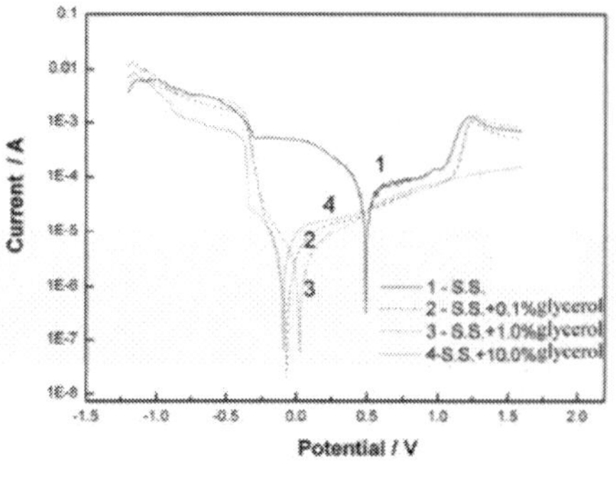

Fig. 1 Potentiodynamic polarization scans of anodic electrolyte

Table 1 Polarization parameters of anodic electrolyte

	βa/mV	-βc/mV	Rp/Ω	Icorr/μA
Non glycerin	86.2	79.9	1654	10.9
0.1% glycerin	64.1	62.3	44744	0.307
1.0% glycerin	54.0	52.9	28617	0.406

Cyclic voltammetry or CV is a type of potentiodynamic electrochemical measurement. Cyclic voltammetry was used to study the electrochemical properties of the glycerine in solution. The results are shown in Fig. 2. When glycerine was added to the anodic electrolyte, the quantity and status of vanadium is changing between the electrolyte and electrode. With the glycerine content increased from 0.1% to 10%, the oxidation peak current decreased. The hydroxy groups on glycerine are bound with the vanadium ions, resulting in the decrease of surface electrode charge, transferring more vanadium ions to the bulk solution.

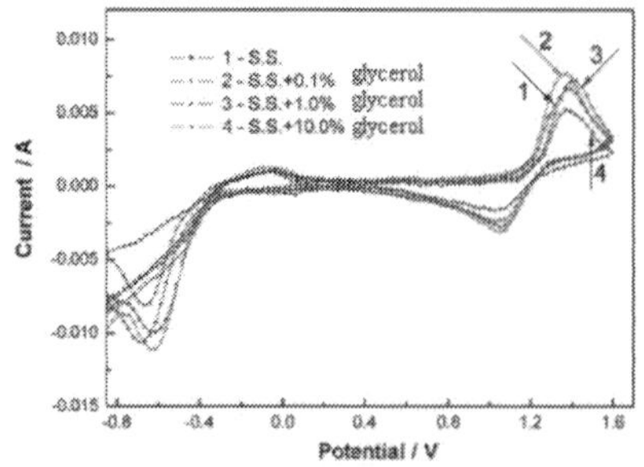

Fig. 2 Cyclic voltammetry scans of anodic electrolyte

EIS reveals information about the reaction mechanism of an electrochemical process: different reaction steps will dominate at certain frequencies, and the frequency response shown by EIS can help identify the rate limiting step. The EIS was employed to investigate the electrode reaction mechanism in anodic electrolyte containing glycerine as additive. The spectrums were simulated and analyzed through the equivalent circuit of $R_{\Omega}(C_d(R_pW))$. The reaction mechanism of glycerine added to electrolyte is shown in Fig 3.

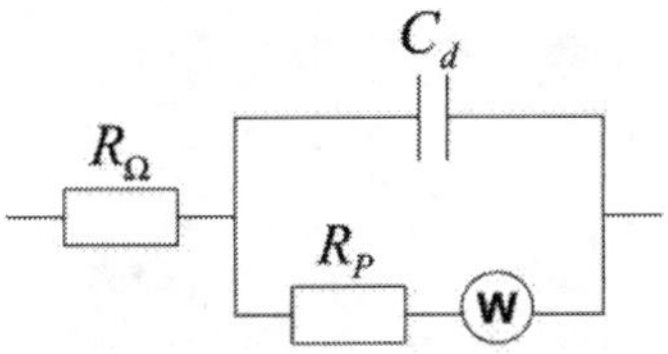

Fig 3. Equivalent circuit using RΩ(Cd(RpW))

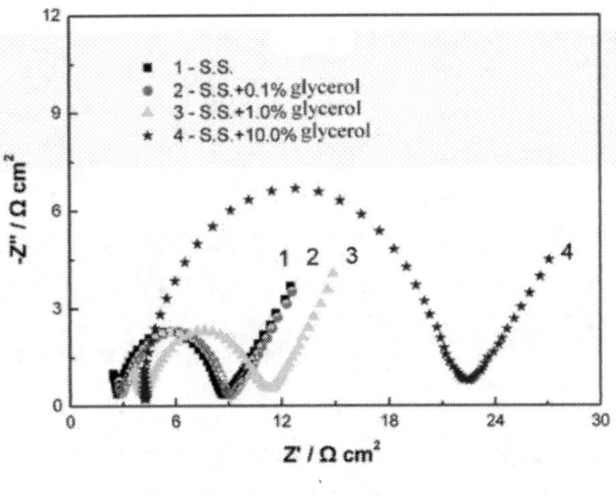

Fig.4 Impedance spectra for anodic electrolytes

The results of EIS are shown in Fig 4. From the impedance curve, with the glycerine concentration increasing, the impedance seems to increase. It means there is an optimal amount for glycerine that can be added to the electrolyte. Therefore, we should conduct the simulation of equivalent circuit to explore the mechanism of reaction with different concentration of glycerine added in. The results show the solution resistance increases by the addition of glycerine. It is difficult for glycerine to ionize in water. When the non-conducting glycerine is added to the electrolyte, the solution resistance should increase. With the increase in glycerine concentration, the resistance of electrolyte increases from 2.795Ω·cm^2 to 4.16 Ω·cm^2. In fact, the molecular structure of glycerine has three hydroxyls,

Table 2 Parameters of the equivalent circuit

	RΩ/(Ω·cm^2)	Cd/(S·cm^2)	Rp/(Ω·cm^2)	W/(S·cm^2)
Non glycerin	2.572	7.735	6.106	0.2478
0.1% glycerin	2.795	8.809	6.25	0.2676
1.0% glycerin	3.855	4.04	7.413	0.245

which are prone to bind with vanadium ions. With the glycerine content increasing from 0% to 10%, the polarization resistance increased. It means the glycerine added to electrolyte is unfavorable for reaction VO_2^+ to VO^{2+}. When the concentration exceeds 1%, the unnecessary glycerine might occupy the reaction site which can catalyze the redox reaction of vanadium ions. Therefore, when more glycerine is added to electrolyte, the reaction impedance increases. It means less glycerine is helpful for VO_2^+ to react to VO^{2+}. The results are in agreement with potentiodynamic test and Cyclic voltammetry. It indicates that glycerine contents when increased beyond certain limits will decrease the reaction rate.

4 . CONCLUSIONS

The results show that glycerine added to anodic electrolyte greatly improved oxidation peak current, reaching a maximum when glycerine was about 1%(v/v). Meanwhile, with the increase in glycerine content, the solution resistance ($R_Ω$), the double electric layer capacitor (C_d) and the polarization resistance (R_p) reached their minimum. However the concentration polarization reached its maximum when glycerine added was about 1%(v/v).

The addition of glycerine increased solution resistance ($R_Ω$) which is disadvantageous to the vanadium ion movement in electrolyte solution. Moreover, the glycerine could greatly decrease the corrosion of electrolyte materials. With the increase in additive, the anti-corrosion tends to decrease. When the addivties was up to 0.1%(v/v), the anti-"corrosive ability" was the best.

ACKNOWLEDGEMENTS

The work was funded by Tsinghua University Initiative Scientific Research Program (201010800623), State Key Laboratory for Automotive Safety and Energy(KF10132).

REFERENCES

1) Bard, Allen J.; Larry R. Faulkner (2000-12-18). Electrochemical Methods: Fundamentals and Applications (2 ed.). Wiley. ISBN 0471043729.

2) http://en.wikipedia.org/wiki/Electrochemical_impedance_spectroscopy

Study on preparation and characterization of the composite bipolar plate for Vanadium Redox Flow Battery

XIE Xiao-Feng[1], WANG Wen-Pin[2] , WANG Shu-Bo[1], WANG Jin-Hai[1], LV Ya-Fei[2], MATHUR V. K[3]

1 Institute of Nuclear and New Energy Technology, Tsinghua University, Beijing 100084, China

2 College of Material Science And Engineering, Beijing University of Chemical Technology, Beijing 100029,China

3.Department of Chemical Engineering, University of New Hampshire, NH 03824, USA

Bipolar plate, as one of the main components of all-vanadium redox flow battery (VRB), can collect the current generated by electrochemical reaction, and can separate positive and negative electrolyte. The electrolyte is a sulfuric acid solution containing vanadium ions (V(IV/V,II/III)/ V(V)). Among these ions, the V(II) has the most strong oxidation ability. Therefore, the bipolar plate not only should have good conductivity, but also have high resistance to chemical and electrochemical corrosion.

Bipolar plate is a carbon composite material, which is a mixture of carbon materials and a polymer. The polymers used are polyethylene, nylon, polypropylene (PP), rubber modified polypropylene, polytetrafluoroethylene and polyvinylidene fluoride, etc. Polymer is helpful to form the conductive network, and improve the corrosion resistance of bipolar plate.

Maleic anhydride grafted polypropylene(g-PP) is a non-polar molecules in the main chain with a strong polar side groups maleic anhydride. It can increase the bonding and the compatibility between polar materials and non-polar materials for enhancing the dispersion between the conductive filler and the polymer matrix, and to improve conductive property and corrosion resistance of the bipolar plate.

In this work, composite bipolar plates, by polypropylene (PP), polypropylene-maleic anhydride graft copolymer (g-PP) as polymer matrix, and flake graphite(GC) as conductive filler, were prepared using a solution intercalation method. Four probes method, potentiodynamic experiments, potentiostatic and cyclic voltammetry(CV) were used to investigate the conductivity, corrosion resistance and electrochemical performance of bipolar plates. The results show that g-PP made the graphite uniformly dispersed in the polymer, which significantly improved compatibility between the conductive filler and the polymer matrix. The conductivity of the composite bipolar plate was about 77.8 $S{\cdot}cm^{-1}$. When the voltage was 1.2 V, the corrosion current was about $5.74\times10^{-2} A{\cdot}cm^{-2}$. The prepared bipolar plate could meet the requirements of VRB.

EXPERIMENTALS

The conductivity of samples (5 cm×1.5 cm) was measured by a four probe resistance tester. Other tests carried out were: Cyclic voltammetry: -1.5~2.0V, scanning speed

10mv·s-1; potentiodynamic polarization curves: -1.5~2.0V, scanning speed 5v·s-1; potentiostatic test (1.2 V, scan time 3600 s). Electrochemical measurements used three-electrode cell with the sample as the working electrode. A Pt electrode as the counter electrode and a saturated calomel electrode as reference electrode (SCE) were used. A homemade VRB electrolyte was used for the experiments.

A certain amount of GC, PP, g-PP and xylene were placed in three-necked bottle, stirred, heated to reflow temperature and kept under reflux for a period of time. The product was precipitated by a solvent, leached, dried, and hot-pressed to form the composite conductive plate.

RESULTS AND DISCUSSION

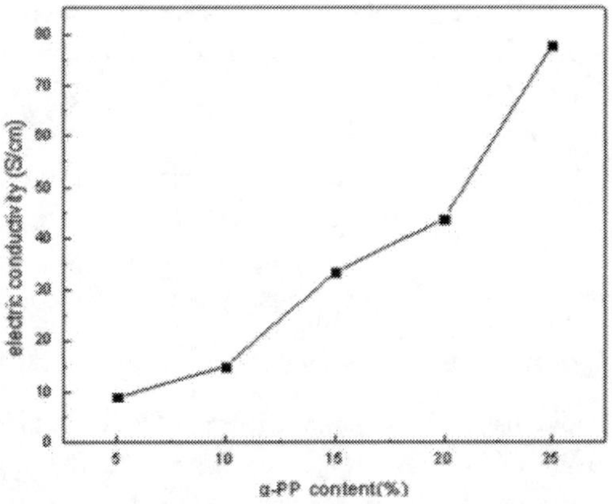

Fig.1 Effect of g-PP contents on conductivity

GC is the most familiar of the natural graphite materials. It has good conductivity. For a GC content of 70wt%, bipolar plates were prepared by using the g-PP content 10wt%, 15wt%, 20wt% or 25wt%. The effect of g-PP contents on conductivity of bipolar plates are shown in Fig. 1. It can be seen that with an increase in g-PP content, the conductivity of bipolar plate increases steeply. It means that g-PP addition contributes to the uniform distribution of the GC in the matrix, resulting in the conductive network structure formation and thus conductive property improvements. Polyacrylonitrile graphite felt (PAN-based) as electrode materials is widely used in VRB. Its conductivity commonly is about 50 S·cm^{-1}. It is preferred to have the g-PP content greater than or equal to 20 wt% for collecting current.

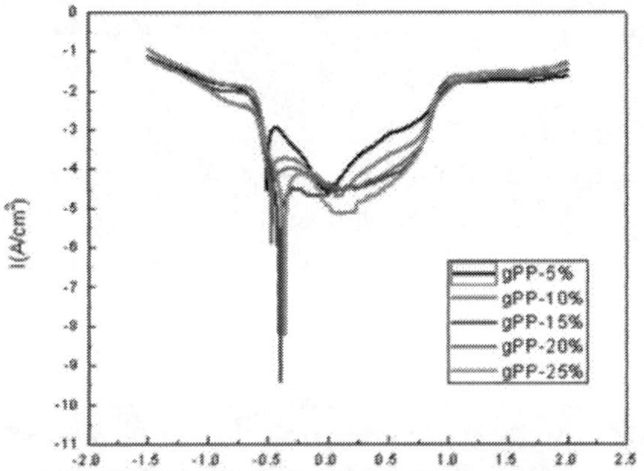

Fig.2 Potentiodynamic polarization curves

Fig.2 shows the potentiodynamic polarization curves of bipolar plates. It can be seen that as the voltage increases, anode passivation of bipolar plate occurs. That means the corrosion of bipolar plate is mainly due to oxygen corrosion. As the g-PP content increases, the corrosion current of bipolar plate decreases first and then increases. When the g-PP content of bipolar plate is 20 wt%, the corrosion current is the lowest.

The effect of g-PP contents on corrosion current are showed in Fig.3. It can be seen that with the increase in g-PP content, the electrochemical corrosion current of bipolar plate decreases rapidly, and then runs through the transition state into the steady-state. When g-PP contents is about 5%, the corrosion current density is 1.8×10^{-2} A/cm^2. With the g-PP content increasing, the corrosion current density reduces to 9.1×10^{-3} A/cm^2. Moreover, once corrosion occurs, it will cause intergranular corrosion or pitting corrosion, and then the corrosion rate further increases. When the g-PP content is 20wt%, the corrosion current of bipolar plate is close to the minimum, which is approximately 8.9×10^{-3} A·cm^{-2}.

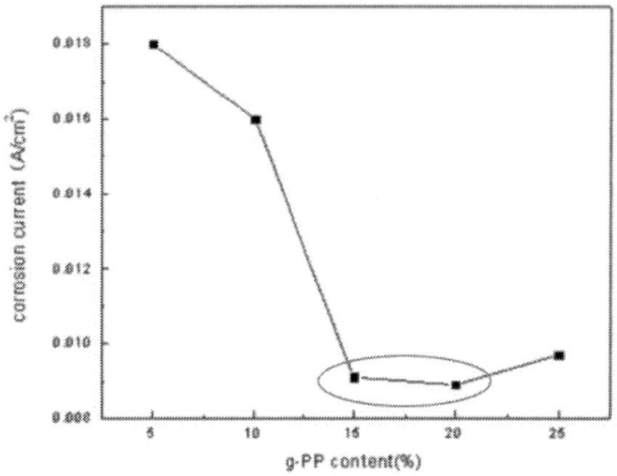

Fig.3 Effect on g-PP contents on corrosion current

Fig.4 shows cyclic voltammetry of bipolar plates under various g-PP contents. With the g-PP content increasing, more V(II) and V(III), V(IV) and V(V) transformation peaks were observed with reversibility. When g-PP is about 10%, it has the high activity. If it was used as bipolar plate in battery, the bipolar plate was easily corroded. For oxygen evolution peak of bipolar plate at anode, the higher the oxygen evolution potential, the more stable the battery is. When the g-PP content is 20wt%, there is low activity and high oxygen evolution potential of the bipolar plate providing the best the corrosion resistance of the bipolar plate.

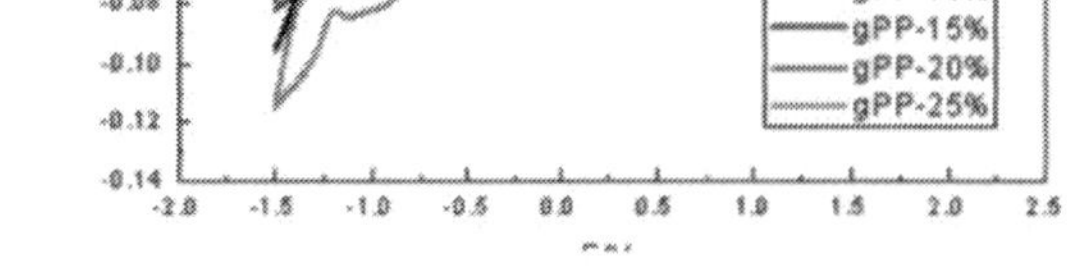

Fig.4 Cyclic voltammetry of bipolar plates

With the increase of g-PP content, the dispersion of GC in the polymer matrix is more uniform, the holes in the bipolar plate surface are reduced, and vanadium ions have difficulty in attaching to the surface of bipolar plate. As the g-PP content increases to 25wt%, the functional group COO^- content of the bipolar plate surface increases. COO^- can contribute to vanadium ion adsorption from solution to the surface. With the reaction of vanadium ion adsorption and desorption, the charge transfer should obviously change. It makes the electrochemical reaction of the vanadium ions on the surface of the bipolar plate to easily happen. These results are in agreement with the previous investigation.

CONCLUSIONS

Analysis of potentiodynamic, potentiostatic experiments and cyclic voltammetry shows when the GC content is constant, within a certain range, conductivity of bipolar plate is proportional to the g-PP content. The g-PP is added to make the dispersion between GC and the polymer matrix and improve the compatibility between the conductive filler and the polymer matrix significantly. Addition of g-PP leads to improving the corrosion resistance of bipolar plate. Conductivity and corrosion resistance of composite bipolar plate can meet the requirements of VRB.

ACKNOWLEDGEMENTS

The work was funded by Tsinghua University Initiative Scientific Research Program (201010800623), State Key Laboratory for Automotive Safety and Energy(KF10132), State Key Lab of Power Systems(SKLD10KZ03).

REFERENCES

[1] WANG Wenpin , WANG Jinhai, WANG Shubo, XIE Xiaofeng , LV Yafei. Properties of Conductive Bipolar Plate for All-Vanadium Redox Flow Battery [J]. The Chemical Industry and Engineering Society of China, 2011, 62(SI): 203-207

[2] **http://www.asbury.com/Natural-Flake-Graphite.html**

[3] Xia Liping, Lin Changwu, Luo Dongmei, Zhang Aimin. Study on properties of fluoro rubber composite current collector for vanadium redox battery [J]. Power Sources, 2010, 134(7): 672-675

[4] Chen maobin, Zhang Shengtao, Li Xiaobing, Liu Lian, Liu Xiaojiang, Meng Fanming. Preparation and performance of polytetrafluoroethylene conducting plastic on vanadium redox flow battery collector [J]. Polymer Materials Science And Engineering, 2009, 25(6): 121-124

[5] Chen Maobin, Lan Wei, Zhang Shengtao, Cui Yanhua, Meng Fanming, Liu Xiaojiang. Performance of Conductive Polyethylene Composite [J]. Applied Chemistry, 2009, 26(5): 619-621

(429b) Supercritical CO2 Extraction for Algal Biodiesel Production

Prafulla Dinkarrao Patil[1], Harvind Reddy[1], Tapaswy Muppaneni[1], Tanner Schaub[2], Peter Cooke[3], Peter Lammers[4] and Shuguang Deng[1], (1)Chemical Engineering, New Mexico State University, Las Cruces, NM, (2)College of Agricultural, Consumer and Environmental Sciences, New Mexico State University, Las Cruces, NM, (3)Core University Research Resource Laboratory, New Mexico State University, Las Cruces, NM, (4)Energy Research Laboratory, New Mexico State University, Las Cruces, NM

SC-CO2 is cleaner green technology for extraction of algal lipids/oils for algal biodiesel production. This work offers an optimized method for the direct conversion (One-step) of algal biomass to biodiesel using supercritical CO_2 extraction and supercritical methanol transesterification. SC-CO_2 uses moderate temperature and pressure so that it can be useful for the extraction of thermolabile compounds which may decompose at the temperature at which thermal methods are carried out. This study examines the effects of pressure, temperature and algae to methanol ratio on fatty acid methyl ester (FAME) yield from algal biomass. The extraction with SC-CO2 allows to obtain a higher amount of long chain fatty acids. The algal biomass characterization and algal biodiesel analysis were performed using ATR-FTIR, SEM-EDS, TEM, TGA and GC-MS.

See more of this Session: Alternative Fuels and Enabling Technologies I
See more of this Group/Topical: Fuels and Petrochemicals Division

(278d) Power Controlled Study of Microwave-Accelerated In-Situ Transesterification of Algal Biomass

Prafulla Dinkarrao Patil[1], Harvind Reddy[1], Tapaswi Muppaneni[1], Aravind Mannarswamy[1], Tanner Schuab[2], Peter Lammers[3], Nirmala Khandan[4] and Shuguang Deng[1], (1)Chemical Engineering, New Mexico State University, Las Cruces, NM, (2)Director, Chemical Analysis and Instrumentation Laboratory, New Mexico State University, Las Cruces, NM, (3)Energy Research Laboratory, New Mexico State University, Las Cruces, NM, (4)Civil Engineering Department, New Mexico State University, Las Cruces, NM

The main objective of this study was to investigate the power controlled effect on microwave-accelerated simultaneous extraction and transesterification (in situ transesterification) of dry algal biomass. Experimental runs were designed using a response surface methodology and the process parameters such as dry algae to methanol ratio, reaction time and catalyst concentrations were optimized for power controlled process. The major benefits of this microwave-accelerated process are rate enhancement, increased yield and improved purity of the product.Proper application of microwave energy may result in greater benefits in terms of energy efficiency and reaction product quality. It was observed that this direct conversion technique has the potential to provide energy-efficient and economical routes for biodiesel production from algal biomass. The algal biomass characterization and algal biodiesel analysis were performed using ATR-FTIR, SEM-EDS, TGA and GC-MS. Nile Red method has been implemented for Lipid identification in microalgae using Confocal Microscopy and Fluorescence Spectroscopy.

See more of this Session: Catalytic Biofuels Refining II
See more of this Group/Topical: Fuels and Petrochemicals Division

(39f) Development of a Scalable Bottom-up Nanomanufacturing Platform for Highly Efficient Photovoltaics

Blayne Phillips and **Peng Jiang**, Chemical Engineering, University of Florida, Gainesville, FL

Solar energy is clean, abundant, and renewable, yet it is vastly under-utilized in the current world. Although crystalline silicon solar cells are the premier candidate for renewable and environmentally friendly solar energy conversion, high manufacturing cost has prevented such cells from economically competing with the combustion of fossil fuels as a source of electricity on a large scale. We have developed a scalable bottom-up nanomanufacturing platform for mass-producing highly efficient photovoltaics by exploring effective photon management using surface-plasmon-enabled transparent conducting electrodes and subwavelength antireflective coatings. In this methodology, self-assembled colloidal crystals with remarkably large domain sizes are used as templates to create periodic metallic nanohole arrays and subwavelength moth-eye antireflection coatings. The resulting metallic nanohole arrays exhibit high and tunable optical transmission and high electrical conductivity, promising as transparent conducting electrodes for flexible photovoltaics. The subwavelength-structured moth-eye antireflection coatings show much improved broadband antireflective properties than traditional quarter-wavelength coatings. This novel nanomanufacturing platform combines the simplicity and cost benefits of bottom-up self-assembly with the scalability and compatibility of top-down microfabrication.

See more of this Session: Nanomaterials for Photovoltaics I
See more of this Group/Topical: Topical 5: Nanomaterials for Energy Applications

(498d) Accelerated Thermal Decomposition of Ammonia Borane by Carbon Dioxide

Junshe Zhang[1], Yu Zhao[2], Daniel L. Akins[2] and Jae W. Lee[1], (1)Department of Chemical Engineering, The City College of New York, New York, NY, (2)Department of Chemistry, The City College of New York, New York, NY

This talk presents the enhanced dehydrogenation of ammonia borane (AB) at 85 ^oC under CO2, an abundant and naturally occurring material. The presence of CO_2 significantly accelerates the thermal decomposition and it kinetics becomes faster as the CO_2 pressure increases from 1.70 to 3.67 bar. Around 4 bar of CO_2, 1.33 equivalents of hydrogen is released from AB in 1 hour. The accelerated AB thermolysis is attributed to CO_2 reacting with AB, or intermediates like diammoniate of diborane (DADB), or primary thermolysis products including linear dimers of aminoborane (LDAB), which accelerates AB thermolysis, and this reaction has been confirmed by ATR-FTIR, Raman, and ^{11}B and ^{13}C solid sate MAS-NMR measurements. The results open up a potential avenue to convert CO_2 to other useful chemicals under very mild conditions.

See more of this Session: Aspects of Hydrogen Storage Materials
See more of this Group/Topical: International Congress on Energy 2011

TORREFACTION OF SWITCHGRASS AS A PRETREATMENT FOR CATALYTIC UPGRADING OF BIO-OIL

Brandon Smith, Trung Pham, Shaolong Wan, Sarah Zhang, Prof. Lance Lobban, Prof. Daniel Resasco and Prof. Richard Mallinson

School of Chemical, Biological and Materials Engineering, University of Oklahoma, Norman, OK 73071

Abstract:

The objective of this work is to use thermal pretreatment to separate the complex bio-oil mixture to produce less complex vapor streams of products from 1. light oxygenates (small aldehydes, acids) , 2. furan derivatives, and 3. phenolics. The selective production of products by thermal treatment is beneficial for catalytic upgrading of bio-oil vapors since different types of products require different catalytic upgrading and would be the first stage of a multi-stage pyrolysis process. Thus more selective and stable catalysts can increase the yields of desired products and lower the cost. In this study, torrefaction of switchgrass has been carried out at various temperatures lower than that for fast pyrolysis (500°C). Because biomass components, including cellulose, hemicellulose and lignin, break down over different temperature ranges, the enrichment of the product vapors to focus their composition by controlling their evolution may be achieved efficiently. While cellulose and hemicelluloses completely decompose in the range of 240-360°C, lignin completely decomposes at higher temperature although starting out at lower temperature as observed from Figure 1.

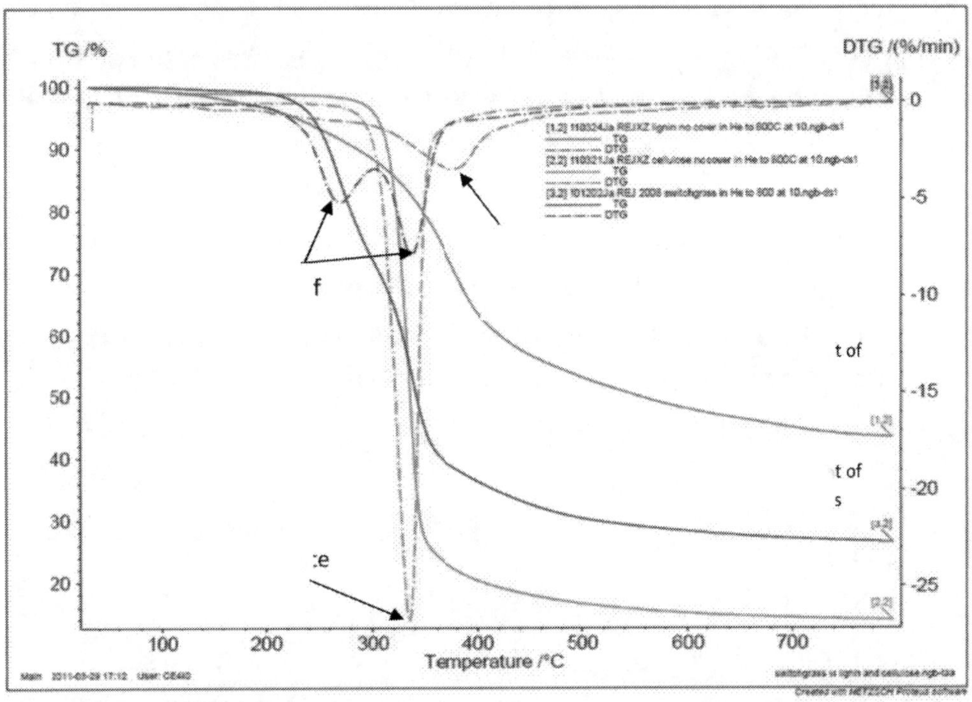

Figure 1: TG plot for thermal decomposition of lignin, cellulose and switchgrass in helium.

The TG experiments have been used to screen decomposition of whole switchgrass, lignin, and cellulose, certain temperatures can be selected to test the thermal pre-treatment, torrefaction of switchgrass using the same reactor as in pyrolysis experiments. The TG systems incorporates online analysis by MS and IR of the evolving vapors. Thus it is the product evolution, not the weight loss (or assumptions about what is decomposing) that provides value to this screening. Examination of the results suggests that temperature zones might be broken down into three zones: 150-270°C, 270-380°C, and 380-500°C, though this may not be suggestive of an optimal, or practical, process. The first stage is mostly to remove mainly water and light oxygenates, the second stage is desired for removal of sugar derived compounds such as furans while the third is enriched in phenolics that come from lignin. The catalytic upgrading strategy is different for each stage. While the light oxygenates in the first stage can be upgraded via condensation reactions to, for example, build up carbon chain length[1,2], the last stage of rich phenolics can be directly deoxygenated with methoxy methyl transfer to aromatic hydrocarbons[3].

We have performed in-situ catalytic upgrading of vapors from a first stage torrefaction. For this stage, various temperatures were screened and 200°C was selected wherein the major organic product in the stream is acetic acid, with no phenolics and low amount of furfural with substantial water. The vapor is then passed through a packed bed of $CeZrO_x$ where the acetic acid can be converted to acetone via ketonization. From the top of the graph for the non-catalytic run, acetic acid is observed as a major product. For the catalytic experiment, the initial sampling of products with time on stream shows complete conversion of acetic acid to acetone. In this case acetone could be the most desirable product; to be hydrogenated to propanol and used as na alkylation agent. Previous work has also shown that ZrO_2 can be used to make larger, less oxygenated, products via aldol condensation.

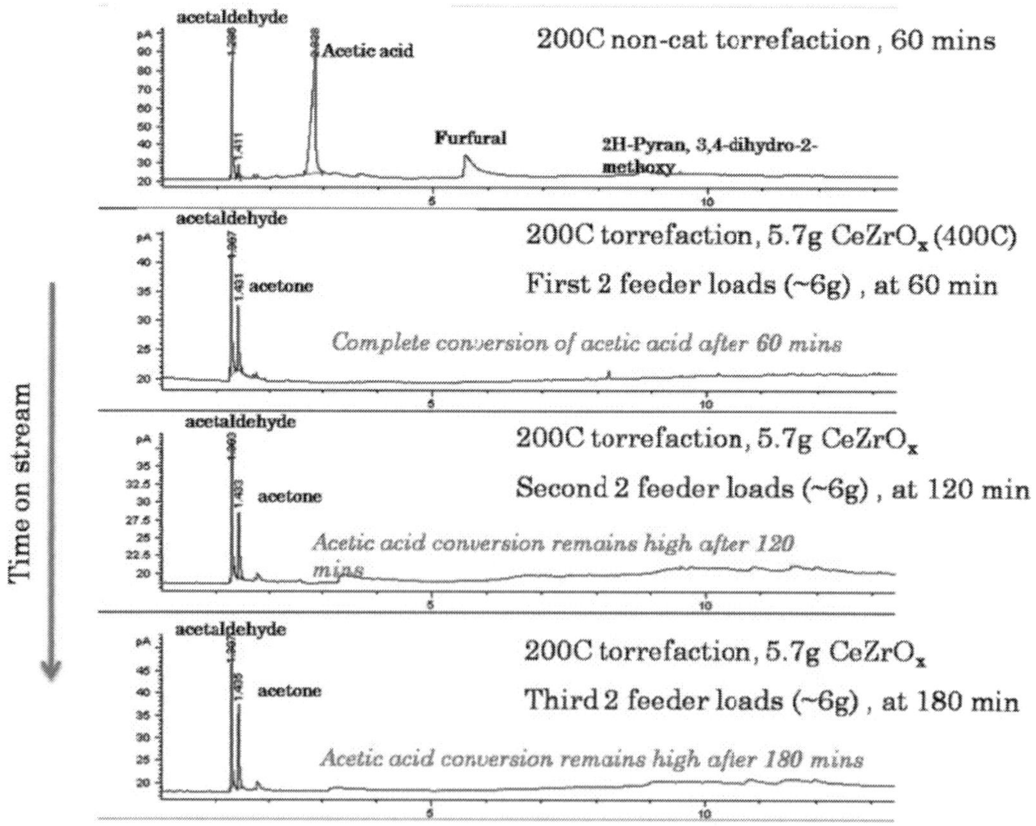

Figure 2: FID chromatograph of the first stage for non-catalytic and catalytic time-on-stream treatment using $CeZrO_x$.

References:

[1] Gangadharan, A; Shen, M; Sooknoi, T; Resasco, D.E; Mallinson, R.G; *Applied Catalysis A*, 385, 2010, 80.
[2] Deng, L; Fu, Y; Guo, Q; *Energy & Fuels*, 23, 2009, 564.
[3] Zhu, X; Lobban, L.L; Mallinson, R.G; Resasco, D.E; *Journal of Catalysis*, 281, 2011, 21.

Direct Coupled Catalytic Upgrading of Multi-stage Pyrolysis Bio-oil Vapors

Shaolong Wan, Trung Pham, Taiwo Omotoso, Sarah Zhang, Prof. Lance Lobban, Prof. Daniel Resasco and Prof. Richard Mallinson

School of Chemical, Biological and Materials Engineering, University of Oklahoma, Norman, OK 73071

Abstract:

A two-stage pyrolysis was conducted in this work. The purpose of the multiple stages is to provide a segregated product evolution in which the vapor products maybe be more selectively treated by catalysts to improve product quality and catalyst stability in comparison to one step pyrolysis in which all components of the bio-oil vapor are present, complicating the upgrading requirements to achieve high yield. This particular two-stage configuration is designed to remove much of the light oxygenates (acetic acid, water, acetol, acetaldehyde) and sugar derived components (furfurals, levoglucosan, pyrans, etc.) in the first stage, with a more concentrated lignin derived phenolics stream produced from the high temperature fast pyrolysis stage. The phenolics rich vapor from the high temperature stage is treated by a dual function metal and oxide catalyst bed with hydrogen at atmospheric pressure coupled to the exit of the pyrolysis reactor.

Dried switchgrass samples were first torrefied under He flow in a drop feed fluidized reactor at 350 °C for 10 minutes, after which the solid products were immediately blown out of the reactor by a large gas flow from the reactor bottom and collected by a cyclone installed at the exit of the reactor. Once the torrefied solid was obtained, it was then transferred into a continuous auger feeder for the second stage – fast pyrolysis with catalytic vapor upgrading. This two-step method allows the use of the same reactor, while providing a temperature-time profile as if it was a continuous staged system. During the fast pyrolysis second stage, the solid sample was continuously fed into the reactor tube preheated at 500°C, where the produced vapor passed through either an inert bed or a catalytic bed at 400 °C and then was condensed by a train of two traps: ice water and liquid nitrogen. For the catalytic run, He gas was still used as the main pyrolysis carrier gas while an independent H_2 flow was introduced prior to the catalytic bed. The catalyst bed consists of 5 g of 5% ruthenium supported on TiO_2, aimed at methyl transfer to the ring (from methoxy) and deoxygenation of the phenolic vapor products.

As shown in Figure 1 below, aromatic compounds including benzene, toluene, ethyl benzene, xylene etc. were produced in the catalytically upgraded oil, whereas none of those compounds were detected in the non-catalytic oil. Guaiacol, one of the major phenolics components in the non-catalytic oil, was not observed after catalytic treatment. These preliminary results clearly show that the Ru/TiO_2 catalysts can effectively convert the phenolics components into aromatics. More systematic work and in-depth analysis in this direction is under way.

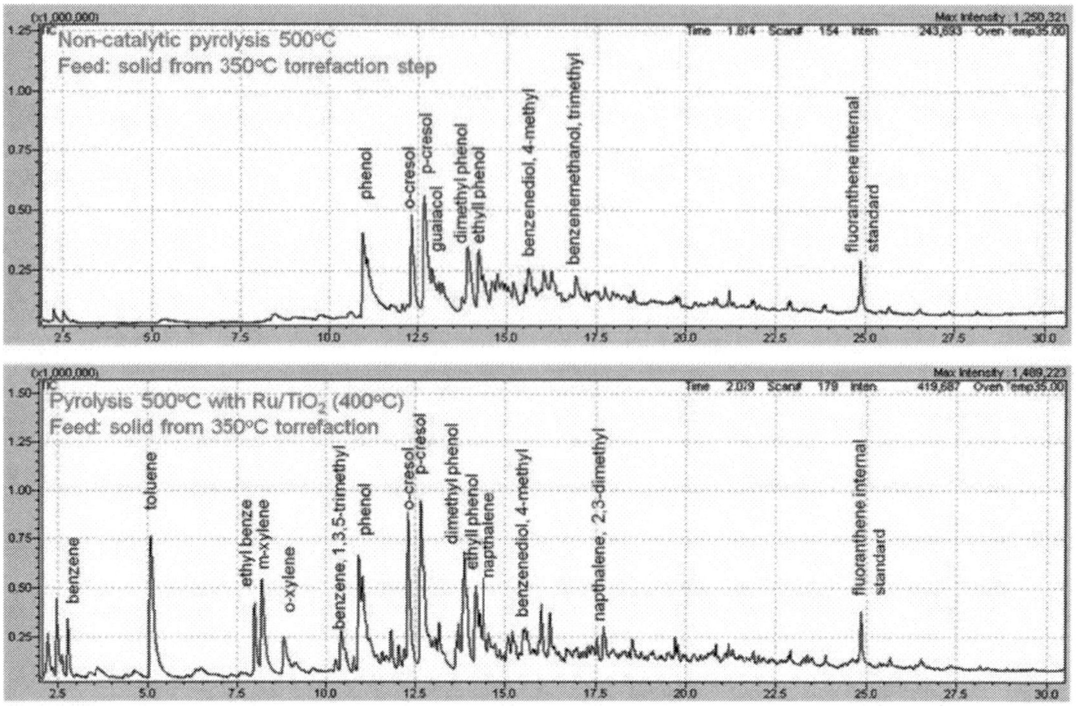

Figure 1: GC-MS spectrum for the oil phase collected in the second trap from non-catalytic and catalytic run respectively.

(496b) Combinational Multifeedstock Biorefining

Marc Privitera, PreProcess,Inc., Walnut Creek, CA

Single feedstock, single product systems combined into multifeedstock, multi-process biorefineries allow the economic optimization on the system with the change in feedstock cost and product value. The use of common utilities and the ability to handle a wide range of the variability inherent in waste material feedstocks is presented and discussed.

See more of this Session: Alternative Fuels and Enabling Technologies II
See more of this Group/Topical: Fuels and Petrochemicals Division

(35b) Graphitic Material with Engineered Graphene Units for High-Power Li Ion Battery Electrodes

Cary M. Hayner, Xin Zhao, Mayfair C. Kung and Harold H. Kung, Chemical and Biological Engineering, Northwestern University, Evanston, IL

Li ion batteries capable of delivering a high current without sacrificing voltage or energy density are highly desirable, since they can be recharged faster and provide longer operation time when used in devices that demand high power. In the current generation of batteries using graphite as the negative electrode material, the rate of Li ion insertion into and migration out of the graphite is a major factor that limits the power delivered by the battery. We have found that by carefully engineering the graphene units, which are the primary building block of graphite, it is possible to increase the Li ion diffusivity in the material significantly, which results in greatly increased charge capacity during high current operation. We report here a solution chemical method to produce the engineered structure that is readily scalable and add minimal costs to the manufacturing process. Electrodes prepared with the graphite formed from such engineered graphene sheets maintained a charge capacity of 180 mAh/g even when discharged at a rate of 2 A/g. Without the engineered structure, the charge capacity would be less than 70 mAh/g. The electrode prepared with the engineered material exhibits excellent cycling stability with no detectable capacity loss after >100 cycles. The nature of the engineered graphene sheets was investigated using high resolution electron microscopy, and the results show that there is an optimum size and density of the engineered structure.

See more of this Session: Lithium Battery Technology and Materials
See more of this Group/Topical: Fuels and Petrochemicals Division

Study on the Pyrolysis Characteristics of Biomass Compositions in Medical Wastes

DENG Na*, WANG Weiwei, ZHANG Yu-feng, MA Hong-ting

(School of Environmental Science & Engineering, Tianjin Uninversity, Tianjin 300072, China)

Abstract: To obtain pyrolysis characteristics of biomass compositions in medical wastes, thermogravimetric study of cotton swabs, toilet paper, gauze, absorbent cotton was carried out using the thermogravimetric analyser (TGA) with N_2. The heat change in pyrolysis process was analyzed and the properties of pyrolysis residues are reported. The overall mathematics model with two-step and four-reaction was established to simulate the pyrolysis processes. The results show that:(1) The decomposition processes of toilet paper, gauze and absorbent cotton appear one obvious stage in 260~420°C, whose pyrolysis mechanisms are in agreement with cellulose, their main ingredient. The pyrolysis of cotton swabs occurs at lower temperature than other three samples, with the weight loss at the temperature range of 207~417°C. A shoulder peak is observed on the left of the main DTG (differential thermogravimetric) peak of cotton swabs corresponding to semi-cellulose decomposition. (2) Due to the influence of the solvent used during the papermaking process, the maximum rate of weight loss of toilet paper takes place at 371°C lower than 383°C of the other three ones. (3) All the biomass medical wastes present endothermic pyrolysis reaction. The peak temperatures of DTA (differential thermal analysis) are a little higher than that of DTG for the delay of heat transfer. (4) All of the pyrolysis residues of the samples are black fluffy carbon-like solid matter, which could reflect the traits of original material. (5)The established model could satisfactorily describe the weight loss and differential process of the above four biomass samples.

Keywords: medical waste; biomass; pyrolysis characteristics; kinetic model

1. *Introduction*

Energy demand is growing up rapidly with the industrialization progress and the increasing of world population. The exploitation and utilization of biomass energy will relieve fossil fuel crisis and environmental pollution, which is beneficial for sustainable development in the world. Biomass pyrolysis technology has been paid great attentions during these years, which can convert low-quality biomass to high-quality liquid fuel or gas and some valuable chemical materials, and give zero net carbon dioxide emission.

Biomass compositions in medical waste contain infectious wastes and will pose a great health risk to the public if they are not properly disposed of. There are many technologies such as autoclaving, gas sterilization, and chemical disinfection, but many unavoidable disadvantages exist in each method, such as tail gas pollution and exhaust of dioxins to the environment in waste incineration. Now the novel pyrolysis technology for the treatment of medical waste is coming into use [1], which could reduce waste mass, make it innocuous, and finally convert it into fuel; the technology has been proposed for the treatment of municipal solid waste (MSW) and biomass by many scientists [2-3].

In the field of biomass energy, pyrolysis characteristics had been further studied into the reaction mechanism. Researchers also applied many kinds of kinetic models to get kinetic parameters based on the reaction mechanism they proposed. Some models put forward were based on independent pyrolysis of the three components in biomass: cellulose, hemicellulose and lignin [4-8]. Some only focus on cellulose pyrolysis behavior, whose content in biomass materials reaches up to almost half of the total amount. The corresponding kinetic models recently proposed can be divided into three principal categories: single-step global reaction models, multiple-step models, semi-global models, and so on [9-14]. However, the more complicated models give more difficulty for engineering use. So, it is pressing that to look for a more appropriate model to be used for medical biomass waste, being mechanistic and conveniently serving for the engineering.

* Corresponding author. Tel: 86-22-27890131
E-mail address: denglouna@tju.edu.cn; denglouna@163.com

In order to make clear the pyrolysis characteristics of medical biomass waste, the study turns to thermogravimetric analysis, which is one of the techniques for studying the primary reaction in the decomposition of solids, and has been widely used to study the mechanism and kinetics of the thermal decomposition of solid wastes.

Thermal decomposition of typical medical biomass product - cotton swabs, toilet paper, gauze, absorbent cotton with DTG equipment has been studied in the present work. The detail information about the process of medical biomass waste decomposition was obtained. The heat change in pyrolysis process was analyzed and the properties of pyrolysis residues are reported. The overall mathematics model was established to simulate the pyrolysis processes. The final purpose of the study is tried to afford theoretical direction and basic information to practical engineering project of fuel processing and waste treatment for biomass composition in medical waste.

2. *Experimental*

2.1. *Materials*

Medical waste refers to any waste generated from the healthcare industry such as hospitals and medical laboratories. The samples studied in this work contained 4 materials, which are cotton swabs, toilet paper, gauze and absorbent cotton, supplied by Tianjin General Hospital. Only the thermal physical chemical properties were cared about, not the virus and pathogenic bacteria, so the unused samples were selected for experiment. The main constituent and ultimate analysis of the samples are listed in **Table 1**.

To meet medical application, medical wastes were dealt with special physical and chemical treatment in the manufacturing process. The special technological process in production of a medical product may affect its degradation behavior and kinetics, which has been observed in later research.

Table 1. Ultimate analysis and proximate analysis of the test samples

Samples	**Major constituent**	**Ultimate analysis (wt.%)**					**Proximate analysis (wt.%)**				**HHV (MJ/kg)**
		C_{daf}	H_{daf}	O^*_{daf}	N_{daf}	S_{daf}	W_{ar}	A_{ar}	V_{ar}	C_{ar}	
Cotton swabs	Cellulose,hemicelluloses lignin	46.26	6.43	47.09	n.a.	0.22	8.36	0.26	84.05	7.34	17.2
Toilet paper	Cellulose	41.63	5.71	52.35	n.a.	0.31	6.34	0.18	89.19	4.29	15.1
Gauze	Cellulose	42.51	6.35	51.01	n.a.	0.13	5.79	0.08	90.91	3.22	16.5
Absorbent cotton	Cellulose	42.90	6.69	50.19	n.a.	0.22	5.29	0.11	91.30.	3.3	18.0

—: not tested.
n.a.: tested but not available.
O^*: obtained by mass balance, containing impurity content

2.2. *Apparatus and procedure*

The equipment used to carry out the pyrolysis of medical waste was a SHIMADZU DTG-60H thermo-balance. The crucible was pyro-ceramic material and the reference substance was an appropriate amount of inert material, Al_2O_3 (aluminum oxide) powder. The dynamic (non-isothermal) runs were performed in an atmosphere of N_2 for all biomass samples. Firstly, high purity N_2 was purged for 40 min at a flow rate of 75 ml/min, before starting the heating of the samples, to be sure of an atmosphere free of oxygen. The nitrogen used in obtaining the results was at a flow rate of 20ml/min. The samples were heated up to the desired temperature 800°C at the pre-selected heating rate of 20°C/min. The mass of sample spread in the crucible is about 2.2 mg. The samples were cut into very small pieces, about 0.5mm in diameter, and distributed all over the surface of the crucible, to ensure a representative sample from the different materials and a reaction situation mainly controlled by chemical kinetics, free of heat transfer and mass transfer. A check of the repeatability of experiments was performed for all samples. The results showed the TG and the DTG curves were almost identical, consequently giving very low standard deviations (less than 1.86%) on the obtained calculated kinetic parameters. The results presented here are a set of ones among those experiments which meet the above-mentioned conditions.

3. *Results*

3.1. *The results of TG and DTG*

The TG, DTG and DTA curves of all the analyzed samples (cotton swabs, toilet paper, gauze, absorbent cotton) at heating rate of 20°C /min are all shown in **Fig.1.** The thermal analyses data are presented in **Table 2**.

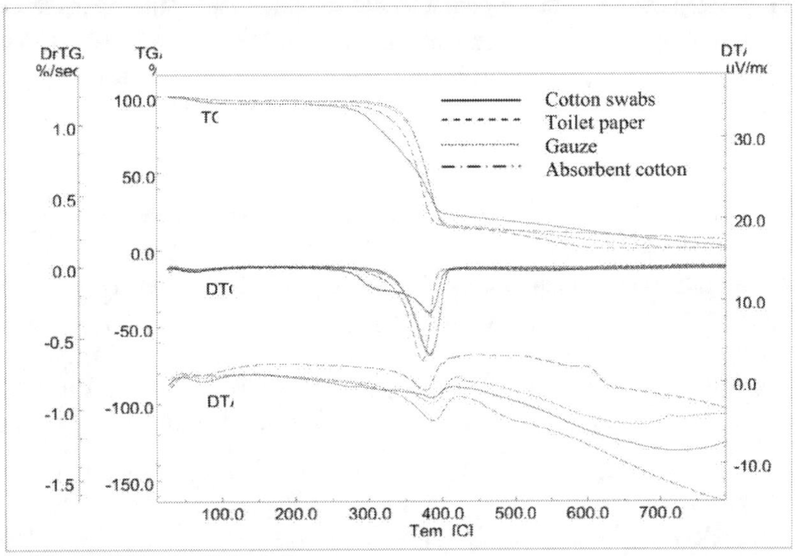

Fig. 1. TG/DTG/DTA curves of biomass compositions in medical waste, β=20°C/min

Biomass material mainly consists of three components: cellulose, hemicellulose and lignin. According to product specification, it can be known cotton swabs included all of the three components. While, the major constituent of toilet paper, gauze, absorbent cotton are cellulose. Song Chuncai[15] reported cellulose pyrolysis occurred at the temperature range of 300~400°C, the DTG peak was very sharp; while the degradation of hemicellulose occurred earlier mainly at the temperature range of 250~350°C, then ended almost after 370°C. The decomposition of lignin had the widest temperature region at 110~900°C [16], with the maximum rate of weight loss at the temperature range of 320~360°C [17]. From **Fig.1.**, it can be observed the initial and final temperatures of the decomposition of four samples are some different. This is attributed to the difference of pyrolysis characteristics of different components in experimental materials. The results show that:

1) Below 110°C, four biomass compositions show slight weight loss due to water evaporation. This phenomenon can be observed in the DTG and DTA curve correspondingly.

2) The decomposition processes of toilet paper, gauze and absorbent cotton appear one obvious stage in 260~420°C, whose pyrolysis mechanisms are in agreement with cellulose, their main ingredient. And similar DTG peaks are observed for all three samples. Due to the influence of the solvent used during the papermaking process, the TG curve of toilet paper shifts to lower temperature than those of gauze and absorbent cotton, the maximum rate of weight loss of toilet paper takes place at 371°C lower than 383°C of the other two ones.

3) The pyrolysis of cotton swabs occurs at lower temperature than other three samples, with the weight loss at the temperature range of 207~417°C. A shoulder peak is observed on the left of the main DTG (differential thermo-gravimetric) peak of cotton swabs corresponding to semi-cellulose decomposition.

4) Up to 600°C, Four samples almost finish their pyrolysis process. The quantity of pyrolysis residues of cotton swabs is more than those of the other three ones under 600°C.

It can be concluded that the main peak of pyrolysis of four samples should be caused by the decomposition processes of cellulose, and the shoulder peak of cotton swards was corresponding to hemicellulose pyrolysis, the large amount of char residue showed that cotton swards contain lignin compement[17]. Orfao[16] also considered hemicelluloses played the important role in the production process of char. This paper considered it was affected by both reasons. Raw material of gauze and absorbent cotton is cotton fiber after being disinfection treated (the content of fiber reaches 98%) , being kept the characteristic of cellulose. Although toilet paper is produced from wood pulping, during the boiling process of chemical pulping, either mechanically or chemically, hemicelluloses and lignin are gradually removed and the proportion of cellulose increase more and more[18]. Finally, almost only the cellulose component was left. Meanwhile, the pyrolysis reactivity of cellulose was enhanced due to the influence of the solvent (NaOH and Na_2S) used during the papermaking process, resulting in the whole TG curve of toilet paper shifted to lower temperature than the others.

Table 2 shows the key temperature points of the TG and DTG curve of every test specimen, containing initial temperature T_i, extrapolated onset temperature T_{onset}, extrapolated final temperature T_f and temperature at the peak T_P of every stage weight loss of decomposition. $\alpha_{\Delta T}$ is the weight loss between $T_i \sim T_f$. The weight loss α of test samples at 600°C is up to 89.65- 98.93%, which proves that pyrolysis technology can remarkably reduce the mass of waste, and provides data for the design of the pyrolytic reactor. The most active reaction step is the research focus on thermo-gravimetric analysis, reflecting strong pyrolysis behavior, which can be briefly represented by the point, maximum rate of weight loss $(\frac{d\alpha}{dt})_p$, with corresponding temperature at T_P. As shown in **Table 2**, all T_P locate within the temperature range of 371–383°C. Thereby, it was of consideration to extend the residence time of the feedstock in the above time interval, to ensure reaction to the maximum extent.

Table 2 Thermal analysis data of biomass compositions in medical waste at heating rate of 20°C/min

Samples	T_i	T_{onset}	T_P	T_f	$\alpha_{\Delta T}$	$(\frac{d\alpha}{dt})_p$
	(°C)	(°C)	(°C)	(°C)	(%)	(%/s)
Cotton swabs	206.65	330.52	381.64	417.38	72.77	-0.34
Toilet paper	261.37	358.95	370.88	412.01	79.24	-0.66
Gauze	259.56	352.00	382.73	429.44	81.42	-0.52
Absorbent cotton	265.47	356.76	382.57	426.08	83.21	-0.63

3.2 The heat change in pyrolysis process

As shown in **Fig.1.**, DTA curves of all biomass samples are similar except that the one for cotton swabs looks like more gently. **Table 3** reports the DTG/DTA peak temperatures for the four biomass compositions (β=20□/min) . The weak endothermic peaks of DTA curves representing the release of water appear at 66~76°C for all biomass samples. Furthermore, corresponding to DTG peak, obvious endothermic peaks appear on the DTA curves, meaning all the biomass medical wastes present endothermic pyrolysis reaction. However, there are some literatures that proved exothermic behavior in the biomass pyrolysis [19]. From **Table 3**, we can see the peak temperature of DTA is a little higher than that of DTG for the delay of heat transfer.

Table 3 The DTG/DTA peak temperatures for biomass compositions in medical waste (β=20□/min)

Samples	T_P^{DTG} (℃)	T_P^{DTA} (℃)
Cotton swabs	381.64	73.51-383.89
Toilet paper	370.88	66.73-375.60
Gauze	382.73	75.87-385.89

Absorbent cotton	382.57	75.14-387.25

3.3 The properties of pyrolysis residues

As all known, the pyrolysis product of biomass samples is gas, liquid and carbon solid residue. In TGA experiments, the small molecule gases (water and fuel gas) and the large and medium molecule condensable volatile are released into the air with purge gas; solid residues were detained at the bottom of the crucible. In order to get a clear residue characteristics, the large amount of biomass samples are used in experiments.

Fig.2. shows the photograph of the pyrolysis residue of biomass compositions in medical waste. From the results of experiments, we can see the traits of pyrolysis residues of samples are quite similar, which are black fluffy carbon-like solids with low density, automatically shrinking and staying in the central crucible. After a slight touch by the tweezers, the residues would be scattered. Moreover, the char residue could reflect the traits of original material. For cotton swabs, we can find a little amount of small black fibers residue in addition to the powder residue, resulting from the existence of wood fiber material even after being ground. Pyrolysis residue of toilet paper samples slightly curled still representing the original state of paper scrapes. Absorbent cotton and gauze samples remain downy surface residue, likewise, reflecting the characteristics of the raw samples. When pyrolysis residues of these samples continued to decompose at high temperatures, the mass would be reduced, but the shape would not change.

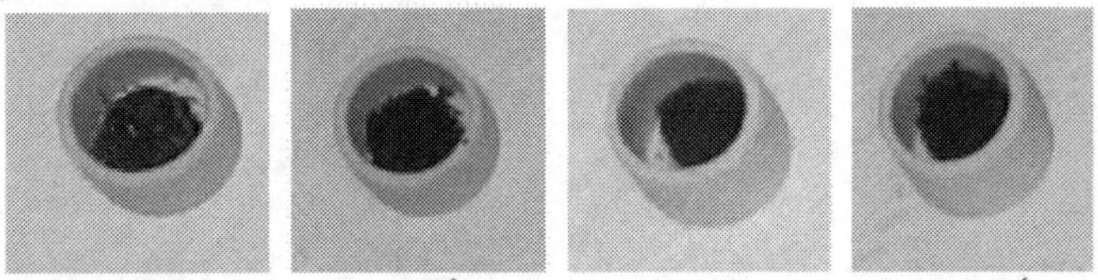

a. b. c. d.

Fig.2. The photograph of the pyrolysis residue of biomass compositions in medical waste
(a. for Cotton swabs; b. for Toilet paper; c. for Gauze; d. for Absorbent cotton)

4 . The global 'two-step four-reaction kinetic model' and the results

4.1. The establishment of the model

The characteristic of a good model is that it should require the least number of parameters for the engineering design of the reactor and that it should be mechanistic. On the other hand, biomass degradation is a complex phenomenon, and then the expected model should express the principle of chemical reaction kinetics, especially for those more complex materials containing kinds of additives. So, the more reasonable models should be established.

The main ingredient of the four biomass samples here is not the same. Cotton swabs contain cellulose, hemicellulose and lignin; while, the other three samples are only mainly composed of cellulose. Hopefully, the established kinetic model can be used to describe the pyrolysis behavior of the all medical biomass samples, even for all medical waste compositions. That is meaningful for engineering design.

Gupta[20] ever reported a consecutive reaction mode for paper pyrolysis assuming the existence of an hypothetical intermediate product. Ahuja[21] also proposed a consecutive-competitive model via. primary and heterogeneous secondary reactions of wood pyrolysis. The reaction scheme of the typical Brodio-Shafizadeh kinetics model[22] was thought to correspond with the formation of active cellulose (AC), followed by two competitive reactions to produce char and volatile. Based on this, no matter how many parallel reactions and chain reactions occur in the process or how the additives influence the procedure, the 'two-step four-reaction kinetic model' as follows will be proposed in the paper. It simplifies and lumps the complicated reactions nets of biomass compositions pyrolysis to two respective steps, which step corresponds to two parallel reactions.

The reaction scheme is as (1) shown. It is assumed that the four reactions follow the Arrhenius. Firstly, S (biomass samples) pyrolysis via two parallel reactions (the constant of the reaction rate are k_1 and k_2, separately) to produce corresponding primary product, volatile V_1 and solid product S_1. Then, V_1 escapes with purge gas. At the same time, solid product S_1 begins to pyrolysis during its formation, to produce secondary volatile V_2 and solid product S_2, also via two parallel reactions whose reaction rate constant are k_3 and k_4 , separately. S_2 is the final solid residue. The proposed reaction coincides with the biomass pyrolysis mechanism. S_1 can be considered as 'hypothetical intermediate product' or 'active cellulose'. It is appropriate for describing the weight loss behavior of feedstock happening in less than or equal two steps.

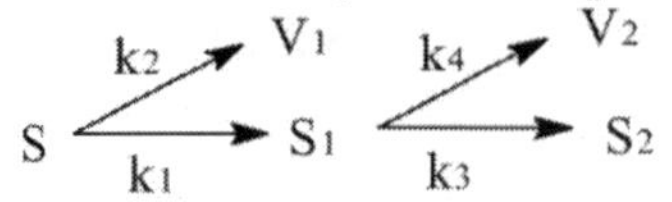

Reaction scheme (1)

The following assumption is needed for the model:

1) Neglecting minor reactions;
2) Considering the volatile product escapes with flow gas, not reacts with solid product;
3) Neglecting the heat transfer between sample and crucible.

The ordinary differential equations of the overall kinetic model become:

$$\begin{cases} \dfrac{dS}{dt} = -k_1 S^{n_1} - k_2 S^{n_2} & (2) \\ \dfrac{dS_1}{dt} = k_1 S^{n_1} - k_3 S_1^{\ n_3} - k_4 S_1^{\ n_4} & (3) \\ \dfrac{dS_2}{dt} = k_3 S_1^{\ n_3} & (4) \end{cases}$$

with the initial condition $[S, S_1, S_2]_{t=t0} = [1,0,0]$; there in, $k_i = A_i \exp(-E_i / RT)$ $_{i=1\cdots4,}$ represents the rate constant of the four reactions; $T = \beta t + T_0$, when the heating rate is β; the solid residual weight fraction in the crucible is w , given by the sum: $w_{cal} = S + S_1 + S_2$. All the optimized kinetic parameters of $x = [A_1, A_2, A_3, A_4, E_1, E_2, E_3, E_4, n_1, n_2, n_3, n_4]$ are identified by comparing the experimental curves with those resulting from solving Eqs. (2), (3) and (4) using iterative optimization, which minimize the values of the objective function OF given below (Eq. (5)).

$$\text{OF} = \sum_{j=1}^{m} \left(w_{cal}(t_j, x) - w_{exp}(t_j) \right)^2 , \qquad (5)$$

Where, w_{cal} is the calculated value of the solid residual weight fraction in the crucible; w_{exp} is the experimental value; x is the kinetic parameters; m is the total number of experimental points; j is the experimental point of j; t_j is the time for the experimental point of j.

The systems of differential equations corresponding to the model were solved using a 2nd and 3th order Runge Kutta algorithm. The Levenberg-Marquardt optimization method was used in order to calculate the kinetic parameters, which has little request for initial values and fast convergence speed.

4.2 The result of the model

The 'two-step four-reaction kinetic model' is used to fit the experimental data at β=20°C/min. The kinetic parameters for all medical biomass samples obtained are shown in **Table 4**. The validity of the models was tested by the value of the variation coefficient (VC), calculated as Eq. (6): where D (m) and P are the number of data and parameters fitted, respectively, and $\overline{w_{exp}}$ is the average of the experimental weights. **Fig.3.~Fig.6.** indicate the calculated and experimental results, with satisfactory agreement. Especially, as **Fig.3.b.** shown, the asymmetry of DTG curve can be described by the proposed model for cotton swabs. The facts also show that the model can be used for the pyrolysis process of all of the medical waste compositions.

$$\mathrm{VC}(\%) = \frac{\sqrt{OF/(D-P)}}{\overline{w}_{\exp}} \times 100 \qquad (6)$$

Table 4 Kinetic parameters obtained by 'two-step, four-reaction, comprehensive model' (β=20□/min)

Samples	**Reaction number**	*E*	*A*	*n*	**D**	$\overline{w}$	**OF**	**CV**
		(kJ/mol)	**/min^{-1}**				**/10^{-3}**	**%**
Cotton swabs	1	86.709	1.43×10^{8}	2.30	158	0.683	17.65	1.61
	2	60.544	7.59×10^{5}	0.33				
	3	80.742	1.03×10^{5}	1.01				
	4	88.382	1.25×10^{5}	1.80				
Toilet paper	1	157.42	1.85×10^{14}	0.556	43	0.430	1.99	1.86
	2	153.08	4.83×10^{14}	0.602				
	3	173.98	7.69×10^{13}	1.87				
	4	177.21	6.21×10^{13}	1.03				
Gauze	1	184.37	7.24×10^{14}	0.96	35	0.463	1.02	0.96
	2	160.05	7.85×10^{14}	0.59				
	3	185.02	6.99×10^{14}	1.01				
	4	184.02	7.40×10^{14}	0.74				
Absorbent cotton	1	163.5	1.98×10^{14}	0.80	90	0.4598	0.80	0.70
	2	181.03	1.53×10^{16}	0.80				
	3	64.67	2.31×10^{13}	1.05				
	4	152.38	1.10×10^{19}	0.86				

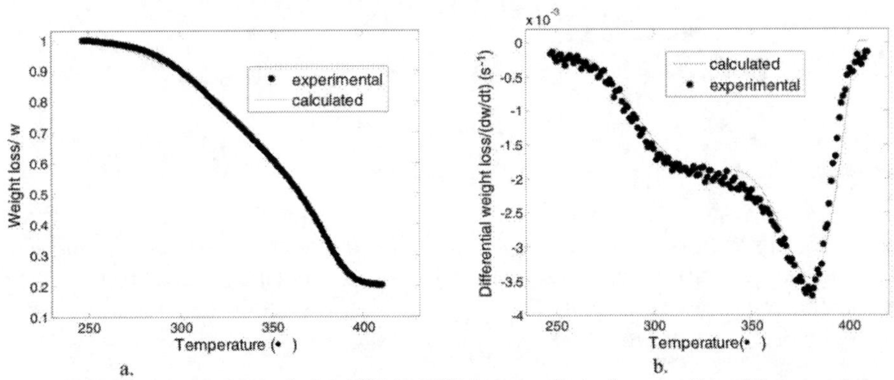

Fig.3. Experimental and calculated TG (a)-DTG (b) curves in the decomposition of Cotton swabs

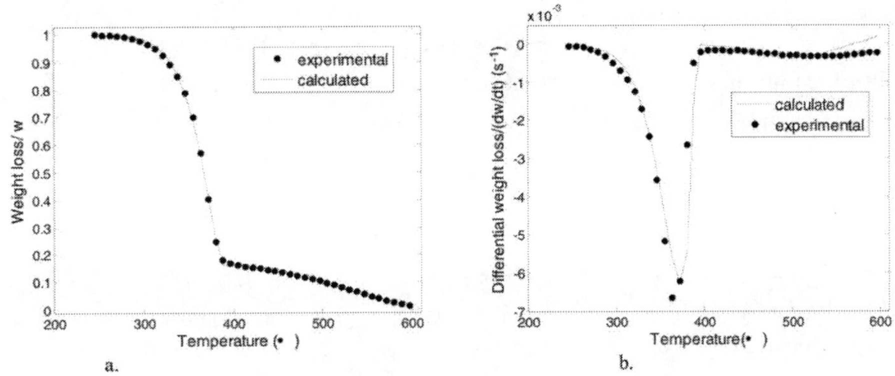

Fig.4. Experimental and calculated TG (a)-DTG (b) curves in the decomposition of Toilet paper

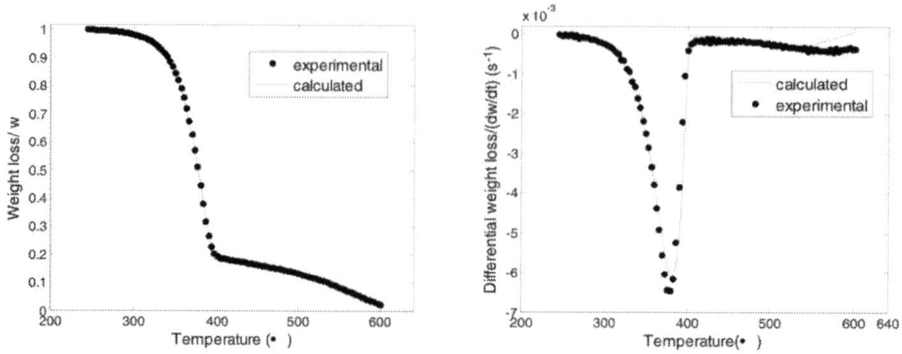

a. b.

Fig.5. Experimental and calculated TG (a)-DTG (b) curves in the decomposition of Gauze

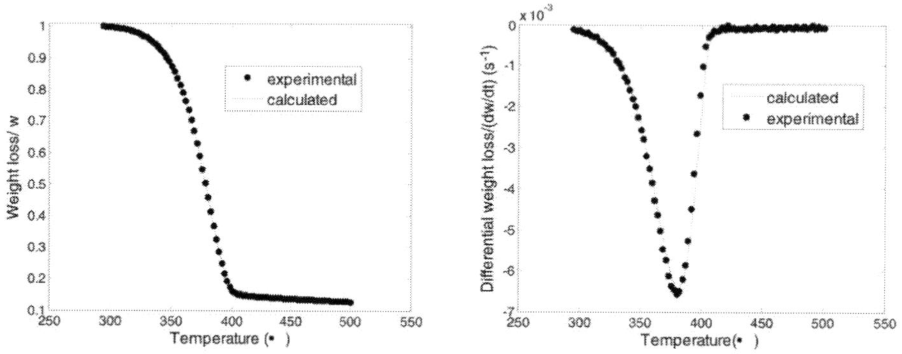

a. b.

Fig.6. Experimental and calculated TG (a)-DTG (b) curves in the decomposition of Absorbent cotton

4.3. Discussion of the kinetic results

All the four samples belong to biomass material. The kinetic parameters obtained were similar. Among them, cotton swabs separated from the others by having the cellulose, hemicellulose and lignin; while, toilet paper, gauze and absorbent cotton contained only cellulose as their main compositions. So the kinetic parameters for the latter three ones are very close.

The author also applied Coats-Redfern method, the Ozawa method and the Kissinger method to compare the results of 'two-step four-reaction kinetic model'. Apparent activation energy obtained for each method is as follows: 84.45, 140.86, and 131.71 kJ/mol for cotton swabs; 170.11, 164.84, and 153.72 kJ/mol for toilet paper; 185.85, 146.10 and 139.79kJ/mol for gauze; 198.23, 139.48 and 134.37 kJ/mol for absorbent cotton, respectively. The results from all of the four methods indicate good agreement.

Reina et al [23]reported kinetic parameters for waste woods(furniture, forest and pallet) obtained by dynamic experiment(2℃/min) at the same conversion(0.75). They were 127.61-136.18 kJ/mol for E and $1.23\text{-}3.38\times10^{6}s^{-1}$ for A, little higher than those of cotton swabs. Kinetic values found in the literature[21] for Accia wood and Eucalyptus concerning a model for primary and heterogeneous secondary reaction were following: E_1 of 91.5 and 93.1 kJ/mol, A_1 of 5.87×10^{4} and $1.28\times10^{5}g^{-0.1}$ /s, E_2 of 85.9 and 86.7 kJ/mol, A_2 of 6.57×10^{3} and $1.16\times10^{4}g^{-0.1}$ /s, E_3 of 80.5 and 80.3 kJ/mol, A_3 of 1.045×10^{6} and $1.329\times10^{6}g^{-0.1}$ /s, E_4 of 76.0 and 78.3 kJ/mol, A_4 of 1.015×10^{6} and $1.264\times10^{6}g^{-0.1}$ /s. The above values almost equal to those of a cotton swabs due to identical physical and chemical composition and structure in test samples.

Kinetic parameters for cellulose had been given by many researchers. Activation energies of 153-238 kJ/mol, pre-exponential factor of 6.06×10^{9}-2.52×10^{16} s^{-1}, and reaction order of 0.46-1, which are close to those of cellulose substance here(toilet paper, gauze and absorbent cotton), were found in the previous work [24].

5. *Conclusions*

In this study, thermogravimetric study of cotton swabs, toilet paper, gauze, absorbent cotton was carried out using the thermogravimetric analyser (TGA) with N_2. The following conclusions are deduced:

(1) The decomposition processes of toilet paper, gauze and absorbent cotton appear one obvious stage in 260~420°C, whose pyrolysis mechanisms are in agreement with cellulose, their main ingredient. The pyrolysis of cotton swabs occurs at lower temperature than other three samples, with the weight loss at the temperature range of 207~417°C. A shoulder peak is observed on the left of the main DTG peak of cotton swabs corresponding to semi-cellulose decomposition. The reason that cotton swabs separated from the others by having the cellulose, hemicellulose and lignin.

(2) the pyrolysis reactivity of cellulose was enhanced due to the influence of the solvent (NaOH and Na_2S) used during the papermaking process, resulting in the whole TG curve of toilet paper shifted to lower temperature, with the maximum rate of weight loss takes place at 371°C lower than 383°C of the other three ones.

(3) All the biomass medical wastes present endothermic pyrolysis reaction. The peak temperatures of DTA (differential thermal analysis) are a little higher than that of DTG for the delay of heat transfer.

(4) All of the pyrolysis residues of the samples are black fluffy carbon-like solid matter, which could reflect the traits of original material.

(5) It is advisable that the global 'two-step four-reaction kinetic model' can satisfactorily describe the weight loss and differential process of the four biomass samples.

Acknowledgements

This project received funding from Independent Innovation Project of Tianjin University and Natural Science Foundation of Tianjin City(09JCYBJC08100).

References

1. Zhang Y F, Deng N, Ling J H, et al(2003), "A New Pyrolysis Technology and Equipment for Treatment of Municipal Household Garbage and Hospital Waste," *Renewable Energy*, December, pp. 2383–2393.
2. Sorum L, Gronli M G and Hustad J E(2001), "Pyrolysis Characteristics and Kinetics of Municipal Solid Wastes," *Fuel*, July, pp. 1217–1227.
3. Garcı'a A N, Marcilla A and Font R(1995), "Thermogravimetric Kinetic Study of The pyrolysis of Municipal Solid Waste," *Thermochimica Acta*, April, pp. 277–304.
4. Soltes E J, Wiley A T and Lin S C K(1981), "Biomass Pyrolysis-towards an Understanding of Its Versatility and Potentials," in *Biotech & bioengineering symposium*.
5. Williams P T, Besler S(1993), "The Pyrolysis of Rice Husks in a Thermogravimetric Analyser and Static Batch Reactor," *Fuel*, February, pp. 151-159
6. Nunn T R, Howard J B, Longwell J P, et al(1985), "Product Compositions and Kinetics in the Rapid Pyrolysis of Sweet Gum Hardwood," *Ind. Eng. Chem. Process*, July. pp. 836-844
7. Miller R S, Bellan J A(1997), "A Generalized Biomass Pyrolysis Model Based on Superimposed Cellulose, Hemicellulose and Lignin Kinetics," *Combust Science and Tech*, pp. 97-137.
8. Janse A M C,Westerhout R W J and Prins W(2000), "Modelling of Flash Pyrolysis of a Single Wood Particle". *Chemical Engineering and Processing*, May, pp. 239-252.
9. John E, White W, James Catallo, et al(2011), "Biomass Pyrolysis Kinetics: A Comparative Critical Review with Relevant Agricultural Residue Case Studies," *Journal of Analytical and Applied Pyrolysis*, May, , pp. 1–33.
10. Agrawal R K(1988), "Kinetics of Reactions Involved in Pyrolysis of Cellulose I. The Three Reaction Model," *Canadian Journal of Chemical Engineering*. June, pp. 403–412.
11. Tsamba W. Yang W. Blasiak(2006), "Pyrolysis Characteristics and Global Kinetics of Coconut and Cashew Nut Shells," *Fuel Processing Technology*. June, pp. 523–530

12. Lanzetta M, Di Blasi C(1998), "Pyrolysis Kinetics of Wheat and Corn Straw," *Journal of Analytical and Applied Pyrolysis*, January, pp. 181–192.
13. Lv P, Chang J, Wang T, et al(2004), "A Kinetic Study on Biomass Fast Catalytic Pyrolysis," *Energy & Fuels*, November, pp. 1865–1869.
14. Senneca O, Chirone R, Salatino P, et al(2007), "Patterns and Kinetics of Pyrolysis of Tobacco under Inert and Oxidative Conditions," *Journal of Analytical and Applied Pyrolysis,*May , pp. 227–233.
15. Song C C(2003), "Biomass Pyrolysis and Its Kinetic Parameters with Different Methods," Doctoral dissertation. Dalian University of Technology.
16. Orfao J J M, Antunes F J A and Figueiredo J L(1999), "Pyrolysis Kinetics of Lignocellusosic Materials-Three Independent Reactions model". *Fuel*, February, pp. 349-358
17. Zhao Ming(2000). "Pyrolysis Kinetic of Rice Straw," Master's dissertation. *University of Shanghai for Science and Technology*, December.
18. Sørum L, Grønli M G and Hustad J E(2001), "Pyrolysis Characteristics and Kinetics of Municipal Solid Wastes," *Fuel*, July , pp. 1217-1227
19. Hu S, Andreas J and Xiang J(2007), "Kinetic Study of Biomass Pyrolysis Based on Different Three-pseudocomponent Models," *Journal of Chemical Industry and Engineering (China)*, October, pp.: 2580-2586
20. David C, Salvador S, Dirion J L, et al(2003), "Determination of a Reaction Scheme for Cardboard Thermal Degradation Using Thermal Gravimetric Analysis," *Journal of Analytical and Applied Pyrolysis*, May, pp.: 307-323
21. Ahuja P, Kumar S and Singh P C(1996), "A Model for Primary and Heterogeneous Secondary Reactions of Wood Pyrolysis," *Chemical Engineering and Technology*, June, pp.272–282.
22. Bradbury A G W, Sakai Y and Shafizadeh F(1979), "A Kinetic Model for Pyrolysis of Cellulose," *Journal of Applied Polymer Science*" June. pp. 3271–3280.
23. Reina J, Velo E and Puigjaner L(1998), "Thermogravimetric Study of The Pyrolysis of Waste Wood," *Thermochimica Acta, November,* pp.161–167.
24. Wu C H, Chang C Y, Lin J P, et al (1997), "Thermal Treatment of Coated Printing and Writing Paper in MSW: Pyrolysis Kinetics," *Fuel*, 76(12), pp. 1151–1157.

(278b) Preparation of Mesoporous Silica-Supported Palladium Catalyst for Biofuel Upgrade

Ling Fei[1], Joshua Hill[1], Qianglu Lin[1], Yun Xu[2], Shuguang Deng[2] and Hongmei Luo[2], (1)Chemical Engineering Department, New Mexico State University, Las cruces, NM, (2)Chemical Engineering, New Mexico State University, Las cruces, NM

Preparation of Mesoporous Silica-supported Palladium Catalyst for Biofuel Upgrade
Ling Fei, Joshua Hill, Qianglu Lin, Yun Xu, Shuguang Deng, Hongmei Luo
Department of Chemical Engineering, New Mexico State University, Las Cruces, NM 88003
A large number of reactions in organic chemistry (such as the Heck and hydrogenation) are facilitated by catalysis with palladium compounds. Due to its highly selective chemical transformations in interfacial reactions, palladium plays an important role in petroleum cracking and biofuel upgrading. In this report, Pd catalyst based on mesoporous silica and/or carbon nanotubes (CNT) support has been prepared, such as Pd/CoMo/SiO2 and Pd/CNT/CoMo/SiO2. The structure and morphologies of catalysts are studied by X-ray diffraction (XRD), scanning electron microscopy (SEM) and transmission electron microscopy (TEM) analysis. The BET surface area is tested by ASAP nitrogen adsorption and desorption. We will study the catalyst activity for biofuel upgrade, and report the influence of mesoporous silica with different pore size and surface area, CNT, and metals from different synthetic approach on catalyst activity.

See more of this Session: Catalytic Biofuels Refining II
See more of this Group/Topical: Fuels and Petrochemicals Division

(508e) Conversion of Furfural Over Supported Metal Catalysts

Surapas Sitthisa, School of Chemical, Biological and Materials Engineering, University of Oklahoma, Norman, OK and Daniel E. Resasco, Cbme, University of Oklahoma, Norman, OK

The conversion of furfural was studied over silica-supported Cu and Pd catalysts. The results showed that Cu is highly selective for the hydrogenation of furfural to furfuryl alcohol, with 2-methyl furan only appearing at high temperatures. In contrast, Pd is highly active for decarbonylation yielding furan as the dominant product. This contrasting behavior indicates that the reaction pathways and product selectivities may strongly depend on the nature of the metal, which opens an opportunity for molecular engineering of biofuels based on specific catalyst selection. However, the exact reaction pathways by which the oxygenated molecules are converted on difference metal surfaces are not fully understood. In this work, kinetics, DFT (density functional theory) and DRIFTS studies have been combined to identify the adsorbed species on difference metal surfaces and propose possible reaction pathways. Based on this knowledge, tailored metal surfaces have been developed taking into consideration the relative stabilities of the surface-bounded reaction intermediates to achieve the desired product.

See more of this Session: Catalytic Biomass Conversion to Chemicals I
See more of this Group/Topical: Fuels and Petrochemicals Division

(173d) A Facile Way to Fabricate MoO3 /Graphene Composite Anode for Lithium Ion Battery

Yun Xu, Qianglu Lin, Ling Fei, Joshua Hill, Shuguang Deng and Hongmei Luo, Chemical Engineering Department, New Mexico State University, Las cruces, NM

A Facile Way to Fabricate MoO_3 /Graphene Composite Anode for Lithium Ion Battery

Yun Xu, Qianglu Lin, Ling Fei, Joshua Hill, Shuguang Deng, Hongmei Luo

Department of Chemical Engineering, New Mexico State University, Las Cruces, New Mexico 88003

MoO_3 is capable of Li-$^+$ insertion/extraction in excess of 6 Li-$^+$ per formula unit, resulting in a significantly reversible capacity. Recently, chemically modified graphene materials have been used to form hybrid materials with SnO_2, Fe_2O_3, and Mn_3O_4 in attempts to improve electrode capacity and cycling stability. However, an anode combining MoO_3 with graphene has never been reported. MoO_3/ graphene composite was prepared through a facile two-step synthesis by homogeneous coprecipitation and subsequent reduction of the graphite oxide with hydrazine. The synergic effect of graphene and MoO_3 improves the electrochemical properties, like cycling performance and rate capability. SEM and TEM images show that MoO_3 uniformly dispersed on the surface of graphene. The synthesis approach presents a promising route to produce large amount of high performance electrode materials for Lithium ion batteries.

See more of this Session: Nanomaterials for Energy Storage III
See more of this Group/Topical: Topical 5: Nanomaterials for Energy Applications

(508d) The Role of Acidic Catalyst Support On the Catalytic Upgrading of Bio-Oils

Sunya Boonyasuwat, Chemical Biological and Material Engineering, CBME - The University of Oklahoma, Norman, OK and Daniel E. Resasco, Cbme, University of Oklahoma, Norman, OK

Catalytic conversion of bio-oil vapor has attracted considerable interest as a promising pathway for upgrading. In this work, the reaction of anisole and guaiacol, phenolic model compounds commonly existing in bio-oils, were studied over an acidic silica-alumina catalyst and at varying space times (W/F) and time-on-streams. The reactions were carried out in gaseous phase fixed bed reactor under hydrogen at atmospheric pressure and 300 ^{o}C. It is found that transalkylation is the predominating pathway, while insignificant deoxygenation activity was detected under the investigated reaction conditions. Deactivation of the catalysts is rapid and more severe in the case of guaiacol feed. As a result, conversion of anisole is generally higher as compared to that of guaiacol. It is suggested that the strong adsorption of the phenolic compounds is responsible for the catalyst deactivation. In this contribution, Density Functional Theory (DFT) calculation was performed to simulate the adsorption of the phenolic compounds on acid sites. Based on this present calculation, it exhibited the adsorption energy of guaiacol is significantly higher than that of anisole. This trend is consistent well with the experimental observation.

See more of this Session: Catalytic Biomass Conversion to Chemicals I
See more of this Group/Topical: Fuels and Petrochemicals Division

(294a) Ultra-Thin Ptfe-Nanowires As Durable Electrocatalysts for Oxygen Reduction Reaction

Zhiyong Zhang[1], Wenzhen Li[1], Meijun Li[2] and Zili Wu[3], (1)Chemical Engineering, Michigan Technological University, Houghton, MI, (2)Oak Ridge National Laboratory, Oak Ridge, TN, (3) Chemical Science Division and Center for Nanophase Materials Sciences, Oak Ridge National Laboratory, Oak Ridge, TN

Abstract:The sluggish cathode reaction rate and poor catalyst durability are two mainhurdles that hamper the large-scale application of proton exchange membranefuel cells (PEMFCs). Recently, we prepared an ultra-thin Pt_xFe_y-nanowire(NW) catalyst with a diameter (2 - 3 nm) and a large surface area (52 m^2/g), through a solution-phase reduction method. The Pt_xFe_y-NWcatalyst demonstrated a higher oxygen reduction reaction (ORR) activity andbetter electrochemical durability than conventional Pt/C catalyst in athree-compartment-cell test. After 1000 cycles of 0 - 1.3 V (vs. RHE)accelerated durability test, the relative electrochemical surface area (ECSA)of Pt_2Fe_1-NW drops down to 46%, which is two times betterthan commercial Pt/C catalyst. In addition, the mass activity at 0.85 V (vs.RHE) on Pt_1Fe_1-NW is 39.9 mA/mg_{Pt} afterdurability test, which is twice that of Pt/C catalyst(18.6 mA/mg_{Pt}).

See more of this Session: Fuel Cell Technology I
See more of this Group/Topical: Fuels and Petrochemicals Division

(294b) Non-Pt Cathode Catalysts for Anion Exchange Membrane Fuel Cells

Le Xin, Zhiyong Zhang, Zhichao Wang and Wenzhen Li, Chemical Engineering Department, Michigan Technological University, Houghton, MI

Abstract: Anion exchange membrane fuel cells (AEMFC) haverecently attracted enormous attention, due to their **high energyconversion efficiency, low emission and low cost. Non-Pt catalysts can beused in AEMFC cathode due to improved kinetics (high activity) and lesscorrosive environment (long term stability). We recently developed a wetchemical method to prepare Pd, Ag and their alloy nanostructures for AEMFCcathode. The Ag particle size can be controlled within 2-9 nm. The RRDE testshows that oxygen reduction on Ag catalyst in liquid alkaline electrolyteundergoes a 4-electron-transfer pathway. In a single AEMFC test, the Ag/C-basedAEMFC has demonstrated a peak power density** of190 mW/cm^2 at 80^oC,2 atm, which is justslightly lower than the Pt/C-based AEMFC (**250** mW/**cm^2**). **The V-type durability tests show that theAg/C-based AEMFC is more stable than the Pt/C-based AEMFC. Our latest progressin this area will also be presented.**

See more of this Session: Fuel Cell Technology I
See more of this Group/Topical: Fuels and Petrochemicals Division

(496c) Nuclear Magnetic Resonance and Small-Angle Neutron Analysis of Native and Deuterium Enriched Biomass for the Detailed Analysis of Lignocellulosic Breakdown

Marcus Foston[1], Art Ragauskas[1], Sai V. Pingali[2], Volker S. Urban[2], Hugh O'Neill[2] and Barbara R. Evans[2], (1)Department of Chemistry and Biochemistry, Georgia Institute of Technology, Atlanta, GA, (2)Chemical Sciences Division, Center for Structural Molecular Biology, Oak Ridge National Laboratory, Oak Ridge, TN

In an effort to better understand the complex mechanisms of lignocellulose recalcitrance and breakdown during pretreatment and enzymatic hydrolysis various potential energy crops have been enriched with deuterium for small-angle neutron scattering (SANS) and nuclear magnetic resonance (NMR) studies. Laboratory grown deuterium-labeled plants including: duckweed, switchgrass, rye grass, and a commercially deuterated kale, were analyzed for deuterium incorporation levels and distribution by two novel applications of NMR. The first methodology uses solid state NMR and the second is based on whole plant cell wall dissolution by an ionic liquid followed by liquids $^{2}H/^{1}H$ NMR both verifying the control deuterated kale showed 33-35% deuterium incorporation. Preliminary SANS results on fully protonated biomass to characterize morphological changes in switchgrass during dilute acid pretreatment demonstrate an increase in the cross-sectional radius of the crystalline cellulose fibril. This change is accompanied by the formation of R_g ~ 135 Å lignin aggregates.

See more of this Session: Alternative Fuels and Enabling Technologies II
See more of this Group/Topical: Fuels and Petrochemicals Division

(560e) One-Step Production of Lactate and Isobutanol From Pretreated Biomass by Recombinant Cellulolytic Bacillus Subtilis

Xiaozhou Zhang, Biological Systems Engineering, Virginia Polytechnic Institute and State University, Blacksburg, VA, Chun You, Biological Systems Eng, Virginia Polytechnic Institute and State University, Blacksburg, VA and **Y.-H. Percival Zhang**, Biological Systems Engineering Department, Virginia Polytechnic Institute and State University, Blacksburg, VA

Consolidated bioprocessing (CBP) is low-cost cellulosic biomass bioconversion by integrating cellulase production, cellulose hydrolysis, and sugar fermentation into a single step. Although significant efforts had been made to create recombinant cellulolytic microorganisms, real recombinant cellulolytic microorganism that can produce sufficient secretory active cellulases, hydrolyze cellulose, and utilize soluble sugars for supporting cell growth and cellulase synthesis without any other organic nutrient is not available yet. We demonstrated that over-expression of *Bacillus subtilis* endoglucanase BsCel5 by more than 30-fold enables non-cellulose-utilizing *Bacillus subtilis* to grow on amorphous cellulose and pretreated lignocellulosic biomass as the sole carbon source without addition of costly organic nutrients. After directed evolution and screening on amorphous cellulose, both the expression/secretion level and specific activity of Cel5 and Cel9 had been increased successfully. An advanced recombinant cellulolytic *Bacillus subtilis* strain has been constructed with engineered cellulase and the genes knock out/knock in. Besides these, we also achieved the co-expression and secretion of family 5, 9 and 48 cellulases in a single *B. subtilis* strain. The one-step production of lactate and isobutanol from cellulose and pretreated biomass in the minimal medium without adding any organic nutrient had been demonstrated. We are reconstituting *in vivo* mini-cellulosome on the cell surface of *B. subtilis*. The recombinant cellulolytic B. subtilis would be an ultra-low-cost platform for producing biofuels and other value-added products from non-food biomass.

See more of this Session: Advances In Biofuels: DOE Bioenergy Research Centers II
See more of this Group/Topical: Sustainable Engineering Forum

(231a) A Novel SHF Process for Conversion of AFEX-CS to Ethanol Featuring High Ethanol Productivity, Enzyme Recycling, and Yeast Cells Reuse

Ming Jie Jin, Christa Gunawan, Nirmal Uppugundla, Shishir Chundawat, Bruce Dale and **Venkatesh Balan**, Chemical Engineering and Material Science, Michigan State University, East Lansing, MI

There are two major problems in current cellulosic ethanol production process: low ethanol productivity and high enzyme loading. Based on the current Separate Hydrolysis and Fermentation (SHF) process, it takes at least 96h to perform enzymatic hydrolysis to convert AFEX-CS to fermentable sugars at 6% glucan loading. It will then take another 96-168h to ferment those sugars to ethanol. By simply integrating enzymatic hydrolysis and fermentation, which is SSCF (simultaneously saccharification and co-fermentation), it still requires at least 168h to reach comparable results as SHF. Here we will present a novel SHF process. In this process, the easily hydrolyzed part of biomass is hydrolyzed first to fermentable sugars in 24h. The hydrolysate was then separated from residue solids and fermented to ethanol in 24h using high yeast cell inoculation. The residual solids (difficultly hydrolyzed part of biomass (DHPB)), which bound enzymes, are then transferred to the next cycle of hydrolysis. This way, enzymes are largely reused and enzyme loadings in the next cycles can thus be reduced. The yeast cells after 24h fermentation are also recovered and reused for the next cycle of hydrolysate fermentation. The time for each cycle hydrolysis/fermentation is 24h/24h. After five cycles, the difficult hydrolyzed part is collected and enzymatic hydrolysis is performed without adding additional enzymes. Since enzymes are accumulated to a relatively high concentration and part of the decomposition products are removed with liquid during the five cycles, the hydrolysis of DHPB would be easier to achieve high sugar conversion. Profiles of enzyme recycled in each step were characterized.

See more of this Session: Integrated Processes for Biochemical Conversion of Renewable Feedstocks to Fuels and Chemicals I
See more of this Group/Topical: Sustainable Engineering Forum

(491a) Glycome Profiling Using Glycan-Directed Monoclonal Antibodies: Applications to Plant Cell Wall/Biomass Characterization

Sivakumar Pattathil[1], Michael G. Hahn[2], Heather McCormick[1], Jeffrey Miller[2], Virginia Brown[2], Ajaya Biswal[2], Trina Saffold[2], Malcolm O'Neill[2], Debra Mohnen[2], William S. York[2], Jaclyn D. DeMartini[3], Charles Wyman[4], Michael W.W. Adams[5], Irina Kataeva[6], Richard A. Dixon[7], Huanzhong Wang[7] and Fang Chen[7], (1)Complex Carbohydrate Research Center, University of Georgia, Athens, GA, (2)University of Georgia, Athens, GA, (3)Chemical and Environmental Engineering, University of California Riverside, Riverside, CA, (4)University of California Riverside, Riverside, CA, (5) Biochemistry and Molecular Biology, University of Georgia, Athens, GA, (6)Department of Biochemistry, University of Georgia, Athens, GA, (7)Samuel Roberts Noble Foundation, Ardmore, OK

Plant biomass is the major source for lingo-cellulosic bioenergy. Plant cell walls form the major part of the plant biomass. Research on plant cell wall/biomass is challenging owing to the complexity of the cell walls. Thus, one of the focus areas of bio-energy research has been the development of advanced tools that enable easy, rapid, high-throughput and reliable characterization of plant cell walls/biomass. Cell wall-directed monoclonal antibodies have now emerged as powerful probes to assist in the detailed structural analysis of plant cell walls. The worldwide collection of plant cell wall glycan-directed monoclonal antibodies is now sufficiently large and comprehensive to be used to monitor the majority of cell wall glycans. Here we report an ELISA-based method, Glycome Profiling, that employs a toolkit of 155 monoclonal antibodies (recognizing most major classes of plant polysaccharides) for characterizing diverse plant cell walls. The method involves sequential extraction of cell wall materials with increasingly harsh reagents to obtain fractions enriched in diverse wall polysaccharides, such as pectins, arabinogalactans, xylans, xyloglucans, and mannans. Each of these fractions is subjected to ELISA against the toolkit of antibodies and the binding responses are depicted as heat maps. The studies presented here outline multiple applications of Glycome Profiling to plant cell wall/biomass characterization and utilization. [This work is supported by a grant from US Department of Energy BioEnergy Science Center and grant DE-PS02-06ER64304. The BioEnergy Science Center is a U.S. Department of Energy Bioenergy Research Center supported by the Office of Biological and Environmental Research in the DOE Office of Science. Generation of the monoclonal antibodies used in this work was supported by the NSF Plant Genome Program (DBI-0421683)]

See more of this Session: Advances In Biofuels: DOE Bioenergy Research Centers I
See more of this Group/Topical: Sustainable Engineering Forum

(373a) Nature-Inspired Gas Distributor Design with Application to Low-Temperature PEM Fuel Cells

Jeffrey Marquis, Chemical & Biological Engineering, Rensselaer Polytechnic Institute, Troy, NY and Marc-Olivier Coppens, Isermann Department of Chemical and Biological Engineering, Rensselaer Polytechnic Institute, Troy, NY

Abstract: This work investigates a novel design for a low-temperature proton exchange membrane fuel cell (PEMFC) flow field[1]. Typical, existing flow field designs suffer from the same fundamental deficiency, namely the non-uniform distribution of reactant gases over the surface of the catalytically active area. This leads to inefficient usage of the expensive, platinum-based catalyst material within the fuel cell.

The proposed, new design is inspired by the architecture of the human lung, which has been shown to be an extremely efficient gas transport network as a result of displaying equipartition of entropy production[2, 3]. Furthermore, the fractal, self-similar geometry of the human lung is capable of performing two other key tasks: evenly distributing oxygen throughout the lung, and slowing down the gas velocity, so that, when the oxygen reaches the alveoli, the oxygen's convective flux is equal to its diffusive flux. These two properties are highly desirable in a fuel cell gas distributor as well.

By evenly distributing reactant gases over the surface of the catalytically active material all the platinum catalyst can be efficiently used. As a result, less platinum can be used to achieve the same power output. The novel gas distributor design is guided by the geometry of the bronchial tree. It has one inlet and 4^n outlets, where n is the number of branching generations. Similar fractal geometries have been used in applications ranging from fluidization and irrigation to electronics cooling[4-6]. A variety of product (liquid water, excess air) removal flow paths are also being studied, including a second fractal network, as well as more conventional serpentine and parallel flow fields.

Simulations of this gas distributor geometry have been conducted using COMSOL Multiphysics 3.5a. The results of these simulations have shown that 8 branching generations are needed to achieve an equality of convective and diffusive fluxes at the outlet of the gas distributor when operating at standard conditions. However, as few as 4 generations, in conjunction with a typical gas diffusion layer will provide a nearly uniform distribution of reactant gases over the catalytically active surface.

As a result of these simulations a prototype design of the fractal gas distributor has been generated using the computer-aided-design package NX 7.5. This design has been successfully prototyped by using a high resolution sterolithography machine. These gas distributors are assembled along with other necessary fuel cell components to form a test cell. The performance of this gas distributor is tested using the fuel cell test equipments available within the Center for Automation Technologies & Systems (CATS) at Rensselaer. Key performance characteristics, including average current density, average power output, and platinum usage, will be discussed, and compared to systems that utilize more conventional flow field geometries (serpentine, parallel, and parallel-serpentine).

References:

1) S. Kjelstrup, M.-O. Coppens, J. G. Pharoah, P. Pfeifer, *Energy & Fuels* 24, 5097 (2010).

2) S. Gheorghiu, S. Kjelstrup, P. Pfeifer, M.-O. Coppens, *Fractals in Biology and Medicine*, Vol. 4 (2005).

3) E. Johannessen, S. Kjelstrup, *Chemical Engineering Science* 60, 3347-3361 (2005).

4) M.-O. Coppens, *Ind. Eng. Chem. Res.* 44, 5011-5019 (2005).

5) W. Escher, B. Michel, D. Poulikakos, *International Journal of Heat & Mass Transfer* 52, 1421-1430 (2009).

6) L. Lou, D. Tondeur, *International Journal of Thermal Sciences* 44, 1131-1141 (2005).

See more of this Session: Fuel Cell Technology II
See more of this Group/Topical: Fuels and Petrochemicals Division

(560f) Development of Alkene Biofuels

Harry R. Beller, Joint BioEnergy Institute (JBEI), Emeryville, CA and Ee-Been Goh, Fuels Synthesis, Joint BioEnergy Institute (JBEI), Emeryville, CA

Aliphatic hydrocarbons are appealingtargets for advanced cellulosic biofuels, as they arepredominant components of petroleum-based gasoline and diesel fuels and thuswould be compatible with existing engines and fuel distribution systems. Long-chainalkenes are useful as feedstocks that can be crackedto form diesel- or gasoline-range hydrocarbons. We have studied alkenebiosynthesis in the bacterium *Micrococcus luteus*,a close relative of which was reported in the 1960s to biosynthesize *iso*- and *anteiso*-branched,long-chain alkenes. The underlyingbiochemistry and genetics of alkene biosynthesis werenot elucidated in those studies. In fact, until very recently, there were no known sequences forprokaryotic enzymes catalyzing the synthesis of non-isoprenoidaliphatic hydrocarbons (> C_2). We have shown that heterologousexpression of a three-gene cluster from *M. luteus*(Mlut_13230-13250) in a fatty-acid overproducing *E. coli* strain resultedin production of long-chain alkenes, predominantly 27:3 and 29:3 (no. carbon atoms: no. C=C bonds). Heterologous expression ofMlut_13230 (*oleA*) alone produced no long-chainalkenes but unsaturated aliphatic monoketones,predominantly 27:2, and *in vitro* studies with the purified Mlut_13230protein and tetradecanoyl-CoA produced the same C_{27}monoketone. Gas chromatography-time of flight (GC-TOF) mass spectrometry confirmedthe elemental composition of all detected long-chain alkenes and monoketones (putative intermediates of alkenebiosynthesis). Negative controlsdemonstrated that the *M. luteus* genes wereresponsible for production of these metabolites. We have proposed a metabolic pathway for alkene biosynthesisstarting with acyl-CoA (or -ACP) thioestersand involving decarboxylative Claisencondensation as a key step, which we believe is catalyzed by OleA. Suchactivity is consistent with our data and with the homology of Mlut_13230 (OleA) to FabH, which catalyzes decarboxylative Claisencondensation during fatty acid biosynthesis. We will discuss how understanding of the alkenebiosynthetic pathway can lead to enhanced production.

See more of this Session: Advances In Biofuels: DOE Bioenergy Research Centers II
See more of this Group/Topical: Sustainable Engineering Forum

(650c) Modeling of Asphaltene Deposition In a Vertical Tubing

Dmitry Eskin, John Ratulowski, Kamran Akbarzadeh and Simon Andersen, DBR Technology Center, Schlumberger, Edmonton, AB, Canada

Asphaltene depositionphenomena are investigated both theoretically and experimentally. A Couette device, where the inner cylinder rotates andparticles deposit on the outer wall, is used for laboratory studies of thedeposition process. The modeling approach, recently proposed by the authors(Eskin et al.[1]), is improved. Thedeposition model is composed of the sub-models describing asphalteneagglomerate growth and their transport to the wall. The agglomerate growthmodel is simplified without reducing its accuracy and the more accuratemodel of particle transport to the wall is employed. According to thedeposition model, tuned by the experimental data, only sub-micron asphaltene particles can deposit on the pipe wall and onlya very small fraction of particles touching the wall sticks to itssurface. Those phenomena are explained based on an analogy between thedeposition and water in oil emulsion stabilization by asphaltenes,and on an analysis of interaction of asphaltenemolecules. The deposition model performance is illustrated by modeling ofoil production, accompanied with asphaltenedeposition, from a cylindrical reservoir through vertical tubing. Thecalculations (see figure below), performed for a reservoir depleting over time,demonstrate an excellent visual agreement with the filed datareported in literature[2].

References

1. Eskin D., Ratulowski J., Akbarzadeh K., Pan S. Modeling Asphaltene Deposition in Turbulent Pipeline Flows. Canadian Journal of Chemical Engineering. 2011(in Press).

2.Haskett CE and Tartera M. Apractical solution of the problem to the asphaltenedeposits - Hassi MessaoudField, Algeria. Journal of Petroleum Technology. 1965;April: 387-391.

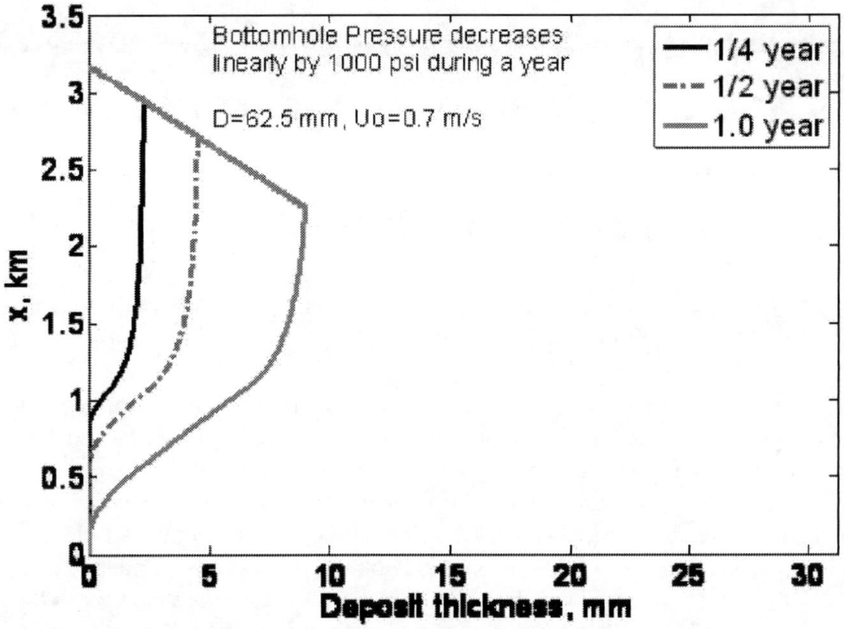

See more of this Session: Heavy Oil and Flow Assurance I
See more of this Group/Topical: Energy and Transport Processes

(647b) Optimization Framework for the Process Synthesis and Simultaneous Heat and Power Integration of a Thermochemical Coal, Biomass, and Natural Gas to Liquids Facility

Richard Baliban, Josephine A. Elia and Christodoulos A. Floudas, Chemical and Biological Engineering, Princeton University, Princeton, NJ

Thermochemical conversion of biomass or coal proceeds via either direct liquefaction which produces a synthetic crude oil or indirect liquefaction which produces synthesis gas (syngas) that can be converted to liquid hydrocarbons directly via the Fischer-Tropsch (FT) process [1]. Thermochemical conversion of natural gas generally occurs through indirect liquefaction using an autothermal reactor. A coal, biomass, and natural gas to liquids (CBGTL) process was recently developed [2,3] that utilizes input hydrogen to have an almost-100% conversion of the carbon in the feedstock to liquid fuels. While the use of hydrogen in the CBGTL process can produce near-zero CO_2 quantities without the need for carbon sequestration, it remains unclear if that is the most cost-effective option to produce liquid fuels. To facilitate the transition from petroleum-based transportation fuels to new carbon-based systems, it is essential to determine the process topology that can deliver the fuels at the lowest levelized cost while using the limited supply of coal, biomass, and natural gas available throughout the country.

To address the above problem, a process synthesis model is proposed to find the CBGTL refinery system topology that provides the lowest levelized cost of transportation fuel production [4]. The model utilizes a combination of continuous variables to model stream flow rates, process operating conditions, and thermodynamic properties and binary variables to model the logical use of a process unit or a process stream. Several constraints dictating the conservation of mass and energy, equilibrium of reactor species, split fractions, and extents of reaction are imposed to ensure that a feasible solution is found. The model is formulated as a mixed-integer non-linear non-convex optimization problem (MINLP). The MINLP model includes simultaneous heat and power integration utilizing heat engines to recover electricity from the process waste heat. Four case studies are presented to investigate the effect of CO_2 sequestration (CCS) and GHG reduction targets on the process topology along with detailed parametric analysis on the role of biomass and electricity prices. Optimal process topologies, the complete heat and power integration, and the overall costs of each case study are compared.

[1] National Academy of Sciences, National Academy of Engineering, and National Research Council. Liquid Transportation Fuels from Coal and Biomass: Technological Status, Costs, and Environmental Issues. Prepublication. Washington, D. C., EPA, 2009.

[2] R. C. Baliban, J. A. Elia, and C. A. Floudas. Toward Novel Biomass, Coal, and Natural Gas Processes for Satisfying Current Transportation Fuel Demands, 1: Process Alternatives, Gasification Modeling, Process Simulation, and Economic Analysis. Ind. Eng. Chem. Res., 49:7343-7370, 2010.

[3] J. A. Elia, R. C. Baliban, and C. A. Floudas. Toward Novel Biomass, Coal, and Natural Gas Processes for Satisfying Current Transportation Fuel Demands, 2: Simultaneous Heat and Power Integration. Ind. Eng. Chem. Res., 49:7371-7388, 2010.

[4] R. C. Baliban, J. A. Elia, and C. A. Floudas. Optimization Framework for the Simultaneous Process Synthesis, Heat and Power Integration of a Thermochemical Hybrid Biomass, Coal, and Natural Gas Facility. Comp. Chem. Eng., doi:10.1016/j.compchemeng.2011.01.041,2011.

See more of this Session: Energy Systems Design and Alternative Energy Sources I
See more of this Group/Topical: Computing and Systems Technology Division

(339c) Effect of Storage Method and Duration On the Bioprocessing of Lignocellulosic Biomass

Arun Athmanathan, ABE/Lorre, Purdue University, West Lafayette, IN and Nathan S. Mosier, Agricultural and Biological Engineering & Laboratory of Renewable Resources Engineering (LORRE), Purdue University, West Lafayette, IN

Lignocellulosic biorefineries require an efficient system of biomass storage between harvesting and processing. This study reports the effect of storage method and duration on the composition and processing performance of sweet sorghum bagasse. Sorghum bagasse was harvested at 85% moisture content, crushed to extract the juice, and then stored. Four storage conditions were used: frozen after crushing (control), ensiled after crushing (67% moisture content), or dried in the field and baled at high (27% by weight) or low (11% by weight) moisture contents. After 24 weeks of storage indoors at 20°C, 5.5% dry matter was lost from low moisture content bales, 29.4% from high moisture bales, and 7% from silage. Compositional analysis and mass balances determined that mass loss was primarily carbohydrates. Pretreatment and SSF at high solids loading (15% w/w) were then carried out for the materials. Distinct differences in performance were noted, with lower enzymatic digestibility of the glucan observed for dry storage compared to wet storage.

To examine non-compositional impacts of storage method, bagasse samples were enzymatically hydrolyzed at 1% w/w solids loading without drying, air-dried (20°C), and oven-dried (45 °C). Release of glucose stopped after <5 hours for dried bagasse with final yields being 70% (air-dried) and 57% (oven-dried) of total glucan. By comparison, non-dried samples continued to hydrolyze beyond 5 hours, resulting 87% glucan yield after 24 hours. These results indicate that hornification during drying may also negatively affect enzyme accessibility to the cellulose and lower bioprocessing performance.

See more of this Session: Sustainable Biomass Feedstock Production and Supply for the Emerging Biorefinery Industry
See more of this Group/Topical: Sustainable Engineering Forum

(38c) Novel Quaternary Ammonium Ionic Liquid Electrolyte for Stable, High Capacity Si Nanowire Based Li-Ion Battery

Vidhya Chakrapani, Florencia Rusli, Michael A. Filler and Paul A. Kohl, School of Chemical & Biomolecular Engineering, Georgia Institute of Technology, Atlanta, GA

*Email: paul.kohl@chbe.gatech.edu

In recent years, there has been interest in silicon nanowires as an anode for Li-ion battery because of its large gravimetric capacity of 4200 mAh/g. 1D nanostructure are essential because they permit radial strain relaxation of mechanical stress which occurs due to the volumetric expansion (>300%) as result of Li insertion. However, commercialization of silicon nanowire-based material has been impeded by number of key issues such as severe capacity fade, limited cycle life and mechanical degradation as result of structural changes. Overcoming these challenges requires optimization of anode structure and choice of better electrolyte than those commonly used. The electrolyte should enable the formation of a stable solid-electrolyte-interface (SEI). Furthermore, the use of organic electrolyte in lithium batteries poses a flammability issue and limits its operating temperature range.

In this work we show a novel quaternary ammonium based ionic liquid as an electrolyte for high capacity and high stability silicon nanowire based anode. Battery testing in these electrolytes showed steady reversible capacity of 2000 mAh/g and columbic efficiency ~97% even after 50 cycles. The efficiency for the first cycle was 91% with very little loss in the discharge capacity indicating good SEI formation in these ionic liquids. Results are discussed in the terms of the nature of this SEI layer and the role of additives that promotes its formation. New strategies were developed to minimize the capacity fade and obtain battery life > 600 cycles.

See more of this Session: Nanomaterials for Energy Storage I
See more of this Group/Topical: Topical 5: Nanomaterials for Energy Applications

(560c) Dissecting the Cellulosome Enzyme Complex: Assembly and Function

Yannick J. Bomble[1], Mark Nimlos[2], Michael E. Himmel[1] and Michael F. Crowley[1], (1)Biosciences Center, National Renewable Energy Laboratory, Golden, CO, (2)National Bioenergy Center, National Renewable Energy Laboratory, Golden, CO

Most bacteria and fungi use free enzymes to degrade plant cell walls. However, a small number of bacteria have developed a fundamentally different approach, where enzymes are tethered to a large protein scaffold forming a multi-domain complex known as, the cellulosome. The study of these large protein assemblies is an ongoing research topic that has already yielded numerous breakthroughs. However, the mechanism of assembly of the enzyme subunits onto the natural scaffoldin; as well as the modes of action of the cellulosome and its enzymatic components, are currently not well understood.

In this study, we focused on the cellulosome-integrating protein (CipA) of C. thermocellum; as well as cellulosomal enzymes from families 5, 9 and 48. These three enzymes are representative of the diversity of enzymes secreted by C. thermocellum. This work not only aims at understanding the mechanisms involved in the sequential binding of the cellulosomal enzymes to the CipA scaffold of C. thermocellum, but also the binding of CipA to secondary scaffolds, such as OlpB. We focused on the physical properties of the binding of component enzymes to the scaffoldin. The modularity of the enzymes was found to be one of the main influences on the cellulosome assembly process.

Additionally, we examined several modules from the multi-modular enzyme complex, CbhA. Several of these modules have undefined functions, such as the X1 modules found closely linked to the catalytic domain. We hypothesize that the X modules could serve as tethers between several modules in CbhA to fine-tune the positioning of this enzyme during cellulose deconstruction. We conducted steered molecular dynamics simulations (directly comparable to AFM pulling experiments) to evaluate the energy and the profile for unfolding these modules. We found that the energy required to partially unfold this domain is accessible by the cellulosome and the unfolding pathways are similar in the multiple simulations conducted.

See more of this Session: Advances In Biofuels: DOE Bioenergy Research Centers II
See more of this Group/Topical: Sustainable Engineering Forum

(491d) Molecular Simulation of Lignocellulosic Biomass Components

Loukas Petridis, Center for Molecular Biophysics, Oak Ridge National Laboratory, Oak Ridge, TN

Lignocellulosic biomass comprises the vast majority of biomass on Earth and has the potential to play a major role in generation of renewable biofuels if cost-effective conversion can be achieved. Largely composed of plant cell walls, lignocellulosic biomass is a complex biological composite material that shows significant recalcitrance towards the structural deconstruction and enzymatic hydrolysis into sugars that is necessary for fermentation to bioethanol. Here, we report on studies combining molecular dynamics simulation and small angle neutron scattering to better understand structural properties of biomass components cellulose and lignin. Lignin, a major polymeric component of plant cell walls, forms aggregates in vivo after pretreatment of lignocellulosic biomass for ethanol production. The aggregates are thought to reduce ethanol yields by inhibiting enzymatic hydrolysis of cellulose. Extensive atomistic molecular dynamics simulations were used to demonstrate that the surfaces of the aggregates are highly irregular and characterized by a surface fractal dimension that is invariant across length scales from ~1–1000Å. The simulations reveal extensive water penetration of the aggregates and heterogeneous chain dynamics corresponding to a rigid core with a fluid surface. The detailed multiscale structure revealed here should aid in understanding biomass recalcitrance to hydrolysis and in feedstock engineering efforts to improve biofuel yield.

See more of this Session: Advances In Biofuels: DOE Bioenergy Research Centers I
See more of this Group/Topical: Sustainable Engineering Forum

(206b) Anisole and Guaiacol Deoxygenation Over Pt/Sn On Monolith Catalyst Supports

Miguel Angel Gonzalez Borja and Daniel E. Resasco, School of Chemical, Biological and Materials Engineering, University of Oklahoma, Norman, OK

Upgrading of pyrolysis oil vapors requires a catalytic reactor able to treat high reactant flow rates without generating a large pressure drop. Therefore, monolithic catalysts appear as good candidates for this purpose. In this contribution, low-surface-area Inconel monoliths were coated with in-situ-grown carbon nanofibers and subsequently impregnated with catalytic species (Pt and Sn) both as monometallic and bimetallic particles. These monoliths were tested for the deoxygenation of guaiacol and anisole (products lignin pyrolysis), two of the most deactivating compounds present in pyrolysis bio-oil. The main products obtained with these feeds were phenol and benzene. Coating with carbon nanofibers provided increased surface area and anchoring sites for the active species (Pt, Sn), thus increasing the yield of desired products. The bimetallic Pt/Sn catalyst have shown the highest activity and stability compared to monometallic Pt and Sn catalysts. These tests indicate that monoliths of Pt/Sn CNF-Inconel are potential catalysts for the upgrading of lignin fractions present in bio-oil.

See more of this Session: Catalytic Biofuels Refining I
See more of this Group/Topical: Fuels and Petrochemicals Division

(436f) Formation of Liquid and Solid Products of Liquid Phase Pyrolysis

Nikolaus Schwaiger[1], Roland Feiner[2], Kerstin Zahel[2], Angela Pieber[2], Verena Witek[2], Peter Pucher[2], Hartmuth Schrttner[3], Edgar Ahn[2] and Matthus Siebenhofer[1], (1)Department of Chemical Engineering and Environmental Technology, Graz University of Technology, Graz, Austria, (2)R&D, BDI BioEnergy International AG, Grambach, Austria, (3)Institute of Electron Microscopy, Graz University of Technology, Graz, Austria

Formation of Liquid and Solid Products ofLiquid Phase Pyrolysis

Schwaiger, N. Feiner*, R. Zahel*, K. Pieber*, A. Witek, V.* Pucher*,P. Ahn*, E. Schroettner[+], H. Siebenhofer,M.

GrazUniversity of Technology Institute for Chemical Engeneeringand Environmental Technology, * BDI-BioEnergy InternationalAG, [+]Graz University of Technology Institutefor Electron Microscopy, Austria

Lignocellulosic feed is expected to contributesignificantly to production of liquified and solid combustibles.The aim of this project is the production of high quality biocharand a liquid energy carrier. Therefore the pyrolytic degradation of softwood duringliquid phase pyrolysis was investigated. The processwas carried out in a semi-batch reaction vessel under isothermal conditions atvarious temperatures. Process pressure was ambient. For optimum heat transfer pyrolysis was carried out in liquid heat carrier phasewhich provides high heat conductivity. Liquid phase pyrolysisis an exothermic process which produces 25-28% liquid CHO products, independentof biomass particle size within a range of 630m and 10mm. The enthalpy ofreaction is -86425 kJ/kg at T=350°C. To gain data on kinetics of formation of liquidproducts and the solid residue, degradation of soft wood and product formationwas recorded.

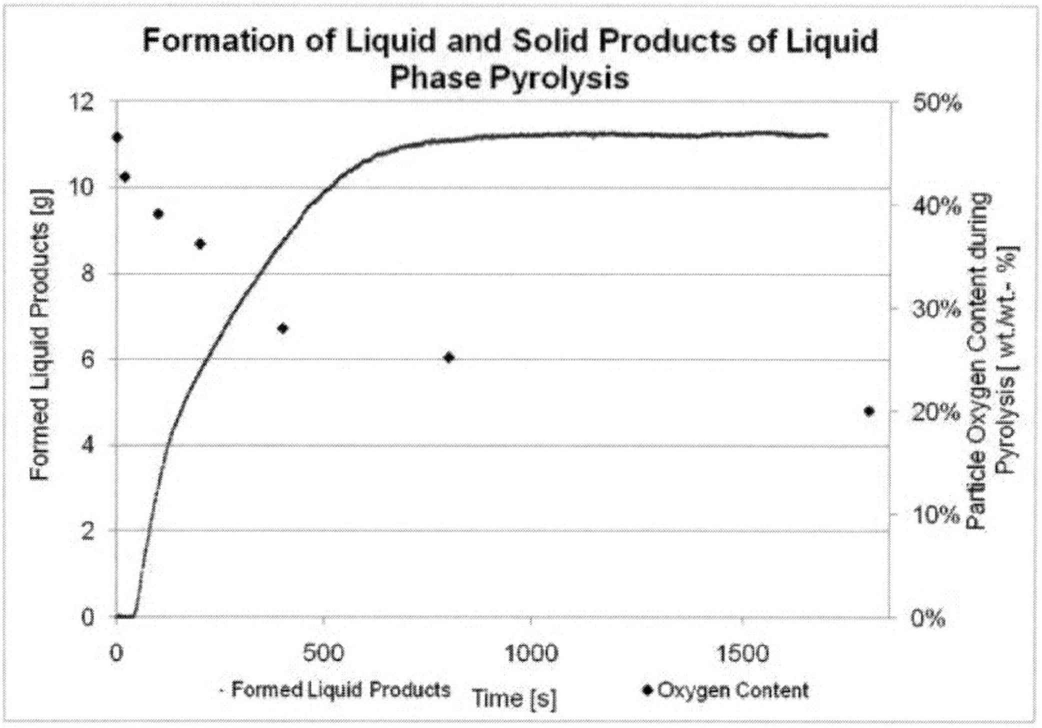

During pyrolysis theCarbon to Oxygen ratio of the solid phase was tracked. The measurements weremade with energy dispersive x-ray spectroscopy, elemental analysis and ATR-IR. Figure1 shows three distinct phases of product formation in liquid phase pyrolysis at T=350°C and ambient pressure. The feedundergoes rapid change in the C:O ratio in two stepsand finally stabilizes at nearly constant ratio but ongoing loss of solidsmass.

See more of this Session: Biomass Pyrolysis II
See more of this Group/Topical: Fuels and Petrochemicals Division

(278a) A New Generation of Zirconia Supported Metal Oxide Catalysts for Converting Low Grade Renewable Feedstocks to Biodiesel

Manhoe Kim[1], **Craig DiMaggio**[1], Shuli Yan[2], Steven Salley[3] and K. Y. Simon Ng[1], (1)Department of Chemical Engineering and Material Science, Wayne State University, Detroit, MI, (2)Chem. Eng., Wayne State University, Detroit, MI, (3)Chemical Engineering and Materials Science, Wayne State University, Detroit, MI

A new class of zirconia supported metal oxide catalysts has demonstrated the ability to convert low quality bio-oils, high in FFA, to biodiesel. Operating at mild to moderate reaction conditions, 195°C and 300psi, these catalysts can perform simultaneous esterification and transesterification reactions to achieve FAME yields as high as 90% in a single reaction step. This suggests that a wide range of crude bio-oil feedstocks, with varying FFA content, including palm oil (FFA 12%), residual corn oil(15%) from the distiller dried grains (DDGs), animal fats (2~35%), yellow grease (<15%), and brown grease (50~100%), can be converted into FAME with this catalyst technology. The heterogeneous nature of the ZrO_2 formulations also simplifies the production process while avoiding the problems inherent in homogeneous processes related to separation and washing, corrosive catalyst, and excessive methanol usage.

See more of this Session: Catalytic Biofuels Refining II
See more of this Group/Topical: Fuels and Petrochemicals Division

(496e) A New Generation of Zirconia Supported Metal Oxide Catalysts for Esterification of Free Acid Feedstocks

Manhoe Kim[1], Craig DiMaggio[1], Shuli Yan[2], Steven Salley[3] and K. Y. Simon Ng[1], (1)Department of Chemical Engineering and Material Science, Wayne State University, Detroit, MI, (2)Chem. Eng., Wayne State University, Detroit, MI, (3)Chemical Engineering and Materials Science, Wayne State University, Detroit, MI

The acid catalyst promotes esterification by reacting the carboxylic group of the FFA with methanol to form methylesters and water. In this reaction mixture, the sulfuric acid is more miscible in the water product phase than the methanol phase leading to a pronounced water inhibition effect that decreased biodiesel yield due to reverse hydrolysis. In order to compensate for this effect and increase the biodiesel yield, greater sulfuric acid (10~25% of fat weight) concentrations, higher excess methanol (Methanol/oil =30), and much longer reaction times (24 hrs) were typically required(Dias, Alvim-Ferraz et al. 2009). In addition, the high levels of sulfuric acid were very corrosive, difficult to recover, and required neutralization at the end of the process. These disadvantages also increased operational and capital costs.

A new class of zirconia supported metal oxide catalysts has demonstrated the ability to convert large molecules of free fatty acids such as brown grease. Zirconia supported catalysts can perform simultaneous esterification and transesterification reactions to achieve FAME yields as high as 98% in a single reaction step. This suggests that a wide range of bio-oil feedstocks containing a lot of FFAcan be converted into FAME with this catalyst technology. The heterogeneous nature of the ZrO_2 formulations also simplifies the production process while avoiding the problems inherent in homogeneous processes related to separation and washing, corrosive catalyst, and excessive methanol usage.

See more of this Session: Alternative Fuels and Enabling Technologies II
See more of this Group/Topical: Fuels and Petrochemicals Division

(39a) Towards Quantum-Dot Solar Cells From Abundant Nontoxic Materials

Ankur Khare[1], Andrew W. Wills[2], Boris D. Chernomordik[1], David J. Norris[3] and **Eray S. Aydil**[1], (1) Chemical Engineering and Materials Science, University of Minnesota, Minneapolis, MN, (2) Department of Chemistry, University of Minnesota Twin Cities, Minneapolis, MN, (3)Optical Materials Engineering Laboratory, ETH Zurich, Zurich, Switzerland

Quantum confinement of electrons and holes in nanometer size crystals (quantum dots or QDs), endows them with properties that may be advantageous for efficient solar-to-electric energy conversion. First, electronic energy levels and optical absorption in QDs can be manipulated by changing their size. This allows the optimization of their optical absorption for maximum overlap with the solar spectrum. Second, the ability to manipulate energy levels through size raises the possibility to make inexpensive multijunction solar cells by judiciously layering different size QDs. Third, it has been suggested that quantum confinement may slow energy dissipative electron and hole relaxation rates such that two new physical processes, multiple exciton generation and hot electron extraction may now compete with relaxation and lead to higher photocurrents or higher photovoltages, respectively. Finally, QDs can be prepared in large quantities as stable colloidal solutions under mild conditions and deposited on surfaces of various planar or nanostructured substrates as thin films through inexpensive high-throughput coating processes to form photovoltaic devices. For these reasons, solar cells based on QDs may have the potential to achieve high power conversion efficiencies at low cost and are promising candidates for third generation photovoltaic devices.

Most quantum dots made to date are based on PbSe and PbS QDs because we know the most about their photophysical properties and synthesis. Lead causes irreversible neurological, reproductive, renal and cardiovascular damage to humans and its use has been banned from many products. Quantum dots based on nontoxic and abundant materials are needed. Copper zinc tin sulfide (Cu_2ZnSnS_4 or CZTS) is emerging as a promising new sustainable semiconductor for photovoltaics. CZTS has a high absorption coefficient in the visible range of the electromagnetic spectrum and an electronic bandgap (~1.5 eV) that is ideal for photovoltaics. We have developed a novel and facile synthesis method for making CZTS nanocrystals and achieved quantum confinement in nanocrystals with diameters less than 3 nm. Our synthesis method is based on decomposition of metal dithiocarbamates in an organic solvent such as dodecene. Presence of oleylamine in the reaction mixture reduces the decomposition temperature of copper, zinc and tin dithiocarbamates to a narrow temperature range (170-220 °C). Rapid injection of oleylamine and tin diethyldithiocarbamate into a hot mixture of copper and zinc dithiocarbamate in dodecene initiates a burst of nanocrystal nucleation followed by growth. We show that using this approach avoids the formation of binary and ternary sulfide impurities such as zinc sulfide or copper tin sulfide. Moreover, we have been able to control the sizes of CZTS nanocrystals from 2-7 nm by varying the experimental conditions such as the synthesis temperature and time. In this talk we will present our latest results towards making CZTS quantum dot solar cells.

See more of this Session: Nanomaterials for Photovoltaics I
See more of this Group/Topical: Topical 5: Nanomaterials for Energy Applications

(218c) Component and Performance Analysis of Vanadium Redox Flow Batteries: Experimental and Modeling Studies

Emin Caglan Kumbur, Chris R. Dennison, Kevin W. Knehr and Ertan Agar, Mechanical Engineering and Mechanics, Drexel University, Philadelphia, PA

One major issue limiting the implementation of renewable energy sources is the lack of efficient, cost-effective, and reliable energy storage technologies. The vanadium redox flow battery (VRFB) is a flow-assisted electrochemical system which shows great promise for grid-scale energy storage due to its high energy efficiency (up to 85%), long lifetime (>12,000 cycles) and low maintenance. Despite these many advantages, the current state of VRFB technology possesses several unresolved issues which affect the efficiency, power density, and power output of the system. To address these limitations, a macroscopic 2-D, transient performance model of a VRFB has been developed to understand the role of components, operating conditions and electrolyte flow strategies.

This model is constructed on five main domains to simulate the physical phenomena in components of a vanadium flow cell, namely: the current collectors, the positive and negative porous electrodes, and the membrane. The unique feature of the present model is the incorporation of species crossover through the membrane and utilization of a more complete description of the Nernst equation to accurately predict the open circuit voltage and electrode potentials. Input parameters for the various domains are obtained by direct measurement of the properties of actual cell components. For instance, the microstructure of the graphite felt electrodes is quantified using x-ray microcomputed tomography, and the resulting properties (*i.e.* porosity, specific surface area, etc.) are incorporated into the model. The 2D temporal distributions of local reaction current density, overpotential, species concentrations and cell potential are predicted. Simulations are performed using experimental data for a variety of components and their impact on cell performance is evaluated to provide guidance for component and cell design efforts.

See more of this Session: Flow Batteries for Energy Storage (In Memory of Bernard Lee)
See more of this Group/Topical: Fuels and Petrochemicals Division

(470c) Molecular and Hybrid Solution Processible Thermoelectrics

Rachel A. Segalman[1], Shannon Yee[2], Boris Russ[2], Nelson Coates[2] and Jeffrey Urban[3], (1)Department of Chemical and Biomolecular Engineering, University of California, Berkeley, Berkeley, CA, (2)University of California at Berkeley, Berkeley, CA, (3)Lawrence Berkeley National Laboratories, Berkeley, CA

Thermoelectric materials for energy generation have severaladvantages over conventional power cycles including lack of moving parts,silent operation, miniaturizability, and CO2 freeconversion of heat to electricity. Excellentthermoelectric efficiency requires a combination of high thermopower(S, V/K), high electrical conductivity (σ, S/cm), and low thermalconductivity (κ, W/mK). To date the bestmaterials available have been inorganic compounds with relatively low earthabundance and highly complex, vacuum processing routes (and hence greaterexpense), such as Bi_2Te_3. Molecular materials andhybrid organic-inorganics bring the promise ofinexpensive, solution processible, mechanicallydurable devices. While highlyconductive polymers are now common place, theygenerally demonstrate low thermopower. Our work onmolecular scale junctions demonstrates that nanostructuringof organics allows them to act as thermionic filters between inorganicjunctions which can lead to enhanced thermoelectric properties. We have takeninspiration from this fundamental understanding to design material systems inwhich we combine a high electrical conductivity, low thermal conductivitypolymer with a nanoparticle that contributes high thermopower. Additionally, the work functions of the two materials are well-aligned which introduces the possibility of thermionicfiltering at the interface and an additional boost to the power factor. The combination of these effects results in a new hybrid,solution processible material with a thermoelectricfigure of merit within an order of magnitude of the Bi_2Te_3. In this talk, I will discuss both the use ofthermoelectric measurements to gain insight to molecular junctions and how thisinsight translates to design principles for polymer and hybrid thermoelectrics.

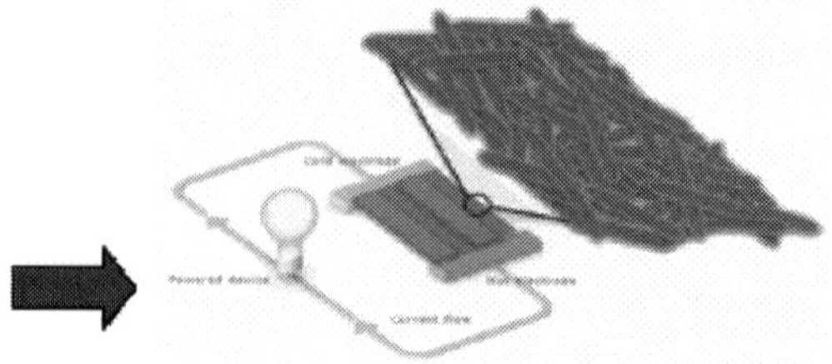

See more of this Session: Nanomaterials for Thermal-to-Electric Conversion I
See more of this Group/Topical: Topical 5: Nanomaterials for Energy Applications

(13b) Mixed Rare Earth Oxides (REOs) for Desulfurization and Tar Removal From Gasifier Effluents

Rui Li, Joseph Bridges, Weishi Kong and Kerry.M. Dooley, Chemical Engineering, Louisiana State University, Baton Rouge, LA

The removal of H_2S, NH_3 and tars from syngas generated in biomass or coal/biomass gasification plays an important role in further processing of syngas. We have developed materials capable of both H_2S adsorption and tar reforming at high temperatures (~600-650ºC). These are mixed rare earth oxides (REOs) such as CeO_2/La_2O_3, prepared by templated sol-gel methods, and then impregnated with transition metals (Mn, Fe, etc.). Desulfurization tests have been carried out using model syngas mixtures. Successive desulfurization-regeneration cycles some of these materials (e.g., $Mn/Ce/LaO_x$) demonstrated reasonable sulfur capacities in the presence of water and CO_2, along with regenerability. Naphthalene and propane have been used as models for tar reforming/cracking, and the effects of H_2S, CO and CO_2 on the model reactions were examined. H_2S and CO_2 are strong inhibitors, while CO is not. Fresh, used and regenerated sorbents and catalysts have been characterized by BET, TPR, XRD, XANES and EXAFS.

See more of this Session: Catalytic Conversion of Renewable Resources to Synthesis Gases and Pyrolysis Oils
See more of this Group/Topical: Sustainable Engineering Forum

(491f) The Binding Properties of Cellulases On Allomorphs Cellulose and Pretreated Biomass During Enzymatic Hydrolysis

Dahai Gao, **Shishir Chundawat**, Venkatesh Balan and Bruce Dale, Chemical Engineering and Material Science, Michigan State University, Lansing, MI

The depolymerization of cellulose to monomeric glucose is the key step for cellulosic ethanol biorefinery process. Cellulases binding on biomass is the essential first step for effective breakdown of biomass to sugars. A new fast flow liquid chromatography (FPLC) based method has been developed to quantify endoglucanase I (EG I, GH family 7B), cellobiohydrolase I (CBH I, GH family 7A) and cellobiohydrolase II (CBH II, GH family 6A) individually in the heterogeneous reaction environment.

The binding level of individual enzymes during 48 hours hydrolysis was studied on different cellulose allomorphs: micro crystal cellulose Avicel (cellulose Iβ), liquid ammonia treated cellulose (cellulose III), sodium hydroxide treated cellulose (cellulose II) and phosphoric acid swollen amorphous cellulose (AC). The digestibility rank is AC>cellulose III>cellulose II>cellulose I. However, the AC has the highest initial binding capacity while cellulose III has the lowest. After the substrate solubilized, most of the cellulase returned to supernatant. But CBH II is less stable. Only around 60% CBH II can be recovered.

The time course binding studies were also performed under pretreated biomass. Ammonia Fiber Expansion (AFEX) treated corn stover (CS), dilute acid treated CS and ionic liquid pretreated CS are compared. The results show the existence of lignin is responsible for a significant amount of unproductive binding. For AFEX-CS, the initial binding level for those cellulases are >80%. However, after 48 hours digestion, only around 50% of CBH I, 28% of CBH II and 19% of EG I are detected in the supernatant. The remaining cellulases are believed to be bound unproductively onto lignin. Similar results were found on ACID-CS. The results are critical for improving our understanding of enzyme synergism, productive/unproductive enzyme binding and the role of pretreatment on enzyme accessibility to lignocellulosic plant cell walls, as well as help to engineer novel low unproductive binding enzymes and develop an economic enzymes recycle system.

See more of this Session: Advances In Biofuels: DOE Bioenergy Research Centers I
See more of this Group/Topical: Sustainable Engineering Forum

(560d) Construction of Efficient Gram-Negative Biocatalysts for Lignocellulose Conversion

David Keating, Michael Schwalbach, Jeffrey Gardner, Mary Tremaine, Patricia Kiley and Robert Landick, Great Lakes Bioenergy Center, University of Wisconsin, Madison, WI

Research within the Great Lakes Bioenergy Research Center aims to better understand the bottlenecks associated with bioconversion of lignocellulose to liquid transportation fuels, knowledge which will then be used to improve the efficiency of this process. Much of these studies are currently focused on the Gram-negative bacterium *Escherichia coli,* due to its sophisticated genetics, well-understood physiology, and widespread industrial use.

To better understand the limitations associated with microbial biocatalysis of lignocellulosic biofuels, we have engineered *E. coli* strains capable of high-level ethanol production and tolerance. To develop efficient *E. coli* ethanologens, we have constructed mutations that reduce alternative pathways of election flow, and introduced the pyruvate decarboxylase and alcohol dehydrogenase genes from *Zymomonas mobilis*. In addition, we have carried out extensive directed evolution of *E. coli*, which has resulted in strains capable of growth in the presence of 10% ethanol. These strains are currently being subjected to a systems-level microbial analysis during lignocellulosic bioconversion. In collaboration with the Dale laboratory, we have generated enzymatic hydrolysates of AFEX-pretreated corn stover at high solid loadings. Furthermore, we have developed methodology to measure extracellular amino acids, metabolic endproducts, intracellular metabolites, and global transcription. We have used these techniques to characterize the physiological state of our ethanologenic strains during fermentation in lignocellulosic hydrolysates. Our initial analysis has already generated several surprising findings. First, lignotoxins within the hydrolysate appear to strongly limit the production of biomass, but have a more modest effect the conversion of glucose to ethanol. However, these inhibitors appear to strongly limit the conversion of xylose to ethanol. Second, microbial growth in lignocellulosic hydrolysates is a dynamic process, in which global transcription and overall physiology change dramatically during the course of fermentation. Our systems-level analysis of these strains has identified transcriptional and metabolic changes that provide insight into the mechanism of action of the lignotoxins. We are using this information, as well as data from similar experiments with lignotoxin-tolerant mutants, to generate new ethanologens with improved productivities under these conditions. Furthermore, we expect these studies will produce an improved molecular understanding of the current limitations of ethanologenesis in these strains, which we expect will be applicable to diverse ethanologenic microorganisms.

See more of this Session: Advances In Biofuels: DOE Bioenergy Research Centers II
See more of this Group/Topical: Sustainable Engineering Forum

(491c) Rheological Properties of Biomass Deconstruction Process Using Ionic Liquid

Alejandro G. Cruz[1], Jeff Mentel[2], Seema Singh[3] and Blake Simmons[3], (1)Deconstruction, Joint BioEnergy Institute, Emeryville, CA, (2)Malvern Instruments LTD, MA, (3)Joint BioEnergy Institute, Emeryville, CA

Lignocellulosic biomass must be pretreated prior to enzymatic hydrolysis to improve sugar yields and saccharification efficiency. Ionic liquid (IL) pretreatment has shown great potential as a novel pretreatment technology with sugar yields as high as ~100%. However, to improve process economics, higher biomass loading is necessary. Biomass slurry in IL are highly viscous. To decrease production cost, the biomass slurry should be pumpable and exhibit least resistance to flow. Here we investigate rheological properties of neat IL, biomass slurry in ionic liquid through dissolution and regeneration processes. Complex viscosities and modulus (both elastic and viscous) measurements provide a detail insight of biomass solubilization and cellulose regeneration process and pinpoint the onset of cellulose regeneration event. By generating small-angle oscillatory frequency sweeps of pretreated switchgrass at various increased loading percentages we can characterize the nature of the slurry where the material has strong intermolecular association. For a viscous material such as in the 3% pretreated switchgrass sample, the viscous modulus is dominant over the elastic modulus. Just a slight increase in percent loading makes the slurry behave like a gel like structure and the elastic modulus is dominant over the viscous modulus. All percent loading pretreated switchgrass show frequency dependence and shear thinning behavior.

See more of this Session: Advances In Biofuels: DOE Bioenergy Research Centers I
See more of this Group/Topical: Sustainable Engineering Forum

(491e) Microfluidic Assays for Biofuels Research

Rajiv Bharadwaj[1], Aarthi Chandrasekaran[2], Chieh Chang[3], April Wong[4], Blake Simmons[2], Paul D. Adams[2] and Anup Singh[2], (1)Sandia National Laboratories, Joint BioEnergy Institute, Livermore, CA, (2)Joint BioEnergy Institute, Emeryville, CA, (3)Sandia National Laboratories, Joint BioEnergy Institute, Livermore, (4)Sandia National Laboratories, Livermore, CA

We are developing next-generation assays for biofuels research using microfluidic technologies to provide significant improvement over conventional platforms in throughput, sensitivity, multiplexing and speed of analysis. To develop these microscale assays we utilize technologies covering a wide spectrum of complexity based on the requirements of the assays. For example, using off-the-shelf components we have developed a microtiter plate-based assay for screening cellulases against heterogeneous substrates. This approach relies on in-situ biomass

regeneration in micro-volumes to volumetrically meter biomass (sub-mg loading) and also precisely control the amount of residual IL for screening activities of novel IL-tolerant cellulases. We have also developed more sophisticated platforms for screening of lignocellulolytic enzymes using microfluidic chip for electrophoretic analysis of glycans. These chips are being used to screen genetically-engineered enzymes, assess performance of pretreatment processes, and for discovering cellulase activities in microbial communities. We are also developing high throughput microchips to perform combinatorial DNA assembly for optimization of biofuel synthesis pathways.

See more of this Session: Advances In Biofuels: DOE Bioenergy Research Centers I
See more of this Group/Topical: Sustainable Engineering Forum

(378d) Removal of Heat From a Hydrogen Storage Media by a Novel Microscale-Based Heat Exchanger

Christopher Loeb[1], **Agnieszka Truszkowska**[1], Bruce Hardy[2], Richard Chahine[3] and Goran N. Jovanovic[1], (1)Chemical, Biological and Environmental Engineering, Oregon State University, Corvallis, OR, (2)Computational Sciences, Savannah River National Laboratory, Aiken, SC, (3) Hydrogen Research Institute, Trois-Rivieres, QC, Canada

High volumetric storage density of hydrogen gas is a major challenge in the progress towards hydrogen-fueled automobiles. Of the various storage methods, pressurized gas tanks, liquid hydrogen storage, liquid chemical systems, and solid-state storage, activated carbon adsorbents show great potential. A reliable thermal management system within the storage tank is vital to provide quick, safe, and efficient charging and a controlled release of hydrogen gas on demand.

The current work investigates room temperature adsorption of nitrogen gas (model gas for hydrogen) in activated carbon bed and compares the temperature distribution within the storage media without thermal management, and with the addition of a microscale-based heat exchanger. Experimentally obtained temperature distribution data is used to assess the effect of charging time, and removal of generated heat within the system. A mathematical model is used to numerically simulate the experimental process.

The maximum height of activated carbon bed positioned between two heat exchange devices, while providing adequate thermal management, is investigated by varying the height of the adsorption bed and monitoring nitrogen-charging time and temperature distribution.

A novel microscale-based heat exchanger design takes advantage of 250 μ m high features, which facilitate the flow of cooling fluid. The heat exchanger is fabricated by diffusion bonding of patterned stainless steel shims, resulting in an overall 250 μ m thick resistance to thermal transport. The heat exchanger design also provides uniform gas distribution into the bed of storage media. The device's ability to provide a uniform temperature at the contacting surface is verified using infrared thermal imaging.

Experimental results indicate a dramatic improvement in charging time due to the fast and efficient removal of generated heat.

See more of this Session: Hydrogen Storage Systems: Designs, Analysis, and Safety Aspects
See more of this Group/Topical: International Congress on Energy 2011

(583d) Feedstock Mixture Effects On Sugar Monomer Recovery Following Dilute Acid Pretreatment and Enzymatic Hydrolysis

Michael J. Brodeur-Campbell, Jordan Klinger and David Shonnard, Chemical Engineering, Michigan Technological University, Houghton, MI

Feedstock MixtureEffects on Sugar Monomer Recovery Following Dilute Acid Pretreatment andEnzymatic Hydrolysis

MichaelBrodeur-Campbell[1], Jordan Klinger[1], and DavidShonnard[1,2]

[1] Department of Chemical Engineering,Michigan Technological University

[2] Director Sustainable Futures Institute,Michigan Technological University

Increaseduse of lignocellulosic biomass for bioenergyproduction holds promise for improved domestic energy security, reducedgreenhouse gas (GHG) emissions, rural economic development, and developing amore sustainable energy infrastructure [Chum and Overend2001; Perlack, Wright, et al. 2005; U.S. DOE 2006]. Transportation fuels are especially difficultto replace as their replacement has to have a high energy density, be compatiblewith existing fuel distribution infrastructure, and be compatible with existingautomobile technology. In 2009, 97% oftransportation fuels were derived from fossil sources [EIA 2008], andtransportation accounted for one-third of carbon dioxide emissions in 2008 [EIA2009]. Liquid fuels derived from biomassare particularly attractive as a replacement for fossil-derived transportationfuels. Cellulose is the most abundantbiopolymer on Earth, and lignocellulosic feedstocks represent one of the fewnatural resources available in sufficient quantity to replace a significantfraction of liquid transportation fuel.

In acommercial biorefinery feedstocks are likely to come from a variety of sourcesand include many different materials, such as softwoods, hardwoods, agriculturalresidues, and herbaceous or woody energy crops.Much work has been done to understand and optimize the pretreatment andenzymatic hydrolysis conditions for pure feedstocks, but little is knownabout mixture effects on enzymatic hydrolysis.A primary concern in processing woody biomass to produce a range ofbiofuels is to maximize the production of desired intermediates, in this case,determining the quantity of monomer sugars that are recoverable. The biochemical platform as described by theNational Renewable Energy Laboratory (NREL) was the process studied. Briefly, lignocellulosic biomass which iscomposed primarily of hemicellulose, cellulose, and lignin is treated withdilute sulfuric acid to hydrolyze the hemicellulose into soluble sugars. The solid residue remaining after dilute acidpretreatment is then treated with cellulose-degrading enzymes to hydrolyze thecellulose into soluble sugars. Theremaining lignin may be utilized as a high energy co-product stream to provideheat and power for the cellulosic ethanol conversion processes.

This studyseeks to clarify the effects of processing mixtures of feedstocks on overall sugarrecovery. We wish to test the hypothesisthat sugar yields from mixed feedstocks can be predicted by a simple linearinterpolation model based on the behavior of the pure species. Experiments were performed on pure species(aspen, balsam, and switchgrass), and on 50/50 blends of aspen/balsam andaspen/switchgrass. Aspen was chosen as amodel species due to its high suitability for processing in a commercialbiorefinery. Balsam was chosen to mixwith aspen to study the effects of high lignin content on mixed feedstockprocessing, and switchgrass was chosen to mix with aspen to investigate theeffects of a high ash content energy crop.

One-halfgram (0.5 g) of dry sample was weighed and placed into one of a set of 9stainless steel Swagelok cylindrical reactor tubes, and 4.5 ml of dilutesulfuric acid (0.5% w/v) was added to achieve 10% (wt) solids loading in thereactor. Reactors were rapidly heated to160°C in a silicon oil bath (Dow Corning 550 fluid), and three tubes wereremoved from the oil bath at previously determined time intervals. From previous research involving pure samplesof aspen, balsam, and switchgrass (Jensen et al. 2010), the time to achievemaximum yields of monomer and oligomer sugars from combined dilute acidhydrolysis and enzymatic hydrolysis was determined; the "optimum" time. Threereactors were removed at one-half the optimal time, three reactors were removedat the optimal time point, and the last three were removed at two times theoptimal time. Reactor tubes removed fromthe oil bath were immediately submerged in an ice-water bath to quench thereaction. One milliliter of liquid fromeach reactor was removed, filtered, neutralized, and analyzed on an Agilent1200 series HPLC for glucose and xylose concentration. One milliliter of liquid from each reactorwas filtered and then further acidified to 4% (w/v) sulfuric acid, and treatedat 121°C for one hour tohydrolyze any soluble oligomers into monomer form. After neutralization these samples were analyzedby HPLC as above. Results from these twoliquid samples were combined into total soluble (monomer plus oligomer) sugarrecovery from dilute acid pretreatment.Solids were washed with excess distilled water on a glass-fiber filter(VWR) and carried on to enzymatic hydrolysis.

Solidresidues containing cellulose and lignin were subjected to enzymatic hydrolysisusing Accelerase 1500 (Genencore). Each of the three reactors from the samepretreatment time point were loaded with a different enzyme concentration 0.1, 0.25, and 0.5 ml Accelerase 1500 per gram ofcarbohydrate in the feedstock in order to study the interaction betweenpretreatment severity and enzyme loading on sugar yields. Reaction conditions were 50mM citrate buffer,pH 4.8, 50°C,250 RPM orbital shaking. Liquid sampleswere removed at 0, 1, 2, and 3 days, filtered, and analyzed for sugar concentrationsby HPLC as above.

Maximumyields of xylose from pretreatment were seen to vary with both time andfeedstock composition in the expected manner. Maximum glucose yields following enzymatichydrolysis were seen to be strongly affected by both pretreatment severity andenzyme loading. Results are summarizedin Table 1 below.

Table 1: Experimental sugar recoveries for Dilute AcidHydrolysis (DAH), Enzymatic Hydrolysis (EH), and experimental and predictedrecoveries for Total Sugar (TS) yield as percent of theoretical.

	DAH		EH		TS		TS (Predicted)	
	Xylose	**Glucose**	**Xylose**	**Glucose**	**Xylose**	**Glucose**	**Xylose**	**Glucose**
Balsam	72%	7%	0%	2%	72%	9%		
Aspen/Balsam	82%	5%	8%	27%	90%	32%	85%[1]	29%
Aspen (1)	83%	3%	4%	47%	87%	50%		
Aspen/Switchgrass	79%	5%	4%	53%	83%	58%	86%	60%
Switchgrass	69%	12%	5%	59%	74%	71%		
Aspen (2) w/ β-Glucosidase	93%	4%	6%	61%	100%	65%		

[1]Predicted value based on Aspen (2) pretreatment results.

Xyloseyields from pretreatment varied from 69-93% of the theoretical maximum. Pure switchgrass gave the lowest yield andpure aspen the highest. Each purespecies and mixture showed a peak hydrolysis yield near the expected optimaltime. Earlier time points showedincomplete hydrolysis with greater quantities of soluble oligomericsugars in the hydrolyzate. Later timepoints yielded sugar decomposition products such as furfural and hydroxymethylfurfural (HMF) in the hydrolyzate. Longer pretreatment reaction times alwaysresulted in higher glucose yields, but due to the degradation of xylose atlater time points total sugar yields were generally lower for extendedpretreatment times.

Glucoseyields following enzymatic hydrolysis varied from 2-61% with pure balsam givingthe lowest yield and pure aspen with supplemental β-glucosidase giving thehighest. The first aspen experiment wasrepeated due to incomplete pretreatment (indicated by lower than expected sugarrecoveries), and subsequently treated with the highest Acceleraseconcentration plus supplemental β-glucosidase to study the effect ofadditional enzyme loading on yield.

Using asimple linear interpolation model to predict sugar recoveries for mixedfeedstock streams based on the known yields obtainable for pure species gives aresult that is accurate to within about 2% of the experimental data for bothglucose and xylose recovery following dilute acid pretreatment, and to within3% for glucose recovery following enzymatic hydrolysis. Xylose recovery following enzymatichydrolysis for blends containing balsam could not be predicted accurately fromthese data, but since that is only a very small amount of the total sugar yieldthe discrepancy did not affect the model results for total sugar yield. Total sugar yield was predicted to within 5%of experimental results.

References:

Chum, Helena L. and Ralph P Overend (2001). *Biomass and renewable fuels*, FuelProcessing Technology 71, 187-195.

Perlack,R. D., L. L. Wright, et al. (2005). Biomass as Feedstock for a Bioenergy and Bioproducts Industry: The Technical Feasibility of aBillion-Ton Annual Supply. U. S. D. o. A. (USDA) and U. S. D. o. E. (DOE).

U.S. DOE (2006). *Breakingthe Biological Barriers to Cellulosic Ethanol: A Joint Research Agenda*,DOE/SC-0095, U.S. Department of Energy Office of Science and Office of EnergyEfficiency and Renewable Energy (www.doegenomestolife.org/biofuels/).

EIA (2008). *Emissionsof Greenhouse Gases in the United States*. U.S. Energy Information Administration Office of Integrated Analysis andForecasting (http://www.eia.doe.gov/oiaf/1605/ggrpt/carbon.html).

EIA (2009). *Annual Energy Review 2009*. U.S.Energy Information Administration, (http://www.eia.doe.gov/aer/pecss_diagram.html).

Jensen, Jill R., Juan E. Morinelly, Kelsey R. Gossen,Michael J. Brodeur-Campbell, and David R. Shonnard (2010). Effectsof dilute acid pretreatment conditions on enzymatic hydrolysis monomer andoligomer sugar yields for aspen, balsam, and switchgrass. BioresourceTechnology 101, 2317-2325.

See more of this Session: Developments In the Pretreatment of Lignocellulosics for Bioconversion III
See more of this Group/Topical: Sustainable Engineering Forum

(305b) Lifecycle Assessment and Policy: Implications of the Renewable Fuel Standard for Upper-Midwest Energy Supply

Michael J. Brodeur-Campbell[1], Jordan Klinger[1], Kathleen Halvorsen[2] and David Shonnard[1], (1)Chemical Engineering, Michigan Technological University, Houghton, MI, (2)School of Forest Resourses and Environmental Science and Department of Social Sciences, Michigan Technological University, Houghton, MI

LifecycleAssessment and Policy: Implications of the Renewable Fuel Standard for

Upper-MidwestEnergy Supply

Michael Brodeur-Campbell[1], Jordan Klinger[1],Kathleen Halvorsen[2,3], and David Shonnard[1,4]

[1] Department of ChemicalEngineering, Michigan Technological University

[2] Department of SocialSciences, Michigan Technological University

[3] School of Forest Resourcesand Environmental Science, Michigan Technological University

[4] Director Sustainable FuturesInstitute, Michigan Technological University

The use of biofuels derived from lignocellulosic biomass has receivedsignificant public attention with Congressional policies such as the EnergyPolicy Act of 2005 (EPAct), updated in the EnergyIndependence and Security Act of 2007 (EISA) which included the Renewable FuelStandard (RFS). The RFS mandates theblending of ethanol and advanced / cellulosic biofuels into the U.S.transportation fuel supply for each year until 2022. "The Renewable Fuel Standard has a lifecycleperspective and contains lifecycle greenhouse gas emissions reductionrequirements (relative to petroleum gasoline) for the production ofbiofuels." Conventional ethanol derivedfrom corn starch must meet a 20% reduction level and is capped at 15 billiongallons/yr. "°Cellulosic" ethanol mustmeet a 60% reduction standard and is expected to make up 16 billion gallons/yr.by 2022. "Advanced" biofuel that meet a50% reduction must make up the remaining5 billion gallons/yr mandated by 2022 (36 billion gallons/yr total ethanol). In addition to demanding an accounting for bothdirect and indirect land use change CO_2 emissions, the policy alsospecifies several other environmental and social metrics to consider inevaluating biofuel production.Environmental metrics specified include airquality, water quantity and quality, wetlands and ecosystem health, andwildlife habitat. Other indirect effectson society include energy security, commercial fuel production andinfrastructure, consumer fuel prices, job creation, agriculture impacts, ruraleconomic development, and future food prices.

While many studies have been performed on agricultural residues such ascorn stover (Sheehan, Aden et al. 2004, Hsu 2010) and wheat straw (Kabel, Bos et al. 2007), muchless attention has been given to woody feedstocks. For this LCA, several regionally importantpotential woody feedstocks are analyzed (hybrid poplar, hybrid willow, andmixed hardwood logging residues) as well as two herbaceous feedstocks (switchgrassmonoculture, and a diverse prairie grass ecosystem). This LCA is directly driven by therequirements of the RFS and includes metrics for water and air quality, as wellas global warming potential. Eutrophicationpotential was chosen as the most important measure for water quality, as theeffects of fertilizer use in biomass cultivation generally dominate allcategories of water quality.

Particulatematter emissions were chosen as the most important measure for air quality, asthis category has the greatest implication for human health.

SimaPro 7.2 and the EcoInventdatabase were used to model the process and assign environmental burdens. A co-product credit is assigned to therenewable electricity generated from the non-fermentable portion of thebiomass. This electricity is modeled todisplace grid electricity at the regional mix for the states of Michigan,Wisconsin, Minnesota, Iowa, and Illinois, which is derived approximately 70%from coal-fired power plants. Inputs andoutputs for the conversion were based on the National Renewable EnergyLaboratory Technical Report NREL/TP-510-32438 (Aden, Ruth, et al. 2002). The ratio of fuel to electricity producedfrom each feedstock was modified based on typical values for cellulose,hemicellulose, lignin, ash, and extractives for each feedstock. Table 1 below summarizes the results forGlobal Warming Potential, Eutrophication Potential, and Particulate MatterEmissions for the feedstocks analyzed.

Table1: Summary of LCA Results for Cellulosic Ethanol

	Logging Residue	**Prairie**	**Switchgrass**	**Willow**	**Poplar**
Global Warming Potential (g CO_2 eq. / MJ)	9.0	-11.0	0.9	-2.4	-7.9
Eutrophication Potential (g N eq. / MJ)	0.0	0.0	0.3	0.1	0.1
Particulate Matter <2.5 m (mg / MJ)	0.4	-1.4	-1.1	-2.8	-3.1

All fuel pathways analyzed meet 60% greenhousegas reduction requirements for a cellulosic biofuel, even before any co-productcredit is applied. With co-productcredit some pathways result in net negative carbon emissions due to coaldisplacement. Eutrophication potentialis increased in all cases due to feedstock production activities; the highestis for switchgrass which is the most fertilizer-intensive feedstock, while thelowest is for multispecies prairie. Particulateemissions are reduced for all cases except logging residues, largely due toavoided coal power. Particulate matteremissions for logging residues are high because of the low productivity peracre and long transportation distances required to supply enough material forconversion.

Future work will focus on refining ourfeedstock conversion emission estimates using the ASPEN Plus process simulationsoftware, and on including land use change estimates for carbon flows.

References

Aden, A., M. Ruth, K. Ibsen, J. Jechura, K. Neeves, J. Sheehan,B. Wallace, L. Montague, A. Slayton, and J. Lukas (2002). *Lignocellulosic Biomass to Ethanol ProcessDesign and Economics Utilizing Co-Current Dilute Acid Prehydrolysisand Enzymatic Hydrolysis for Corn Stover.* National Renewable EnergyLaboratory, Golden, Colorado.

Hsu,D., D. Inman, G.A. Heath, E.J. Wolfrum, M.K. Mann,and A. Aden (2010). *Life CycleEnvironmental Impacts of Selected U.S. Ethanol Production and Use Pathways in2022.* Environmental Science &Technology; 44, 5289-5297.

Kabel, M.A., Bos,G., Zeevalking, J., Voragen,A.G.J., Schols, H.A., (2007). *Effect of pretreatment severity on xylansolubility and enzymatic breakdown of the remaining cellulose from wheat straw.*Bioresource Technology; 98, 20342042.

Sheehan J, Aden A, Paustian K, Killian K,Brenner J, Walsh M, et al. (2004) *Energyand environmental aspects of using corn stover for fuel ethanol.* Journal of Industrial Ecology;7(4):117e46.

See more of this Session: Life Cycle Analysis of Renewable Feedstock-Based Processes and Products I
See more of this Group/Topical: Sustainable Engineering Forum

(278c) Biojet Production From Hydrocracking of Vegetable Oil

Huali Wang, Shuli Yan, Manhoe Kim, Steven Salley and Simon K. Y. Ng, Chemical Engineering and Materials Science, Wayne State University, Detroit, MI

The development of fuels from alternative and renewable sources to replace commercial petroleum products has been a subject of increasing research in recent years. At present, there are few economically feasible processes to produce jet fuel from plant oils and animal fats, especially with non-sulfided catalysts. In our research, bifunctional catalysts, NiMo/HY carbide and nitride catalysts were developed for hydrocracking of vegetable oils to obtain hydrocarbons in the jet fuel range. Around 2 g of the catalyst was loaded in a BTRS flow reactor and the catalysts were then reduced in a hydrogen flow of 30 mL/min at 450 °C for two hours. The reactor pressure was increased up to 650 psi and the reactions were carried out at the temperature range of 360-450 °C. Quartz beads with a size about 200 μ m were used to dilute the catalyst bed (1:1 v/v) to improve the mass and heat transfer. This study is the first application of noble metals and transitional metal carbide/nitride on hydrocracking of soybean oil to produce biofuels. A comparable yield of jet fuel (about 16-20 wt %) was obtained under a much lower pressure (650 psi) compared to about 17wt% kerosene jet yield under 2000 psi over the commercial hydrocracking catalyst. A 20-29 wt% diesel yield was also obtained during the process. Compared with other catalytic cracking processes with vegetable oils, a more stable continuous flow reaction was obtained by using this process.

See more of this Session: Catalytic Biofuels Refining II
See more of this Group/Topical: Fuels and Petrochemicals Division

(456e) A Computational Study of Steam Hydrolysis of NaBH4

Ping Li, University of Pittsburgh, Pittsburgh, PA, Karl Johnson, Chemical and Petroleum Engineering, University of Pittsburgh, Pittsburgh, PA and Michael A. Matthews, University of South Carolina, Columbia, SC

Complex hydrides are promising hydrogen storage materials and have received significant attention due to their high hydrogen storage capacity. $NaBH_4$ has high gravimetric (10.7 wt%) and large volumetric (115 kg of H_2 m^{-3}) hydrogen content. The hydrolysis reaction of $NaBH_4$ releases hydrogen with both fast kinetics and high extent of reaction under technical conditions by using steam deliquescence of $NaBH_4$. This catalyst-free reaction has many advantages over traditional catalytic aqueous phase hydrolysis. However, the reaction mechanism and kinetic barriers associated with steam hydrolysis have not yet been elucidated. We are developing a fundamental molecular level understanding of the steam hydrolysis reaction of $NaBH_4$ using a combination of experiments and first-principles density functional theory (DFT) calculations. We have applied finite temperature *ab initio* molecular dynamics (AIMD) to investigate the reaction mechanism. AIMD provides clues to elementary reaction steps. We have studied the effect of $NaBH_4$ concentration on the initial reaction rate, and found that high $NaBH_4$ concentrations lead to high initial reaction rates. The reason for this is appears to be due to a cooperative interaction between two BH_4^- groups. We have computed the free energy profiles of possible reaction pathways observed in AIMD using a cluster approach with an implicit solvent model and explicit water molecules. Explicit water is critical because proton transfer is a key step in the hydrolysis reaction.

See more of this Session: Hydrogen Storage Systems and Materials: Modeling, Simulation, and Optimization:
See more of this Group/Topical: International Congress on Energy 2011

(456a) Temperature Distribution In Activated Carbon Bed During Adsorption of Nitrogen (model gas for hydrogen); Experiment & Mathematical Model

Agnieszka Truszkowska[1], **Christopher Loeb**[1], Richard Chahine[2], Bruce Hardy[3] and Goran Jovanovic[4], (1)Chemical, Biological and Environmental Engineering, Oregon State University, Corvallis, OR, (2)Hydrogen Research Institute, Trois-Rivieres, QC, Canada, (3)Computational Sciences, Savannah River National Laboratory, Aiken, SC, (4)Oregon State University, Corvallis, OR

Temperaturedistribution in activated carbon bed during adsorption of nitrogen (model gasfor hydrogen); experiment & mathematical model

Agnieszka Truszkowska[1],Christopher Loeb[1], Richard Chahine[2], Bruce Hardy[3]and

Goran N. Jovanovic[1]

[1]Oregon State University,Department of Chemical, Biological, and Environmental Engineering, Corvallis, OR 97330, USA[2]Hydrogen Research Institute at University du Quebec a Trois-Rivires, PO BOX 500, Trois- Rivires, QC, Canada, G9A 5H7

3•
Savannah RiverNational Laboratory•
ProcessModeling and Computational Chemistry Group •
Building 999-2W, Rm. 122 •
Aiken, SC 29808, USA •
•

Emails: truszkoa@onid.orst.edu, loebc@onid.orst.edu, goran.jovanovic@oregonstate.edu, bruce.hardy@srnl.doe.gov, richard.chahine@uqtr.ca

Providinghigh volumetric and mass storage density of hydrogen gas is a major challengein the progress towards hydrogen-fueled automobiles. Of the various storage media,liquid chemical systems, and metal hydrides, activated carbon adsorbents havesubstantial advantages in cost, weight, ease of handling and safety.

In this paperwe present experimental data (temperature and pressure) obtained during adsorptionof nitrogen (model gas for hydrogen) in an activated carbon bed. We alsopresent the results of a numerical simulation of the same process.

Experimentalapparatus used in this study consists of a cylindrical activated carbon bed (2in diameter x 1.5 in high) placed into a sealed high-pressure cylindricalvessel. The carbon bed, consisting of densely packed activated carbon materialand binding agent, is instrumented with six thermocouples placed at suitablelocations throughout the bed. In addition, a thermocouple, and three pressuretransducers are placed into the high-pressure vessel outside the bed. Dynamic change of temperature andpressure during the charging process are recorded, corresponding to compressionand adsorption phenomena. The charging process starts at 1.0 bar and ends at50.0 bar pressure. A parametric study of filling time is performed for time intervalsbetween 10 and 30 seconds, to observe its effect on the charging process.

The mathematicalmodel is developed for the carbon bed pressure vessel system, which enablesnumerical simulation of the charging process. Initially, numerical simulationsare used to analyze experimental data and to explore characteristic parametersof the carbon bed. The same numerical tool will be used in the future todesign carbon bed with optimal characteristics imposed by requirements forhydrogen storage media in motor vehicles.

See more of this Session: Hydrogen Storage Systems and Materials: Modeling, Simulation, and Optimization:
See more of this Group/Topical: International Congress on Energy 2011

(305a) Life Cycle Assessment of Advanced Regional Biomass Processing Depots Integrated with Sustainable Cellulosic Feedstock Landscapes

Pragnya L. Eranki, Bryan Bals, Seungdo Kim and Bruce E. Dale, Department of Chemical Engineering, Michigan State University, Lansing, MI

Lignocellulosic biomass can help fulfill escalating demands for liquid fuels and mitigate the environmental impacts of petroleum-derived fuels. Cellulosic biofuels also offer potential solutions to many socio-economic and sustainability problems caused by petroleum dependence. However, the large biorefineries which process cellulosic biomass to fuels face many logistical problems because they are fully integrated, centralized facilities in which all units of the conversion process are present in a single location. Biomass logistical problems are further exacerbated by physical characteristics of feedstocks such as low bulk density, compositional variability, seasonality and their tendency to decompose.

A potential solution to these problems might be a network of distributed processing facilities called "Regional Biomass Processing Depots" (RBPDs). In their simplest configuration these depots procure, preprocess /pretreat, densify and deliver feedstock to the biorefinery while providing a single co-product animal feed to end-users. However, the RBPDs can be configured into other, more complex setups by employing multiple technologies such as the ammonia fiber expansion pretreatment (AFEX), Leaf Protein Concentrate (LPC) extraction, leaf/stem separation, pyrolysis and anaerobic digestion, as dictated by feedstock characteristics and local conditions, to produce a variety of valuable co-products and for energy generation via synergies among these technologies.

In a previous analysis we considered the differences between distributed processing and centralized processing combined with a variety of landscape scenarios. This study established that combining high yielding perennial grasses with distributed processing that includes densification of AFEX treated biomass prior to transportation to the biorefinery may generate greater energy yields and lower greenhouse gas emissions compared to centralized processing. In this study we aim to extend this analysis to watershed- scales in various landscape configuration scenarios using more sophisticated geospatial information system tools as well as by including a wider range of feedstocks and sustainable management practices, including the use of marginal lands. On the processing side, we evaluate the energy balances and environmental impacts of advanced RBPDs and their associated transport systems. We then intend to integrate the various elements within this overall system and conduct a life cycle analysis (LCA) using the Gabi software.

Through an integrated systems-wide analysis and by developing flexible models we seek the most sustainable landscape-processing-transportation arrangements for cellulosic biofuel production. We believe that greater energy and environmental benefits can be added to biofuel production by not merely aiming at improved logistics but by reimagining agricultural systems and landscapes, minimizing the challenges associated with biomass supply and its processing and by investigating multiple, potentially synergistic processing technologies and related complexities.

See more of this Session: Life Cycle Analysis of Renewable Feedstock-Based Processes and Products I
See more of this Group/Topical: Sustainable Engineering Forum

(388e) α-Olefin Biosynthesis In the Cyanobacterium *Synechococcus Sp.* PCC7002

Daniel Mendez-Perez[1], Matthew B. Begemann[2] and Brian F. Pfleger[1], (1)Department of Chemical and BIOLOGICAL Engineering, UNIVERSITY OF WISCONSIN-MADISON, Madison, WI, (2) Microbiology Doctoral Training PROGRAM, UNIVERSITY OF WISCONSIN-MADISON, Madison, WI

Biological hydrocarbons are a promising alternative to petroleum-based liquid transportation fuels as they would be compatible with current engines and distribution systems, yet be sourced from renewable substrates. Cyanobacteria have received significant attention because the direct conversion of sunlight and carbon dioxide to lipids will by-pass recalcitrant intermediates found in terrestrial biomass. Although it has been known for decades that some species of cyanobacteria, including *Synechococcus sp.* PCC7002, can synthesize hydrocarbons of different lengths, very little is known about their biosynthesis. In this study, we analyzed the hydrocarbon profile of *Synechococcus sp.* PCC7002 and demonstrate the involvement of a gene in the biosynthesis of two C19 α-olefins. The gene encodes a large multi-domain protein with homology to Type I polyketide synthases, suggesting a new route for hydrocarbon biosynthesis from fatty acids via an elongation-decarboxylation mechanism. The combinatorial nature of polyketide synthesis raises the possibility of using this type of genes to produce a wide range of industrially relevant α-olefins and high value chemical building blocks.

See more of this Session: Microbial Engineering for the Synthesis of Bulk and High-Value Products
See more of this Group/Topical: Sustainable Engineering Forum

(738a) Optimization of IGCC with CFD Based ROM

Yi-dong Lang, Department of Chemical Engineering, Carnegie Mellon University, Pittsburgh, PA, Stephen E. Zitney, U.S. Department of Energy, National Energy Technology Laboratory, Morgantown, WV and Lorenz T. Biegler, Chemical Engineering, Carnegie Mellon University, Pittsburgh, PA

Integrated gasification combined cycle (IGCC) plants have significant advantages for efficient power generation with carbon capture. Moreover, with the development of accurate CFD models for gasification and combined cycle combustion, key units of these processes can now be modeled more accurately. However, the integration of CFD models within steady-state process simulators, and subsequent optimization of the integrated system, still presents significant challenges. This study describes the development and demonstration of a reduced order modeling (ROM) framework for these tasks. The approach builds on the concepts of co-simulation and ROM development for process units described in earlier studies. Here we show how the ROMs derived from both gasification and combustion units can be integrated within an equation-oriented simulation environment for the overall optimization of an IGCC process. In addition to a systematic approach to ROM development, the approach includes validation tasks for the CFD model as well as closed-loop tests for the integrated flowsheet. This approach allows the application of equation-based nonlinear programming algorithms and leads to fast optimization of CFD-based process flowsheets. The approach is illustrated on two flowsheets based on IGCC technology.

See more of this Session: Energy Systems Design and Alternative Energy Sources II
See more of this Group/Topical: Computing and Systems Technology Division

(62d) Medium Optimization for Enhanced Lipid Production by Chlorella Protothecoides UTEX25

Kuan-Chen Cheng[1], Ming Ren[1] and Kimberly Ogden[2], (1)Chemical and Environmental Engineering, The University of Arizona, Tucson, AZ, (2)University of Arizona, Tucson, AZ

Algal biodiesel as the third generation biofuels has received considerable attention in recent years because of its renewable, biodegradable, and nontoxic fuel properties. Algal also exhibit several advantages over other energy crops, such as higher photosynthetic efficiency, higher biomass production and lipid content, and faster growth rate. The composition of algal-produced hydrocarbons vary among the microalgae species resulting in many different lipids, hydrocarbons, and other complex oils.

In the present study, a fresh water microalgae, *Chlorella protothecoides* (UTEX25), was evaluated for its lipid productivity through medium optimization. A one-factor-at-a-time approach was applied to determine the influence of each ingredient of proteose medium on both biomass and lipid production. $MgSO_4$°E7H_2O, $NaNO_3$, $CaCl_2$°E2H_2O, K_2HPO_4 and KH_2PO_4, and NaCl were removed individually from the culture medium, and the results were compared with the one obtained from the original medium. After 11-day cultivation, the results demonstrated that the removal of MgSO4°E7H_2O has positive effect on both biomass growth and lipid production. The biomass reached 0.95 g/L (the biomass production obtained from original medium was 0.54 g/L). The lipid content also increased to from 4.6 to 9.5 % when MgSO4 was removed. On the other hand, the removal of $NaNO_3$ and proteose adversely reduced biomass growth. The biomass production decreased from 0.54 to 0.37 and 0.29 g/L, respectively, when $NaNO_3$ and proteose was removed. Therefore, MgSO4°E7H_2O, $NaNO_3$, and proteose were chosen for further evaluation.

A statistical experimental design, response surface methodology (RSM), was applied to determine the optimal concentration of the three nutrients for both biomass and lipid production. The predicted results of these three nutrients are validated in comparison with experimental data. The change of the lipid profiles obtained from original and optimized medium are evaluated using GC-FID. Results are applied to minimize nutrient requirements in large-scale pond cultivation.

See more of this Session: Advances In Algal Biorefineries I
See more of this Group/Topical: Sustainable Engineering Forum

(583c) Pretreatment of Corn Stover for Bioethanol Production Using Low-Moisture Anhydrous Ammonia (LMAA); Ammoniation and Pretreatment

Chang Geun Yoo, Agricultural and Biosystems Engineering, Iowa State University, Ames, IA, Nhuan-John Nghiem, Crop Conversion Science and Engineering, USDA-ARS-ERRC, Wyndmoor, PA, Kevin B. Hicks, Eastern Regional Research Center, Agricultural Research Service, Wyndmoor, PA and **Tae Hyun Kim**, Department of Agricultural and Biosystems Engineering, Iowa State University, Ames, IA

In the past several decades, various pretreatment methods have been suggested to enhance the enzymatic digestibility and fermentability of lignocellulosic biomass. Although a few of them may be effective, several cost barriers which prohibit scale-up exist including high chemical input and excessive water use. In order to solve these problems, a simpler pretreatment method using anhydrous ammonia was developed. In this method, corn stover with 30~70% moisture was used, which was contacted with anhydrous ammonia in a packed bed reactor. After the ammoniation step, biomass was treated at different reaction temperatures (40-120°C) and reaction times (24-96 h). Pretreated biomass was directly saccharified and fermented "with no additional washing step." Therefore, it is anticipated that this method can reduce energy cost substantially due to the extremely low liquid loading and moderate conditions applied. With the treated solids, we have obtained significantly improved ethanol fermentability (>85%) using *E.coli* KO11 strain.

In this investigation, it is speculated that ammonia swells cellulose structure effectively and ammoniation prevents methoxyl groups on lignin from absorbing cellulase enzymes, thus it enhances the enzyme hydrolysis rate and ethanol fermentation yield of lignocellulosic biomass. Ammoniation pretreatment can also supply assimilable nitrogen (up to 1.25 wt.% of the weight of dry biomass) for microorganism growth in the fermentor.

In this paper, various pretreatment conditions such as moisture levels, residual ammonia concentrations, reaction time, and reaction temperature were explored to reduce the pretreatment cost and its pertinent costs. Enzyme digestibility and ethanol fermentation test results will also be presented.

See more of this Session: Developments In the Pretreatment of Lignocellulosics for Bioconversion III
See more of this Group/Topical: Sustainable Engineering Forum

(39d) Electrostatic Stabilization of Silicon Nanocrystal Colloids for Solution-Processed Photovoltaics

Lance M. Wheeler and Uwe R. Kortshagen, Mechanical Engineering, University of Minnesota, Minneapolis, MN

The tunable optoelectronicproperties of semiconductor nanocrystals (NCs) have garnered significant attention for photovoltaicdevice integration. Additionally,solution processing of semiconductor nanocrystals (NCs) provides a low-cost,scalable technique for the fabrication of third-generation solar cells. Group II-VI and IV-VI compoundsemiconductors have dominated the field of NC photovoltaics, as solution-basedsynthesis processes are well established. The process results in a stericallystabilized colloid from which ligand-capped NCs will naturally self-assemblewhen cast into a film. The ligandscan then be removed or exchanged to form a dense, electronically-coupled NCfilm.

Solution synthesis of Silicon NCshas proven difficult due to the high temperatures needed. Nonthermal flow-through plasmasynthesis offers an effective alternative. Typical plasma processes using silane as the siliconprecursor lead to free-standing, hydrogen-terminated Si NCs that will not,however, stabilize in solution. A hydrosylilation reaction will attach alkylligands to the surface to achieve a stable colloid similar to group II-VI andIV-VI NCs, but post-processing of cast films to remove or exchange the ligandsis problematic.

Here we report nonthermal plasmasynthesis of Si NCs with an alternative silicon precursor, silicontetrachloride. The presence of Clduring synthesis allows for tuning of the surface chemistry of the Si NCs. FTIR analysis reveals Cl present on thesurface, which facilitates a novel colloidal stabilization in aprotic polarsolvents. The Si NCs are readilydispersed and stabilized in aprotic dipolar solvents. The solutions are optically transparent and have remainedstable for over 8 months. Webelieve the electronegativity of the chlorine surface leads to solvation of theNCs by electrostatic interaction with solvent molecules.

Solutions of Cl-terminated Si NCsare spin- or drop-cast into self-assembled films that do not require a post-treatmentto remove ligands. Electronmicroscopy and Scanning Probe Microscopy (SPM) reveal the smooth and continuousfilm morphology needed for photovoltaic device structures. Electrical characterization reveals photoconductiveSi NC films with differential dark conductivities of 1.09×10^{-7}S/cm. This is competitive with nano-and micro-crystalline Si filmsdeposited using high-vacuum techniques such as PECVD.

This work was supported in partsby the National Science Foundation under grant CBET-0756326 and by DOE underthe EFRC Center for Advanced Solar Photophysics.

(564c) Production of Middle Distillate Range Transportation Fuels Via Fischer Tropsch Synthesis with Integrated Upgrading Under Supercritical Phase Conditions

Sihe Zhang, Rui Xu, Ed Durham and Christopher B. Roberts, Department of Chemical Engineering, Auburn University, Auburn, AL

Fischer-Tropsch (FT) synthesis involves surface catalyzed polymerization reactions that inherently yield a high selectivity towards methane and a broad product distribution of gaseous light products, middle distillate hydrocarbons (olefins + paraffins), and heavy wax, as well as CO_2 and water. Subsequent product upgrading via catalytic oligomerization and hydrocracking/isomerization can be employed to convert light olefin FT hydrocarbons into middle distillate range liquids, while also converting the long-chain hydrocarbons of heavy wax into fuel range products. This results in a reduction in olefin selectivity, generation of aromatics, suppression of heavy hydrocarbons selectivity and an enhancement of cold-flow properties.

In an effort to further direct the FT product distribution towards middle distillate transportation fuel range products, we have undertaken a series of experiments where we have sequentially incorporated catalytic oligomerization and hydrocracking/isomerization stages subsequent to FT in single pass operation. A multiple fixed bed reactor system was designed and utilized in this investigation. A precipitated iron-based FT catalyst was applied due to its low price and high tendency to yield olefins, which can be upgraded in the oligomerization, hydrogenation and isomerization reactions within the sequential reaction beds. The reaction temperatures were maintained differently in each bed to provide optimal reaction performance, while the system pressure was kept uniform in all three reaction beds.

Moreover, we have utilized the enhanced heat transfer properties of a supercritical fluid reaction medium in the FT and subsequent upgrading stages in order to reduce the selectivity towards CH_4 and CO_2. The use of the supercritical solvent provides prolonged catalyst life and activity maintenance. Distinguishable enhancement of propagation probability was observed under the supercritical phase operation. In particular, this study involves a systematic investigation into the influence of the supercritical fluid reaction medium on the overall product distribution and selectivity towards fuel range products.

See more of this Session: Alternative Fuels and Enabling Technologies III
See more of this Group/Topical: Fuels and Petrochemicals Division

An electrokinetic model for combined fouling during cross flow filtration of electrolyte solution and charged Colloids

Md Abdullaha Al Mamun, Reeshav Chatterjee, Subir Bhattacharjee,
Department of Mechanical Engineering, University of Alberta, Edmonton, AB, Canada T6G 2G8
Email: subir.b@ualberta.ca
Sirshendu De
Department of Chemical Engineering, Indian Institute of Technology, Kharagpur 721 302, India

Abstract

Colloidal fouling, formation of a cake due to deposition of colloids on to a membrane, is commonly encountered in pressure driven filtration processes, such as nanofiltration and reverse osmosis. The influence of these deposited charged colloids on the ion concentration polarization, ion rejection and the overall permeate flux is the focus of the present work. A common situation encountered during membrane filtration is the combined influence of charged colloids and the associated electrolyte on the flux decline and rejection. In this context, coupled model of concentration polarization (CP), cake layer formation by charged colloidal particles and ion transport through the interstitial spaces of the cake layer has been developed to predict the permeate flux decline and observed salt rejection during cross flow filtration of an electrolyte and colloidal mixture. The suspension structure of the cake layer is considered as a swarm of colloidal particles, represented using the Kuwabara cell model. For this system, the electroosmotic flow is calculated analytically using the Levine-Neale model. The analysis is then combined with the standard film theory of cross flow membrane filtration to predict the flux decline, observed rejection, cake layer thickness and resistance, and osmotic pressure difference. The model illustrates the combined influence of electrical and physical properties of colloids and electrolytes, mass transfer coefficient or crossflow condition and membrane resistance on the separation process. This approach provides a comprehensive insight into the mechanism of cake enhanced concentration polarization for charged colloids.

Keyword: Cake Enhanced Concentration Polarization, Kuwabara Cell Model, Levine-Neale Electrophoretic Model, Film Theory

1 Introduction

Membrane separation processes are increasingly being used in industries as a treatment process for separation of colloids, macromolecules, solutes, and ions. This is often achieved through processes like nanofiltration, whereby a single unit operation is used to remove multiple molecular weight and size species. The major operating costs associated with membranes include creating the pressure driving force for permeation, and mitigation of fouling. Both these operating costs are directly related to the overall energy consumption of membrane filtration. Several methods for fouling prevention and minimization has been suggested over the last few decades. But, most studies on fouling have typically ignored the concept of microscale manipulation of the fouling deposit itself to modify its permeability, or its ability to selectively adsorb ions or small organic contaminants. Therefore study of the fouling mechanism is important to develop a better understanding of the relation between the permeate flux, salt rejection, cake formation and to manipulate the fouled layer to enhance the preformece of the filtration process.

Hoek and Elimelech [1] conducted colloid fouling experiments during cross flow RO and NF, and reported cake enhanced osmotic pressure (CEOP) as the mechanism of flux decline. They proposed a model of CEOP, where hindered back-diffusion of ions within colloidal deposit layers was mentioned as the main mechanism of concentration polarization, resulting in an increase of osmotic pressure and decrease of pemeate flux.

Kim and co-workers [2] presented a detailed model of three component combined fouling to represent flux decline behavior of the system during dead end filtration. The model however introduces several parameters that need to be determined independently, such as, cake layer tortuosity, diffusive tortuosity, and hindered diffusivity of electrolyte through the cake layer. Additionally the model does not incorporate the effect of zeta potential of colloids and electrolyte solution conductivity in the flux decline behavior. However, the model presents a comprehensive mechanism for multi-component fouling from classical filtration concept.

Sim and co-workers [3] investigated the coupled effect of cake enhanced osmotic pressure and colloidal fouling in RO, using cross flow sampler-modified fouling index ultrafiltration (CFS-MFI_{UF}). They used CFS-MFI_{UF} measurement to determine the deposition rate of colloidal mass and cake porosity, which then incorporates the CEOP effect in RO fouling. They also considered the hindered diffusivity of ions within the cake layer to model CEOP. Though, the technique provides good estimation of RO fouling profile, the model depends on the CFS-MFI_{UF} or dead-end RO experimental data.

De and Bhattacharjee [4] proposed an electrokinetic model for flux decline during combined filtration of charged colloids and electrolyte. They assumed the cake layer as a bundle of charged cylindrical nano-pores similar to the membrane. Their model considered the electroviscous effect created by the charged nano-pores, which results in additional resistance of the cake layer. Though their model represents an improvement, the actual flow model should consider the flow of electrolytes around the charged colloids in a cake layer to be more realistic.

In this paper, a coupled model of concentration polarization, cake layer formation by charged colloidal particles and ion transport through the interstitial spaces of cake layer has been developed to predict the permeate flux decline and observed rejection due to fouling during cross flow nanofiltration of an electrolyte and colloidal suspension. The model cosiders the cake layer as a swarm of charged spherical particles instead of cylindrical nanopores, which is more realistic, and it can predict the flux decline and observed rejection without considering any correction parameter except the cake tortuosity. The model illustrates the combined influence of colloid mass transfer coefficient, electrical and physical properties of colloids and electrolytes, crossflow geometry and membrane resistance on the cake enhanced concentration polarization of the electrolyte.

2 Mathematical Model

During combined fouling of colloidal particles of radius a, surface potential ψ_c and electrolyte feed concentration of $C_{i,f}$ in a crossflow membrane filtration operation, the colloidal particles form a cake layer on the membrane and the electrolyte solution forms a concentration polarization layer on the cake surface as shown in Fig. 1. The dimensionless permeate flux through the cake layer and membrane can be represented by the following modified Darcy's equation [4]:

$$v_w^* = \frac{v_w}{v_w^0} = \frac{1-\Delta\pi_m^*}{1+R_g^*} \tag{1}$$

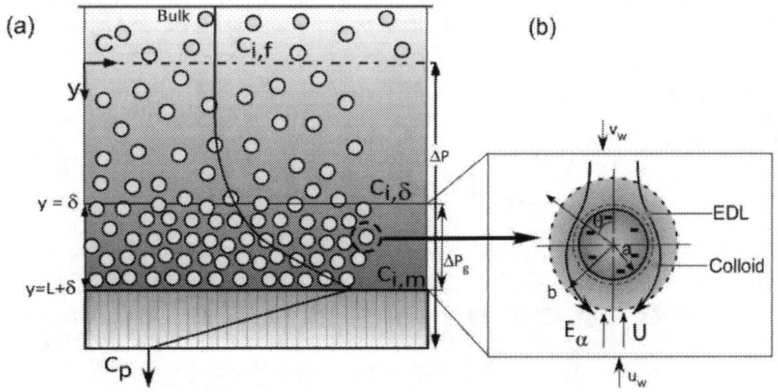

Figure 1: (a) Cake enhanced concentration polarization (b) Electrophoretic flow model around a charged colloid according to the Levine-Neale and Kuwabara cell model

Here, $v_w^o = \Delta P/\mu R_m$ is the pure water flux, $\Delta\pi_m^*$ is the dimensionless osmotic pressure across the membrane and R_g^* is the dimensionless resistance. The cake layer resistance R_g^* has been formulated based on the Kuwabara cell model and Levine-Neale electrophoretic flow model [5] of charged particles in a concentrated colloidal suspension as shown in Fig. 1b. The model considers the combined influence of hydrodynamic and electrokinetic resistance to the transport through the cake layer. According to the Levine-Neale model, electroosmotic flow (U) through the interstices of the cake layer due to induced streaming potential is given by [5]:

$$U = E_\infty \left(\frac{\varepsilon\psi_c}{\mu}\right) g_2(\kappa a, \varphi_p) \tag{2}$$

The expression of induced electric field (E_∞) in Eq. 2 at zero current condition, in accordance with Onsager principle of reciprocity for irreversible phenomena and for thin Debye layer, is [5]:

$$E_\infty = \frac{\Delta P_g}{L}\left(\frac{\varepsilon\psi_c}{\mu\sigma^\infty}\right) \tag{3}$$

Thus, the expression of electroosmotic flux (u_w) from the continuity assumption using Eqs. 2 and 3 is given as follows:

$$u_w = U(1-\varphi_p) = \frac{\Delta P_g}{L\sigma^\infty}\left(\frac{\varepsilon\psi_c}{\mu}\right)^2 (1-\varphi_p)g_2(\kappa a, \varphi_p) \tag{4}$$

Therefore, the total flux across the cake layer at no current condition, due to effective pressure across the cake ΔP_g and reverse electroosmotic flow is given as follows:

$$\begin{aligned} v_w &= \frac{\Delta P_g}{\mu r_g L} - u_w \\ \text{or, } v_w &= \frac{\Delta P_g}{\mu r_g L}\left[1 - 9\varphi_p(1-\varphi_p)A_K \frac{\beta g_2(\kappa a, \varphi_p)}{2(\kappa a)^2}\right] \\ \text{or, } v_w^* &= \frac{\Delta P_g^*}{L^*/Mg} = \frac{\Delta P_g^*}{R_g^*} \end{aligned} \tag{5a}$$

$$r_g = \frac{9\varphi_p A_K}{2a^2}; \quad M = \frac{2a^2 R_m}{9\varphi_p A_K d_e} \quad \text{and} \quad g = \left[1 - 9\varphi_p(1-\varphi_p)A_K \frac{\beta g_2(\kappa a, \varphi_p)}{2(\kappa a)^2}\right] \tag{5b}$$

Here, r_g is the hydrodynamic resistance of the cake layer, $A_K = 1/(1 - 9\varphi_p^{1/3}/5 + \varphi_p - \varphi_p^2/5)$ accounts for the presence of neighbouring particles, $\beta = (\varepsilon\psi_c\kappa)^2/(\mu\sigma^\infty)$ and $L^* = L/d_e$. The function g accounts for the electroosmotic effect in a swarm of colloidal particles, depends on dimensionless Debye length κa, surface potential of colloidal particles ψ_c, solution conductivity σ^∞ and volume fraction of the cake layer, φ_p.

The dimensionless osmotic pressure $\Delta\pi_m^*$, for simple electrolyte (1:1) can be written from the vant Hoff equation:

$$\Delta\pi_m^* = \frac{2R_{ug}T}{\Delta P}(C_{i,m} - C_{i,p}) = \alpha\bar{C}_{i,m}R_r \tag{6}$$

where, $\alpha = 2R_{ug}TC_{i,f}/\Delta P$ and $\bar{C}_{i,m}$ is dimensionless salt concentration at the membrane surface. The expression of $\bar{C}_{i,m}$, derived by integrating the mass balance of electrolytes within the concentration polarization and cake layer with appropriate boundary conditions is:

$$\bar{C}_{i,m} = \frac{C_{i,m}}{C_{i,f}} = \frac{\exp\left[v_w^*(R_i + L^*R)\right]}{R_r + (1-R_r)\exp\left[v_w^*(R_i + RL^*)\right]} \tag{7}$$

Here, $R_i = v_w^o/k_i$, $R = v_w^o d_e/D_i^*$. The observed rejection of the membrane can be written as:

$$R_o = 1 - \frac{C_{i,p}}{C_{i,f}} = 1 - \bar{C}_{i,m}(1-R_r) \tag{8}$$

The dimensionless governing equation for colloidal cake layer growth, obtained by integrating the mass balance equation of colloids in the cake layer with appropriate boundary conditions, is given by [6]:

$$\frac{dL^*}{d\tau} = v_w^* \frac{C_{c,m}^* - \exp(v_w^* R_c)}{1 - \exp(v_w^* R_c)} \tag{9}$$

Here, $\tau = v_w^o C_{c,f} t/(d_e \rho_c)$, $R_c = v_w^o/k_c$, $C_{c,m}^* = C_{c,m}/C_{c,f}$. The dimensioless pressure drop with respect to operating pressure can be written as:

$$\Delta P_m{}^* + \Delta\pi_m^* + v_w^* R_g^* = 1 \tag{10}$$

Here, $\Delta P_m{}^*$ is dimensionless pressure drop across the membrane.

3 Numerical Methodology

Equations 1 and 9 are solved simultaneously in conjunction with the Eqs. 6 and 7, using a differential algebric equation solver DDASSL. For a 50 nm particle with surface potential -45 mV and cake volume fraction 0.64 for random-closed-packing, in 0.01 M NaCl solution, the value of g is 0.67. The value of M depends on particle size, cake volume fraction and hydrodynamic radius of crossflow channel and is less than unity for RO or NF. The volume fraction can be as high as 0.72 for hexagonal packing. The viscosity and relative permitvity of water is 0.89 mPa.s and 80.0, respectively. The electrolyte flow turbulence parameter R_i, is considered to be in the order of 1.0 becuase pure water flux and mass transfer coefficient of electrolyte both are of the order 10^{-5}. The corresponding value of R_c is considered to be of the order 10.0, because the diffusivity of colloidal particle is at least one order magnitude less than the electrolytes. For the silica colloidal particles, the bulk concentration is 0.2 kg/m^3, whereas the cake concentration is about 173 kg/m^3 [7]. Typical equivalent diameter of the cross flow channel is 10^{-3} m. For colloid feed concentration 0.2 kg/m^3, with particle density 2317 kg/m^3, τ= 0.01 represents 6.5 hr of operation in real time. The range of dimensionless parameters are listed in table 1.

Table 1: Range of dimensionless parameters used in simulations

Parameter	Range
κa	1 – 100
g_2	0 – 0.9
β	1.0 – 5.0
g	0.05 – 0.95
M	0.005 – 0.5
R_i	0.05 – 1.5
R_c	10.0 – 60.0
R	100 – 1000
α	0.09 – 0.9
$C_{c,m}^*$	800 – 900
R_r	0.8 – 0.9

4 Results and Discussion

Each of the dimensionless parameters listed in Table 1 affect the flux decline and observed rejection during cross flow filtration process. For the present parametric study, the effect of four parameters M,

g, R and R_c are considered and explained. The physical properties of the colloidal particle, cake layer and membrane are incorporated in the membrane and colloidal system interaction parameter M. The value of M is proportional to the square of particle size, i.e., for small particles, the value of M is small resulting in higher resistance since specific resistance is inversely proportional to M. As a result, the permeate flux decline is higher as shown in Fig. 4a. The observed rejection also increases when M is small, this is due to lower convection of salt compared to back diffusion across the cake layer. Therefore, concentration at the membrane surface is less significant for M=0.05 resulting in lower osmotic pressure as shown in Fig. 4a. Therefore, permeate flux and observed rejection can be changed by changing the size of the cake forming particles.

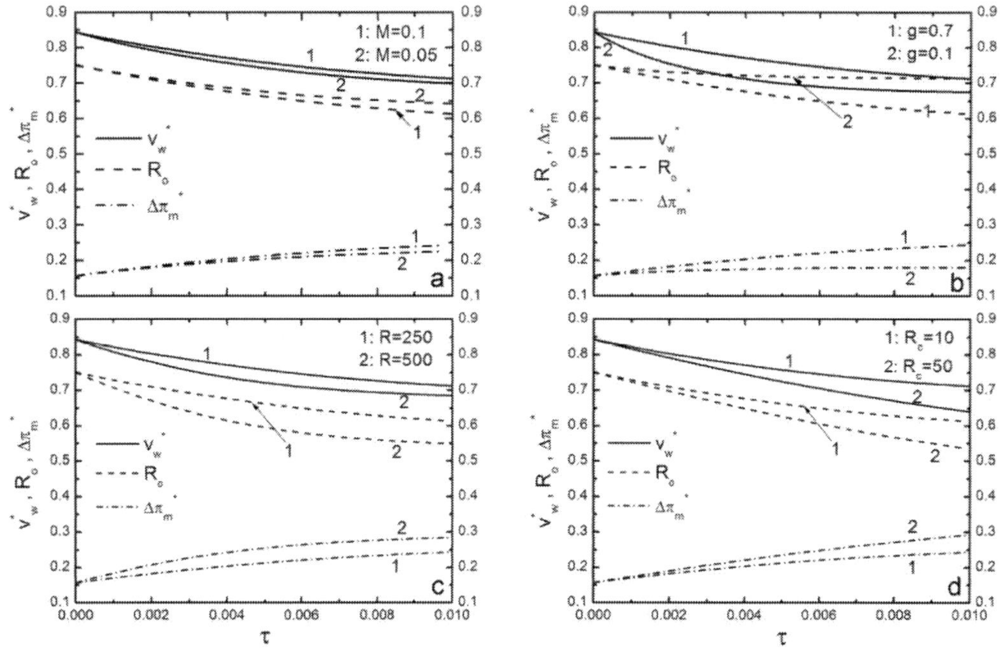

Figure 2: Variation of normalized permeate flux (v_w^*), observed rejection (R_o) and osmotic pressure $\Delta\pi_m^*$. (a) g=0.7, R=250.0, R_i=1.0, R_c=10.0, R_r=0.875, α=0.09 and $C_{c,m}^*$=850.0; (b) M=0.1, R=250.0, R_i=1.0, R_c=10.0, R_r=0.875, α=0.09 and $C_{c,m}^*$=850.0; (c) M=0.1, g=0.7, R_i=1.0, R_c=10.0, R_r=0.875, α=0.09 and $C_{c,m}^*$=850.0; (d) M=0.1, g=0.7, R=250.0, R_i=1.0, R_r=0.875, α=0.09 and $C_{c,m}^*$=850.0

The effect of electroosmotic parameter g, induced due to the presence of charged colloids and electrolyte solution, is shown in Fig. 4b. In absence of surface potential, the value of g goes to unity, implying absence of any induced electroosmotic flow. In presence of charged particles and salt solution the value of g is always less than 1. The parameter g is a strong function of κa and it decreases with increase of κa or with increase of electrolyte feed concentration. The parameter g is also a function of surface potential ψ_c and decreases with increase of ψ_c. Therefore, increase of surface potential of colloidal particles or increase of feed electrolyte concentration causes decrease of g. Since cake resistance is inversely proportional to g, increase of surface potential and feed concentration increases the electroviscous resistance of cake layer. As a result, permeate flux decline is more severe due to higher reverse electroosmotic flow induced from charged colloids. Due to the reverse electroosmotic flow of electrolyte, electrolyte concentration is less at the membrane surface, increasing the observed rejection and decreasing the transmembrane osmotic pressure as shown in Fig. 4b. Therefore, smaller value of g or higher value of κ and ψ_c, decreases the permeate flux and increases the salt rejection siginificantly. Thus, formation of cake layer with nano colloidal particles of high surface potential in concentrated salt solution can enhance the observed rejection of nanofiltration membranes or can enhance concentration polarization on cake surface significantly, which means the cake layer can be modeled as a secondary low salt rejection membrane of nano size pores with high surface potential.

The effect of parameter R on permeate flux and observed rejection is shown in Fig. 4c. The parameter R is defined as the ratio of pure water flux through the membrane to the diffusive flux through the cake layer. Higher value of R indicates that the convective flux is dominant compared to back diffusion through the cake layer which results in enhanced accumulation of electrolyte within the cake layer and increase in transmembrane osmotic pressure. As a result, the permeate flux decline is comparatively higher and observed rejection is lower for large value of R, according to Fig. 4c. Therefore, higher value of R enhances salt concentration at membrane surface and declines the permeate flux and observed rejection.

The parameter R_c is defined as the ratio of pure water flux to colloids mass transfer coefficient. The mass transfer coefficient solely depends on the geometry of cross flow channel and velocity. Since the model is for cross flow filtration lower value of R_c indicates higher mass transfer coefficient or, higher degree of turbulence and shear at the cake surface. Due to higher shear rate at cake surface, the cake layer can not grow much which leads to lower cake layer thickness and resistance, resulting in higher permeate flux as shown in Fig. 4d. Therefore, convective flux of electrolyte is higher for small value of R_c. Though the convective flux is higher, observed rejection is higher and transmembrane osmotic pressure is lower, due to smaller of ratio of convective flux to diffusive flux within the thinner cake layer for small R_c. Thus, higher mass transfer coefficient or small R_c enhances the permeate flux and observed rejection.

5 Concluding Remarks

The electrokinetic model presented in this paper is a simple coupled model for quantifying the combined effect of colloid-electrolyte system during crossflow nanofiltration. Due to reverse electroosmotic flow, the permeate flux and salt transport through the cake layer decrease. The salt concentration at the cake surface increases, which enhances the concentration polarization on cake surface. The cake enhanced concentration polarization will be severe for nano-colloids with higher surface potential. A detailed electrokinetic model of ion transport through the cake layer of nano-colloids and membrane will lead to a better understanding of the combined fouling process for high salt rejection membranes. Consideration of the electrokinetic interaction between the electrolyte and the colloid is important to develop a detailed model of the inter-relations between the permeate flux, salt rejection and cake formation. The study also includes that manipulation of the meso-scale structure of the cake will allow us to alter the porosity of cake layer, extent of adsorption of different species, and alter the salt rejection of the membranes.

Nomenclature

ΔP	Applied operating pressure, Pa
R_m	Membrane resistance, 1/m
R_r	Real rejection of membrane
d_e	Hydraulic diameter of cross flow channel, m
μ	Solvent viscosity, Pa.s
ε	Solvent electrical permitivity
R_{ug}	Universal gas constant, J/mole K
T	Absolute temperature, K
C_f	Solute concentration at feed (i.e., i= ions, c=colloids)
C_m	Solute concentration at membrane surface (i.e., i= ions, c=colloids)
ρ_c	Density of colloids, kg/m^3
L	Cake thickness, m
$D = \frac{k_B T}{6\pi\mu r}$	Brownian diffusion coefficient of solutes (i.e., i= ions, c=colloids), m^2/s
$D_i^* = \frac{(1-\varphi_p)*D_i}{[1-\ln(1-\varphi_p)^2]}$	Hindered diffusivity of ions within the cake layer, m^2/s
$k = 1.86\,\frac{D}{d_e}\left(\frac{u_c d_e^2}{DL_c}\right)^{\frac{1}{3}}$	Mass transfer coefficient co-relation of solute (i.e., i= ions, c=colloids) for laminar flow in thin channel [8], m/s

References

[1] E.M.V. Hoek and M. Elimelech. Cake-enhanced concentration polarization: A new fouling mechanism for salt-rejecting membranes. *Environmental Science and Technology*, 37:5581–5588, 2003.

[2] A.S. Kim, A.E. Contreras, Q. Li, and R. Yuan. Fundamental mechanisms of three-component combined fouling with experimental verification. *Langmuir*, 25:7815–7827, 2009.

[3] L.N. Sim, Y. Ye, V. Chen, and A.G. Fane. Investigations of the coupled effect of cake-enhanced osmotic pressure and colloidal fouling in ro using crossflow sampler-modified fouling index ultrafiltration. *Desalination*, 273:184–196, 2011.

[4] S. De and S. Bhattacharjee. Flux decline during cross flow membrane filtration of electrolytic solution in presence of charged nano-colloids: A simple electrokinetic model. *Journal of Colloid and Interface Science*, 353:530–536, 2011.

[5] J. Masliyah and S. Bhattacharjee. *Electrokinetic and Colloid Transport Phenomena*. Wiley Interscience, 2006.

[6] S. De and P. K. Bhattacharya. Modeling of ultrafiltration process for a two-component aqueous solution of low and high (gel-forming) molecular weight solutes. *Journal of Membrane Science*, 136:57 – 69, 1997.

[7] S.K. Zaidi and A. Kumar. Experimental analysis of a gel layer in dead-end ultrafiltration of a silica suspension. *Desalination*, 172:107–117, 2005.

[8] S. Kim and E.M.V. Hoek. Modeling concentration polarization in reverse osmosis processes. *Desalination*, 186:111–128, 2005.

(411c) Gas Flow Behavior In FCC Disengagers Vessel with Different Outlet Tube Length of Primary Cyclone Separator

Jiangyun Wang, Yu Mao, Juan Wang and Meili Liu, College of Chemical Engineering, China University of Petroleum,Beijing, Beijing, China

To deeply understand the effect of primary cyclone separator's outlet tube length on gas oil flow behavior in fluid catalytic cracking disengager space, the gas oil flow in a full-size disengager is numerically studied on the platform of a commercial computational fluid dynamics (CFD) code Fluent 6.1.22. The Reynolds stress model (RSM) is applied to simulate the gas oil flow field. As no geometry simplification is made in generating the structured mesh for the space both inside and outside the two-stage separator system, the whole flow behavior of the gas oil in the disengager is obtained. Coupled with the RSM, the stochastic tracking mode is used to quantify the gas oil into the secondary cyclone separator. It is shown that the growth of the outlet tube length of primary cyclone separator contributed to oil gas flowing into the inlet of secondary cyclone separator, effectively reducing the gas oil quantity into the disengager and the gas oil residence time, which avoids unwanted overreaction of hydrocarbon.

See more of this Session: Solids-Liquid Separation Processes for the Petroleum Industry
See more of this Group/Topical: Separations Division

(305c) Optimal Design and Operations of Cellulosic Biofuel Supply Chains Under Uncertainty

Fengqi You, Dept of Chemical Engineering, Carnegie Mellon University, Pittsburgh, PA and Belinda Wang, Department of Chemical and Biological Engineering, Northwestern University, Evanston, IL

Concerns about climate change, energy security, and the diminishing supply of fossil fuels are causing our society to search for new renewable sources of transportation fuels. Domestically available biomass has been proposed as part of the solution to our dependence on fossil fuels. Biofuels, especially liquid transportation fuels produced from cellulosic materials, have the benefits of significantly reducing greenhouse gas (GHG) emissions and leading to new jobs and greater economic vitality in rural areas. Renewable Fuels Standard (RFS), part of the Energy Independence and Security Act of 2007, establishes a target of 16 billion gallons of cellulosic biofuel annual production by 2022 [1, 2], although the United States produced less than 1 billion gallons of liquid fuels from cellulosic materials in 2010 [3, 4]. In observance of this mandatory production target, many new cellulosic biomass-to-biofuels supply chains will be designed and developed in the coming decade for better economic, environmental and social performances. However, uncertainty resulting from supply and demand variations may have significant impact on the biofuel supply chain. Therefore, an efficient optimization strategy is urgently needed to for the design and operations of sustainable and robust cellulosic biofuel supply chains.

In this work, we address the optimal design and planning of cellulosic biofuel supply chains under supply and demand uncertainty. A two-stage stochastic mixed-integer linear programming (SMILP) model combined with Monte Carlo sampling and the associated statistical analysis [4, 5] is proposed to deal with different types of uncertainty, and it is incorporated into a multi-period planning model that takes into account the main characteristics of the biofuel supply chains, such as seasonality of feedstock supply, biomass deterioration with time, geographical diversity and availability of biomass resources, feedstock density, diverse conversion technologies and byproducts, infrastructure compatibility, demand distribution, regional economic structure, and government incentives. In the two-stage framework, the supply chain network design and capacity planning decisions are made "here-and-now" prior to the resolution of uncertainty, while the production, transportation and storage decisions for each time period are postponed in a "wait-and-see" mode. The SMILP model integrates decision making across multiple temporal and spatial scales and simultaneously predicts the optimal network design, facility location, technology selection, capital investment, production operations, inventory control, and logistics management decisions. In order to solve the resulting large scale SMILP problems effectively, a decomposition algorithm based on sampling average approximation [5] and multi-cut L-shaped method [6, 7] is proposed by taking advantage of the problem structure.

In addition to the economic objective of minimizing the annualized net present cost, the SMILP model is also extended to integrate with life cycle assessment (LCA) and regional economic input-output (EIO) analysis through a multiobjective optimization scheme to include two other objectives: the environmental objective measured by life-cycle greenhouse gas emissions and the social objective measured by the number of accrued local jobs resulting from the construction and operation of the cellulosic biofuel supply chain. The multiobjective optimization framework allows the model to establish tradeoffs among the economic, environmental, and social performances of the cellulosic biofuel supply chains in a systematic way. The multiobjective optimization problem is solved with an ε-constraint method and produces Pareto-optimal curves that reveal how the optimal annualized cost and the supply chain network structure change with different environmental and social performance of the entire supply chain [10, 11].

The proposed optimization model and solution method is illustrated through county-level case study for the state of Illinois. Three major types of biomass, including crop residues, energy crops, and wood residues, and three major conversion pathways, including biochemical conversion, gasification followed by Fischer-Tropsch synthesis and fast pyrolysis followed by hydroprocessing are considered. Uncertainty information is generated from the time series analysis [8] based on the historical data of biomass feedstock supply [9] and liquid fuel demand [1]. County-level results will be presented that provide regionally-based insight into transition pathways of biomass production and conversion. Computational results also demonstrate the effectiveness of the proposed decomposition algorithm for the solution of large-scale SMILP problems.

References:

[1] Biomass Program Multi-Year Program Plan 2010; EERE, U.S. DOE, March 2010.

[2] National Biofuels Action Plan; Biomass Research and Development Board: U.S. U.S. Energy Information Administration.

[3] National Renewable Fuel Standard Program for 2010 and Beyond; U.S. EPA, February 2010

[4] Department of Agriculture and U.S. Department of Energy: 2008.

[5] Shapiro, A.; Homem-de-Mello, T., A simulation-based approach to two-stage stochastic programming with recourse, Mathematical Programming, 1998, 81, 301-325

[6] Birge, J.R.; Louveaux, F., Introduction to Stochastic Programming, Springer Verlag, New York, 1997

[7] You, F.; Wassick, J. M.; Grossmann, I. E., Risk management for global supply chain planning under uncertainty: models and algorithms. AIChE Journal 2009, 55, 931-946.

[8] Enders, W., Applied Econometric Time Series. Wiley: Hoboken, NJ, 2004.

[9] National Agricultural Statistics Service.

[10] You, F.; Tao, L.; Graziano, D. J.; Snyder, S. W., Optimal Design of Sustainable Cellulosic Biofuel Supply Chains: Multi-objective Optimization Coupled with Life Cycle Assessment and Input-Output Analysis. AIChE Journal 2011, In press, DOI: 10.1002/aic.12637.

[11] You, F.; Wang, B., Life Cycle Optimization of Biomass-to-Liquids Supply Chains with Distributed-Centralized Processing Networks. Industrial & Engineering Chemistry Research 2011, Submitted.

See more of this Session: Life Cycle Analysis of Renewable Feedstock-Based Processes and Products I
See more of this Group/Topical: Sustainable Engineering Forum

A Future Energy Supply Based on Liquid Organic Hydrogen Carriers (LOHC)

Daniel Teichmann, Wolfgang Arlt, Peter Wasserscheid
University of Erlangen-Nuremberg, Germany

Introduction

Our current energy system is mainly based on the availability of fossil fuels that can be extracted from the ground to a comparably cheap price and that are also excellent energy carriers in terms of storage density and manageability. The on-going transition towards renewable sources is a big challenge for our energy system, mainly due to the highly intermittent character of many renewable energies like wind or solar.

While hydrogen is often considered as a suitable energy vector in a future sustainable energy system [1,2], the weight and volume-efficient storage of sufficient amounts of hydrogen is a difficult task due to its low density. A comprehensive hydrogen infrastructure does not exist yet and makes high investments necessary.

Liquid Organic Hydrogen Carriers (LOHC)

This contribution describes a concept for the establishment of a competitive energy distribution network based on Liquid Organic Hydrogen Carrier (LOHC) compounds [3]. These compounds are characterized by the fact that they can be loaded and un-loaded with considerable amounts of hydrogen in a cyclic process. In contrast to other hydrogen storage methods (like compressed gaseous or cryogenic hydrogen) LOHCs can be safely stored under ambient conditions in big quantities. As they have many physico-chemical similarities to Diesel, LOHCs can make use of the existing energy infrastructure (e.g. storage tanks or road tankers).

Many different substances are discussed in the literature as potential LOHC compounds [4,5]. In this contribution we concentrate on heterocyclic aromatic hydrocarbons such as N-ethylcarbazole (Fig. 1) which were first considered by the company AirProducts [6]. The fully hydrogenated substance perhydro-N-ethylcarbazole offers a hydrogen capacity of up to 5.8 wt%. The hydrogenation reaction for this particular LOHC system is typically carried out at 130–160 °C with 70 bars pressure using supported noble metal catalysts like ruthenium on aluminum oxide. The dehydrogenation reaction proceeds usually in the presence of supported Pt or Pd catalysts. The reaction is typically carried out at 200–230 °C and pressures slightly above ambient conditions [7-9].

Energy storage and mobility applications

In times of high loads from intermittent sources like wind or solar energies, excess energy can be used to locally produce hydrogen by water electrolysis, which is efficiently stored in the form of LOHC. The so-produced energy carrier can be stored at the spot of production and re-converted to electricity in times of high demand.

Alternatively it can be easily transported to be used in various off-site applications like mobility (as discussed in the following), heating, local hydrogen production, etc.

Compared to other energy storage technologies like pumped hydroelectric or compressed air energy storage [10], LOHCs show a high energy storage density, thus enabling a safe and loss-free storage of renewable energy with only limited land use. These reasons make LOHCs a promising technology especially for seasonal energy storage where large amounts of energy have to be stored over extended periods of time.

Another promising field of application for LOHCs is mobility [11]. To allow cruising ranges of 500km, about 5 kg of on-board hydrogen are necessary which correspond to about 100 liters of carrier compound. The hydrogen which is produced from the LOHC on-board in a catalytic converter can either be used in a combustion engine or a fuel cell (see Fig. 2). At the gas station the unloaded carrier compound is returned in exchange to fresh material and can be reloaded with hydrogen for further use.

Conclusion
The proposed concept links the technical challenge of storing temporary and local energy over-production from regenerative sources with the vision of a sustainable, hydrogen-based mobility. Due to their beneficial handling and storage properties we consider LOHCs as an attractive way to provide wind and solar energy for mobility applications.

Figures

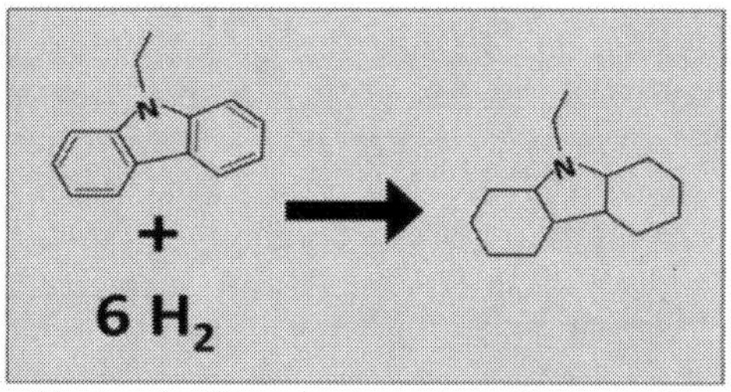

Figure 1: Chemical energy storage via hydrogenation of a LOHC compound [3]

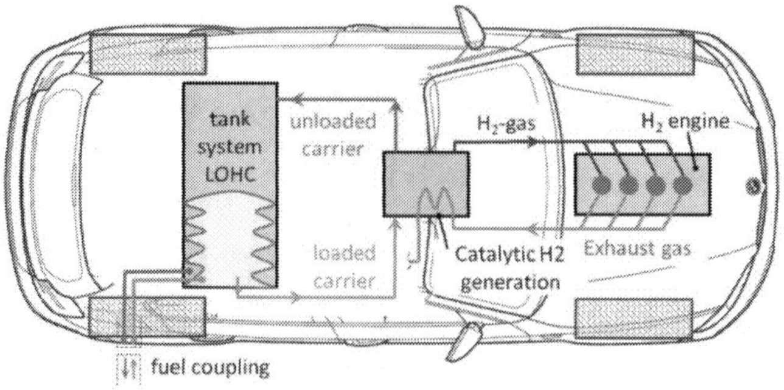

Figure 2: Concept of a hydrogen car with liquid organic hydrides [3]

References

1. Sartbaeva, V. L. Kuznetsov, S. A. Wells and P. P. Edwards, Energy Environ. Sci., 2008, 1, 79–85.
2. U.S. Department of Energy, Hydrogen Posture Plan, http://www.hydrogen.energy.gov.
3. D. Teichmann, W. Arlt, P. Wasserscheid, R. Freymann, Energy Environ. Sci., 2011, 4 (8), 2767 – 2773
4. G. W. H. Scherer, E. Newson, A. Wokaun, Int. J. Hydrogen Energy, 1999, 24, 1157–1169.
5. R. H. Crabtree, Energy Environ. Sci., 2008, 1, 134–138.
6. Pez et al., World Pat., WO 2005/000457A2, 2005.
7. M. Sobota et al., Chem.–Eur. J., DOI: 10.1002/chem.201101311.
8. F. Sotoodeh and K. J. Smith, Ind. Eng. Chem. Res., 2010, 49, 1018–1026.
9. K. M. Eblagon, D. Rentsch, O. Friedrichs, A. Remhof, A. Züttel, A. J. Ramirez-Cuesta and S. C. Tsang, Int. J. Hydrogen Energy, 2010, 35, 11609–11621.
10. Verband der Elektrotechnik, Elektronik, Informationstechnik (VDE), Energiespeicher in Stromversorgungssystemen mit hohem Anteil Erneuerbarer Energieträger, Frankfurt, 2009.
11. J. von Wild, R. Freymann and M. Zenner, Potentiale von alternativen Wasserstoffspeicherungstechnologien, VDI-Berichte: Innovative Fahrzeugantriebe, 2008, vol. 2030, pp. 273–298.

(278e) Value Added Products From Lignocellulose Based Biochar

Roland Feiner, Nikolaus Schwaiger, Susanne Lux and Matthaeus Siebenhofer, Department of Chemical Engineering and Environmental Technology, Graz University of Technology, Graz, Austria

Biochar-to-Liquid (B_ctL) – a pathway for producing energy carriers

The European fuel market is facing two major challenges. One challenge is the rising need of motor fuels which is at present satisfied with rising imports. The second challenge is to meet the directives of the European Commission to increase the amount of biofuels within the field of motor fuels to 10% until 2020 [1].

This target cannot be achieved with industrially established technologies such as production of biofuels (biodiesel, bioethanol). New complementary technologies are required to aim this goal. One of the main suggestions is the usage of whole plant energy which is not in competition with the food industry. Possible feedstock with respect to the total usage of the plant is lignocellulose biomass like wood or straw.

Gasification of the feedstock and liquefaction by Fischer-Tropsch-Synthesis is an established pathway utilized for example by Coren Company (GER). Although gasification and liquefaction by Fischer-Tropsch Synthesis are very complex processes with an overall efficiency of 50% [2], it is favoured at present.

Another promising route of producing liquid energy carriers from biomass is Liquid-Phase-Pyrolysis. Preliminary research work in this field was accomplished with promising results [3]. Beside condensibles biochar is a major product of Liquid-Phase-Pyrolysis with a broad field of applications.

The specific surface area of biochar gives access to chemical modification even including intermediates for fuel synthesis. In comparison with Fischer-Tropsch synthesis direct liquefaction of biochar needs less upgrade of the feed. The direct liquefaction is a heterogeneous catalytic high temperature/high pressure process. Several catalysts and coal based feed was investigated in the past [4].

Focus of the present project is the validation of liquefaction processes with biochar and with selected catalysts. Liquefaction by catalytic hydrogenation is carried out in a batch reactor of 1100 ml volume. The reactor can be operated at a maximum pressure of 20 MPa and a maximum temperature of 400 °C. Batch size is 25g to 50g of biochar dissolved in n-alkane carrier. Prior to liquefaction biochar was carefully mixed/impregnated with the catalyst (5 wt% at maximum) and dried to minimize physically dissolved water. Overall kinetics of liquefaction is deduced from hydrogen consumption records.

With molybdenum oxide based catalysts and hydrogenation at T=370 °C and P = 18 MPa the formation of a wide range of alkanes <C30 as well as cycloalkanes is observed. Latter substances are assumed to be derivatives of the lignin structure. Asphaltenes could not yet be identified but are still subject of investigations, because adsorption on the solid residues can not yet be excluded.

Beside oxide based catalysts sulphide based catalysts are subject of catalyst screening. In both cases the recycling of catalysts as well as product poisoning is governing direction of investigations.

[1] **2009/28/EC, EU Directive.** Directive on the promotion of the use of energy from renewable sources. Strasbourg : European Union, 2009.

[2] **Coren, Company.** Information & Press. [Online] 24. 01 2011. http://www.choren.com/informationen/faq/.

[3] **Verena, Mertlitz.** *Dissertation: Flüssigphasen-Pyrolyse biogener Edukte.* Graz : TU Graz, 2010.

[4] **Sol, Weller et.al.** Coal Hydrogenation Catalysts Batch Autoclave Tests. *Industrial & Engineering Chemistry.* 42, 1950, 330-334.

See more of this Session: Catalytic Biofuels Refining II
See more of this Group/Topical: Fuels and Petrochemicals Division

(523c) Alkaline Deacylation of Corn Stover to Reduce Toxicity of Prehydrolyzates and Improve Enzymatic Hydrolysis

Urvi D. Kothari and Y. Y. Lee, Chemical Engineering, Auburn University, Auburn, AL

Hydrolysates from dilute acid pretreatment of biomass contain various biomass degradation products which are inhibitory to enzymes and toxic to microorganisms. One of the main inhibitory compounds is acetic acid which is released from the lignin-hemicellulose matrix in the biomass. Acetic acid at the level of 1 g/L inhibits cellulase activity by 10%. In most pretreatment liquors its concentration is in the range of 3-10 g/L and inhibition of cellulase activity by acetic acid at these concentrations is 20-60%. In this study, sodium carbonate was investigated as a deacylation reagent that can selectively remove part of the acetyl components in corn stover and reduce the acetyl concentration in the pretreatment hydrolysate. Sodium carbonate was chosen since it is a mild alkali that can preserve carbohydrates under the deacylation process conditions. It is inexpensive and can easily be recovered through established Kraft recovery process. Sodium carbonate of 5-10 wt% was used over 30-60°C for deacylation. A dilute-acid pretreatment was then applied under the conditions of 0.5-1% H_2SO_4 at 60-110°C in order to assess the effects of deacylation. The two-step process operated under optimized conditions led to reduction of the final acetyl concentration by 50-80%, and minimal loss of hemicellulose at 2-12%. The combined process further improved the glucan and xylan hydrolysis yields of the pretreated corn stover over those of the dilute-acid pretreatment alone.

See more of this Session: Developments In the Pretreatment of Lignocellulosics for Bioconversion II
See more of this Group/Topical: Sustainable Engineering Forum

(13d) Catalytic Cracking of Oak Pyrolytic Vapors Via Bench-Scale In-Situ Fixed-Bed Catalysis

David J. Mihalcik, Akwasi A. Boateng, Charles A. Mullen and Neil Goldberg, Eastern Regional Research Center, Agricultural Research Service, USDA, Wyndmoor, PA

Pyrolysis oils derived from lignocellulosic feedstocks by way of fast pyrolysis offer the possibility of a distinct alternative to the use of petroleum-based fuels. However, an inherent drawback associated with pyrolysis oil its high oxygen content which is due to the nature of the lignocellulosic biomass from which it derives. Extensive efforts are being put forth to treat pyrolytic vapors in situ using zeolite catalysts. For our experiment, the catalytic conversion of oak pyrolysis vapors was performed at 500°C using a 2.5 kg/hr fluidized-bed fast pyrolysis reactor that was fitted with a re-heated (425°C) split-stream fixed-bed catalytic cracker containing zeolite-type catalyst. The placement of the split-stream fixed-bed catalytic reactor was varied within the condenser train to determine the effect of the removal of increasing amounts of water and oxygenated components from within the pyrolytic vapor stream. The cracked pyrolytic vapors resulting from the fixed-bed catalyst were collected by way of immediate contact with a dry-ice acetone bath followed by secondary collection using a methanol spray condenser. Karl Fisher titration and elemental and quantitative GC/MS analyses were performed to determine relevant trends within the experiment. The recovered liquid products that passed through the fixed-bed catalytic reactor displayed a remarkable decrease in the amount of oxygen containing species and a significant increase in the presence of aromatic hydrocarbons.

See more of this Session: Catalytic Conversion of Renewable Resources to Synthesis Gases and Pyrolysis Oils
See more of this Group/Topical: Sustainable Engineering Forum

A Novel Reactive Distillation Approach for Producing 1,3-Dihydroxyacetone from Glycerol

Xi Hong, Carl T. Lira, and Dennis J. Miller
Department of Chemical Engineering and Materials Science, Michigan State University
East Lansing, Michigan 48824

Introduction

Glycerol is a viable renewable feedstock that has emerged as a consequence of triglyceride conversion to biodiesel. In addition to glycerol conversion to commodity chemicals such as propylene glycol and epichlorohydrin, formation of higher-value specialty chemical products such as 1,3-dihydroxyacetone (DHA) will provide income for the biorefinery.

DHA is the active ingredient in sunless tanning products. Currently, DHA is industrially produced by fermentation of glycerol, but this microbial approach is challenging because of low production rates and high recovery costs [1]. As a result, the high selling price of DHA limits access of sunless tanning products in the general population.

Recently, scientists have attempted conventional chemical oxidation processes to make DHA from glycerol using metal catalysts with air or molecular oxygen. The goal is to achieve direct oxidization of the secondary hydroxyl group in glycerol to a carbonyl bond. Unfortunately, this reaction is always accompanied with primary alcohol oxidation and a variety of products (i.e. glyceroaldehyde, glyceric acid, etc.) are produced. The highest yield obtained from this method is in the range of 40-60% [2-4].

We present here a novel chemical process to convert glycerol to DHA that greatly improves the efficiency and reduces cost. The essential feature of this process is the selective formation of a 6-membered acetal of glycerol via reactive distillation, followed by selective oxidation of the secondary alcohol to its corresponding ketone product using ruthenium metal and a chemical oxidant. The ketone product is then hydrolyzed to obtain DHA.

A continuous process of making DHA from glycerol has been designed and will be presented. Reactions involved in the process are rapid and can be carried out under mild conditions. Yields of individual process steps are all in excess of 90%, suggesting that this process has good potential to reduce DHA production costs.

Experiments

Materials

Anhydrous glycerol (99.5%), 1-propanol (99.9%) and 1-butanol (99.9%) were obtained from J.T. Baker. Ethanol (99.5%), 1,1-diethoxyethane (99%) and sodium hypochlorite (15% aqueous solution) were purchased from Pharmco and Aldrich, respectively. Sodium carbonate (> 99%), sodium bromate (> 99%), potassium bromate (> 99.8%), calcium hypochlorite (> 98%), sodium periodate (> 99%), sodium iodide (> 99%) were purchased from Sigma-Aldrich. Amberlyst-15 cationic exchange resin (16-50 mesh), purchased from Sigma-Aldrich, was investigated as the catalyst. Before being used in reactions, it was pretreated at 90°C in a vacuum oven for 2 hours. Ruthenium dioxide hydrate (99.9% trace metals basis, 28% water content) and ruthenium trichloride hydrate (reagent plus, 46% Ru content) were obtained from Aldrich. Hydrogen (99.999%) and helium (99.999%) were used for GC analysis; both were purchased from Linde Gas, Inc.

Analysis

Gas Chromatography Analyses

1-Butanol and 1-propanol were used as internal standard and solvent, respectively, for this analysis. To prepare samples for analysis, 0.16 ml 1-butanol and 0.2 ml of reaction sample were added to a volumetric flask and the mixture was diluted with 1-propanol to 10 ml of solution. A Perkin-Elmer Sigma 2000 gas chromatograph with FID and a DB-Wax capillary column (15m × 0.53mm × 1.0 μm) was used for analysis. Ultra high purity helium was used as the carrier gas; the column temperature was ramped from 40°C (3min) at a rate of 30°C/min to 220°C and held for 1 min. The total program time was 10 min. On column injection was applied to analyze the samples. Injector and detector chamber temperatures were set at 60°C and 300°C, respectively.

GC-MS analysis

The structure of some products was identified using GC-MS analysis (GC-Agilent 6890, MS-Agilent 5973). A 1μL sample was separated in a DB-wax capillary column (30m × 0.25mm × 0.25 μm). The initial column temperature was 40 °C (for 5min), the temperature was then increased at 20 °C/min to 250 °C and held for 10 min. Total program time was 24.5 min. Injector temperature was set at 240 °C. Electron impact (EI) mass spectra were monitored from 20 to 300 m/z. The MS source was set at 230 °C.

Results and discussion

Synthesis of glycerol acetal

GA isomers were synthesized at 45 °C by mixing glycerol (3.1 kg, 33.9 mol), DEE (4.0 kg, 33.9 mol), and Amberlyst-15 cationic exchange resin (120 g) and stirring for 4 h. Following reaction and removal of volatile species (DEE and ethanol) by rotary evaporation, GC-MS analysis of the product mixture showed the presence of four different structures corresponding [5, 6] to the four GA isomers displayed in Table 1.

Table 1. Structure and IUPAC name of GA isomers

	Structure	IUPAC Name	Designation
a)	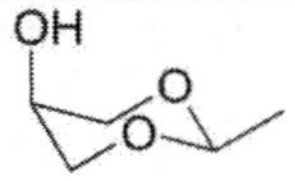	cis-5-hydroxy-2-methyl-1,3-dioxane	Cis-6
b)	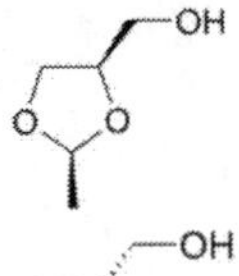	cis-4-hydroxymethyl-2-methyl-1,3-dioxolane	Cis-5
c)	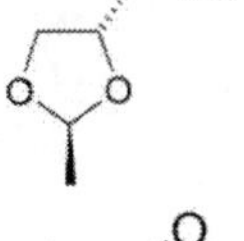	trans-4-hydroxymethyl-2-methyl-1,3-dioxolane	Trans-5
d)	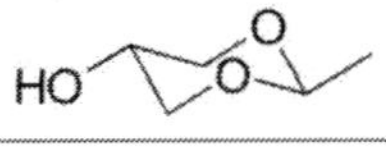	trans-5-hydroxy-2-methyl-1,3-dioxane	Trans-6

GA isomerization equilibrium

In the presence of acid catalyst [7, 8], the four GA isomers interconvert and reach an equilibrated composition. To demonstrate this equilibration, three samples obtained from the distillation experiments below with widely differing isomer compositions (cis-6 from 0% to 84%, cis-5 from 2% to 54%, trans-5 from 2% to 41%, and trans-6 from 0% to 63%) were stirred for 5 hr at 25 °C in the presence of 5 wt% (relative to GA) Amberlyst-15 cationic exchange resin. Identical isomer compositions were observed in the products from the three samples: cis-6, cis-5, trans-5 and trans-6 were 37±1%, 17±1%, 13±1%, and 33±1%, respectively. The molar percentages of 6-membered isomers (cis-6 + trans-6) and 5-membered isomers (cis-6 + trans-5) are approximately 70% and 30%, respectively, at equilibrium.

Rapid isomerization requires vigorous agitation (to ensure mass transfer between solid catalyst and solution) and elevated temperature. For example, Sample 1 took 12 h to reach equilibrium at 0 °C, but only one hour at 80 °C. Increasing temperature also shifts the equilibrium composition. For Sample 1, increasing temperature from 0 °C to 80 °C decreased the sum of cis-6 + trans-6 isomers from 78% to 54%. However, the total quantity of the cis- form isomers (cis-6 + cis-5) and trans- form isomers (trans-6 + trans-5) isomers remained essentially constant over all temperatures at 55±2% and 45±2%, respectively.

Production of the cis-6 GA isomer

A glass vacuum distillation column was constructed to separate the four GA isomers. The column (height = 3.6 m, i.d. = 50 mm) was packed with 0.6 cm 316 Stainless Steel PRO-PAK distillation packing (Scientific Development Co.). The column reboiler, consisting of a two-liter round-bottom flask, was heated with a 750 W electric heating mantle. Vapor temperature was measured at the top of the column.

In the initial distillation, 896 g (800 ml) of a GA mixture containing 35% cis-6, 18% cis-5, 15% trans-5, and 32% trans-6 isomers was placed in the reboiler. To avoid isomer interconversion, the solution was maintained at a weakly basic state by adding 50 g of sodium carbonate to the mixture. This experiment was carried out at a pressure of 20.5 ± 1.0 mm Hg (except for Sample 1, where the pressure was temporarily reduced to 9.0 mmHg).

Twelve 60-mL distillate samples were collected over the course of the distillation and analyzed by GC. Ninety percent of the initial charge to the still pot was recovered as distillate in these samples (Table 2). The material remaining in the still pot consisted primarily of the trans-6 isomer and glycerol; inclusion of this material gave essentially complete (100±2%) closure of the material balance for the experiment.

Table 2. Summary of GA isomers recovered in the initial distillation

	Initial GA charge		Product recovery in distillate		Percent recovered in distillate (%)
	Mass (g)	Composition (wt%)	Mass (g)	Percent of total	
Cis-6	311	35	288	36	93
Cis-5	166	18	152	19	92
Trans-5	132	15	120	15	91
Trans-6	287	32	246	30	86
Total	896	100	806	100	90

Figure 1 gives the reboiler liquid temperature, vapor temperature at the top of the column, and composition (via GC) of the distillate samples taken during the distillation. The results clearly show the order of decreasing volatility of the GA isomers as cis-6, cis-5, trans-5, and finally trans-6. As the more volatile isomers are removed from the still pot, liquid and vapor temperature gradually increase.

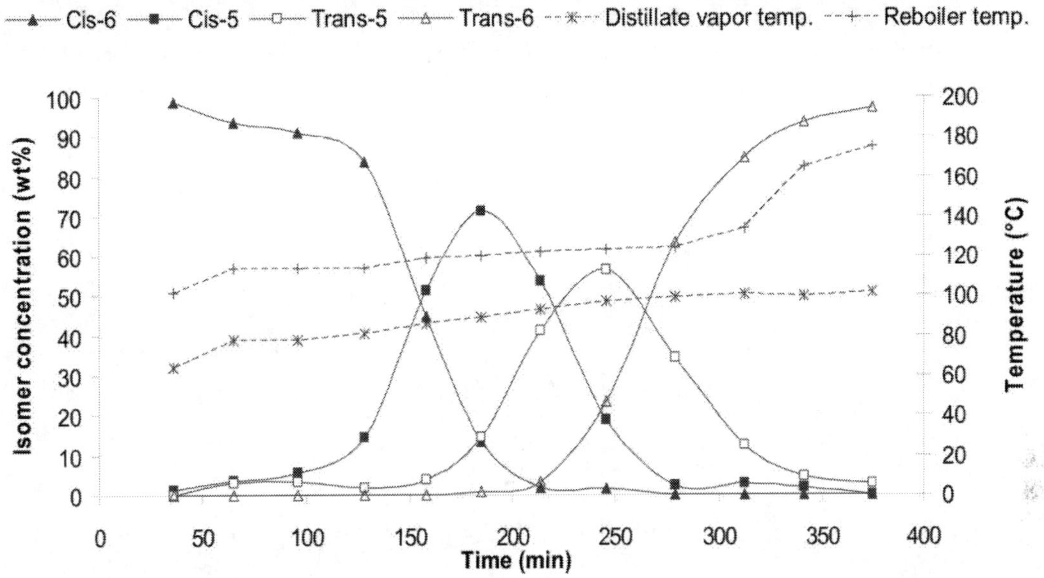

Figure 1. Initial distillation of GA isomer. Column pressure was 20.5±1.0 mmHg (except for the first sample, where the pressure was 9.0 mmHg).

Based on the result of the initial distillation and the fact that GA isomers can interconvert to each other and reach an equilibrated isomer composition in the presence of acid catalyst, a continuous reactive distillation experiment was then carried out to produce the cis-6 isomer as the sole product. In this run, 504 g (450 mL) of a mixture of 36 wt% cis-6, 16 wt% cis-5, 14 wt% trans-5, and 34 wt% trans-6 GA isomers was charged to the reboiler pot of the glass column along with 50 g of Amberlyst-15 cationic exchange resin. For this experiment, the column pressure was maintained at 2.9±0.2 mmHg. Once the distillate product started to condense at the top of the column, a continuous feed of the same GA isomer mixture was introduced to the column reboiler to maintain its liquid level constant. The feed rate was adjusted to match the distillate rate from the column, which was approximately 1.8 mL/min. During the distillation, 16 distillate samples were collected with an average volume of 59 mL.

As shown in Figure 2, this experiment produced samples with high concentration of cis-6 isomer (from 90% to 96%) continuously during operation of the column. The reboiler temperature and distillate vapor temperature were essentially constant during operation, indicating that the reboiler composition was constant and that the column was operating at steady state. The total quantity of distillate obtained was 1028 g, more than twice the amount of the initial charge to the reboiler, demonstrating that continuous production of the cis-6 isomer from the GA acetal mixture is feasible. Based on these results, it is fully expected that an appropriately designed commercial-scale column with sufficient separation capacity would provide essentially pure cis-6 isomer.

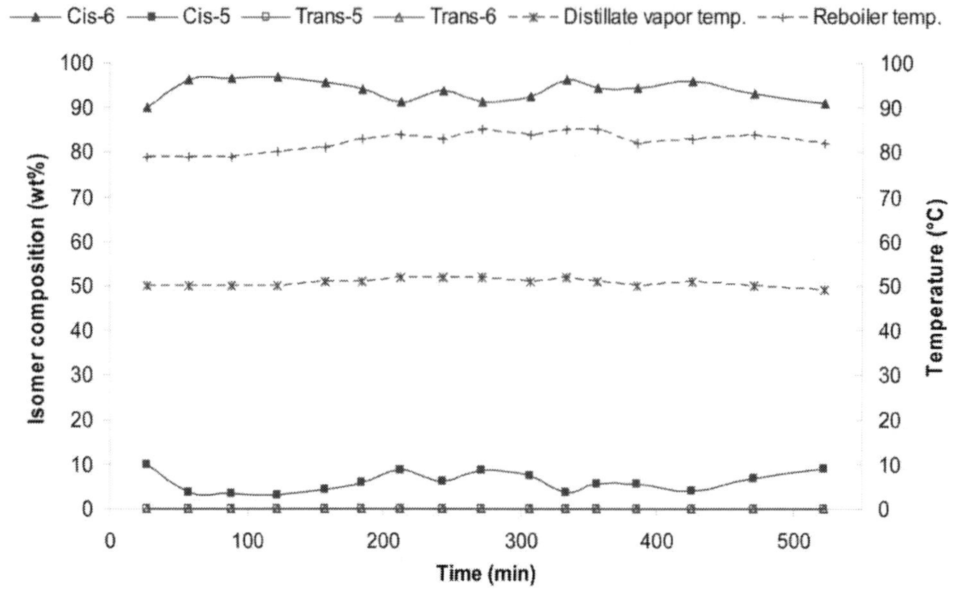

Figure 2. Continuous reactive distillation of GA isomers. Column pressure = 2.9±0.2 mm Hg.

Oxidation of cis-6 to the precursor of DHA

With the availability of cis-6 isomer, a new synthesis pathway can be developed to obtain valuable chemicals. For instance, by oxidizing the remaining free hydroxyl group in cis-6 and hydrolyzing the corresponding ketone product (2-methyl-1,3-dioxan-5-one, G-tone), 1,3-dihydroxyacetone (DHA) can be produced [7].

For this purpose, a variety of oxidation reagents including pyridinium chlorochromate (PCC), ruthenium tetroxide, hydrogen peroxide, molecular oxygen, and ozone were tested to convert cis-6 into its ketone product.

The most effective method is the ruthenium tetroxide-catalyzed oxidation. The reagents used in this reaction involves a ruthenium species (i.e. $RuCl_3$ and RuO_2) and another oxidizing salt (i.e. NaClO, $NaBrO_3$, $NaIO_4$, $Ca(ClO)_2$, $KBrO_3$, etc.) During reaction, the ruthenium species is first oxidized to RuO_4, and then RuO_4 oxidizes cis-6 to the corresponding ketone product. The co-product is a black precipitate of RuO_2, which can be re-oxidized to RuO_4 by the oxidizing salt in the aqueous solution. The regenerated RuO_4 then enters the organic phase again for another cycle of oxidation. In this way, the alcohol substrate can be continuously oxidized to the desired product with catalytic amounts of RuO_4 while the secondary oxidant is consumed. When the reaction is finished, the residue RuO_4 species in the organic phase can be easily recovered by adding isopropanol to make it RuO_2 precipitate [9].

Table 3 presents the results of cis-6 oxidation experiments catalyzed by ruthenium-based substances. It is clear that this type of reaction very efficient. In some reactions over 90% yield of G-tone was achieved within 30 min.

Table 3. Ruthenium tetroxide catalyzed oxidation of cis-6 at 25°C, 1.0 atm.

Reaction	Cis-6 (g)	Water (g)	Catalyst ($RuO_2 \cdot xH_2O$) loading (g)	Secondary oxidant solution/mass (g)/concentration	Time (min)	G-tone yield (%)
1	0.41	10	0	NaClO/4/10%	30	0
2	0.42	10	0	$KBrO_3$/10/saturated	1440	0
3	4.0	50	0.40	NaClO/100/10%	30	80
4	0.39	11	0.31	NaClO/4/10%	30	76
				NaClO/5/10%	60	95
5	0.45	10	0.05	$Ca(ClO)_2$/10/ saturated	20	87
6	0.40	10	0.30[a]	$NaIO_4$/10/saturated	75	88
7	0.42	10	0.26	$KBrO_3$/10/ saturated	45	> 99
8	0.43	10	0.32[a]	$KBrO_3$/10/saturated	75	95
9	4.1	10	0.40	$KBrO_3$/100/saturated	20	97
10	0.30	10	0.41	$NaBrO_3$/10/10%	20	96
11	10.0	10	0.51	$NaBrO_3$/100/20%	40	90
12	3.0	10	0.10	$NaBrO_3$/12.5/20%	20	95

a. $RuCl_3 \cdot xH_2O$ was used as catalyst.

Hydrolysis of G-tone to DHA

Hydrolysis of G-tone to DHA is very fast and can be carried out at mild conditions (Scheme 1). In one experiment, 5 wt% (weight to substrate) of Amberlyst-15 cationic exchange resin was added to the final mixture of Reaction 12 in Table 3. When the mixture was heated to 30°C, complete hydrolysis of G-tone was achieved in 10 min.

Scheme 1. Processes of catalytic conversion of glycerol into useful chemicals.

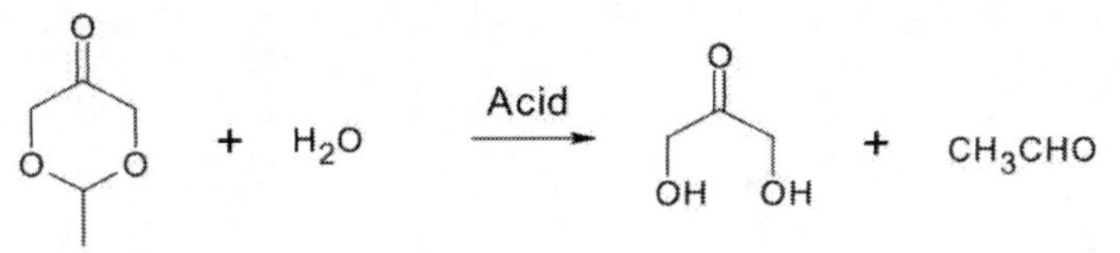

A novel process of making DHA from glycerol

Using cis-6 as an intermediate provides a relatively quick, environmentally safe and economical route to 1,3-dihydroxyacetone (DHA) from glycerol. Based on the experimental results obtained, a novel process of making DHA from glycerol is developed and described here.

Figure 3 presents the overall scheme of a novel DHA production process based on acetaldehyde or 1,1-diethoxyethane (DEE) as the glycerol acetal forming reagent. The process may also work with other aldehydes or ketones. Selection of the acetal forming reagent is dependent on price, availability and process feasibility.

In Figure 3, glycerol and acetaldehyde are introduced to Reactor 1 to form cyclic glycerol acetals. The reaction mixture, which contains four isomers of glycerol acetal, is then evaporated to get rid of water and unreacted acetaldehyde for recycling. After that, the mixture of glycerol acetal isomers is introduced to a vacuum distillation column, in which the cis-6 GA isomer is formed, distilled, and thus obtained from the top of the column. At the same time, other isomers, which have higher boiling points, react in the reboiler to re-establish isomerization equilibrium. The continuous recycling of the less-volatile glycerol acetal isomers through the equilibration-distillation procedure can transform all of the products into the desired 6-membered structure.

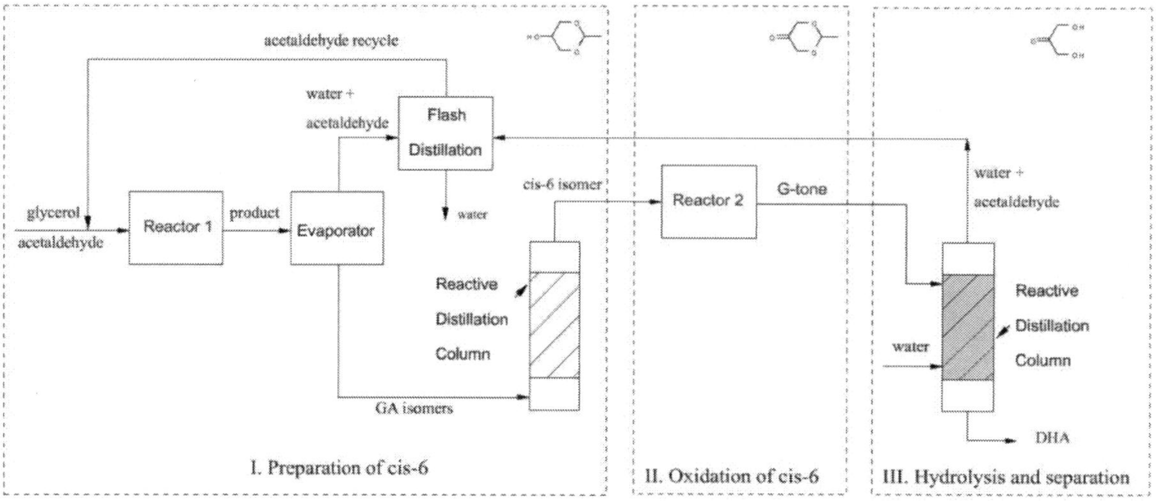

Figure 3. A novel process to produce DHA from glycerol.

In the second stage of this design, cis-6 GA isomer is introduced to the oxidation Reactor 2 to form 2-methyl-1,3-dioxan-5-one (G-tone). A preferred operation for this stage is to use ruthenium tetroxide-related oxidation.

In the last stage, G-tone and an excess amount of water are reacted, perhaps in a reactive-distillation column, to hydrolytically decompose G-tone to the desired product DHA and acetaldehyde, which can be recycled or reformed into DEE. The possible application of reactive distillation along with a large amount of water will overcome any chemical equilibrium limitations of this reaction, such that a high yield of DHA can be achieved from this process.

Summary

A novel process to make 1,3-dihydroxyacetone (DHA) from glycerol is presented in this paper. The process involves three major steps: 1) formation of glycerol acetal (GA) and reactive distillation to get the 6-membered GA isomer 5-hydroxyl-2-methyl-1,3-dioxane (cis-6); 2) oxidation of cis-6 GA isomer to the corresponding ketone product 2-methyl-1,3-dioxan-5-one (G-tone); 3) conversion of G-tone to DHA via hydrolysis. Experimental results show that in each step, over 90% yield of the desired product can be achieved. Thus, the new method holds a good potential for commercial success.

References

1. Mishra, R, Jain, S. and Kumar, A. (2008). Microbial production of dihydroxyacetone. Biotechnology Advances. 26, 293-303.
2. Garcia, R, Besson, M. and Gallezot, P. (1995). Chemoselective catalytic oxidation of glycerol with air on platinum metals. Applied Catalysis A: General. 127, 165-176.
3. Hu, W, Knight, D, Lowry, B. and Varma, A. (2010). Selective Oxidation of Glycerol to Dihydroxyacetone over Pt-Bi/C Catalyst: Optimization of Catalyst and Reaction Conditions. Industrial & Engineering Chemistry Research. 2010, 49, 10876–10882.
4. Kimura, H. (1993). Selective oxidation of glycerol on a platinum-bismuth catalyst by using a fixed bed reactor. Applied Catalysis A: General. 105, 147-158.
5. Da Silva Ferreira, A, Barbe, J. and Bertrand, A. (2002). Heterocyclic acetals f from glycerol and acetaldehyde in port wines: Evolution with aging. Journal of Agricultural and Food Chemistry. 50, 2560-2564.
6. Câmara, J, Marques, J, Alves, A. and Da Silva Ferreira, A. (2003). Heterocyclic acetals in Madeira wines. Analytical and Bioanalytical Chemistry. 375, 1221-1224.
7. Deutsch, J, Martin, A. and Lieske, H. (2007). Investigations on heterogeneously catalysed condensations of glycerol to cyclic acetals. Journal of Catalysis. 245, 428-435.
8. Aksnes, G, Albriktsen, P. and Juvvik, P. (1965). Studies of cyclic acetal and ketal isomers of glycerol. Acta Chemica Scandinavica. 19, 920-930.
9. Tojo, G. and Fernandez, M. (2006). Oxidation of alcohols to aldehydes and ketones. NY: Springer.

(498e) Hydrogen Production From Deliquescence of Ammonia Borane

Chang-Chen Chou[1], Cheng-Hong Liu[1], **Bing-Hung Chen**[1] and Duu-Jong Lee[2], (1)Department of Chemical Engineering, National Cheng Kung University, Tainan, Taiwan, (2)Department of Chemical Engineering, National Taiwan University of Science and Technology, Taipei, Taiwan

Chemical hydride is one of the most potential hydrogen storage materials which can effectively increase hydrogen storage capacity. To meet the new goal of U.S. department of energy (DoE), 4.5wt% and 5.5 wt% of gravimetric hydrogen density capacity by 2010 and 2015, respectively, ammonia borane (AB) is an attractive material due to high intrinsic hydrogen density (i.e. 19.6 wt%). Such a high intrinsic hydrogen density increases the possibility to apply to be the on-board power source of vehicle system, not only for 3C electric product.

In this study, hydrogen evolution from hydrolysis of ammonia borane in limiting water conditions is studied. The solid composite of ammonia borane hydrogen fuel is synthesized with pulverized catalyst through the high energy ball milling process. Various parameters such as the amount of water dosage, temperature, loading of catalysts, on hydrogen production characteristics will be discussed. Kinetic behavior of the hydrolysis reaction in such limiting water supply conditions will also be investigated. Based on experimental outcomes, optimal conditions on hydrogen production from this hydrogen storage system will be proposed.

See more of this Session: Aspects of Hydrogen Storage Materials
See more of this Group/Topical: International Congress on Energy 2011

(349b) Conversion of Algal and Duckweed Biomass to Liquid Fuels Through Catalytic Fast Pyrolysis

Alejandrina Campanella and Michael P. Harold, Department of Chemical and Biomolecular Engineering, University of Houston, Houston, TX

Microalgae and duckweed constitute a potentially viable alternative for replacing fossil fuels, given their economic and environmental benefits. In this study fast pyrolysis of microalgae was studied to generate a third-generation bio-fuel, with the goal of producing a stable liquid fuel that has acceptable energy content and composition. The impacts of the relevant process variables on the pyrolysis of microalgae and duckweed process were studied in detail. The pyrolysis reactions were carried out in out in a continuous fixed bed reactor in which an auger screw feeder was used to feed biomass particles into the hot reaction zone set at a prescribed temperature. Product vapors were condensed and collected for analysis employing a GCMS and elemental analyzer, while the post-condenser gas phase was evaluated by a combination of FTIR, GC, and FID. The yields and chemical species distribution were shown to be dependent on both the feedstock type primarily and the pyrolysis processing conditions secondarily. The analyses reveal that the pyrolytic liquid has a large fraction of oxygenated compounds which undermine the liquid chemical stability. To this end, upgrading of the bio-oil is necessary to improve stability and the heating value. Pyrolysis coupled with catalytic upgrading was utilized to achieve a pyrolitic oil with improved properties. Several metal-exchanged MFI framework zeolites (e.g. Fe-ZSM-5, Cu-ZSM-5 and Ni-ZSM5) and H-ZSM-5 were compared in terms of the enhancement in the stability. The catalyst performance studies showed an increase in the fraction of hydrocarbons and decrease of oxygenates, with H-ZSM-5 having the best performance. Notably, this catalyst produce a higher bio-oil fraction compared to the other zeolites and, also, exhibited the least coking. The effects of biomass weight hourly space velocity, and the comparison between zeolite powder and monolithic catalysts (i.e. synthesized with one or more layers of washcoated films containing the active catalyst) will also be presented.

See more of this Session: Biomass Pyrolysis I
See more of this Group/Topical: Fuels and Petrochemicals Division

(647d) An Integrated Biofuels Strategy: Catalytic Conversion of Lignocellulosic Biomass to Liquid Hydrocarbon Fuels

Sercan Murat Sen[1], Carlos A. Henao[2], Elif I. Gurbuz[1], David Martin Alonso[1], James A. Dumesic[1] and Christos Maravelias[3], (1)Chemical and Biological Engineering, University of Wisconsin-Madison, Madison, WI, (2)Chemical and Biological Engineering, University of Wisconsin - Madison, Madison, WI, (3)University of Wisconsin, Madison, WI

An integrated biofuels strategy: Catalytic conversionof lignocellulosic biomass to liquid hydrocarbon fuels

Replacement of fossil fuels with new sustainableresources becomes crucial due to depletion of petroleum reserves, increasingglobal energy demand and arising environmental concerns. Lignocellulosicbiomass provides sustainable and environmentally friendly ways of producingchemicals and fuels as an alternative for fossil fuels. One critical step isthe conversion of lignocellulosic biomass to versatile intermediates such aslevulinic acid (LA), which can be transformed into liquid fuels, fuel additivesand even other specialty chemicals. In this respect, we studied a LA-basedcatalytic process to convert lignocellulosic biomass into liquid hydrocarbonfuels for use in the transportation sector. Using experimental results for allassociated reactions, we synthesized an integrated biomass-to-fuels strategythat has a number of advantages over existing strategies.

The first step of this process is the removal ofhemicellulose fraction of the biomass by using dilute acid pretreatment. After thehemicellulose fraction (that mainly contains xylose) is removed, the celluloseportion of the biomass is converted to levulinic acid (LA) and formic acid in acellulose deconstruction reactor using sulfuric acid. The insoluble lignin and C_5-sugarpolymers are sent to a boiler/ turbogenerator to produce heat and electricity,covering the utility requirements of our process. Excess electricity is sold tothe grid. LA is converted to γ-valerolactone (GVL) over a Ru/C or RuRe/Ccatalyst in the presence of sulfuric acid. GVL is extracted from the sulfuricacid and GVL aqueous solution using butyl acetate solvent and sulfuric acid isrecycled back to the cellulose deconstruction reactor. Finally, GVL isseparated from butyl acetate by using a distillation column and purified GVL isconverted to butene and to butene oligomers.

To determine the economic potential of thisstrategy, we carried out detailed process simulation (based on experimentalresults) and capital/operational cost calculations. A comparison with alignocellulosic ethanol production facility reveals the potential feasibilityof the LA-based catalytic approach. Various feedstock alternatives (cornstover, sugarcane, wheat straw, hybrid poplar, switchgrass, loblolly pine andaspen wood) were analyzed and compared for cost-effective processing. Loblollypine was identified to be the most cost-effective feedstock due to its high C_6sugar content. Using loblolly pine as the feedstock, the minimum sellingprice (MSP) of butene oligomers was calculated as $4.31 per gallon of gasolineequivalent. We also performed sensitivity analysis studies for several processparameters (e.g., feedstock cost, equipment cost etc.) as well as economicparameters (e.g. equipment lifespan, income tax rate, return on investmentdiscount rate) to determine the bottleneck of the base case design. Feedstockprice appears to be the major cost driver: a 20%change results in slightly more than 8% change in the MSP of oligomers . Itwas also shown that the MSP of alkenes are sensitive to variations in the costof the turbogenerator and economic parameters (equipment life span and the ROIdiscount rate). Finally,we present the results of an energy efficiency analysis for the proposedprocess. The energy efficiency of biofuel production from C_6 sugarcontents was equal to 44.5%.

The presence of sulfuric acid in the GVLproduction step causes a dramatic decrease in the catalytic activity of the Ru/Ccatalyst which is preferred because it is cheaper than the RuRe/C catalyst. Toaddress this shortcoming, we considered the production of hydrophobic esters (Sec-Butyl Levulinate and Sec-ButylFormate) from the reaction of levulinic acid and formic acid with butene. Thesource of reacting butene is mainly provided from the conversion of GVL tobutene and CO_2 by catalytic decarboxylation over an acid catalyst ina later step. Hydrophobic esters can be separated from sulfuric acid withoutneed of an energy intensive distillation and they are converted to GVL over adual-catalyst-bed system. This sulfuric acid management strategy provides downstream catalytic processing in theabsence of sulfuric acid and therefore with higher yields. We investigate theeconomic impact of integration of the proposed reactive extraction strategywith the original process.

See more of this Session: Energy Systems Design and Alternative Energy Sources I
See more of this Group/Topical: Computing and Systems Technology Division

(378b) Out-Gassing Study and Species Measurement for Cryogenic Carbon Fiber Hydrogen Storage Vessels

Joseph W. Reiter, Rajeshuni Ramesham and Alexander W. Raymond, Jet Propulsion Laboratory, Pasadena, CA

Out-gassingStudy and Species Measurement for Cryogenic Carbon Fiber Hydrogen Storage Vessels[1]

Joseph Reiter, Rajeshuni Ramesham, and AlexanderRaymond

JetPropulsion Laboratory, California Institute of Technology, Pasadena, CA 91109,USA

Cryo-adsorption andcryo-compressed hydrogen storage are two promising technologies for vehicularhydrogen storage. These low-temperature storagemethods have the benefits of increasing hydrogen density and materialadsorption capacity, which enhance both the gravimetric and volumetricperformance of these systems. In bothcases, carbon-fiber wrapped tanks appear to be preferable to metal pressurevessels for vehicular applications because of their higher gravimetriccapacity; however, maintaining adequate thermal isolation for sustaineddormancy at cryogenic temperatures is challenging with carbon compositevessels. Cryogenic pressure vessels aretypically surrounded by a vacuum jacket and multi-layer insulation (MLI.) The effectiveness of MLI depends on theextent to which high-vacuum is maintained in the vacuum jacket. Even moderate vacuum degradation can resultin significant gas conduction across the separated layers and will render theMLI ineffective. Carbon compositeout-gassing is studied in this preliminary investigation with relation to itseffects on vacuum insulation degradation.Species and total out-gassing rates will be presented on both agravimetric and on an areal basis using the sampleporous surface area provided by the BET (**Brunauer-Emmett-Teller**) method. Temperature dependence of the out-gassingrate will be deduced by conducting experiments at cryogenic operatingtemperature (77 K), room temperature (298 K), and elevated temperature (358 K). A discussion of the allowable out-gassingrates will be presented along with potential mitigation strategies.

Acknowledgments:The researchwork described in this paper has been carried out at the Jet Propulsion Laboratory,California Institute of Technology, Pasadena, CA. We acknowledge theDepartment of Energy (DOE) for supporting this effort.

See more of this Session: Hydrogen Storage Systems: Designs, Analysis, and Safety Aspects

(560b) Discovery and Characterization of Cellulolytic Enzymes From the Wood Wasp Symbiont Streptomyces Sp. ActE

Taichi E. Takasuka, Adam J Brook, Cameron R. Currie and Brian G. Fox, Great Lakes Bioenergy Center, University of Wisconsin, Madison, WI

One of the major hurdles for producing cost effective biofuels is the enzymatic conversion of lignocellulosic biomass into fermentable sugars. The US Department of Energy has issued the challenge of discovering and designing enzymes with novel biomass-deconstruction capabilities. This study focuses on the discovery of better enzymes and accessory proteins for biomass deconstruction from a novel bacterial species. *Streptomyces* sp. ActE, an isolate from wood-boring wasps of the genus *Sirex*. Its genomic sequence has been determined and initial physiological studies suggested high cellulolytic potential. We analyzed the genome-wide gene expression of ActE grown on insoluble substrates such as cellulose, xylan, chitin and others as the sole carbon source in the culture medium, and compared these expression profiles with glucose-grown control ActE cells. The secretomes obtained from growth on different substrates were analyzed by shotgun LC-MS/MS and enzymatic assays. From the combination of the genomic, transcriptomic, and secretomic analyses, we discovered ~200 genes that have carbon source-specific regulation of gene expression or protein secretion. This work provides the first extensive systems biology characterization of the capabilities of the genus *Streptomyces* for wide variety of polysaccharide utilization.

This work was funded by the DOE Great Lakes Bioenergy Research Center (DOE BER Office of Science DE-FC02-07ER64494).

See more of this Session: Advances In Biofuels: DOE Bioenergy Research Centers II
See more of this Group/Topical: Sustainable Engineering Forum

(573b) Selective Dehydration of Polyols to Commodity Chemicals

Michael Nolan, School of Chemical and Biological Engineering, Iowa State University, Ames, IA and Brent H. Shanks, Chemical and Biological Engineering, Iowa State University, Ames, IA

Synthesis of commodity chemicals from biorenewable sources is an important component of a sustainable society. Due to the very different chemical nature of bio-based feedstocks, a fundamental understanding of the chemical conversion of bio-derived molecules will be necessary for production of biorenewable chemicals. One reaction of interest for biorenewable chemicals is the selective removal of oxygen from polyols. In this presentation, acid-catalyzed dehydration of a biorenewable model compound, 1,2,6-hexanetriol, to selectively produce precursors of 1,6-hexanediol, which is a large volume commodity chemical, will be discussed. To investigate the fundamental chemistry pathways leading to the desired product, the model compound was converted via gas phase dehydration over a wide variety of heterogeneous acid catalysts. Acid strength was found to have an important effect on activity regardless of catalyst morhphology, but did not necessarily impact selectivity. Interestingly, a pathway to ε-caprolactone was found, which is proposed to be through an intermediate generated in the reaction. This talk will include a discussion of catalytst properties leading to selective dehydration of the polyol, proposed reaction mechanisms and initial results from mechanistic studies. Finally, possible catalyst surface effects and self-reactions that could be applied to other polyol systems will be presented.

See more of this Session: Catalytic Biomass Conversion to Chemicals II
See more of this Group/Topical: Fuels and Petrochemicals Division

(62e) Single-Cell Growth Kinetics of Algae, Chlorela Vulgaris

Alim Dewan[1], Jihye Kim[2], Swastika S. Bithi[2], Marci Kerls[2], Siva A. Vanapalli[2] and M. Nazmul Karim[2], (1)Chemical Engineering, Texas Tech University, Lubbock, TX, (2)Chenical Engineering, Texas Tech University, Lubbock, TX

Single-cell growth kinetics of algae, *Chlorela vulgaris*

AlimDewan, JihyeKim, Swastika S. Bithi, Marci Kerls, Siva A.Vanapalli, M. Nazmul Karim*

Departmentof Chemical Engineering, Texas Tech University, Lubbock, Texas 79409, USA.

*E-mail: naz.karim@ttu.edu

Abstract

Traditionally,the growth kinetics of algae is investigated using cell culture equipment thatcontains large number of cells. The kinetic parameters are measured fromrepresentative samples of the cell culture and the measured parametersrepresent the statistical average of all the cells in the samples. In thisstudy, we have investigated the growth kinetics of single-cell of *Chlorellavulgaris* using droplet based PDMS microfluidic chip. The kinetic parameterswere determined by counting the total number of cells under investigationduring the entire period of the experiment. We found that there is significantheterogeneity in the growth and size of single cells and kinetic parameters aredifferent even if the initial cell numbers are the same. We found that thereis significant heterogeneity in the growth and size of single cells and kineticparameters are different even if the initial cell numbers are the same. Figure1exmplifies the heterogeneity of single-cell growth kinetics. We hypothesizethat the heterogeneity occurs due to the difference in initial cell conditionssuch as the age of the cell, and the size.

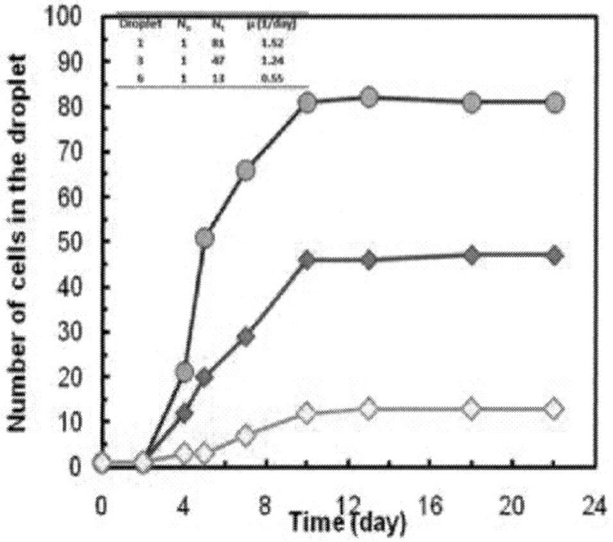

Figure1.Example of single-cell growth curves. Inset shows the initial cell number (N_o),final cell number (N_t) and the specific growth rate ().

See more of this Session: Advances In Algal Biorefineries I
See more of this Group/Topical: Sustainable Engineering Forum

(690a) Development of New Redox Flow Battery Systems

Liyu Li[1], Wei Wang[2], Soowhan Kim[2], Vijayakumar Murugesan[2], Zimin Nie[2], Baowei Chen[2], Qingtao Luo[2], Feng Chen[2], Jun Liu[2], Jianzhi Hu[2] and Gary Yang[2], (1)Hydrocarbon Processing Group, Pacific Northwest National Laboratory, Richland, WA, (2)Pacific Northwest National Laboratory, Richland, WA

Concerns over the environmental consequences of burning fossil fuels and their remaining reserves have led to an increasing use of renewable energy generated from sources such as solar and wind. However, the intermittent, varied nature of the renewable resources makes it difficult to integrate these valuable energies into the electrical grids. A solution to this problem is to employ large-scale electrical energy storage (EES), which is widely considered as an effective approach to improve the reliability, power quality, and economy of these renewable energies. Among the most promising large-scale EES technologies are the redox flow batteries (RFBs). RFBs are special electrochemical systems that can repeatedly convert and store multi-megawatt-hours (MWhs) of electricity energy to chemical energy and convert the chemical energy back to electricity energy when needed. The advantages of RFBs include short response time, low self discharge, long life-time, and independent tunable power and storage capacity, which make it possible to independently scale up the electricity storage capacity and power generation capacity. With the support of US Department of Energy's Office of Electricity, PNNL has carried out extensive efforts in the development of advanced redox flow batteries. Substantial progress has been achieved in the past two years, including two new redox chemistries (V-Fe system and all vanadium mixed acid system), which will be reported in this presentation

See more of this Session: Energy Storage and Electrochemical Processes I
See more of this Group/Topical: International Congress on Energy 2011

(373b) In Situ Electrical Characterization of a High Temperature PEM Fuel Cell Stack At Loads

Ying Zhu, Wenhua H. Zhu and Bruce J. Tatarchuk, Center for Microfibrous Materials, Department of Chemical Engineering, Auburn University, Auburn, AL

PEM fuel cells operated at temperature over 100°C has several advantages over traditional PEM fuel cells (generally operated at 80°C), including better reaction kinetics and higher CO tolerance. In this work, the Serenus fuel cell module was operated under ambient air and pure hydrogen at loads. The module contains a 65 planar cell stack with a nominal power around 1 kW. The area of each cell is ca. 285 cm^2. The operating temperature was varied from 140°C to 170°C and the current density was increased from OCV (open circuit voltage) up to 100 mA/cm^2. A small AC signal was superimposed on the fuel cell and the impedance data of the stack were collected by Gamry FC350 fuel cell monitor. The contributions of the different processes in the stack to the stack impedance have been mechanistically discriminated using the EC (equivalent circuit) diagram derived from the collected impedance. This work is useful for further study on kinetic mechanisms and cell performance of high temperature PEM fuel cells.

See more of this Session: Fuel Cell Technology II
See more of this Group/Topical: Fuels and Petrochemicals Division

(373c) Validation of the Equivalent Circuit Diagram for SOFC Modeling

Ying Zhu, Wenhua H. Zhu and Bruce J. Tatarchuk, Center for Microfibrous Materials, Department of Chemical Engineering, Auburn University, Auburn, AL

AC impedance spectroscopy, combined with equivalent circuit diagram (ECD) simulations, achieves to characterize a chemical reactor system in terms of electronics. Their application on solid oxide fuel cells (SOFCs) provides experimental data and physical models for cell designs, diagnostics, and performance improvements. However, different arrangements and combinations of circuit elements may yield the same dynamic response when three or more elements are involved in one EC diagram. To overcome the ambiguity in EC diagram interpretations, the impedances of the SOFCs were measured under different operating conditions. Temperature, current density and fuel composition were changed to validate the consistency of the derived EC diagram and to mechanistically distinguish the kinetics and processes limiting cell behaviors.

See more of this Session: Fuel Cell Technology II
See more of this Group/Topical: Fuels and Petrochemicals Division

(484a) Modeling of Long-Time Flux Decay In Cross-Flow Filtration

Philip P. Schonewill, Richard C. Daniel, Rick W. Shimskey, Justin M. Billing and Reid A. Peterson, Pacific Northwest National Laboratory, Richland, WA

Recent work at Pacific Northwest National Laboratory has been concerned with understanding how cross-flow filter flux decays with time. Understanding the cause(s) of flux decay is important for assessing long-term behavior of this separation technology, especially when used in challenging processing environments e.g. nuclear waste treatment. Fouling in the filter is a complex issue that is affected by many parameters: the chemistry of the solution being filtered, particle size and morphology, operating conditions, and cleaning methods. Based on filtration work performed at PNNL over the past several years, it is clear that a portion of the fouling can be "reversed" by back-pulsing the filter and another portion is irreversible (usually only rectified via cleaning protocols). However, the sheer number of parameters requiring consideration complicates effective modeling of both the long-time behavior of the cross-flow filter and its response to back-pulses.

To combat this, simple models were developed from first principles based on different suspected fouling mechanisms. The models were tested against long-term data (100-hour tests at a constant solids concentration) collected with PNNL's bench-scale filtration system. The flux was found to be well described by a cake mechanism model at intermediate times (starting at approximately 3 hours elapsed time); at longer times (approximately greater than 20 hours of filtration time) the flux behavior transitions to an intermediate mechanism behavior. The decay constant for these models can be predicted using a single dimensionless parameter, the ratio of the transmembrane pressure to the inertial pressure. Additional testing suggested that the filter could be conditioned by back-pulsing. These observations were used to build a comprehensive model to assist in determining a back-pulse strategy for cross-flow filters that will experience long-term decays in the filter flux.

See more of this Session: Theories and Applications of Cross Flow Filtration
See more of this Group/Topical: Separations Division

(738c) Modeling and Optimization of Cathode Layer Microstructures In Polymer Electrolyte Fuel Cells

Sesha Hari Vemuri[1], Parag Jain[2], Myung S. Jhon[1] and Lorenz T. Biegler[1], (1)Chemical Engineering, Carnegie Mellon University, Pittsburgh, PA, (2)UOP LLC - A Honeywell Company, Des Plaines, IL

The optimal design of catalyst layer (CL) sub-component is critical for achieving commercialization of polymer electrolyte fuel cells (PEFCs). Typically, a CL is designed as a heterogeneous porous composite of Platinum catalyst over carbon support, void space, and polymer electrolyte membrane (PEM) materials, through which heat & mass transport processes occur with electrochemical reaction. To overcome performance limitations within the CL, a detailed understanding of these transport and reaction processes is required. Especially, an accurate modeling & optimization tool that establishes relationship between CL micro-structures to overall PEFC performance can significantly expedite experimental efforts towards developing tailored CL structures. The non-homogeneous media theories [1-3] estimate the effective molecular transport coefficients through averaged phase hold-up and/or tortuosity parameters which do not account for the microscopic structural information. On the other hand, detailed sub-continuum theories may account for detailed CL structural geometry, they however, may not be suitable for systems level analysis.

In our earlier study [4], we successfully constructed a compact modeling and optimization framework comprising a macro-/meso-scale CL agglomerate model capturing essential features transport & reaction processes and linked to our in house state-of-the-art interior point optimization algorithm, IPOPT [5]. This framework was further exploited to examine PEFC model and optimization sensitivity for various CL agglomerate shapes (sphere, cylinder, and plate) and sizes [6].

Here, we analyze a generalized class of spherical shell agglomerate models originating from the approach by Aris [7], with a goal of providing a basis for realistic CL structures, and amenable to systems analysis.

First, we introduce non-homogeneity in distribution of triple phase boundaries (required for electrochemical reaction) by solving multi-zone spherical agglomerate reaction-diffusion problem, possessing radial variation in reaction rates. We examine sensitivity to PEFC performance in both simulation and optimization results for different zone thicknesses. *Second*, instead of the heuristic 'shape-factor' approach in Aris [7], we derive effectiveness of a spherical shell from first principles by introducing a Robin boundary condition, expressed in terms of Biot number (Bi), which plays an important role in estimating the effectiveness of CL layer. This generalized model gives a good agreement with Aris [7] for various values of shell thickness as Bi values approach to zero. Also, the case $Bi = 0$ for a solid sphere reduces to our previous model [4]. The optimization based on maximum current density is obtained in detail for different shell thicknesses and various values of *Bi*. We examine how *Bi* influences the critical system design parameters including the I-V characteristics.

1. J. C. Maxwell, A Treatise on Electricity and Magnetism (Clarendon, Oxford, 1881), Vol. 1, p. 435.
2. D.A.G. Brugemann, *Ann. Phys.*, 24, 637 (1935).
3. J. Wang, J. K. Carson, M. F. North, D. J. Cleland, *Int. J. Heat Mass Tran.*, 49, 3075 (2006).
4. P. Jain, L.T. Biegler, and M.S. Jhon, *Electrochem. Solid St.*, 11, B193 (2008).
5. A. Wächter, L.T. Biegler, *Math. Program.*, 106, 25 (2006).
6. P. Jain, L.T. Biegler, and M.S. Jhon, *J. Electrochem. Soc.*, 157, B1222 (2010).
7. R. Aris, *Chem. Eng. Sci.*, 6, 262 (1957).

See more of this Session: Energy Systems Design and Alternative Energy Sources II
See more of this Group/Topical: Computing and Systems Technology Division

(498b) A New Metal-Decorated Carbon Catalyst for Dehydrogenation of Complex Hydride Hydrogen Storage Materials

Sean S.-Y. Lin[1], Jun Yang[2] and Harold H. Kung[1], (1)Chemical and Biological Engineering, Northwestern University, Evanston, IL, (2)Fuel Cell and Hybrid Electric Vehicle Department, Ford Motor Company, Dearborn, MI

In recent years, many new solid-state hydrides with a high hydrogen capacity that meet DOE requirement (> 7.5 wt.% H_2) have been synthesized as promising on-board hydrogen storage materials.

In these hydrides, however, only a small fraction of hydrogen can be released at a desired on-board operating temperature (<85°C) due to a slow dehydrogenation kinetics of the hydrides. It has been discovered that the kinetics can be improved by adding transition metal catalysts to form metal-hydride intermediates that facilitate hydrogen transportation and desorption. Nevertheless, identification of such catalyst for various hydrides is a challenging task because the catalytic effect is component-specific. Here we report a new form of catalyst for dehydrogenation of complex hydride, using a well-studied $NaAlH_4$ as a test case. Dehydrogenation of $NaAlH_4$ can be greatly facilitated by activated carbon catalysts. The catalytic function can be further enhanced by decorating the carbon with Co, Ni, or Cu nanoparticles. The decomposition temperature was lowered by as much as 100°C using a 3 wt.% Co or Ni-decorated activated carbon, comparable to a Ti-based catalyst, which were the most effective among the metals tested. The catalytic effect is likely due to a combination of hydrogen spillover effect, high contact area between carbon and the hydride, and confinement of the hydride as nano-sized domains in the pores of the carbon matrix. The catalysts were also effective in facilitating rehydrogenation of $NaAlH_4$ under moderate pressure (75.8 bar H_2) and low temperature (120°C), when no rehydrogenation would occur without the catalyst. The fact that this new catalyst system is not specific to any hydride offers many potential applications.

See more of this Session: Aspects of Hydrogen Storage Materials
See more of this Group/Topical: International Congress on Energy 2011

(173e) Graphene-Encapsulated Hollow Fe3O4 Nanoparticle Aggregates As a High Performance Anode Material for Lithium Ion Batteries

Dongyun Chen[1], Ge Ji[1], Yue Ma[1], **Jim Yang Lee**[2] and Jianmei Lu[3], (1)Chemical & Biomolecular Engineering, National University of Singapore, Singapore, Singapore, (2)Department of Chemical and Biomolecular Engineering, National University of Singapore, Singapore, Singapore, (3)Chemical Engineering and Materials Science, Soochow University, Suzhou, China

Graphene-EncapsulatedHollow Fe_3O_4 NanoparticleAggregates as a High Performance Anode Material for Lithium Ion Batteries

Dongyun Chen[1,2], Ge Ji[1], Yue Ma[1], Jim Yang Lee[1] and Jianmei Lu[2]

-

[1]*Department of Chemical& Biomolecular Engineering, Faculty of Engineering, National University ofSingapore, 10 Kent Ridge Crescent, Singapore, 119260; Tel: (65) 65162186; E-mail:cheleejy@nus.edu.sg.*

[2]*Key Laboratory ofOrganic Synthesis of Jiangsu Province, Key Laboratory of Absorbent Materialsand Techniques for Environment, College of Chemical, Chemical Engineering andMaterials Science, Soochow University, Suzhou, China, 215123; E-mail: lujm@suda.edu.cn*

In this work we combined the unique propertiesof graphene sheets and a hollow assembly of nanoparticles to simultaneouslyprovide large reversible Li^+ storage capacity,good rate performance and long cycle life. Specifically graphene-encapsulatedordered aggregates of Fe_3O_4 nanoparticles withnearly-spherical geometry and hollow interior (G-HPM)were synthesized by a simple self-assembly process. TheHPM particles were first prepared by a one-pot hydrothermal synthesis. Theywere then modified by 3-aminopropyltriethoxysilane (KH550) to acquire a positivesurface charge, which enabled electrostatic assembly with negatively chargedgraphene oxide. Finally, a chemical reduction was used to reduce the grapheneoxide to graphene; finalizing the packaging of HPM particles in individualgraphene cages. This assembly could be carried out under relatively mildconditions; thereby minimizing the perturbations to the properties of HPM. Inour design, the open interior structure has adapted well to the volume changein repetitive Li^+ insertion and extraction reactions; and theencapsulating graphene connects the Fe_3O_4 nanoparticleselectronically. The composite particles showed a stable high specificreversible capacity of about 897 mA h g^{-1} which was nearly unvaryingover 50 cycles. These graphene-Fe_3O_4 composites are thereforea capable reversible Li^+ host with high storage capacity and goodcycle life which can be cycled at high rates.

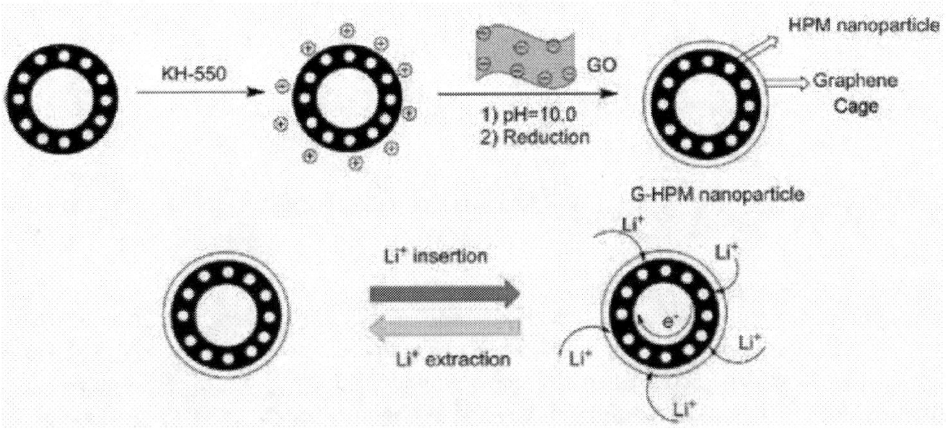

See more of this Session: Nanomaterials for Energy Storage III
See more of this Group/Topical: Topical 5: Nanomaterials for Energy Applications

(294g) Improvement of the PEMFC Performance Under Low Humidification Condition by Using the Hydrophobic/Hydrophilic Double Micro Porous Layer In a Cathode Gas Diffusion Layer

Jeong Hwan Chun, Dong Hyun Jo and Sung Hyun Kim, Department of Chemical & Biological Engineering, Korea University, Seoul, South Korea

Gas diffusion layer (GDL), which consists of a macro porous substrate (GDBL, gas diffusion backing layer) and micro porous layer (MPL), is the crucial component which is closely related with water management in PEMFC due to its basic function that is to drain out liquid water to catalyst layer from flow channels. The MPL, which is composed of carbon powder and a polymer binder, such as polytetrafluoroethylene (PTFE), is a hydrophobic layer that contains several micro pores, and these micro pores serve to improve the water removal ability. Thus, MPL has been recently been used to solve the mass transport problems in PEMFCs, which are caused by water flooding. In this study, a MPL was modified to improve the proton exchange membrane fuel cell (PEMFC) performance under low humidification condition.

We developed a novel hydrophobic/hydrophilic double micro porous layer (MPL) that was coated on a gas diffusion backing layer (GDBL). The carbon slurries for MPL were prepared using the following procedures. The carbon black (Vulcan XC-72, Cabot) was dispersed in a mixture of isopropyl alcohol and deionized water (80:20 volume ratio) using ultrasonication and mechanical stirring and then the polymer binder was added. The hydrophobicity of the MPL was controlled by changing the polymer binder. PTFE was used to fabricate the hydrophobic MPL, while another fluorinated polymer, which had a similar structure to PTFE but contained less fluorine, was used to fabricate the hydrophilic MPL. The hydrophobic/hydrophilic double MPLs (PA2ADL2 and PA2ADL3) were prepared in two steps. First, the hydrophobic (or hydrophilic) MPL slurry was applied onto the wet-proofed carbon fiber paper using screen printing technique. This was followed by a second coating with the hydrophilic (or hydrophobic) MPL slurry. The hydrophilic MPL was located on the surface of PA2ADL2 and in the middle of PA2ADL3. For comparison purposes, a single hydrophobic MPL (PA2AML1) was prepared.

The contact angle of the MPL surface was measured using the Sessile drop method (Phoenix 300, SEO Co.). The water and vapor permeability of GDL was characterized using a home built system. A single cell test was carried out using a 25 cm^2 active area test cell in humidified H_2 / air gases under low humidification conditions, R.H. 50%.

The best cell performance was obtained from the PA2ADL3, and PA2ADL1 resulted in the worst cell performance. In other words, both PA2ADL2 and PA2ADL3 containing hydrophilic MPL showed the better cell performance than PA2AML1, which contained a single hydrophobic MPL. We identified that PA2ADL2 and PA2ADL3 containing a hydrophilic MPL, retained a much larger amount of water than PA2AML1 through measuring water and vapor permeability of GDLs. This result demonstrated that the hydrophilic MPL had the ability to absorb significant amounts of water, like a sponge. PA2ADL3 had a hydrophobic MPL on its surface and a hydrophilic MPL in the middle of the GDL. The hydrophobic MPL surface prevented water flooding and the hydrophilic MPL, which was located inside of the GDL, absorbed water to prevent dehydration of the polymer electrolyte membrane. The polymer electrolyte membrane has to be hydrated for protons to be conducted from the anode to cathode. Under the low humidification condition, the polymer electrolyte membrane can be easily dehydrated, resulting in a decrease in cell performance. As the amount of water generated at the cathode catalyst layer decreases due to low cell performance, dehydration of the polymer electrolyte membrane can be accelerated. In the cathode GDL, air must diffuse from the flow channel to the catalyst layer through the MPL. If the MPL retains water, the air will be the more humidified while diffusing through the MPL. The hydrophilic MPL that was located in the middle of the GDL acted as an internal humidifier. Therefore, PA2ADL3 containing a hydrophilic MPL between the hydrophobic MPL and GDBL showed the best cell performance. In contrast, PA2ADL2 had a hydrophilic MPL on the surface; thus, water produced at the cathode catalyst layer can be easily stagnant between the cathode catalyst layer and surface of the PA2ADL2 when the cell was operated under the low humidification condition. Consequently, the stagnant water could cover up the active site in the catalyst layer, which was the reason why the performance of PA2ADL2 was lower than PA2ADL3

See more of this Session: Fuel Cell Technology I
See more of this Group/Topical: Fuels and Petrochemicals Division

(453c) Infrared Spectroscopy and Thermo Gravimetric Analysis for Cellulose and Residue by Hydrothermal Decomposition

Lidong Geng and Zhongjun Zhu, Chinese Society for Environmental Sciences, Beijing, China

Abstract: Under the background of energy and resource shortage and the environment pollution, the utilization and research of biomass wastes as a kind of resource attracting worldwide attention. Hydrothermal reaction provides an effective method for the reutilization of biomass wastes. The disposal and reutilization of biomass wastes were realized by hydrothermal reaction in batch hydrothermal reactor. By using Fourier transform spectrum analysis, the microcosmic structure of micro crystallite cellulose and its products were compared. The pyrolysis behavior of cellulose was studied by thermo gravimetric analysis (TGA). The experimental results showed that the molecular structure of micro crystallite cellulose changed obviously and the decomposition of micro crystallite cellulose mainly occurred at the temperature range of 250-400°C. The weight loss(TG) and the rate of weight loss (DTG) shift to higher temperature region with the increase of heating-up rate. The thermal dynamics for biomass wastes was coincidence to the first order reaction dynamics by Coat-Redfern analysis. The FTIR and TGA analysis indicated that the hydrothermal decomposition of micro crystallite cellulose was composed by low temperature hydrolysis and high temperature pyrolysis.

See more of this Session: Fundamentals of Biomass Utilization
See more of this Group/Topical: Environmental Division

(573a) Oxidation of Polyols Using Au-Pd and Au-Pt Alloy Supported Nanocrystals

Peter Miedziak[1], Gemma L. Brett[1], Yulia Ryabenkova[1], Nikolaos Dimitratos[1], Jose A. Lopez-Sanchez[1], Meenakshisundaram Sankar[1], Albert F. Carley[1], Ram C. Tiruvalam[2], Andrew A. Herzing[3], Chris K. Kiely[2], David W. Knight[1], Stuart H. Taylor[1] and Graham J. Hutchings[1], (1)Chemistry, Cardiff Catalysis Institute, Cardiff, United Kingdom, (2)Center for Advanced Materials and Nanotechnology, Lehigh University, Bethlehem, PA, (3)Surface and Microanalysis Science Division, National Institute of Standards and Technology, Gaithersburg

Oxidation of polyols using Au-Pd and Au-Pt alloy supported nanocrystals

Peter Miedziak[a], Gemma L. Brett[a], Yulia Ryabenkova[a], Nikolaos Dimitratos[a], Jose A. Lopez-Sanchez[a], Meenakshisundaram Sankar[a], Albert F. Carley[a], Ram C. Tiruvalam[b], Andrew A. Herzing[c], Christopher J. Kiely[b], David W. Knight[a], Stuart H. Taylor[a], and Graham J. Hutchings[a*]

[a] *School of Chemistry, Cardiff University, Cardiff Catalysis Institute,Main Building, Park Place, Cardiff, CF10 3AT, UK*

[b]*Center for Advanced Materials and Nanotechnology, Lehigh University, 5 East Packer Avenue, Bethlehem, PA 18015-3195, USA*

[c]*National Institute of Standards and Technology, Surface and Microanalysis Science Division, 100 Bureau Drive, Mailstop 8371, Gaithersburg, MD 20899-8371, USA*

Graham J. Hutchings: phone +44 (0)29 2087 4059, *fax:* +44 (0)29 2087 4030, *e- mail: hutch@cardiff. ac.uk*

Oxidation of organic compounds with oxidants such as air, molecular oxygen or hydrogen peroxide using supported metal catalysts for synthesising valuable intermediate products for chemical industry is well known. In recent years an increased interest has been focused on catalytic processes because these have demonstrated to be environmentally friendly and economically constructively in comparison to the stoichiometric reactions. The use of bio-renewable resources for the production of materials and chemicals continues to attract significant research attention and is highly desirable from the industry.

Glycerol, a by-product from bio-diesel manufacture is a highly functionalised renewable raw material. Oxidation of glycerol leads to the formation of a number of products, such as glyceric acid, dihydroxyacetone, b-hydroxypyruvic acid, tartronic acid, glycolic acid and oxalic acid, therefore control in the distribution of products by choosing the appropriate conditions and designing the proper catalyst is desirable [1,2,3].

Moreover, glycerol can be used alternatively for the synthesis of 1,2-propane diol which can then be selectively oxidised to lactate, which has immense potential as a monomer for biodegradable polymers. The current production of lactic acid is based on the fermentation and the reaction of acetaldehyde with HCN followed by hydrolysis with sulfuric acid, which is not a desired green process. Therefore, finding an environmentally green process is desired and this can be based on the use of heterogeneous catalysts with O_2 or H_2O_2 as oxidants. We have recently shown that the combination of gold and palladium in an alloy nanoparticle configuration leads to enhanced catalytic activity and selectivity in redox processes, such as alcohol oxidation and hydrogen peroxide synthesis [4,5]. We have shown that gold palladium alloy catalysts can be very effective for the selective oxidation of glycerol and 1,2-propanediol to glycerate and lactate [6,7]. We have extended these studies by using gold-platinum alloy catalysts. We have extensively studied the effect of support, nature of base and Au/Pt ratio in the liquid phase oxidation of glycerol and 1,2-propanediol. In addition, we have carried out the reactions in the absence of base and we have demonstrated successfully the activation of polyols in alkaline-free conditions which is important from industrial point of view. Different type of supports (acidic and basic type) have been studied and the influence of two preparation methods, wet impregnation and sol-immobilisation, are compared. The addition of platinum to gold significantly enhances the activity with respect to the corresponding Au and Pt monometallic catalysts and retains the high selectivity to glyceric and lactic acids using O_2 as oxidant. Detailed STEM-HAADF studies have been carried out and have shown the presence of gold-platinum alloy on the surface of the catalyst. Comparison of the activity for C_3 alcohols shows that the reactivity decreases in the order: glycerol > 1,2-propane diol The use of a sol-immobilisation preparation method as compared to impregnation leads to the highest activity alloy catalysts for glyceric and lactic acids formation from the oxidation of glycerol and 1,2-propane diol respectively and the origins of these activities based on the particle size, gold-platinum metal molar ratio, nature of metal and support are presented.

References

[1] A. Corma, S. Iborra, A. Velty, Chem. Rev. 107 (2007) 2411.

[2] C-H. Zhou, J.N. Beltramini, Y-X. Fan, G.Q. Lu, Chem. Soc. Rev. 37 (2008) 527.

[3] C.D. Pina, E. Falletta, L. Prati, M. Rossi, Chem. Soc. Rev. 37 (2008) 2077.

[4] D.I Enache, J. K. Edwards, P. Landon, B. Solsona-Espriu, A. F. Carley, A. A. Herzing, M. Watanabe, C. J. Kiely, D. W. Knight, G. J. Hutchings, Science 311 (2006) 362.

[5] J.A. Lopez-Sanchez, N. Dimitratos, P. Miedziak, E. Ntainjua, J.K. Edwards, D. Morgan, A.F. Carley, R.Tiruvalam, C.J. Kiely, G.J. Hutchings, Phys. Chem. Chem. Phys. 10 (2008) 1921.

[6] N. Dimitratos, J.A. Lopez-Sanchez, J.M. Anthonykutty, G. Brett, A.F. Carley, R.C. Tiruvalam, A. A. Herzing, C.J. Kiely, D.W. Knight, G.J. Hutchings, Phys. Chem. Chem. Phys. 11 (2009) 4952.

[7] N. Dimitratos, J.A. Lopez-Sanchez, J.M. Anthonykutty, G. Brett, A.F. Carley, S.H. Taylor, D.W. Knight, G.J. Hutchings, Green Chem. 11 (2009) 1209.

See more of this Session: Catalytic Biomass Conversion to Chemicals II
See more of this Group/Topical: Fuels and Petrochemicals Division

Design of the smart community through the use of EV in a university

Yuu Notoji, Kazuya Shimohara, Tsuguhiko Nakagawa,
Graduate school of Okayama Prefectural University, Soja, Japan

Abstract

The CO_2 emissions which is said to be one of the cause of global warming have been researched and developed to reduce all over the world. The smart community that is combined the fossil fuel and the renewable energy to manage the energy supply and demand by the use of IT technology is one of the important system. In this study, the mathematical simulation model which is calculated the energy demand and supply of heat and electric power in the university has been developed on the basis of the more than a year investigation on the hourly demand and changes in the weather. The replacement of the air conditioning apparatuses and the energy demand changes have been calculated when photovoltaic power generation system and a lot of EV are introduced. As a result, if heat pomp type air conditioner is replaced the absorption type, it is possible to reduce 70% of the CO_2 emission. As the feature of Okayama Prefectural university, 80 % of the commutation means is a private car and the average distance is within 30km/day. So the battery capacity of the EV can be changed to a small size. In addition, the photovoltaic power generation system can be generated 200kWh/(m^2·year) of electricity around the university. So it is possible to construct the high cost performance system that is combined the EV with photovoltaic power generation. It has been proved that it is possible to reduce 36% or more CO_2 emission using the proposed integration system.

Introduction

Recently, cutting the CO_2 emission has been the worldwide problem. The Japanese government is promoting various means that are aimed at cutting the country's CO_2 emissions by 60% to 80% on 2050 as compared with 1990. In order to achieve the target, it is important to make the energy demand and supply visibly and the set up to optimize the energy management. As a concrete method, "Smart Community" has been researching all over the world.

In this paper, it is a target to build the Smart Community in the Okayama Prefectural University (hereafter referred to as "OPU"), we have been studied to develop a combination energy system of the energy conservation and the natural energy based on the demand and supply with OPU.

Analysis of the current energy system

Annual energy consumption of OPU

The energy consumption in 2010 is shown in **Table 1.** As can be seen in **Table 1**, 86% of the energy is electric power and 14% is the heavy oil. The OPU has consumed the energy which is 54.1 TJ per year and discharged 2,310 ton CO_2 in 2010.

Table 1 The University's annual energy consumption (2010)

	Amount of energy consumption	**Calorie (TJ/年)**	**CO_2 mission (t-CO_2/年)**
Electric power	4,710MWh/年	46.3	※1 1,760
Heavy oil	202 kL/年	7.8	※2 550
total		54.1	2,310

※1 CO_2 emission coefficient of electric power in Japan 0.3743t-CO_2/MWh

※2 CO_2 emission coefficient of heavy oil 2.71t-CO_2/kL

Innovation of energy system

An image of Smart Community that is a combination energy system of the energy conservation and the natural energy is shown **Fig. 1**. The solid line in the figure shows a current energy network. The short dashed line shows the energy network of the studying object. As shown in **Fig. 1**, we have considered that the future energy system will be using solar cell, fuel cell, biomass fuel and wind power.

In this paper, we discuss the most basic system that is the combination of ①replacing the air-conditioner from the absorption type to a heat pump type, ②introducing the photovoltaic power generation and ③changing the car that is used for commute from the internal-combustion engine type to an electric vehicle (EV).

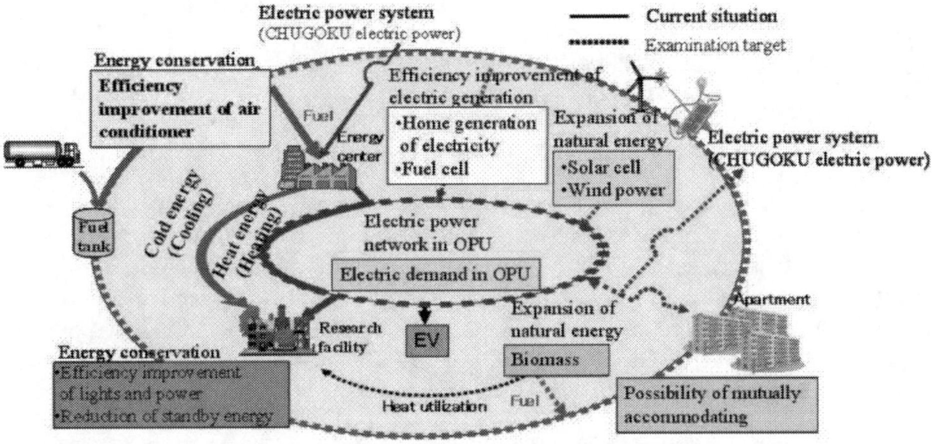

Fig. 1 Basic Smart Community image of OPU

Application of solar power

The capacity of the secondary battery that will need with the substitution rate of photovoltaic power has been presumed. For example, the calculated result that the substitution rate of photovoltaic power will be 100% is shown in **Fig. 2**. As can be seen in **Fig. 2**, the amount of the electricity accumulation becomes the maximum at the end of June. In this case, the battery capacity is calculated at 780MWh, however it is not realistic too huge.

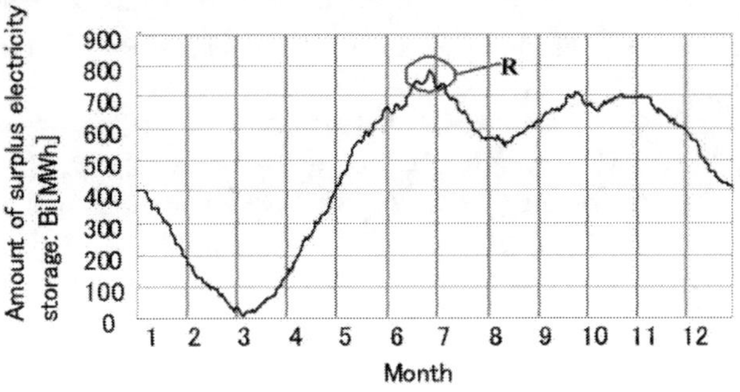

Fig. 2 Necessary capacity of the secondary battery with 100% of the photovoltaic power substitution rate

The relationship between the necessary of the secondary battery capacity with the substitution rate of photovoltaic power is shown in **Fig. 3**.

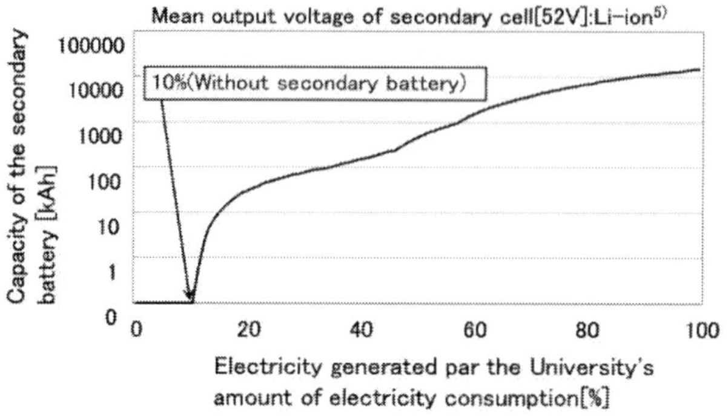

Fig. 3 Relationship between the necessary of secondary battery capacity with the substitution rate of photovoltaic power

As can be seen in **Fig. 3**, it is possible that the substitution rate of photovoltaic power will be 10.5% without using the secondary battery. If the substitution rate of photovoltaic power will increase, the secondary battery capacity will increase in the logarithmic.When the substitution rate of photovoltaic power will be 10.5%, the amount of the CO_2 will be reduced ▲537 [t-CO_2/year] and it means ▲26% reduction compared with 2010. In this case, the photovoltaic power will be generated 0.44MW and the payback period will become 20 years or more. So, we should consider a new system that could cope with both the CO_2 reduction and the payback period.

Combination system use of electric vehicles (EV)

We have thought in a broad concept to reduce the CO_2 emission. That is, we have considered to be related the amount of CO_2 that is emitted by commuting to the university of OPU. In the OPU, the most common commuting method has been driving the car that consumes gasoline and light oil as fuels. These cars will be changed into the electric vehicles (hereafter referred to as EV) and charged the electricity of photovoltaic power generation as the power source.

We have present the method that the generated electricity from photovoltaic power generation has given highest priority to charge EV in weekdays of daytime. In the holidays, the generated electricity has substituted for the basic electricity of OPU. We have studied with the same conditions of **Fig. 3**. In this case, the substitution rate of photovoltaic power is 10.5% and the amount of annual photovoltaic electricity generation is 537MWh/year. The distribution of the photovoltaic power generation is the photovoltaic electricity will be shared of 70% that will charge to EV and 30% that will supply for the demand of OPU.

The final result of all studies is shown in **Table 2.** As can be seen **Table 2**, In the case of replacing the air-conditioner from the absorption type to a heat pump type, the CO_2 emission will be reduced ▲14 %. In addition of introducing the photovoltaic power generation, the CO_2 emission will be reduced ▲20 %. In addition of changing the car that is used for commute from the internal-combustion engine type to an EV, the CO_2 emission will be reduced ▲36 %.

Conclusions

We have studied to develop a combination energy system of the energy conservation and the natural energy based on the demand and supply with OPU.

As a result, it is possible to cut off the 36% or more CO_2 emission with the combination of ①replacing the air-conditioner from the absorption type to a heat pump type , ②introducing photovoltaic power generation and ③changing the car that is used for commute from the internal-combustion engine type to an electric vehicle (EV).

Table 2 Amount of CO_2 emission and reduction ratio each equipment combination

		Current	①	①+②	①+②+③
Electric power	t-CO_2/year	1760	1920	1720	1850
	MWh/year	4710	5120	4590	4950
Heavy oil	t-CO_2/year	550	0	0	0
	kL	202	0	0	0
CO_2 reduction	t-CO_2/year	2310	1920	1720	1850
	%	—	17.0	26.0	20.0
Changing the car (361)	t-CO_2/year	580※1	580※1	580※1	0
	kL	248	248	248	0
Amount of CO_2 reduction	t-CO_2/year	**2890**	2500	2300	**1850**
	%	—	14.0	20.0	**36.0**

※1 CO_2 emission coefficient of gasoline 2.32t-CO_2/kL

①Heat pump type air-conditioning
②Photovoltaic power generation
③EV

References

1 Kazuya Shimohara, Yuu Notoji, Seiji Shiba, Tuguhiko Nakagawa (2010), "A Formation of Distributed Energy Network and Energy Conservation in the University, The 15th National Symposium on Power and Enelgy System," SPES 2010, pp.267-268

2 SANYO Electric Company Ltd (2010), "Progress of Technology of Making to Highly Effective of Solar Battery that Uses Nanotechnology, " pp34-35

3 Yukinori Kuwano (2009), "Gaphic Eplanation All of the Latest Solar Cell, " pp100-105,

(650a) Physical Properties and Rheological Characteristics of Deasphalted Oil Produced From a Solvent Deasphalting Process

Jung Moo Lee, Ji Won Hwang, Sang Chul Shin, Hyun Wook Jung and Ki Bong Lee, Department of Chemical & Biological Engineering, Korea University, Seoul, South Korea

Because of conventional oil being-exhausted and its increasing price, the technologies to use unconventional oil and low value crude residues, which have not been fully exploited, are receiving more and more interest. Unconventional oil contains a lot of asphaltene which hinders utilizability of unconventional oil. As the asphaltene has high viscosity and contains large amounts of heavy metals, it decreases the fluidity of oil and causes catalyst poisoning. Furthermore, the process of desulfurization and denitrification are necessary to use unconventional oil because the asphaltene has a high sulfur and nitrogen content. As the unconventional oil has these drawbacks, an extraction process called solvent deasphalting is commonly used to acquire deasphalted oil(DAO) through the removal of asphaltene. The DAO is generally used for lube base oil, and besides, it is converted into transportation fuel and chemical raw materials by additional refinement. In this study, the physical properties and rheological characteristics of DAO were measured for more efficient utilization of DAO. Pentane and hexane were added into DAO to investigate the effect of the solvents on the viscosity of DAO solution. As the content of pentane and hexane was increased, the viscosity of DAO was decreased, resulting in the improved fluidity of DAO. The degree of decreasing rate in viscosity was declined with increasing amount of solvent. The viscosity of DAO was characterized using a rheometer under various temperature conditions to investigate the effect of temperature on the fluidity. The fluidity of DAO was rapidly increased with increasing temperature and the DAO showed the behavior of a Newtonian fluid above 70 ^{O}C, which is called the softening point. Additionally, the fluid behavior of DAO in a pipe was simulated using computer softwares, GAMBIT and FLUENT. The velocity profile of DAO in a pipe was rapidly changed around the softening point.

See more of this Session: Heavy Oil and Flow Assurance I
See more of this Group/Topical: Energy and Transport Processes

(499g) Effect of Transition Metal Catalyst On Lignin Oxidation During Alkaline Hydrogen Peroxide Pretreatment

Zhenglun Li, Department of Chemical Engineering and Material Science, Michigan State University, East Lansing, MI and David Hodge, Chemical Engineering and Materials Science, Michigan State University, East Lansing, MI

Biomass lignocellulose can be used as feedstock for bioethanol production, and a pretreatment is necessary for achieving high ethanol yield and economical feasibility. As a method to increase enzyme digestibility of biomass by selective delignification, alkaline hydrogen peroxide (AHP) pretreatment has been proposed and studied. The lignin recalcitrance in lignocellulose biomass can be removed by H_2O_2 oxidation and the enzymatic hydrolysis efficiency of the polysaccharides can be greatly improved. However, a big issue with AHP pretreatment is the high cost of hydrogen peroxide during the pretreatment, and the selectivity of the oxidant need to be improved, i.e. to achieve highest pretreatment efficiency with low level of H_2O_2 loading.

In 1890s, Fenton discovered that iron ion can catalyze the oxidation of organic substrate by hydrogen peroxide. Since then, research has been done on the application of transition metal catalysts in catalyzing the phenol contaminant in wastewater and pulp bleaching. It is desired that catalyst can enhance selective electron transfer from lignin substrate to the oxidant and improve the selectivity of delignification reactions to unproductive side reactions.

See more of this Session: Developments In the Pretreatment of Lignocellulosics for Bioconversion I
See more of this Group/Topical: Sustainable Engineering Forum

(641a) Direct Conversion of Lignocellulosic Biomass to Drop-In Hydrocarbon Fuel

Xuejun Pan and Li Shuai, Biological Systems Engineering, University of Wisconisn, Madison, Madison, WI

This research reports an innovative process (HDA) for producing fuel-grade hydrocarbon from lignocellulosic biomass. The process directly converts polysaccharides in lignocellulosic biomass to furan-based precursors for producing fuel-grade hydrocarbons in one step under mild conditions without prior pretreatment and saccharification of the feedstock biomass. Specifically, in the presence of acetone, halide salt, and small amount of mineral acid, the lignocellulosic biomass undergoes hydrolysis of polysaccharides (both cellulose and hemicellulose) to monosaccharides, dehydration of the monosaccharides to furfural (from pentoses) and hydroxymethylfurfural (HMF, from hexoses), and aldol-condensation of HMF and furfural with acetone, and ends up with the formation of furan-derived oxygenates (furfural- and HMF-acetone adducts) with chain length of C5–C15. The resultant furan-based precursors can be subsequently converted to fuel-grade hydrocarbons via hydrogenation. The process integrates the three reactions (Hydrolysis-Dehydration-Aldol condensation) in one step, and is abbreviated as HDA. At the conditions of 120-140°C and 1-2hour, cellulose and hemicellulose in lignocellulosic material (e.g., ground or chipped wood) were completely hydrolyzed into sugars, and more than 80% of the sugars were selectively converted to the furan-acetone adducts. Meanwhile, lignin in the biomass is extensively depolymerized and extracted. Preliminary evaluation indicated the HDA lignin has high purity, low molecular weight with narrow distribution, and good reactivity. The lignin is expected to have great potential in co-products development.

See more of this Session: Chemical and Catalytic Conversions and Processes for Renewable Feedstocks
See more of this Group/Topical: Sustainable Engineering Forum

(294c) A New Approach for Quantification of Morphology-Property Linkages In Fuel Cell Materials: A Case Study for PEM Fuel Cells

Emin Caglan Kumbur[1], Eric A. Wargo[1], Ahmet Cecen[1], Anne Hanna[1] and Surya R. Kalidindi[2], (1) Mechanical Engineering and Mechanics, Drexel University, Philadelphia, PA, (2)Mechanical Engineering, Drexel University, Philadelphia, PA

Due to the complex morphology and small length scales of many fuel cell materials, experimental quantification of the key properties of these materials can be expensive and quite difficult to conduct, if not impossible. As a result, modeling efforts become an imperative approach to evaluate the impact of material structure on fuel cell performance. To date, pore-scale modeling approach is widely used to understand the complex materials-performance relationship in these systems. However, the computational complexity of these models often limits simulation to analyze only a small volume of the material of interest. The model domain selected for the pore-scale models is typically chosen randomly from a much larger microstructure dataset. When considering the complex and heterogeneous internal structure of fuel cell materials, it is highly unlikely that this randomly selected volume would accurately reflect the full material structure and related impact on cell performance.

The objective of this work is two-fold. The first goal is to develop advanced microstructure analysis tools for direct quantification of the key structural properties of the complex fuel cell materials. The second goal is to develop a new approach for intelligently selecting small representative volume elements (RVEs) from much larger material microstructure datasets, which can be confidently used in pore-scale modeling efforts to obtain reliable results regarding the structure-performance relationship. The diffusion media (DM) in polymer electrolyte fuel cells is chosen for initial demonstration of the approach. The microstructure of a dual-layer DM sample (i.e., a micro-porous layer coated on a macro-DM substrate) is quantified using X-ray computed tomography and dual-beam focused ion beam scanning electron microscopy. Computationally efficient algorithms are developed to extract the key structural parameters (e.g., porosity, surface area, phase connectivity) from measured microstructure datasets of these materials. In particular, two novel microstructure analysis techniques are introduced for the quantification of tortuosity and pore size distribution. Using in-house microstructure analysis tools based on n-point statistics and principal component analysis decomposition, sets of small RVEs that accurately represent salient features of tested fuel cell DM samples are selected from the much larger, full datasets. A detailed validation study is performed to assess the reliability of the presented approach.

See more of this Session: Fuel Cell Technology I
See more of this Group/Topical: Fuels and Petrochemicals Division

(609e) A Fundamental Investigation of Electron Transport and Recombination Characteristics of Titania Nanowire/Nanoparticle Hybrid Structures

Venkat Kalyan Vendra[1], Delaina A. Amos[1], Mahendra K. Sunkara[1] and Thad Druffel[2], (1)Chemical Engineering, University of Louisville, Louisville, KY, (2)Conn Center for Renewable Energy Research, Louisville

The electron transport and recombination of long titania nanowires and titania nanoparticle/nanowire composites are investigated. Understanding charge transport and recombination in 1D nanostructures of titania is critical for improving the performance of DSSC employing these 1D nanostructures as photoanodes in dye sensitized solar cells. 1D titania nanostructures are expected to have short transport times and high electron lifetimes. However, most of the 1D titania nanostructures have not shown these desirable properties. For instance, the electron diffusion coefficients and transport time constants in titania nanotubes were found to be very similar to that of titania nanoparticles.[1] But, the recombination time constants for titania nanotubes were 10 times higher than then nanoparticle based DSSCs. In contrast, DSSCs fabricated using titania nanowires showed both shorter transport time scales and higher electron lifetimes.[2] In a seminal paper, Richter and Schmuttenmaer showed that the lower the electron mobility in TiO2 nanotubes was due to the existence of exciton like trap states formed from Ti^{+3} donor states resulting from oxygen deficiencies, fluoride impurities and excess titanium introduced during synthesis.[3] Although thermal annealing is expected to remove the oxygen deficiencies, the structural integrity of the nanotubes would be compromised and hence a new synthesis technique is needed to prepare defect free TiO_2 nanotubes for enhanced DSSC performance.

Several approaches such as electrochemical anodization, solvothermal techniques, hydrothermal methods, thermal evaporation and template based approaches have been widely explored. However, most the methods produce short titania nanotubes (~ 2 μm in length) which is undesirable considering the low dye loading and slow electron transport in these architectures. In addition, all these synthesis techniques discussed above are very time consuming with the reaction time scales ranging from several hours to even days. To our knowledge there have not been any studies which have focused on the fast synthesis of single crystalline titania nanowires. Here we present a novel, scalable approach for making single crystalline titania nanowires arrays for dye sensitized solar cells.[4] Electron transport and recombination in these single crystalline nanowires and in single crystalline titania nanowires coated with titania nanoparticles are discussed.

[1] Zhu, K., Neale, N.R., Miedaner, A., Frank, A.J. Enhanced charge collection efficiencies and light scattering in dye sensitized solar cells using oriented TiO2 nanotubes. Nanoletters 2007, 7, 69-74.

[2] Tetreault, N., Horvath, E., Moehl, T., Brillet, J., Smajda, R., Bungener, S., Cai, N., Peng, W., Zakeeruddin, S.M., Forro, L., Magrez, A., Gratzel, M. High-Efficiency Solid-State Dye- Sensitized Solar Cells: Fast Charge Extraction through Self-Assembled 3D Fibrous Network of Crystalline TiO2 Nanowires. ACS Nano 2010, 4, 7644-7650.

[3] Richter, C., Schmuttenmaer, C.A., Exciton like trap states limit electron mobility in TiO2 nanotubes, Nature Nanotechnology 2010, 5, 769–772

[4] Kumar, V., Kim, J.H. Jasinski, J.B., Clark, E.L., Sunkara, M. Alkali metal assisted gas phase bulk production of titania nanowires. Journal of Crystal Growth and Design 2011, Article Accepted.

See more of this Session: Semiconductor Nanowires and Nanotubes for Next Generation Photovoltaics - From Fundamentals to Devices
See more of this Group/Topical: Topical 5: Nanomaterials for Energy Applications

(738d) Exergy-Based Optimization for Mixed Refrigerant Systems

Jian Zhang, Meiqian Wang, Qiang Xu and Kuyen Li, Dan F. Smith Department of Chemical Engineering, Lamar University, Beaumont, TX

A refrigeration system generally works by indirectly transferring thermal energy from low-temperature sources to high-temperature sinks at the expense of electricity or mechanical work to lower the source temperature to a certain value. In nature gas liquefaction process, mixed refrigerant (MR) system is widely used to chill and condense nature gas from gas to liquid phase for easy shipment. In comparison with pure refrigerant system, an MR system provides refrigerant composite curve that better fits with natural gas cooling curve in the enthalpy-temperature diagram. Thus, MR system needs less shaft work of compressors. To quantify the "fitness" of composite curves, the concept of exergy is employed. Exergy is the maximum useful work possible during a process that brings the system into equilibrium with environment. The exergy of a stream equals to the enthalpy minus the product of entropy and ambient temperature.

In this work, a novel exergy-based optimization model is introduced. To disclose the thermodynamic instincts of a refrigeration system, an exergy-temperature diagram (B-T diagram) is employed, where the refrigerant exergy is divided into phase-exergy, pressure-exergy, and temperature-exergy. Phase-exergy means the exergy change due to solely phase change from liquid to gas. Pressure-exergy means the exergy change due to solely pressure change from current pressure to the ambient pressure P_0. Temperature-exergy means the exergy change due to solely temperature change from current temperature to the ambient temperature T_0. The transition among different types of exergy for a normal refrigeration cycle is perfectly explained. Based on the understanding, a mixed-integer nonlinear programming (MINLP) model is built. The model includes all equipment models used in the MR system and exergy related equations. The objective function is to maximize exergy efficiency, which is the ratio between useful exergy and the total consumed exergy. An industrial case study of liquefied natural gas (LNG) plant is presented. Results show that the optimized result could increase the exergy efficiency by about 10%.

See more of this Session: Energy Systems Design and Alternative Energy Sources II
See more of this Group/Topical: Computing and Systems Technology Division

(38f) Vanadium Oxide Nanowire - Carbon Nanotube Binder Free Flexible Electrodes for Supercapacitors

Kenneth J. Balkus Jr. and Sanjaya D. Perera, Chemistry, University of Texas at Dallas, Richardson, TX

Vanadium pentoxide (V_2O_5) layered nanostructures are known to have very stable crystal structures and high faradaic activity. The low electronic conductivity of V_2O_5 greatly limits the application of vanadium oxide as electrode materials and requires combining with conducting materials using binders. It is well known that the organic binders can degrade the overall performance of electrode materials and need carefully controlled compositions. In this study, we develop a simple method for preparing freestanding carbon nanotube (CNT)-V_2O_5 nanowire (VNW) composite paper electrodes without using binders. Coin cell type (CR2032) supercapacitors are assembled for using the nanocomposite paper electrode as the anode and high surface area carbon fiber electrode (Spectracarb 2225) as the cathode. The supercapacitor with CNT-VNW composite paper electrode exhibits a power density of 5.26 kW Kg^{-1} and an energy density of 46.3 Wh Kg^{-1}. Li-VNWs and CNT composite paper electrodes can be fabricated in similar manner and show improved overall performance with a power density of 8.32 kW Kg^{-1} and an energy density of 65.9 Wh Kg^{-1}. The power and energy density values suggest that such flexible hybrid nanocomposite paper electrodes may be useful for high performance electrochemical supercapacitors.

See more of this Session: Nanomaterials for Energy Storage I
See more of this Group/Topical: Topical 5: Nanomaterials for Energy Applications

(650b) Asphaltene Instability Measured by Small-Angle Neutron Scattering

Michael P. Hoepfner, Chemical Engineering, The University of Michigan, Ann Arbor, MI, Claudio Vilas Boas Favero, Chemical Engineering, Universidade Estadual de Maring, Maring, Brazil and H. Scott Fogler, Chemical Engineering, University of Michigan, Ann Arbor, MI

Asphaltenes are defined as the petroleum fraction that precipitates with the addition of a normal alkane; however, it was only recently discovered that there is a strong time-dependency for precipitation in oil/n-alkane mixtures [1]. It was found that at low precipitant concentrations, asphaltenes would eventually precipitate if one simply waited, sometimes up to several months. Asphaltenes were also recently found to deposit before the precipitation could be detected by optical microscopy, further revealing asphaltene instability [2]. Small-angle neutron scattering (SANS) was selected as a technique that could detect subtle changes in the asphaltene size and shape once a precipitant is added. Optical microscopy requires asphaltene aggregates to grow to ~0.5 μm before the precipitation can be detected. With SANS, changes in size or shape of a few nanometers can easily be measured.

Experiments were performed on the CG-2 SANS instrument at Oak Ridge National Laboratory. Oil samples were diluted with heptane at various concentrations, and were allowed to mix for times ranging from less than 1 day and up to 6 months prior to performing the scattering experiments. The goal was to investigate the asphaltene stability over a wide range of precipitant concentrations and to monitor the aggregation process over time. If asphaltenes are found to be stable at low enough precipitant concentrations, the scattering results will indicate that there is no change in the morphology of an asphaltene nanoaggregate and that any change in the scattering would be due to dilution and scattering contrast effects.

It was found that asphaltenes grew in size with the small addition of precipitant, as low as 10 vol. % heptane, for an oil with an instantaneous onset condition of 40 vol. % heptane. The asphaltenes continued to grow over time. Attempting to detect this instability with optical microscopy may not be possible due to the long-term precipitation/aggregation kinetics [1]. Experiments were also performed on a second oil and a model oil consisting of 3 wt. % asphaltenes dispersed in d8-toluene. Additionally, the asphaltenes remaining in solution were much smaller than those that precipitated, suggesting that the largest of the asphaltene nanoaggregates are the most unstable and precipitate first.

These findings question whether asphaltenes are ever truly stable and a small fraction of asphaltenes may precipitate with any slight perturbation. Future models to understand asphaltene stability need to consider this effect and the assumption of equilibrium may not be valid for a wide range of conditions.

[1] T. Maqbool, A. T. Balgoa, and H. S. Fogler, "Revisiting Asphaltene Precipitation from Crude Oils: A Case of Neglected Kinetic Effects," *Energy & Fuels*, vol. 23, no. 7, pp. 3681-3686, Jul. 2009.

[2] M. Hoepfner and H. S. Fogler, "Effects of Preciptant Concentration on Asphaltene Deposition," presented at the *10th Annual International Conference on Petroluem Phase Behavior and Fouling*, Rio de Janiero, Brazil, 2009.

See more of this Session: Heavy Oil and Flow Assurance I
See more of this Group/Topical: Energy and Transport Processes

(508f) Kinetics of the Water-Gas Shift Reaction In the Liquid Phase

Dhairya D. Mehta[1], M. Cem Akatay[2], Eric A. Stach[2], W. Nicholas Delgass[1] and Fabio H. Ribeiro[1], (1) School of Chemical Engineering, Purdue University, West Lafayette, IN, (2)School of Materials Engineering and Birck Nanotechnology Center, Purdue University, West Lafayette, IN

Water-gas shift (WGS) is one of the hydrogen generating reactions in aqueous phase reforming of biomass derived molecules. We have measured the WGS turnover rate (TOR) at liquid phase conditions and determined the kinetics. The reaction was carried out in a stirred tank reactor at about 40 bar in a semi-batch manner, with continuous flow of gases (Ar, H_2, CO and CO_2) and water pre-loaded along with the catalyst. The percentage Pt loading was measured by ICP-OES and the Pt dispersion was estimated by transmission electron microscopy (TEM). The minimum value of the agitation speed for which external mass transfer was not the controlling step was measured and all experiments were conducted above that speed. The Koros-Nowak criterion was applied for 1% and 5% Pt/C and the TOR* was found to be 5.8×10^{-3} s^{-1} and 4.7×10^{-3} s^{-1} respectively. This test suggests that the actual kinetics was being measured.

The liquid phase WGS reaction (LWGS) was carried out on two different catalysts – Pt/C and PtMo/C. The TOR* obtained for PtMo/C (~$0.05 s^{-1}$) was observed to be 10 times higher than that for Pt/C (~$0.005 s^{-1}$). A factor of about 10 was also observed for both gas-phase WGS and the glycerol reforming reaction. The gas-phase WGS TOR was extrapolated to LWGS conditions and vice-versa in order to compare the two. The TOR obtained in LWGS was found to be about 10 times higher than that for the gas-phase WGS. In LWGS on Pt/C, a higher CO order (~0.5) as compared to gas-phase WGS (~0.2) was observed, indicating lower coverage of CO. This can be explained by competitive adsorption between CO and H_2O. Carbon dioxide was found to be significantly more inhibiting in LWGS (~-0.3) as compared to gas-phase WGS (~-0.1). We attribute this to a lower temperature of operation as well as the higher partial pressure range of CO_2 over which the order was measured. The H_2 orders in LWGS and gas-phase WGS are similar (~-0.33). The kinetic data for WGS at liquid phase reforming conditions are the first reported and the results are consistent with predictions made by extrapolation from gas-phase conditions.

*All the TOR values have been normalized to T=160°C, P_{CO}=4bar, P_{H2}=4bar, P_{CO2}=0.2bar, P_{H2O}=4.7bar

See more of this Session: Catalytic Biomass Conversion to Chemicals I
See more of this Group/Topical: Fuels and Petrochemicals Division

(373f) Macro-Scale Modeling of SOFC with Micro-Structurally Graded Electrodes

Dong Hyun Jo, Jeong Hwan Chun and Sung Hyun Kim, Department of Chemical & Biological Engineering, Korea University, Seoul, South Korea

Solid oxide fuel cell (SOFC) that is operated at high temperature (650~1000 °C), have many advantages such as high efficiency, useful waste heat and fuel flexibility. However, due to its high temperature operation, it is hard to obtain optimal cell design and parameters for improving SOFC performances through experimental works. Mathematical models have been developed to understand and elucidate effects of various designs and operation parameters on SOFC performances, as well as accelerate SOFC developments.

In this work, various electrochemical phenomena of SOFC have been investigated using the change of operating parameters (e.g, temperature, pressure, gas composition, pore size and porosity). The typical operating conditions for P_{H2O}, P_{H2} and P_{O2} were 0.03, 0.97 and 0.21 bar. Butler-Volmer equation, dusty-gas model and Ohm's law were incorporated to determine the activation, concentration, ohmic overpotential. To verify good agreement with experimental data the exchange current density and gas diffusion coefficient have been sought, depending on operational conditions (temperature, pressure and gas composition) and the cell microstructures (porosity and pore size).

As the cell temperature increases, the activation and ohmic overpotentials decrease, whereas the concentration overpotential increases due to the considerable reduction of gas density at the elevated temperature despite the increased diffusion coefficient. Also, increasing the hydrogen molar fraction and operating pressure can further augment the maximum cell output. As the triple phase boundary (TPB) length which is closely related with the pore size, grain size, and overlap of particles principally affects the activation overpotential, activation overpotential can be minimized at maximum TPB length by controlling the pore size and porosity. Furthermore, it is noted that the concentration overpotential decreases with increasing pore size and porosity due to the increased Knudsen diffusion. There were optimal values of electrode pore size and porosity for better cell performance. From the fact that there exists an optimum of electrode pore size and porosity for maximum cell power density, the graded electrode has newly been designed to effectively reduce both the activation and concentration overpotentials. The graded electrode can offer both sufficient reactive area and good gas transport. In typical non-graded electrode case, decreasing pore size and porosity for reducing activation overpotential makes the concentration overpotential greatly larger. Such opposing effect can be eliminated by a well-defined graded electrode. Comparing with non-graded electrode, a micro-structural graded electrode can be classified into two different types, namely porosity grading and pore size grading, respectively. It is assumed that both porosity and pore size increase linearly from the electrode/ electrolyte interface to the electrode surface. The results exhibit 70% improved cell performance than a non-graded electrode. This electrochemical model will be useful to simply understand overpotential features and devise the strategy for optimal cell design in SOFC systems. The simulation data quantitatively agreed with experimental data of anode-supported tubular SOFC.

See more of this Session: Fuel Cell Technology II
See more of this Group/Topical: Fuels and Petrochemicals Division

(288a) Managing Your Career

Wendy Young, GE Water & Process Technologies, El Lago, TX

Powerpoint to be added later

See more of this Session: Developing Leadership Skills
See more of this Group/Topical: Management Division

(737a) Development and Testing of An Asymmetric Supercapacitor Based On a Carbon Foam Electrode Impregnated with Nickel Hydroxide

Wen Nee Yeo[1], Tony N. Rogers[1] and Bahne C. Cornilsen[2], (1)Department of Chemical Engineering, Michigan Technological University, Houghton, MI, (2)Department of Chemistry, Michigan Technological University, Houghton, MI

Supercapacitors have attracted significant attention in recent years due to their potential to power hybrid vehicles, portable audio equipment, and memory back-up devices. The trend in supercapacitors is to combine a double-layer electrode with a pseudo-capacitance electrode in an asymmetric capacitor configuration.[1] The double-layer electrode is usually an activated carbon (AC) since it has high surface area, good conductivity, and relatively low cost.[2, 3] The pseudo-capacitance electrode usually consists of RuO_2, MnO_2, $Ni(OH)_2$, etc., with the most commonly studied material being $Ni(OH)_2$.[4-7]

In this work, a graphitic carbon foam was impregnated with nickel hydroxide active mass using electrochemical deposition to make a nickel-carbon foam electrode. It served as the pseudo-capacitance electrode in an asymmetric supercapacitor. Carbon foam was used as a conductive holder of the active mass because of its electrical conductivity, pore size, and pore connectivity. The objective was to reduce the cost and weight of the supercapacitor while maintaining or increasing its capacitance and gravimetric energy storage density. The specific capacitance of the nickel-carbon foam electrode at different current densities was measured in constant-current flooded cell cycling tests. The measured specific capacitance of 2641 F/g at 5 mA/cm^2 was higher than a reported value of 2080 F/g for an Al-substituted α-$Ni(OH)_2$ electrode.[6] The nickel-carbon foam electrode was then combined with an AC at different ratios of the negative and positive electrode masses to study asymmetric cell performance. The mass ratio for optimum cell discharge capacity was found to be ~3.7.

References

[1] A.I. Belyakov, 3rd European Symposium on Supercapacitors and Applications (ESSCAP) Roma, Italy, 2008.

[2] K. Babel, K. Jurewicz, *Journal of Physics and Chemistry of Solids* 65 (2004) 275-280.

[3] P.-L. Taberna, Geoffroy Chevallier, P. Simon, D. Plee, T. Aubert, *Materials Research Bulletin* 41 (2006) 478-484.

[4] Jong Hyeok Park, Sungwook Kim, O Ok Park, J.M. Ko, *Applied Physics A* 82 (2006) 593-597.

[5] D.-D. Zhao, S.-J. Bao, W.-J. Zhou, H.-L. Li, *Electrochemistry Communications* 9 (2007) 869-874.

[6] J.-W. Lang, L.-B. Kong, M. Liu, Y.-C. Luo, L. Kang, *Journal of Solid State Electrochemistry* 14 (2010) 1533-1539.

[7] Q. Huang, X. Wang, J. Li, C. Dai, S. Gamboa, P.J. Sebastian, *Journal of Power Sources* 164 (2007).

See more of this Session: Energy Storage and Electrochemical Processes II
See more of this Group/Topical: International Congress on Energy 2011

(484b) Control of Packed-Particle Flow Resistances During Cross-Flow Filtration In Microchannels

Nopphon Weeranoppanant, Levy Amar, **Michael Hill** and Edward F. Leonard, Chemical Engineering, Columbia University, New York, NY

Cross-flow filtration of fine suspensions through microsieves is a logical, sometimes necessary step in micro-processing. The observed limitation on steady-state filtrate flux accompanying this operation is generally attributed to particle accumulation upon the sieve. We consider two mechanisms: (1) blockage of filter pores by particles in contact with the filter surface, and (2) formation of a packed bed of particles that intrudes upon and is regulated by shear flow within the microchannel. This study examines the conditions under which a packed bed forms on a sieve, and provides a simple model for predicting its thickness when formed by hard non-interacting microspheres. Extensive experimental data are reported that corroborate the model. The good fit of these data lay a firm basis for the semi-quantitative analysis of the behavior of more complex suspensions.

See more of this Session: Theories and Applications of Cross Flow Filtration
See more of this Group/Topical: Separations Division

(484c) Cross-Flow Filtration of Plasma From Blood From a Microchannel

Xiangnan Ye[1], Levy Amar[1], Michael Hill[1], James Jones[2], Jacob Baggerman[3], Cees van Rijn[3] and **Edward F. Leonard**[1], (1)Chemical Engineering, Columbia University, New York, NY, (2)Department of Medicine -- St. Lukes'/Roosevelt Hospital, Columbia University, New York, NY, (3)Aquamarijn Micro Filtration, Zutphen, Netherlands

Steady cross-flow filtration of plasma through microsieves from blood in a microchannel is an important part of several biomedical applications under development, including a new artificial kidney. Erythrocytes, which comprise the largest volume fraction of all particulates in blood, are believed to create the controlling resistance at the sieve surface. Maximum steady-state filtration flux has been observed to be a function of wall shear rate, as predicted by any conventional cross-flow filtration theory, but to show weak dependence on erythrocyte concentration, contrary to theory based on convective diffusion. Maximum filtration flux has also been observed to be a function of sieve pore size and shape (in spite of their very low flow resistance) as well as sieve surface coating.

Recent experiments have correlated macroscopic measurements (filtration rates, transmembrane pressures) with direct observation of erythrocyte behavior at the filtering surface. At low filtration rates (low transmembrane pressures), erythrocytes roll across the filter surface, but at higher filtration rates (higher transmembrane pressures), erythrocytes are observed sticking to the sieve surface. Post-filtration SEM's, even those obtained at very low transmembrane pressures, reveal significant capture and deformation of erythrocytes in the filter pores. Careful balances among sieve design (which can be precisely defined by photolithography), surface treatments of sieves, and flow conditions, allow filtration fluxes at rates exceeding 0.1 cm^3/cm^2-min.

See more of this Session: Theories and Applications of Cross Flow Filtration
See more of this Group/Topical: Separations Division

(588d) Quantification of [Fe-Fe]-Hydrogenase, MrcA and FTHFS Gene In Mixed Anaerobic Cultures During Dark Hydrogen Fermentation

Subba Rao Chaganti[1], Daniel D. Heath[1] and Jerald A. Lalman[2], (1)Great Lakes Institute for Environmental Research, University of Windsor, Windsor, ON, Canada, (2)Department of Civil and Environmental Engineering, University of Windsor, Windsor, ON, Canada

Energy security, climatic change and decreasing fossil fuel inventories are major issues driving researchers to develop technologies to produce cheap and dependable energy supplies. Hydrogen (H2) is a promising potential energy carrier for meeting future energy needs because it is eco-friendly, it has a high heating value (2.75 times higher energy content than hydrocarbon fuels) and it can be converted into electricity via fuel cells (Benemann, 1996; Momirlan and Vezrioglu, 2002). Currently, H2 is mainly produced from fossil fuel-based feedstock (Oil/naptha reforming, coal gasfication, methane steam reforming etc.,) chemicals because many commercial petrochemical processes are cost effective (Ewan and Allen, 2005). However, the dark fermentation process is emerging as a practical route for producing H2 because of fast reaction rates, technical simplicity and less area (1-5%) requirements compared to the direct photolysis or indirect photolysis or photo fermentation (Levin et al., 2004; Li and Fang 2007).

The successful operation of the dark fermentation process requires a good understanding of how mixed microbial cultures behave under a variety of operating conditions. Identifying microbial communities can provide descriptions of the interaction between different groups; however, this approach may not provide sufficient information about the function of the communities. Noticeable changes in the microbial community composition within the H2 fermenting reactor is expected during operation. Quantifying the gene expression and monitoring fluctuations within the community structure will provide an understanding of the changes linked to the operation of bioreactors. This approach can possibly serve as a rapid tool for bioprocess monitoring.

In the present study, genes responsible for H2 production and H2 consumption were monitored as a means of establishing changes within the microbial community structure. The Fe-Fe hydrogenase gene is monitored because it is related to H2 production. Methanogens and acetogens are major H2 consumers and monitoring includes examination of the McrA gene and formyltetrahydrofolate synthetase (FTHFS) gene, respectively. A monitoring protocol for three different sources of cultures (A, B and C) at pH 5.5 and 37oC was developed using qRT-PCR. Terminal Restriction Fragment Length Polymorphism (T-RFLP) analysis was also performed to characterize changes in the community. The results indicated that the expression levels of [Fe-Fe]-hydrogenase, MrcA and FTHFS gene varied with the culture source. The [Fe-Fe]-hydrogenase gene expression was shown to correlate with H2 production. Increasing MrcA expression as well as increasing FTHFS and [Fe-Fe]-hydrogenase activity was correlated with low H2 yields of approximately 1.65±0.25 mol.mol-1 glucose in culture B. In comparison, cultures A and C with lower FTFHS and higher [Fe-Fe]-hydrogenase activities had higher H_2 yields of 2.4±0.12 and 2.5±0.23 mol.mol-1 glucose, respectively. This data clearly suggest that monitoring the [Fe-Fe]-hydrogenase gene alone is insufficient to gain an understanding of how mixed cultures function and genes related to H_2-consumers (MrcA and FTHFS) should be monitored and down regulated to achieve higher H_2 yields.

References

Bennemann, J. 1996. Hydrogen biotechnology: progress and prospects. Nat. Biotechnol. 14, 1101-1103.

Ewan, B.C.R., Allen, R.W.K. 2005. A figure of merit assessment of the routes to hydrogen. Int. J. Hydrogen Energ. 30 (8), 809-819.

Levin, D.B., Pitt, L., Love, M. 2004. Biohydrogen production: prospects and limitations to practical application. Int. J. Hydrogen Energ. 29 (2), 173-185.

Li. C., Fang, H.H.P. 2007. Fermentative hydrogen production from wastewater and solid wastes by mixed cultures. Crit. Rev. Env. Sci. Tec. 37 (1), 1-39.

Momirlan, M., Veziroglu, T.N. 2002. Current status of hydrogen energy. Renew. Sust. Energ. Rev. 6 (1-2), 141-179.

See more of this Session: Fundamentals of Hydrogen Production and Utilization
See more of this Group/Topical: Environmental Division

Conversion of Cellobiose into Glucose at Near-Neutral pH using a Recyclable Heterogeneous Catalyst

Po-Wen Chung, Oz Gazit, Alexandre Charmot, and Alexander Katz

Department of Chemical and Biomolecular Engineering
University of California, Berkeley, CA 94720

Strong acids under severe reaction conditions are known to rapidly hydrolyze cellulose with moderate selectivity towards glucose; however, enzymes in nature convert saccharides into glucose at higher selectivity using mild conditions dictated by carboxylate acids/carboxylate bases at the active site. Recently we have demonstrated the production of glucose from cellobiose at mild pH and temperature in a homogeneous system, but this system was not recyclable and also is dependent on trace amounts of metals. Herein, we demonstrate the conversion of cellobiose and other oligosaccharides into glucose using a mildly acidic heterogeneous catalyst. We also will also present adsorption results demonstrating the interaction among the surface of the material and various sugars.

ID:232613
PW: 867811

Determination of kinetic parameters of self-ignition wood pellets during storage

Wendi Guo[1], Xiaotao Bi[1], C. Jim Lim[1], Shahab Sokhansanj[1,2]

1. Chemical & Biological Engineering, University of British Columbia, Vancouver, BC V6T 1Z3
2. Environmental Sciences Division, Oak Ridge National laboratory, Oak Ridge TN 37831

1. Introduction

Accidental fires in wood pellets silos happen frequently (Channel 3000 2007; Aines 2008; Bartel 2008). It appears that the oxidative biological reactions are initially responsible for simultaneous gas release and temperature increase. (Walker and Harrison 1977; Lehtikangas 2000; Wihersaari 2005b, 2005a). The off-gassing and self-heating phenomena of wood pellet during storage have been investigated by several research institutions as well as in our Biomass & Bioenergy laboratory at UBC (Svedberg *et al.* 2004; Wadso 2007; Kuang *et al.* 2008; Svedberg *et al.* 2008; Kuang *et al.* 2009a; Kuang *et al.* 2009b; Guo *et al.* 2011). The results to date show that the rate of self-heating decreases with the age of pellets. The reaction rate is also highly dependent on storage temperature. A maximum heat release rate of 0.3 W/(kg pellet) has been recorded when a mass of pellets was stored in a sealed container maintained at 50 °C.

It is postulated that chemical oxidation dominates the self heating process when the storage temperature increases to 80 °C. This process leads to a higher temperature accelerating self heating when a self-ignition may occur (Meijer and Gast 2004). The self ignition of wood and other biomass was widely investigated experimentally under heat flux. Models were developed to predict the thermal ignition for those materials (Bilbao *et al.* 2001; Bilbao 2002; Pauner and Bygbjerg 2007)

The kinetic of wood pellets self heating at higher temperatures and the self-ignition phenomena haven't been fully understood. Experiments were carried out in this work at temperature range of 100 °C to 200 °C to study wood pellets chemical oxidation process. A self-heating model was developed to simulate the self-heating process using the results obtained from the experiment. A numerical method was developed to solve the energy balance and obtain the kinetic parameters for self-heating process.

2. Methodology

Experiments were carried out using a lab size oven (Thermal scientific, 1.7 cu. ft). Cylindrical containers were used as the self heating reactor and were put in the oven (constant temperature) with wood pellets loaded. Assuming the self heating process is a first order reaction and the self heating heat can be described as reaction heat (enthalpy of reaction, $\Delta_r h$) multiplying the reaction rate constant [$k=A\exp(-E/RT)$].

The energy balance of the self-heating process inside a cylindrical container is expressed by

$$\rho C_P \frac{\partial T}{\partial t} = \lambda(\frac{1}{r}\frac{\partial}{\partial r}(r\frac{\partial T}{\partial r}) + \frac{\partial^2 T}{\partial z^2}) + \rho \cdot \Delta_r h \cdot A\exp(-\frac{E}{RT}), \quad (1)$$

where, C_p is the specific heat capacity of the pellets, kJ/kg; λ is the thermal conductivity of the pellets , W/(m K); ρ is the bulk density of the pellets, kg/m^3; T is the temperature of the pellets, K; r is the radius of the container, m; $\Delta_r h$ is the reaction enthalpy of the self-heating reaction, kJ/mol; A is a exponential factor; E is the activation energy of the self-heating reaction, J/mol; R is gas constant, 8.314 J/(molK); t is time of the self heating reaction, s.

A curve-fitting method was developed to extract the kinetic parameters (including $\Delta_r h$, A and E) using the experimental data (temperature vs. time). Eq.(1) was solved by MatLab using the thermal properties obtained from the previous studies along with the boundary conditions shown in Figure 1. The best fitted kinetic parameters were determined by minimizing the residuals between the numerical solution and the experimental temperature profiles.

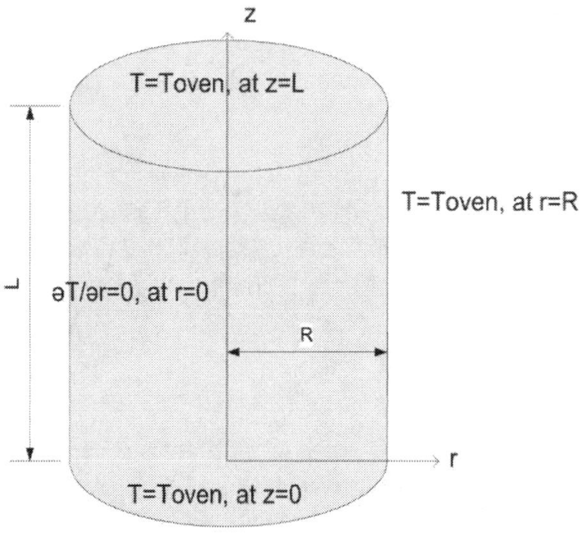

Figure 1 Boundary conditions and assumptions

Assumptions:

Bulk pellets can be treated as homogeneous materials with effective thermal properties (Cp, λ, ρ).

Initial condition t=0, $T=T_0$ (initial temperature)

No air movement inside the container

Fast heat conduction at container wall ($T_{wall}=T_{oven}$)

3. Experiment setup and materials

3.1 Sample

Softwood pellets produced in British Columbia (BC) were used in this research. Pellets with designation A, B and C in this research were manufactured from 3 different wood pellet plants in BC. They were made of whitewood mostly pine; both pellet A and pellet C were manufactured by Princeton Co-Generation Corp. (Princeton, BC, CanadaV0X 1W0). Pellet B was manufactured by Premium Pellet Ltd. (Vanderhoof, British Columbia Canada, V0J 3A0). Pellets were on average 6.2 mm in diameter and 10 to 25 mm in length. All pellets were made using whitewood with no bark. The pellets in this region are made from sawdust and shavings, a by-products of milling spruce, pine and fir for lumber.

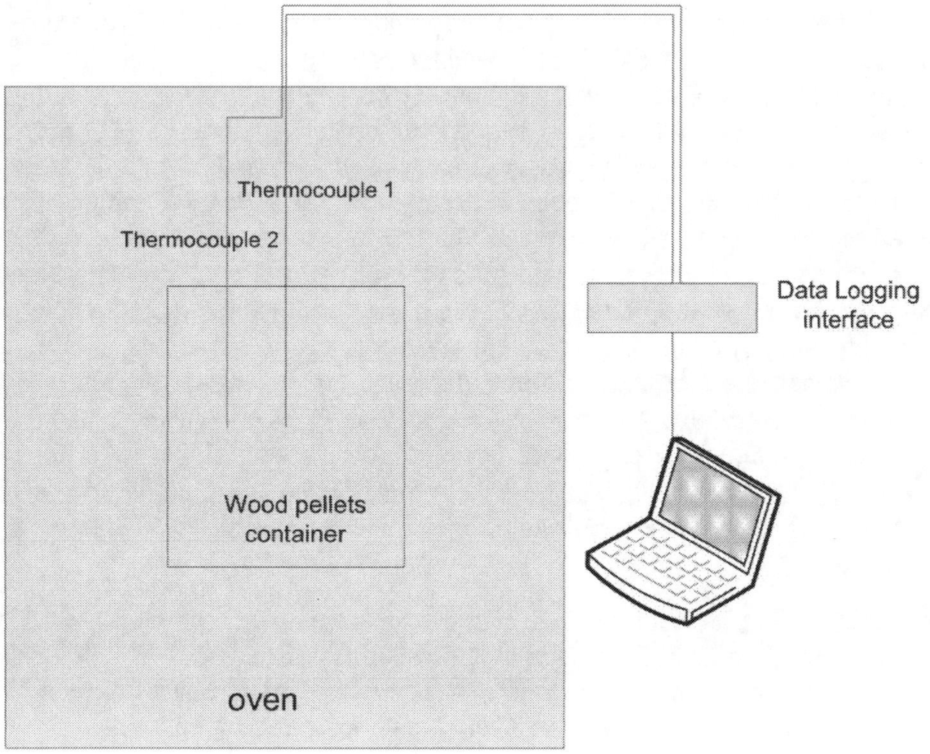

Figure 2 Experimental setup

3.2 Experimental setup

Figure 2 shows the experimental setup. Three sizes of containers made from steel were used in this experiment, small (H 106mm, D 120mm); Medium (H 166mm, D 190mm); Large (H 226mm, D 240mm). Top open short cylinders were used as reactor and two thermocouples were placed inside the container, one right in the middle (T1) and another one half way to the center (T2) (radial). The ends of two thermocouples were kept at the same axial level (same distances to the top and bottom). After placing the thermocouples, the containers were filled up with wood pellets (weight was recorded) and put inside of an oven (maintained at a constant temperature (between 120°C to 200°C); temperatures were recorded using a data acquisition system from the start till after the crossing point.

4 Results and Discussion

4.1 Oven temperature and self-heating

Figure 3 plots the experimental results of two different size containers at different oven temperature. The experimental results show that there is no self heating detected below an oven temperature of 100 °C at all lab size self heating containers which used in this study. Figure 3.a shows the temperature profiles of the large container (L)at 103 °C for 4 days and no self heating was detected. Self-heating was detected when the oven temperature was set at 140 °C for large containers (L), as shown in Figure3b where temperatures increased with time. When the center temperature (T1) crossed over the off-center temperature (T2) at 135°C (the crossing point), both temperature increased rapidly and reached a higher steady state temperature. Here, we consider the CP point (crossing point) as a signal of the detectable self-heating. When the oven temperature increased to 150 °C, the temperature T1 and T2 of the large container (L) continued increasing and reached 158 °C (the peak of T1) a steady state (decrease slightly), as shown in Figure 3.c. On the other hand, for the medium container, the crossing point temperature didn't appear until the temperature reached 147 °C (Figure 3.c.). For the large container, there is finally an uncontrollable thermal run-away which leads to a self-ignition when the oven temperature increased to 170 °C (shown in Figure3.d), while the temperature of medium container remains at a steady state (around 175 °C) . For the large container, temperatures kept running away after the oven heat was turned off and led to a self-ignition at 200 °C. Figure 3.f shows the burning wood pellets. The ignition happens at a point near the center.

It can be concluded that the heat release of the oxidization at temperature 100-200 °C is significant enough to be detected using a lab technique and to cause a self-ignition. The crossing point (detectable point of self-heating) and the thermal run-away temperatures are highly dependent on the size of the containers and the temperature of the oven.

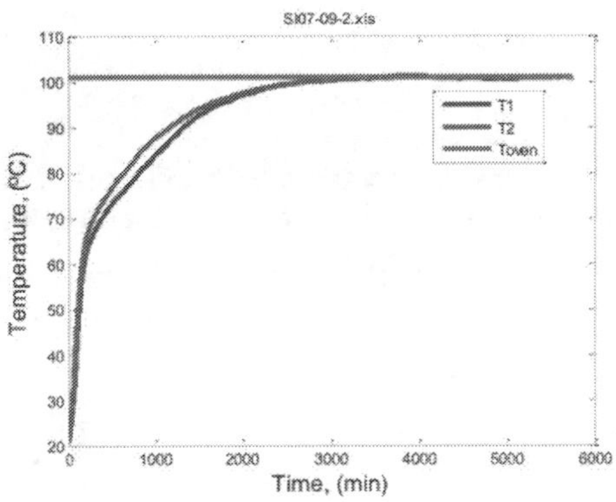

a.Oven temperature 103 °C, no crossing point, no thermal run away.

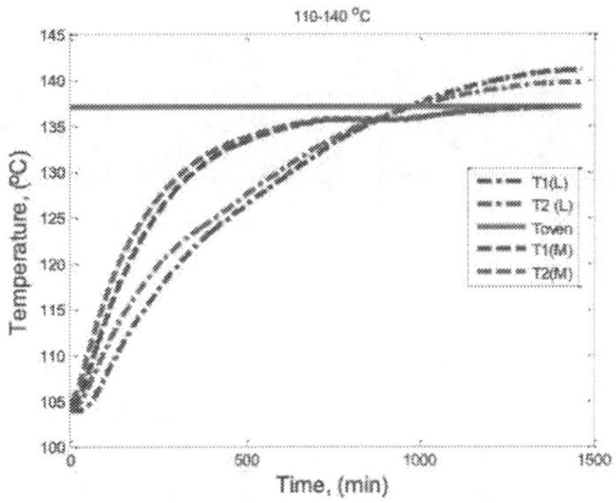

b. Oven temperature 140 °C crossing point after 10 minutes (Large and Medium container)

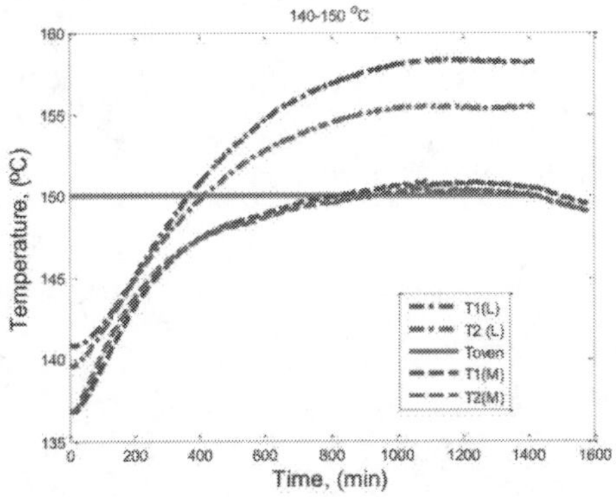

c. Oven temperature 150 °C (L&M)

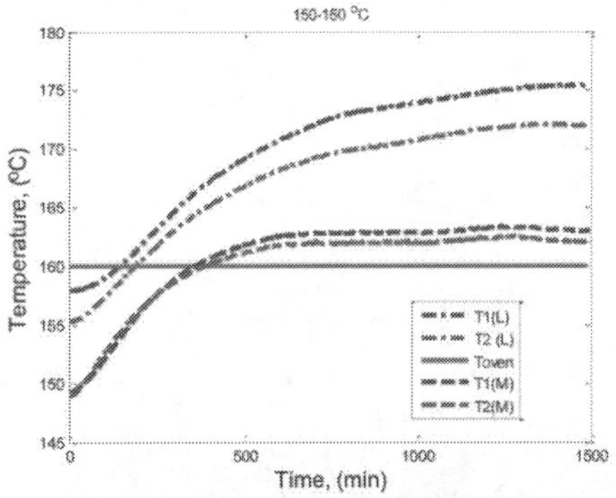

d. Oven temperature 160 °C (L&M)

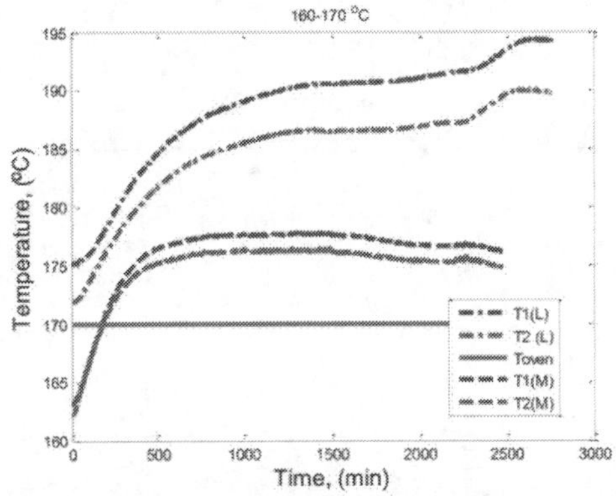

e. Oven temperature 170 °C and self-ignition (L&M).

f. Burning wood pellets.

Figure 3 Self-heating and self-ignition -Temperature profiles at different oven temperatures.

4.2 Kinetic parameters from curve fitting

A curve-fitting method was developed by solving the energy balance equation Eq (1) based on the boundary conditions shown in Fig. 1. The kinetic parameters were obtained by searching the best fit using the curve-fitting method and both data from large containers and medium containers. Figure 4 shows the typical best fitted curves in comparison with the experimental data.

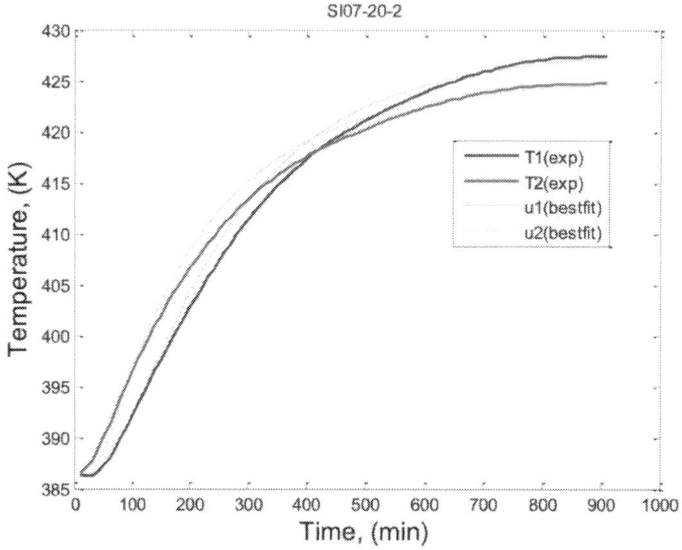

Figure 4 Comparison of the experimental data and best fitted curves of the kinetic parameters. (Experimental results of 'SI07-20-2' comparing to the numerical results of parameter E=62.5 kJ/mol, $\rho\Delta_r hA=5.7*10^7$ J.K/(m^3.s)

Further analysis is in progress.

5. Summary

Experiments were carried out to investigate the self-heating and self-ignition process of wood pellets at temperatures ranging from 100°C to 200°C. Three sizes of containers were used. The results show that the self-heating speed and thermal run-away temperature are highly dependent on the oven temperature and the size of the reactors. A curve-fitting method is developed to extract the kinetic parameters.

Aines, D. 2008. Fire reported for second time in six weeks at wood stove pellet plant. Retrieved 08 July 2010, from http://www.herald-mail.com/?cmd=displaystory&story_id=201753&format=html.

Bartel, R. 2008. Fire at westwood fibre products Retrieved 08 July 2010, from http://www.silk.fm/node/763122.

Bilbao, 2002. A model for the prediction of the thermal degradation and ignition of wood under constant and variable heat flux. Journal of Analytical and Applied Pyrolysis **62**(1): 63.

Bilbao, R., Mastral, J.F., Aldea, M.E., Ceamanos, J., Betr, M., Lana, J.A., 2001. Experimental and theoretical study of the ignition and smoldering of wood including convective effects. Combustion and Flame **126**(1-2): 1363-1372.

Channel 3000. 2007. 2 injured in explosion, fire at wood pellet warehouse. Retrieved 08 July 2010, from http://www.channel3000.com/news/10947450/detail.html.

Guo, W., Trischukb, K., Bi, X., Lim, C.J., Sokhansanj, S., 2011. Measurements of wood pellet self-heating kinetic parameters using isothermal calorimetry. Submitted to Biomass & Bioenergy. **Under Review**.

Kuang, X., Shankar, T.J., Bi, X.T., Lim, C.J., Sokhansanj, S., Melin, S., 2009a. Rate and peak concentrations of off-gas emissions in stored wood pellets—sensitivities to temperature, relative humidity, and headspace volume. Annals of Occupational Hygiene **53**(8): 789-796.

Kuang, X., Shankar, T.J., Bi, X.T., Sokhansanj, S., Jim Lim, C., Melin, S., 2008. Characterization and kinetics study of off-gas emissions from stored wood pellets. Annals of Occupational Hygiene **52**(8): 675-683.

Kuang, X., Shankar, T.J., Sokhansanj, S., Lim, C.J., Bi, X.T., Melin, S., 2009b. Effects of headspace and oxygen level on off-gas emissions from wood pellets in storage. Annals of Occupational Hygiene **53**(8): 807-813.

Lehtikangas, P., 2000. Storage effects on pelletised sawdust, logging residues and bark. Biomass Bioenerg. **19**(5): 287-293.

Meijer, R., Gast, C.H., 2004. Spontaneous combustion of biomass: Experimental study into guidelines to avoid and control this phenomena. KEMA Power Generation & Sustainables.

Pauner, M., Bygbjerg, H., 2007. Spontaneous ignition in storage and production lines: Investigation on wood pellets and protein powders. Fire and Materials **31**(8): 477-494.

Svedberg, U.R.A., Hogberg, H.E., Hogberg, J., Galle, B., 2004. Emission of hexanal and carbon monoxide from storage of wood pellets, a potential occupational and domestic health hazard. Annals of Occupational Hygiene **48**(4): 339-349.

Svedberg, U.R.A., Samuelsson, J., Melin, S., 2008. Hazardous off-gassing of carbon monoxide and oxygen depletion during ocean transportation of wood pellets. Annals of Occupational Hygiene **52**(4): 259-266.

Wadso, L., 2007. Measuring chemical heat production rates of biofuels by isothermal calorimetry for hazardous evaluation modelling. Fire and Materials **31**(4): 241-255.

Walker, I., Harrison, W., 1977. Self-heating of wet wood. 1. Exothermic oxidation of wet sawdust. New Zealand Journal of Science **20**: 191-200.

Wihersaari, M., 2005a. Evaluation of greenhouse gas emission risks from storage of wood residue. Biomass Bioenerg. **28**(5): 444-453.

Wihersaari, M., 2005b. Greenhouse gas emissions from final harvest fuel chip production in finland. Biomass Bioenerg. **28**(5): 435-443.

The Attractiveness of Ammonia Borane to Generate Hydrogen for PEM Fuel Cell Vehicles

Ahmad Al-Kukhun, Hyun Tae Hwang, and Arvind Varma

School of Chemical Engineering, Purdue University,
480 Stadium Mall Drive, West Lafayette, IN 47907

Current promising methods to release hydrogen from Ammonia Borane (NH_3BH_3, AB; 19.6 wt% H_2) including neat thermolysis, thermolysis in ionic liquid bmimCl with or without proton sponge, thermolysis with nano-BN and hydrothermolysis, were investigated for hydrogen yield and ammonia formation. It was found that even trace moisture influences AB dehydrogenation significantly. The hydrothermolysis (13.5 wt% H_2, 1 mol% NH_3) at 85°C and thermolysis in bmimCl with 3 wt% moisture at 110°C (13 wt% H_2, 0.2 mol% NH_3) methods were found to be the most promising. This study demonstrates that AB dehydrogenation is an attractive approach to generate hydrogen for PEM fuel cell vehicle applications.

(429c) High Density Cultivation of a Heat-Resistant Microalga Chlorella Sorokiniana for Biodiesel Production

Tingting Li[1], Yubin Zheng[2] and Shulin Chen[1], (1)Biological Systems Engineering, Washington State University, Pullman, WA, (2)Department of Biological Systems Engineering, Washington State University, Pullman, WA

High density cultivation of aheat-resistant microalga *Chlorella sorokiniana*

forbiodiesel production

Tingting Li, YubinZheng, Shulin Chen

Department ofBiological Systems Engineering,

Washington StateUniversity, Pullman, WA 99164

Abstract

Microalgae areconsidered to be outstanding potential sources for biofuelproduction and thus have attracted growing interest. Heterotrophic algae cultures havethe advantages of fast growth without subjection to light limitation, highbiomass concentration for easily harvesting and high lipid content forbiodiesel production. The green alga *Chlorellasorokiniana* has the capability toaccumulate a high content of lipid heterotrophically, however, the high cell density cultivation in fermentor has not been well examined yet. In this study, we firstly made acomparison between phototrophic and heterotrophic culture of *C. sorokiniana*,and the data showed the heterotrophic cultured algae had twice higher lipidproduction. Then, effects of temperature, carbon source, nitrogen source andC/N ratio on the biomass and lipid production of *C. sorokiniana* were evaluated. Theresults stated that the biomass and lipid reached 20.3 g/L and 5.4 g/L after 72hours cultivation with 40 g/L glucose and 4 g/L potassium nitrate at 37°C,which were approximately 4 and 6 folds respectively higher compared with theculture without optimization. Interestingly, the best culture temperature for *C. sorokiniana*was 37°C since most *Chlorella*species preferred the environmental temperature in the range of 25 to 30°C.Moreover, the temperature had significant effects on the fatty acid profile anda higher content of oleic acid (C18:1) but a lower content of linolenic acid (C18:3) was observed at higher temperature.The low temperature led the algae to accumulate unsaturated fatty acid,especially polyunsaturated fatty acid, because the low melting point was veryimportant for the fluidity of the cell membrane at low temperature. Finally,fed-batch cultivation in fermentor was conducted withthe optimal conditions and the dry biomass and lipid production for *C. sorokiniana*was significantly increased. These results demonstratedthat heterotrophic culture of the microalga *C. sorokiniana*had the great capability for biodiesel production. The finding is verysignificant because a special microalga was presentedin this study with the characteristics of high lipid content andheat-resistant.

See more of this Session: Alternative Fuels and Enabling Technologies I
See more of this Group/Topical: Fuels and Petrochemicals Division

(456c) Molecular Simulation of Hydrogen Adsorption In Porous Material Made up with Silsesquioxane Units and Metal Catalytic Sites

Nethika S. Suraweera[1], Michael E. Peretich[2], Joshua Abbott[2], Craig E. Barnes[2] and David J. Keffer[3], (1)Chemical & Biomolecular Engineering, University of Tennessee, Knoxville, TN, (2)Chemistry, University of Tennessee, Knoxville, TN, (3)Chemical and Biomolecular Engineering, University of Tennessee, Knoxville, Knoxville, TN

The adsorptive hydrogen storage of molecular hydrogen on microporous materials is widely studied as an effort for efficient utilization of hydrogen as a pollution free renewable energy source. Porous structures made up with spherosilicate building blocks and consist of metal catalytic sites exhibit promising results for the ability of adsorption of gases. Experiments are carried out for investigating the hydrogen adsorption ability of these materials. In this study we performed computer simulations to analyze hydrogen adsorption in structures which consist of interconnected silsesquioxane units. This porous material has cubic silicate building blocks (Si_8O_{20}), which are cross-linked by $SiCl_2O_2$ bridges and decorated with $OTiCl_3$ catalytic sites and $SiMe_3$ ends.

The structure for the simulations was developed first using a coarse grain model, followed by an energy optimization and replacing the coarse grains with atoms. The physical properties of the structure such as density, surface area, accessible volume and pore size distribution were compared with experimental vales and they are in good agreement.

Adsorption of hydrogen was modeled by Path Integral Grand Canonical Monte Carlo (PI-GCMC) simulations using standard force fields. From the simulations, hydrogen adsorption isotherms at 300K and 77 K were generated for low pressures up to 10 bars. The energy of adsorption of hydrogen was calculated at each case. Density distribution was developed and analyzed. Classical MD simulations were performed to calculate the self-diffusivity of hydrogen in the porous structure and activation energy was calculated.

Keywords: spherosilicate building blocks, hydrogen adsorption, PI-GCMC simulation, silsesquioxane units, coarse grain model, energy optimization, adsorption isotherm, density distribution, MD simulations, self-diffusivity, activation energy

See more of this Session: Hydrogen Storage Systems and Materials: Modeling, Simulation, and Optimization:
See more of this Group/Topical: International Congress on Energy 2011

A MULTI-SCALE MODEL-BASED APPROACH TO THE SYNTHESIS AND HIGH-THROUGHPUT ANALYSIS OF BIOREFINERIES

Marinella L. Tsakalova, National Technical University of Athens, Greece
Aidong Yang, University of Surrey, United Kindom
Antonis C. Kokossis, National Technical University of Athens, Greece

Abstract

The paper highlights the pivotal role of systems technology to foster innovation, preview options, and support high-throughput computational experimentation, arguing that systems tools are largely under-deployed. Systems-enabled platforms could instead function as powerful environments to generate ideas for integrated designs and offer tremendous services to the complex and large problems produced by the numerous portfolios of feedstocks, unknown portfolios of products, multiple chemistries, and multiple processing paths. A decision support platform is developed to evaluate and screen options for producing chemicals from biomass feedstock in a multi-scale approach. The platform is based on the modelling of the so-called compartmental processing blocks (CPBs) and their connections. It includes a simulation tool and a synthesis/optimisation tool, both based on a central database which holds process and economic information about individual CPBs. The synthesis/optimisation tool performs screening over multiple production routes by means of modelling and optimisation of superstructures that represent possible feedstock, routes, and products in a given case of assessment. The current implementation uses Excel for the database and the simulation tool, while the synthesis/ optimisation tool is based on GAMS. Cases studies of green ethylene, ammonia and propylene glycol are reported to illustrate the application of the platform.

(738b) Optimal Design of Oxy-Fuel Combustion for Carbon Capture and Sequestration

Hussam Zebian, Nicholas D Mancini and **Alexander Mitsos**, Mechanical Engineering, Massachusetts Institute of Technology, Cambridge, MA

Oxy-fuel combustion is a thermodynamically attractive concept that seeks to mitigate the penalties associated with CO2 capture from power plants. The main disadvantage is the need to separate oxygen from atmospheric air. This can be achieved using conventional cryogenic air separation or novel membranes. In either case, careful integration of the air separation, power generation and CO2 treatment is required to achieve high conversion efficiencies. Moreover, in order to obtain meaningful prediction of efficiencies detailed modeling of all processes involved is mandated, but this is often ignored in the literature. The talk describes modeling, design and optimization of various oxycombustion concepts.

In the first part of the talk, a semi-commercial pressurized oxy-coal combustion is considered. Multi-variable optimization is performed involving high-fidelity modeling of the power plant units. In particular, the model accounts for heat losses, flow leaks, pressure drops, irreversibilities, and several technological and economical considerations. Continuous and discrete variables are varied simultaneously to maximize thermal efficiency while satisfying techno-economic constraints. The optimization leads to significant efficiency improvement and more moderate operating conditions (lower pressure) compared to prior studies based on single-variable sensitivity analysis.

In the second part, integrated oxygen ion transport membranes (ITMs) are considered for oxygen separation for the oxycombustion of natural gas. Oxygen separation in an ITM system consists of many distinct physical processes, ranging from complex electrochemical and thermochemical reactions, to conventional heat and mass transfer. The dependence of ITM performance on power cycle operating conditions and system integration schemes must be captured in order to conduct meaningful process flow and optimization studies. An axially distributed, quasi two-dimensional model is presented based on fundamental conservation equations, semi-empirical oxygen transport equations obtained from the literature, and simplified fuel oxidation kinetic mechanisms. Several power cycles from the literature are considered, along with new proposals, including separation in a reactive environment. The detailed modeling demonstrates that several literature claims are overly optimistic since they neglect irreversibilities and operational constraints. Moreover, the model is used to examine the feasibility of novel concepts to overcome some of the challenges associated with the ITM. Finally, the merit of partial CO2 emission concepts is discussed.

See more of this Session: Energy Systems Design and Alternative Energy Sources II
See more of this Group/Topical: Computing and Systems Technology Division

(51c) Design of High Pressure/High Temperature Drilling Fluids

Apurva Samudra and Nick Sahinidis, Chemical Engineering, Carnegie Mellon University, Pittsburgh, PA

One of the important factors that influence drilling costs in oil and gas production, is the drilling fluid used. Drilling fluids, also known as drilling muds, are suspensions with many soluble and insoluble additives in a base fluid like water or diesel. These fluids perform many functions and can be effective in optimizing the drilling operations. Although under high pressure/high temperature (HPHT) conditions oil-based or synthetic fluid-based drilling muds have better performance than water-based fluids, environmental restrictions and prohibitive costs prevent their use in such wells, particularly in offshore drilling operations. At the same time, increasing demand for oil and gas has lead to exploration of deep reservoirs with extreme temperature and pressure conditions and designing base fluids to withstand such demanding conditions is a challenging task.

Computer-aided molecular design (CAMD) is an effective technique used for product design. CAMD techniques based on group contribution have been successfully applied to problems like refrigerant design, solvent replacement design, and polymer design. We have developed a molecular design framework that is efficient, versatile, and allows integration of different property models. In addition to property estimation based on the GC+ model, our framework incorporates molecular modeling methods to predict environmental properties like LC50. Another advantage of the framework is its modular structure and use of mixed-integer optimization in each module to increase the calculation speed by orders of magnitude compared other enumeration-based methods.

This paper presents the use of the CAMD framework to design the components of the drilling fluid for HPHT drilling. We primarily present the efforts to design the base fluid in the drilling mud. The properties that dominate the choice of drilling fluid are included in the design model. These include boiling and melting points, flash point, viscosity, and surface tension. One of the objectives in this work has been to include reliable property models to predict environmental metrics and use them in the design process. Properties like octanol-water coefficient, aqueous solubility, and LC50 are used to guide the molecular design. The targets for molecular design are formulated after an extensive review of deep well conditions and previously used synthetic fluids. We discuss the results of the design problem and compare them to current synthetic fluids like esters, mineral oils, and polyolefins.

Finally, we present the formulation of design targets for various additives in the drilling muds and preliminary results from the design problem.

See more of this Session: Recent Advances In Drilling Under Extreme Conditions

(456b) Optimization of Adsorbent Tanks for Hydrogen Storage for Automotive Applications

Stephen L. Garrison[1], David A. Tamburello[1], Matthew R. Kesterson[1], Claudio Corgnale[2] and Bruce Hardy[1], (1)Computational Sciences, Savannah River National Laboratory, Aiken, SC, (2)Hydrogen Technology Center, Savannah River National Laboratory, Aiken, SC

A key technical hurdle to an energy economy focused on clean-burning hydrogen over non-renewable petroleum is on-board storage of hydrogen for automotive vehicular applications. Adsorbents provide one possible solution to the need for improved gravimetric and volumetric capacities. However, while the heat of adsorption released upon sorption of hydrogen is relatively low, so it the thermal conductivity. The subsequent rise in storage temperature results in decreases in maximum hydrogen storage capacity and loading rate at the short fill times mandated by the Department of Energy (DOE) targets. Novel designs for integrated heat exchangers are needed to maximize the heat transfer out of system to maximize storage rate and capacity.

While many literature studies focus on maximum capacity assuming an (impossible) infinite fill time, full, detailed modeling of the coupled heat and mass transfer is needed. The focus of this presentation will be our work developing detailed numerical models that couple heat and mass transfer via the general purpose finite element solver COMSOL Multiphysics® for various adsorbent media and bed designs and the automatic optimization of the designs by linking the COMSOL models with routines in MATLAB®.

See more of this Session: Hydrogen Storage Systems and Materials: Modeling, Simulation, and Optimization:
See more of this Group/Topical: International Congress on Energy 2011

(110b) Lifetime Investigations of Bi-Functional Air Cathodes Using Electrochemical Impedance Spectroscopy (EIS)

Stephan Weinberger, Katja Frhlich, Franz Bernd Wede, Damijan Movrin and Viktor Hacker, Institute of Chemical Engineering and Environmental Technology, Graz University of Technology, Graz, Austria

Air cathodes were prepared based on various fabrication methods and compositions. Charging and discharging cycles have been performed. EIS measurements have been done on new electrodes, on aged electrodes at certain read-points, and at the end of electrode lifetime.

The experimental data are fitted to an appropriate equivalent circuit. A model is presented which characterizes the ageing of the air electrodes.

See more of this Session: Novel Battery Chemistry and Technology
See more of this Group/Topical: Fuels and Petrochemicals Division

(339b) Agent-Based Modeling Analysis of Biomass Feedstock Production System Dynamics and Resilience

Yogendra Shastri[1], Luis Rodriguez[2], Alan Hansen[2] and K.C. Ting[2], (1)Department of Agricultural and Biological Engineering, University of Illinois at Urbana-Champaign, Energy Biosciences Institute, Urbana, IL, (2)Department of Agricultural and Biological Engineering, University of Illinois at Urbana-Champaign, Urbana, IL

An efficient and sustainable energy crop production system is essential for the success of the emerging biorefinery industry. However, there is currently a significant lack of understanding regarding the development and dynamics of the biomass feedstock market upstream of the biorefinery, which is a critical component of the value chain. This includes assessing the time-scale of its development as well as the impact of various engineering, technological, regulatory and social factors on the dynamics. Various uncertainties and disturbances can impact the functioning of this system, and understanding the resilience of the system in the presence of those is also important. The analysis is especially challenging due to the involvement of multiple stakeholders such as farmers, refinery, transportation companies and storage elevators that must participate and yet compete for a fraction of the profit.

The objective of this work is to conduct such as analysis using the complex adaptive systems theory approach. The work develops an agent-based model that incorporates various stakeholders (agents) within the feedstock production system. The model, developed in MATLAB®, uses an object-oriented programming approach by considering each agent-type as an independent class and a particular realization of that agent-type as an object (instance) of that class. The individual characteristics of an object are defined by a set of attributes. The behavior of each agent as well as its interaction with other agents is based on a set of rules (methods). Since the behavioral and social factors play an important role in the decisions made by individual agents, a mental model incorporating some of these factors is also developed for each agent. The set of decision rules are consequently based not only on the economic and technological rationale, but also on the unique mental models and cognitive processes. The learning/adaptive behavior exhibited by the agents is also modeled which enables the simulation of emergent behavior of the system.

The model has previously been used to study the development of the biomass market under different scenarios. The focus of this work is on understanding the impact of various disturbances and uncertainties on the dynamics of the system, which enables us to study the resilience of the system. Different possible disturbances such as loss of productivity, price fluctuations and infrastructure breakdown will be modeled. Desirable but sudden changes such as the availability of an efficient technology will also be considered. Simulation studies incorporating these disturbances will be conducted to study their impact on the dynamics of the feedstock production system. Preliminary studies using this model have illustrated that the development of this sector is gradual, nonlinear and may take up to 15 years. The impact of the time of disturbance along this development trajectory will therefore be critical and will also be studied. Simulations for different types of energy crops such as switchgrass and Miscanthus will also identify whether the dynamics are a function of the energy crop. The work will also enable us to evaluate different system resilience measures appropriate for analyzing the feedstock production system.

See more of this Session: Sustainable Biomass Feedstock Production and Supply for the Emerging Biorefinery Industry
See more of this Group/Topical: Sustainable Engineering Forum

(609f) Vertically Aligned Silicon Radial p-n Junction Micropillar Array Solar Cells

Kane J. Miller, Chemical Engineering, University of Louisville, Louisville, KY, Kevin M. Walsh, Department of Electrical and Computer Engineering, University of Lousiville, Lousiville, KY and Xiao-an Fu, Chemical Engineering, University of Lousiville, Lousiville, KY

Solar cell applications using silicon have been extensively studied in the past. The theoretical limit of the efficiency of single-crystalline silicon solar cells is well known. Several methods for optimizing silicon solar cells and obtaining the maximum possible efficiency exist in the literature. Among these methods are changes in the geometry and configuration of the solar cells, such as introducing radial p-n junction solar cells.

Micro- and nanoscale radial p-n junction solar cells, in theory, result in increased solar cell efficiency compared to planar p-n junctions by reducing carrier collection length. The geometry of arrays of these radial p-n junctions also introduces internal reflection and light trapping effects, ultimately improving the photon absorption characteristics.

We have fabricated a first-pass vertically-aligned micropillar p-n junction solar cell array with a micropillar diameter of 2 μm, length of approximately 10 μm, and distance of 2 μm between adjacent pillars. These arrays were fabricated using conventional photolithography and deep reactive ion etching on a (100) n-type silicon wafer, followed by boron diffusion using a spin-on dopant. Initial measurements demonstrated an open-circuit voltage of 0.36 V, short-circuit current density of 24 mA/cm^2, a fill factor of 0.6, and efficiency of 5.3%.

Nanopillar p-n junction solar cell arrays of varying pillar diameters and lengths, fabricated using the combination of electron beam lithography and deep reactive ion etching, will be discussed. The test results will be compared to planar silicon solar cells and to theoretical calculations. With various techniques, we expect to increase the efficiencies of our solar cells beyond 10%.

See more of this Session: Semiconductor Nanowires and Nanotubes for Next Generation Photovoltaics - From Fundamentals to Devices
See more of this Group/Topical: Topical 5: Nanomaterials for Energy Applications

(278f) Novel Reactor Design for Conversion of Waste Oils to Biodiesel

Richard A. Cairncross, Colin J. Stacy and Brittany N. Preston, Department of Chemical and Biological Engineering, Drexel University, Philadelphia, PA

Biodiesel is a renewable, carbon-neutral fuel that can ease pressure on the world's dwindling supply of oil and petroleum diesel. This research explores production of biodiesel from alternative low-value feedstocks such as waste oils. Traditional biodiesel reactions mix an alcohol and a refined triglyceride (TAG) are mixed with an alkaline catalyst to create Fatty Acid Alkyl Esters. At Drexel University, a novel method is being developed for the production of biodiesel using an acid catalyst to esterify low-value natural oil and grease feedstocks that contain TAG and high concentrations of free-fatty acids (FFA). This method uses a novel column reactor, in which an alcohol vapor bubbles through the oil at temperatures in excess of 110°C. This method effectively produces biodiesel from unrefined waste oils, which have FFA contents in excess of 50% and are much cheaper than refined vegetable oils. The current reactor method has also shown that a water-diluted alcohol may be nearly as effective as a pure alcohol feed. Because the bubble reactor system is more robust for unrefined waste oils and impure alcohol feedstocks than traditional biodiesel reactors, this reactor could operate at a much lower cost than traditional biofuel processes. This talk analyzes performance of a bubble column reactor using mixed oil feedstocks containing both FFA and TAG which traditionally are reacted in two separate steps with different catalysts. Experimental data shows that in the bubble reactor with an acid catalyst, the FFA reacts quickly to biodiesel in less than 90 minutes and the TAG reacts more slowly. There are two chemical pathways for converting TAG to biodiesel, (1) direct transesterification of TAG and (2) hydrolysis to FFA followed by esterification of FFA; this talk explores the relative importance of these reactions within a bubble reactor.

See more of this Session: Catalytic Biofuels Refining II
See more of this Group/Topical: Fuels and Petrochemicals Division

(641b) A Method for Rapid Screening of Catalysts for Hydrodeoxygenation of Biomass Intermediates

David K. Johnson, Biosciences Center, National Renewable Energy Laboratory, Golden, CO

The focus of biofuels development is shifting from alcohol fuels to advanced biofuels that would be drop-in replacements for petroleum-derived hydrocarbons. There is a wide range of intermediates that can be produced from biomass by chemical or biological routes, but virtually all require further deoxygenation before they can be used as fuel components. Potential routes for decreasing the oxygen content of biomass components include dehydration, hydrodeoxygenation and decarboxylation (possibly after controlled oxidation). Deoxygenation by HDO is preferred over dehydration and decarboxylation reactions, as it is likely to result in higher yields through conservation of carbon, retention of hydrogen, and the production of more energy dense products. Hydrodeoxygenation (HDO) is a hydrogenolysis process that removes oxygen by converting it to water, however, HDO is often performed at high temperatures and pressures that significantly increase the cost of the transformation. We need to develop HDO catalysts that more efficiently convert biomass-derived components into fuel products that are compatible with the existing fuel distribution infrastructure. The feedstocks for the chemical transformations can include sugars, lignins, bio-oils, and chemicals obtained by biological transformations of sugars and biomass. It is clear from previous research that in order to design commercially viable catalysts for HDO upgrading of biomass-derived intermediates, the catalyst system must (1) exhibit high activity under mild conditions (<200°C), (2) contain low-cost materials, (3) demonstrate high stability in an aqueous environment, and (4) have high selectivity for the desired products. To accomplish this, a microwave heated (to 250ºC), pressurizable (to 250 psi), automated reaction system has been assembled to rapidly screen catalyst systems using model compounds. To allow automated operation, hydrogen is provided in-situ from a hydrogen donor material, such as formic acid, *i*-propanol or decalin. The rapid screening not only allows the rapid testing of catalysts, but also of reaction conditions, including temperature, reaction time, solvent, and hydrogen donor. The results from tests with model compounds such as acetovanillone and hydroxymethyl furfural, representing typical lignin and carbohydrate derived intermediates, with heterogeneous catalysts, will be described.

See more of this Session: Chemical and Catalytic Conversions and Processes for Renewable Feedstocks
See more of this Group/Topical: Sustainable Engineering Forum

(173b) Nano-Silicon/Graphene Composite Anode Modifications for Enhanced Performance Lithium Ion Batteries

Rhet Joseph De Guzman[1], Jinho Yang[2], Mark Cheng[2], Steven Salley[1] and K. Y. Simon Ng[1], (1) Department of Chemical Engineering and Materials Science, Wayne State University, Detroit, MI, (2) Department of Electrical and Computer Engineering, Wayne State University, Detroit, MI

Development of lithium ion batteries with improved capacity and cycle performance are in the forefront of efforts to increase the electrification of current-age vehicles. Lately, much attention has been focused in the use of silicon-based anode materials due to their low discharge potential and theoretical capacity (4,200 mAh/g) which is more than 10 fold of the traditional graphite anode. However, due to the increased accommodated lithium ions, silicon's volume varies by 400% causing pulverization and loss of electrical contact. Due to this, strategies have led to the use of nano-sized Si and incorporation of Graphene to be able to mitigate volume variation and therefore produce a stable anode material with improved capacity and cyclability. This project is aimed at the improvement of the dispersion of Si nanoparticles and the optimization of the graphene nanostructure producing a Si nanoparticle/graphene nanosheet composite anode demonstrating improved capacity retention and rate performance.

See more of this Session: Nanomaterials for Energy Storage III
See more of this Group/Topical: Topical 5: Nanomaterials for Energy Applications

(647c) Process Synthesis of Biodiesel Production Plant Using Artificial Neural Networks As the Surrogate Models

Ismail Fahmi, Chemical Engineering, The University of Tulsa, Tulsa, OK and Selen Cremaschi, Department of Chemical Engineering, The University of Tulsa, Tulsa, OK

Biofuel from biomass is very appealing due to its abundance and renewability to replace the usage of the depleting fossil-based fuels. Among many types of biofuels, biodiesel is probably one of the most attractive one because it can be used directly with the current traditional petro diesel engine, either as a substitute or as a blending aggregate [1-3]. Biodiesel is commonly obtained through transesterification of bio oil, either in a form of triglycerides or hydrolyzed fatty acid. There are several alternatives that can be considered for biodiesel production: transesterification with homogeneous catalysts [2, 4] or solid base catalysts [5-7], or by using supercritical methanol without the presence of catalyst [8]. In this work, superstructure optimization is performed to synthesize the optimum biodiesel production plant, i.e., the one that gives the minimum net present sink. Given the high computational resource required to solve the resulting Mixed Integer Nonlinear Programming (MINLP) formulation, using surrogate models is suggested [9, 10]. Among many kinds of surrogate models, Artificial Neural Network (ANN) has been shown to be a powerful tool both for complexity reduction and accurate mapping-function modeling [11, 12]. In this presentation, the feasibility of synthesizing the optimum biodiesel production plant using ANN as the surrogate models for the required unit operations is investigated. The optimum solution of using the alkali-catalyzed reactor with the total cost of USD 41 million was obtained in about 5 CPU seconds. When the solution was modeled in a process simulator, the resulting total cost of the simulation only differs about 1% attesting to the accuracy of the ANNs predictions. However, this comes with a cost of having to provide a large amount of training data for the ANN.

Keywords: Biodiesel Production, Artificial Neural Network, Optimization, Process Synthesis, Surrogate Models

References:

[1] D. Tapasvi, *et al.*, "Process model for biodiesel production from various feedstocks," *Transactions of the American Society of Agricultural Engineers,* vol. 48, pp. 2215-2221, 2005.

[2] Y. Zhang, *et al.*, "Biodiesel production from waste cooking oil: 1. Process design and technological assessment," *Bioresource Technology,* vol. 89, pp. 1-16, 2003.

[3] J. M. Lujan, *et al.*, "Comparative analysis of a DI diesel engine fuelled with biodiesel blends during the European MVEG-A cycle: Preliminary study (I)," *Biomass and Bioenergy,* vol. 33, pp. 941-

947, 2009.

[4] L. F. Bautista, *et al.*, "Optimisation of FAME production from waste cooking oil for biodiesel use," *Biomass and Bioenergy,* vol. 33, pp. 862-872, 2009.

[5] S. Yan, *et al.*, "Oil transesterification over calcium oxides modified with lanthanum," *Applied Catalysis A: General,* vol. 360, pp. 163-170, 2009.

[6] P. D. Patil and S. Deng, "Transesterification of Camelina Sativa Oil Using Heterogeneous Metal Oxide Catalysts," *Energy & Fuels,* vol. 23, pp. 4619-4624, 2009.

[7] D. E. Lopez, *et al.*, "Esterification and transesterification using modified-zirconia catalysts," *Applied Catalysis A: General,* vol. 339, pp. 76-83, 2008.

[8] K. T. Tan, *et al.*, "Production of FAME by palm oil transesterification via supercritical methanol technology," *Biomass and Bioenergy,* vol. 33, pp. 1096-1099, 2009.

[9] C. A. Henao and C. T. Maravelias, "Surrogate-Based Process Synthesis," in *Computer Aided Chemical Engineering.* vol. Volume 28, S. Pierucci and G. B. Ferraris, Eds., ed: Elsevier, 2010, pp. 1129-1134.

[10] N. V. Queipo, *et al.*, "Surrogate-based analysis and optimization," *Progress in Aerospace Sciences,* vol. 41, pp. 1-28, 2005.

[11] H. Seung-Soo and G. S. May, "Using neural network process models to perform PECVD silicon dioxide recipe synthesis via genetic algorithms," *Semiconductor Manufacturing, IEEE Transactions on,* vol. 10, pp. 279-287, 1997.

[12] I. Sabuncuoglu and S. Touhami, "Simulation metamodelling with neural networks: an experimental investigation," *International Journal of Production Research,* vol. 40, pp. 2483-2505, 2002.

See more of this Session: Energy Systems Design and Alternative Energy Sources I
See more of this Group/Topical: Computing and Systems Technology Division

(378e) Mitigating Strategies for Reactive Solids-Based Hydrogen Storage Systems

Joseph W. Pratt[1], Joseph G. Cordaro[2], Daniel E. Dedrick[3], Michael P. Kanouff[1] and Yehia (John) F. Khalil[4], (1)Thermal/Fluid Science and Engineering, Sandia National Laboratories, Livermore, CA, (2) Materials Chemistry, Sandia National Laboratories, Livermore, CA, (3)Hydrogen and Combustion Technology, Sandia National Laboratories, Livermore, CA, (4)Chemical Engineering Group, Physical Sciences Department, United Technologies Research Center (UTRC), East Hartford, CT

Solid-state materials such as metal hydrides and complex hydrides have been shown to be viable candidates for on-board hydrogen storage systems. However, some complex metal hydrides such as sodium alanate ($NaAlH_4$) in the powder form, exhibit unfavorable exothermic reactions when exposed to air or water. Mitigating this reactivity and its associated risks in the event of inadvertent exposure (e. g., tank rupture during a vehicular collision) is an important step for achieving widespread use and public acceptance of these technologies.

Several types of accident scenarios, exposures, and mitigating methods have been investigated in the past. This paper focuses on a specific scenario and mitigating strategy: exposure of the reactive solid to oxygen, and the effect of an active material-polymer composite mixture on its exothermic behavior.

Sodium alanate is selected as the representative of reactive solid hydrogen storage materials that exhibit unfavorable exothermicity, and is the "active material," although the mitigation methods developed are applicable to other similar materials. Previously, we reported on the use and effectiveness of polystyrene polymers synthesized (polymerized) along with the active material to create a composite. This paper presents our work on other polymers using the same methodology, comparing their effectiveness to the active material alone and to the previous work with polystyrene. Additive effectiveness is measured by the amount of heat released during controlled exposure to oxygen. Measurements are done both before and after subjecting the composite material to representative operating conditions of adsorption/desorption cycling, to understand the impacts (if any) of operating conditions on the stability and effectiveness of the composite material. SEM imagery is also used as a tool to better understand the method of mitigation and reasons for degradation (if any) of the composite. Additional tests were performed to investigate the impact of exposing the hydride (both in powder and powder compact forms) to vehicle fluids and road sprays and also to assess the sensitivity of the mitigated material to mechanical impact.

Daniel E. Dedrick, Joseph G. Cordaro, Michael P. Kanouff, Craig L. Reeder, Joseph, W. Pratt, and Y. F. Khalil, "Mitigation Technologies for Hydrogen Storage Systems based on Reactive Solids," presented at AIChE Annual Meeting, Salt Lake City, UT, Nov. 8-12, 2010.

(364a) Impact of Lignin Extraction and Cellulose III Conversion On the Enzymatic Digestibility of Corn Stover

Leonardo da costa Sousa, Shishir Chundawat, Nirmal Uppugundla, Bruce Dale and **Venkatesh Balan**, Chemical Engineering and Material Science, Great Lakes Bioenergy Center, Michigan State University, Lansing, MI

The benefits of delignification on plant cell wall digestibility by hydrolytic enzymes are well described in the literature. The presence of lignin increases non-productive binding of cellulases, resulting in enzyme inhibition and decrease in the sugar yields. Moreover, lignin acts as an anti-microbial agent, reducing cell viability and ethanol yields.

Recently, additional benefits by modifying the crystal structure of native cellulose (cellulose I) to cellulose III were also reported. After this modification, it was possible to double the rate of enzymatic hydrolysis of cellulose. However, this increase was observed for purified cellulose, native from algae (Cellulose Iα), which has a different crystal configuration than the cellulose native from plants (Cellulose Iβ). Therefore, it is not evident that this modification is possible to be obtained in plant cell walls and that it can represent the same kind of benefits observed for purified cellulose. From the literature, cellulose III is obtained by contacting cellulose I to supercritical ammonia or different types of amines in anhydrous conditions. Additionally, water is known to prevent the conversion of cellulose I to cellulose III, however the water concentration limit that allows cellulose III to be formed is not yet clearly defined. For biorefinery applications, plant biomass will contain a certain degree of moisture that could prevent stable cellulose III formation and therefore, it is important to verify in what extent is possible to convert native cellulose to cellulose III in plant biomass.

In this perspective, our goal was to test the benefits of cellulose III conversion under non-anhydrous conditions, using corn stover, and couple this modification with lignin extraction in a single stage process – Extractive AFEX (E-AFEX). These benefits were compared with the traditional non-extractive ammonia based pretreatment method - AFEX.

See more of this Session: Developments In the Pretreatment of Lignocellulosics for Bioconversion I
See more of this Group/Topical: Sustainable Engineering Forum

(429e) Direct Biodiesel Production From Mucor Circinelloides Fungi

Cristiano E. R. Reis, Bo Hu and Jianguo Zhang, Bioproducts and Biosystems Engineering, University of Minnesota, Saint Paul, MN

Biodiesel is one of the most promising alternative biofuels because it is more environmentally friendly and can be directly used in existing diesel engines. Biodiesel is mainly produced from plant oils, and this may have an impact on food supply. Hence, alternative lipid sources, such as oleaginous fungi, needs to be developed. Studies have shown that the dimorphic organism Mucor circinelloides is capable of producing reasonable amount of ethanol during their aerobic growth. It is also known that these filamentous fungi is also forming high content of lipid bodies in their cells, rich in unsaturated fatty acids. This study was about finding a way of reacting the microbial oils generated by Mucor with its own ethanol, in order to direclty produce Fatty acid ethyl esters, also known as biodiesel. The experiments were designed to handle the presence of water in the reaction, therefore, different catalysts were used in the reactions, including the lipase from Mucor miehei, as described in previous studies. The main purpose is funding the best efficiency, leading to a product that fulfills the specifications established by the American standards specifications, and resulting in the minimum environmental impact.

See more of this Session: Alternative Fuels and Enabling Technologies I
See more of this Group/Topical: Fuels and Petrochemicals Division

(110c) Secondary Manganese Dioxide Electrodes for Grid-Scale Batteries

Joshua Gallaway, Nilesh Ingale, Michael Nyce, Yasumasa Ito, Lev Sviridov, Abhinav Gaikwad, Steven Lever, Ali Firouzi, Sanjoy Banerjee and Daniel Steingart, Chemical Engineering, Energy Institute, City College of New York, New York, NY

Secondary Manganese Dioxide Electrodes for Grid-Scale Batteries

Joshua W. Gallaway,Nilesh Ingale, Michael Nyce, Yasumasa Ito, Lev Sviridov, Abhinav Gaikwad, Steven Lever, Ali Firouzi, Sanjoy Banerjee, Daniel A. Steingart

CUNY EnergyInstitute, The City College of New York

160 Convent Ave, NewYork NY 10031

Manganese dioxide (MnO_2) is well-established as aprimary battery cathode, but irreversibility has generally excluded its use insecondary batteries. Repurposingthis chemistry as a rechargeable electrical storage medium make possible a newera of MW-scale advanced batteries to add universal storage capability to theelectrical grid. This would bepossible due to low cost, high availability, and safety of the materials.

There is no universal consensus on the strategy forextending the cycle life of MnO2 electrodes, which are typicallyelectrolyte-filled porous electrodes also containing carbon and a binder. Adding dopant atoms to the MnO_2crystal structure is one. Anotheris low depth of discharge. Thefirst of these renders electrochemically irreversible products (Mn_2O_3)reversible, the second avoids their formation. These methods are not mutually exclusive, and a combinationof them may prove successful. Inany case, the problem is not only one of manganese electrochemistry, but alsocurrent distribution in the electrode and interactions with the conductivecarbon matrix.

Any one of many processes may become limiting in a porousMnO_2-carbon electrode: ionic conduction in pores, electronic conductionin the carbon matrix, interfacial contacts, or charge transfer of the MnO_2reaction itself. Parallel studiesof battery cycling, half-cell cycling, ex situ material characterization, andfailure analysis must be used to identify the limiting processes in such porouselectrodes.[1,2]

For this study, porous manganese dioxide cathodes werepaired with non-limiting cadmium anodes with a base cell size of ~4 Ah. The electrolyte was quiescent 12 M KOH. Electrochemical impedancespectroscopy (EIS) revealed a steady increase in cell impedance with cyclenumber, although battery failure was relatively sudden, usually after 100-200cycles. The EIS results weremodeled to pinpoint which phenomena in the porous electrode were responsiblefor increase in overall impedance.[3,4] Ex situ methods such as XRD were used to track materialchanges in the electrodes.

Recent battery efforts at the CUNY Energy Institute haveincorporated parallel experimentation across many cell sizes, with the targetbeing large cell stacks.[5]

ACKNOWLEDGEMENTS

The authors would like to thank The Wallis Foundation andARPA-E under award number DE-AR0000150 for generous support.

REFERENCES

1. J. S. Newman and C. W. Tobias, *J Electrochem Soc*, **109** (12), 1183-1191 (1962).

2. J. S. Chen and H. Y. Cheh, *J Electrochem Soc*, **140** (5), 1213-1218 (1993).

3. S. W. Donne and J. H. Kennedy, *J Appl Electrochem*, **34**,159-168 (2004).

4. S. W. Donne and J. H. Kennedy, *JAppl Electrochem*, **34**, 477-486 (2004).

5. Y. Ito, M. Nyce, R. Plivelich, M. Klein, D. Steingart, S.Banerjee, *J Power Sources*, **196**, 2340-2345 (2011).

See more of this Session: Novel Battery Chemistry and Technology
See more of this Group/Topical: Fuels and Petrochemicals Division

(388b) Metabolic Engineering of Rhizopus Oryzae for Enhanced Production of Fumaric Acid From Sugars

Baohua Zhang, WILLIAM G. LOWRIE Department of Chemical and Biomolecular Engineering, The Ohio State University, Columbus, OH and Shang-Tian Yang, The Ohio State University, Columbus, OH

Fumaric acid is a dicarboxylic acid used extensively in resins, food acidulants, and other applications including oil field fluids, esters etc. Some strains of *Rhizopus oryzae*, which is a filamentous fungus, can produce fumaric acid from sugars under aerobic conditions but the fumarate yield is relatively low due to the co-production of ethanol and other fermentation byproducts. In this work, metabolic engineering technique is applied to optimize the fermentation for enhanced fumaric acid production. Overexpression of native enzymes fumarase and pyruvate carboxylase showed no significant effect on the fumaric acid production, suggesting that the biosynthesis of fumaric acid is not limited by these two enzymes. On the other hand, overexpression of heterologous enzyme phosphoenolpyruvate carboxylase, which converts pyruvate with CO_2 fixation to oxaloacetate, led to increased fumaric acid production compared to the wild type control. The fermentation kinetics and activities of enzymes in the fermentation pathway of these *Rhizopus* mutants were studied and will be reported in this paper.

See more of this Session: Microbial Engineering for the Synthesis of Bulk and High-Value Products
See more of this Group/Topical: Sustainable Engineering Forum

(583b) Structural Characterization of AHP-Pretreated Biomass

Muyang Li, Chemical Engineering and Material Science, Michigan State University, East Lansing, MI and David Hodge, Department of Chemical Engineering and Material Science, Michigan State University, East Lansing, MI

Alkaline hydrogen peroxide (AHP) pretreatment is exceptionally well-suited to grasses, yielding high digestibilities at low enzyme loadings, while generating relatively few fermentation innhibitors. For AHP pretreatment, the question of how structural and chemical compositional changes within the plant cell wall correlate to pretreatment effectiveness has not been effectively resolved, while knowledge of how selective modification of lignin during pretreatment may improve digestibility may yield insights into both improving pretreatments and tailoring plant cell walls for deconstruction. This study presents a comprehensive chemical and structural characterization of the changes in plant cell walls associated with switchgrass (*Panicum virgatum* cv. Cave-In-Rock), corn stovers (a commercial hybrid and inbred brown midrib lines bm1 and bm3), sugar maple (*Acer saccharum*) and hybrid aspen (*Populus tremula x tremuloides*) that were AHP pretreated at varied severities. Both the remaining solids and solubilized biomass hydrolyzates were subjected to a number of characterizations including total polysaccharide composition (monosaccharides, acetate, uronic acids), lignin content as both Klason and acetyl bromide lignin, and the ratio of H/G/S monolignols as determined by thioacidolysis GC/MS. Solid-state ^{13}C CP/MAS NMR and HSQC 2D NMR were applied to determine changes associated with intra-lignin, carbohydrate-carbohydrate, carbohydrate-lignin linkages and functional groups. Apparent molecular weight distributions of soluble polymeric lignins and hemicellulose aggregates were determined by HP-SEC. Py-GC/MS was utilized to characterize changes in the volatilized compounds before and after pretreatment implying pretreatment-induced alterations in pyrolysis-labile monolignol linkages.

See more of this Session: Developments In the Pretreatment of Lignocellulosics for Bioconversion III
See more of this Group/Topical: Sustainable Engineering Forum

(288d) Successful Networking-Yes, Engineers Can Network Too!

Lori McDowell, Praxair, Inc, The Woodlands, TX

The author will discuss networking.

See more of this Session: Developing Leadership Skills
See more of this Group/Topical: Management Division

(62f) Modeling and Optimization of Algae Based Biodiesel Production

Soumya Yadala, Chemical Engineering, The University of Tulsa, Tulsa, OK and Selen Cremaschi, Department of Chemical Engineering, The University of Tulsa, Tulsa, OK

Processingof biodiesel from algae biomass has attracted much interest in recent times due to many reasons. Significant amongst them are global warming and scarcity in fossil fuels. It has been suggested that this third generation biofuel, a carbon neutral alternative, is capable of meeting the existing demand for transport fuels [1]. However, substantial challenges are encountered in deriving fuels from algae biomass. Primary amongst them is the relatively higher cost of cultivating and harvesting algae species. Due to this reason, currently, algae based biofuels cost much higher than the biofuels from first generation and second generation biofuel feedstock [2].

This work focuses on modeling and optimization of algae based biodiesel production. Design parameters, for cultivation of algal species, at a geographical location are affected by that location's climatic conditions such as annual solar irradiance. Hence, geographical location, at which algae is cultivated, plays a vital role in the biodiesel production. Another significant factor to be considered for producing maximum amount of biodiesel is the selection of algal species. There are a variety of algal species present throughout the world, however, not all algal species are suitable for biodiesel production because of varying amounts of oil content and specific growth rates. Hence, selection of algal species is also an important factor in determining the cost of biodiesel production. There are a number of technologies for each process involved in the production of biodiesel from algae biomass but the question of which technology, for which species, and at which location, to attain maximum profits is still an open challenge to the researchers. Figure 1 presents the tasks that should be completed to produce algae-based biodiesel and options/technologies (which can be used to achieve these tasks) considered in this work. Using Figure 1, a disjunctive programming model is developed to determine the algal species, the location for the cultivationof algae, and the optimum route from growth of algae to production of biodiesel by selecting one technology for each process with the objective of minimum cost of biodiesel production. The resulting Mixed Integer Nonlinear Programming(MINLP) model is solved using GAMS 23.6.5/SBB solver. With the model parameters used, 8000 tons/year of biodiesel can be produced at a cost of about $153 per gallon for a period of ten years using the selected technologies. Accordingly, Tricornutum species was selected to grow at all the given locations by means of artificial growing technique-tubular photo bioreactor. The cultivated algae were harvested through flocculation which was then concentrated using centrifugation. The algae biomass was dried by low pressure shelf drying followed by extraction of algal oil by expeller press and finally, the algal oil was sent to transesterification for biodiesel production.

REFERENCES

1. Y. Chisti, Biodiesel from microalgae, *Biotechnol. Adv.* **25** (2007), pp. 294-306.

2. Sheehan, J., Dunahay, T., Benemann, J., Roessler, P. A Look Backat the U.S. Department of Energy's Aquatic Species Program: Biodiesel fromAlgae; Close-Out Report. U.S. Department of Energy (1998).

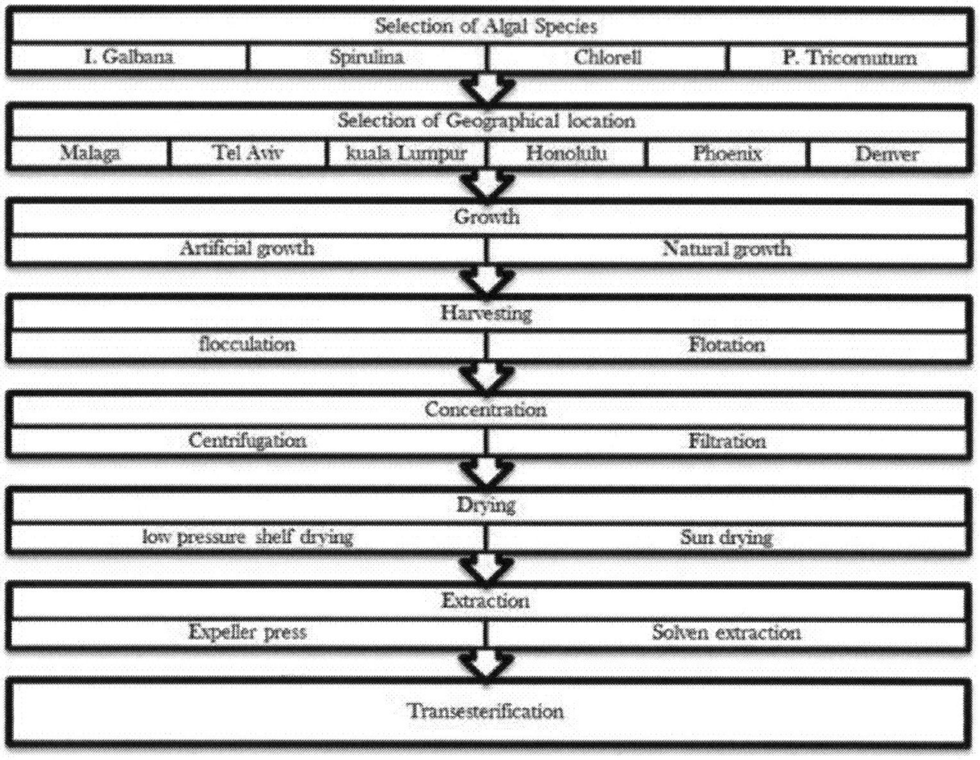

Figure 1. Tasks and corresponding options/technologies for biodiesel production from algae

See more of this Session: Advances In Algal Biorefineries I
See more of this Group/Topical: Sustainable Engineering Forum

(305f) An Integrated Framework for Design of Sustainable Chemical Process Based On Life Cycle Assessment

Masahiko Hirao and Yasunori Kikuchi, Department of Chemical System Engineering, The University of Tokyo, Tokyo, Japan

In this work, we propose a supporting framework for the design of chemical process integrating computer-aided process engineering (CAPE) and life cycle assessment (LCA) tools. LCA has been a useful tool for quantifying the environmental impacts from existing product life cycle and its application to process development enables us to design environmentally conscious product and process. During process development, however, implementation of LCA is difficult due to lack of inventory data. CAPE tools are used for simulation and evaluation of chemical processes, but have potential for the estimation of process inventories used in LCA. Thus, we modeled activities and information flows using integration definition for function modeling (IDEF0) method for the sustainable chemical process development in which CAPE and LCA tools are integrated. Applicability of the proposed activity model is demonstrated and verified by case studies on the process development of bioethanol-derived chemicals.

See more of this Session: Life Cycle Analysis of Renewable Feedstock-Based Processes and Products I
See more of this Group/Topical: Sustainable Engineering Forum

(364b) The Effects of Cationic Polyelectrolyte On the Pretreatment and Enzymatic Hydrolysis of Corn Stover

Shaowen Ji and Ilsoon Lee, Chemical Engineering and Materials Science, Michigan State University, East Lansing, MI

Lignocellulose is the most abundant renewable biomass and the potential substrate for ethanol production. The costly pretreatments and large consumption of enzymes are needed to reach high cellulose conversion. Lignin, as one of major cell wall components, is considered as the barrier to enzymatic hydrolysis by blocking cellulase from productively being adsorbed. The removal or effective modification of lignin is required for efficient pretreatment process. In this study, the effects of cationic polyelectrolyte, poly-(dimethyldiallylammonium chloride) (PDAC), on alkaline pretreated corn stover were investigated. Structural study by Scanning Electron Microscopy (SEM) imaging revealed particles morphology on cell walls of pretreated corn stover. These particles were confirmed to contain lignin by Confocal Laser Scanning Microscopy (CLSM) and Transmission Electron Microscopy (TEM) imaging. Chemical compositions of pretreated corn stover at different alkaline and polyelectrolyte concentrations were analyzed. Higher polyelectrolyte concentration in the pretreatment resulted in higher lignin retention. Evaluation of enzymatic hydrolysis showed the enhancement of cellulose and hemicellulose conversion after polyelectrolyte pretreatment. PDAC was proposed to facilitate cellulose disruption and lignin redistribution via its interaction with lignin, which significantly modified biomass morphological properties, such as increased porosity and improved accessibility of cellulose surface.

See more of this Session: Developments In the Pretreatment of Lignocellulosics for Bioconversion I
See more of this Group/Topical: Sustainable Engineering Forum

(218d) Energy Efficiency of Battery Systems with Forced Convection Between Electrodes

Sun Ung Kim, Department of Chemical Engineering, Department of Mechanical Engineering, University of Michigan, Ann Arbor, MI and Charles W. Monroe, Department of Chemical Engineering, University of Michigan, Ann Arbor, MI

The energyefficiency of an alkaline battery system with forced convective flow betweenthe electrodes has been investigated. Rechargeable alkaline batteries, which dependon the reaction between zinc and manganese dioxide (Zn/MnO_2), havebeen widely used because of their excellent properties such as inexpensiveness,long shelf-life, high-temperature performance, and environmental friendliness comparingto the Leclanche cell or zincchloride type batteries.[1] A full recharge of commercial alkaline AAor AAA type batteries takes, however, more than 10 hours; but, the charge/dischargetime is difficult to be reduced because the limiting current i_L (the maximumcharge-transfer rate) is fixed in a battery system.

In 2010and 2011, Suppes introduced an interesting new battery configuration, which canin principle reduce charge/discharge times by using coaxial liquid flow tocontrol the limiting current, i_L.[2,3]The Suppes system uses a mechanical pump to drive fluid between porouselectrodes as the battery operates, so as to manipulate the concentrationdistributions of ions in the separator region. We applied the Nernst-Planckdilute-solution theory[4-7] to model this battery configuration. Themodel allows the limiting currents, i_L,to be calculated in terms of a Peclet number, Pe_{forced}, whichquantifies the dimensionless rate of forced convection. In this presentation, wewill show quantitatively how a pump can be used to adjust the maximum charge/dischargerate of the Suppes flow battery.

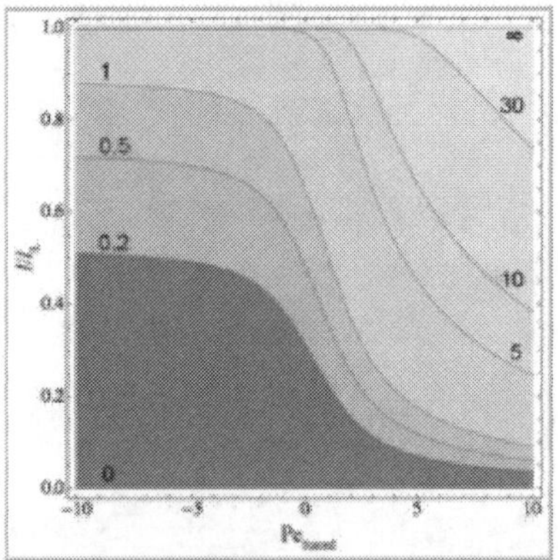

Fig. 1. A contour plot of the dimensionless power loss from diffusion potential $|z_+ F I \Delta\Phi_{diffusion}/(RT)|$ as a function of the dimensionless current, I, and the dimensionless forced convection rate, Pe_{forced}.

Eventhough forced convection can in principle reduce the charge/discharge time as muchas needed, the rate increase is necessarily accompanied by a power loss due toconcentration overpotential, as shown in Fig. 1. The power loss is increased becausethe larger Pe_{forced} leads to a non-uniform concentrationdistribution in the battery, inducing a diffusion potential $\Delta\Phi_{diffusion}$. In addition, an analysis with the Ergun equation[8]shows that significant power may be lost to fluid friction in the porouselectrodes and separator. The fluid friction causes a significant pressure-headloss that increases with the forced-convection rate parameter Pe_{forced};therefore, the use of forced flow to enhance rates may be more effective forincreasing battery charge rates than for increasing rates of discharge.

Once theenergy losses are calculated, the task of evaluating the energy and power efficiencyof the battery presents itself. We investigate the optimal fluid velocity neededto get the highest efficiency from the Suppes flow battery. Also, the comparisonof this battery efficiency with the Carnot efficiency will be discussed.

References

1. K. Kordesh and M. Weissenbacher, *J. Power Sources* **51**(1-2)(1994) 61-78

2. G. J. Suppes, B. D. Sawyer, M. J. Gordon, *AIChE J.* (2010) (in press).

3. B. D. Sawyer, G. J. Suppes, M. J. Gordon, M. G.Heidlage, *J. Appl. Electrochem.* (2011) (online).

4. W. Nernst, *Z. Physik. Chem.* **2** (1888), 613-637.

5. M. Planck, *Ann. Physik*, **275**(2) (1890) 161-186; **276**(8)(1890) 561-576.

6. J. S. Newmanand K. E. Thomas-Alyea, 3rd ed. Wiley-IEEE, (2004) p. 271.

7. A. J. Bard andL. R. Faulkner, 2nd ed. Wiley, (2000) p. 138.

8. S. Ergun,*Chem. Eng. Prog.* **48** (1952) 89.

See more of this Session: Flow Batteries for Energy Storage (In Memory of Bernard Lee)
See more of this Group/Topical: Fuels and Petrochemicals Division

(373e) Molten Mixed Metal Anodes for Solid Oxide Fuel Cells

Ashay D. Javadekar[1], Abhimanyu Jayakumar[2], Sounak Roy[1], Raymond J. Gorte[2], John Vohs[2] and Douglas Buttrey[1], (1)Department of Chemical Engineering, University of Delaware, Newark, DE, (2) Department of Chemical and Biomolecular Engineering, University of Pennsylvania, Philadelphia, PA

Increasing demand for energy calls for efficient use of solid carbonaceous fuels such as coal and waste biomass. Solid oxide fuel cells using molten metal anodes have shown sufficient promise in utilization of solid carbon fuels for energy generation [1,2]. Recently, we have shown that cells with molten Sb anodes have been able to generate electric power densities exceeding 300 mW/cm^2 at 973 K using sugar char as fuel[3]. The research presented in this work is directed toward understanding the chemistry associated with use of a low melting metal mixtures including Sb as anodes in fuel cells in a battery mode (i.e. without added fuel). Various mixtures of Sn, Bi, In and Pb, each with Sb have been used as anodes. In all cases, metals having the higher tendency to oxidize are preferentially oxidized, and the cell behaves much as it would with that metal in pure form as the anode. In cases where the oxide formed from this preferential oxidation has a higher melting point than the operating temperature of the cell (e.g. SnO_2 with T_{mp} = 1903 K or In_2O_3 with T_{mp} = 2185 K in cells operating at around 1000 K or less), the oxide may deposit on the electrolyte-anode interface resulting in degradation of cell performance. In the Sb-Pb system, although PbO melts well above the 973 K operating temperature (T_{mp} = 1161 K), Sb is preferentially oxidized and the cell behaves similarly to that using pure Sb.

Experiments involving extended periods of operation for the cell using mixture of molten Sb and Pb as anode were conducted and it was observed that the cell performs until 100% of the Sb is oxidized. This is a significant improvement compared with a cell using the same amount of pure Sb as the anode, which degrades after 10% oxidation of Sb[3]. Characterization of cross-sections from quenched mixed metal anodes is underway to better understand the nature of the observed improvement in performance.

1) Jayakumar, A.; Lee, S.; Hornes, A.; Vohs, J. M.; Gorte, R. J. A Comparison of Molten Sn and Bi for Solid Oxide Fuel Cell Anodes. *J. Electrochem. Soc.* **2010**, *157*, B365–B369.

2) Jayakumar, A.; Vohs, J. M.; Gorte, R.J. Molten-Metal Electrodes for Solid Oxide Fuel Cells, *Ind. Eng. Chem. Res.* **2010**, *49,*10237-10241.

3) Jayakumar, A.; Küngas, R.; Roy, S.; Javadekar A.; Buttrey D. J.,; Vohs, J. M.; Gorte, R.J. Electrical Power Generation Using Solid Fuels in a Solid Oxide Fuel Cell with a Molten Antimony Anode, *Submitted.*

See more of this Session: Fuel Cell Technology II

(523f) Economics of Pretreatment for Biological Processing

Ling Tao, Andy Aden and Richard T. Elander, National Bioenergy Center, National Renewable Energy Laboratory, Golden, CO

Economics of Pretreatment for Biological Processing

Ling Tao[1], Andy Aden[1] and Richard T. Elander[1]

[1] NationalRenewable Energy Laboratory, 1617 Cole Blvd., Golden, CO 80401, USA

Abstract:

A broad understanding of pretreatment economics within anoverall conversion context will be presented. The importance of pretreatmentin the overall economics of the biorefinery is addressed along with the keyparameters that influence the pretreatment economics. A large array ofpretreatment technology options exist and are currently being researched anddeveloped in this study. There appear to be many performances and costtradeoffs for these varying technologies. Only through continued R&D willthese tradeoffs be more fully understood. Process economics, especially thepromising pretreatment technologies studied in the Consortium for AppliedFundamentals and Innovation (CAFI) projects, are analyzed and compared indetails in this work. The six pretreatment technologies vary greatly in termsof their process design and projected total capital investment. Pretreatmentarea capital costs are based on past research, existing databases andengineering judgments, and include not only pretreatment reactor costs, butalso pretreatment chemical recovery and recycle equipment costs for theprocesses that use high pretreatment chemical loadings. More accurate costestimates for each pretreatment process will require specific vendor quotationsbased on pretreatment process conditions and any pretreatment chemical recoveryrequirements. Overall ethanol yield, which is largely based on achieved overallsugar yield achieved in pretreatment and subsequent enzymatic hydrolysis steps,is one of the single-most important factors in determining projected ethanolcosts for each process. Many key variables and parameters can have significantimpact on the pretreatment economics. While many are discussed within this talk, continuing R&D will generate data that further guide the development of thebiomass conversion industry.

See more of this Session: Developments In the Pretreatment of Lignocellulosics for Bioconversion II
See more of this Group/Topical: Sustainable Engineering Forum

(411b) Alternatives to Tailings Ponds for Oil Sand Tailings Disposal

Yuming Xu and Tadeusz Dabros, CanmetENERGY, Natural Resources Canada, Devon, AB, Canada

ABSTRACT

Oil sand resources are playing an increasingly important role in maintaining energy supplies for North America. Bitumen is produced from oil sand, mainly by two processes: in situ recovery by SAGD (steam assisted gravity drainage) for deep reservoirs and open-pit mining operations for shallow deposits. In open-pit mining operations, bitumen is separated from sand grains by a hot-water extraction and flotation process that produces large volumes of tailings. In the existing commercial plants, tailings from the extraction operation are pumped into large settling ponds. The coarse solids settle out quickly as a sand beach while the fine particles settle very slowly and finally form mature fine tailings (MFT), a stable suspension containing about 30% solids. Over the years, large volumes of MFT have accumulated in the ponds and the continuing accumulation is a cause of growing economic and environmental concern.

In order to address these concerns we are actively investigating alternative methods of dry and stackable tailings disposal such as flocculation and thickening, filtration, and centrifugation. Coarse oil sand tailings are flocculated using polymeric flocculants and the sediments are either disposed in the mined-out pit or further filtered. Various factors affecting flocculation and filtration performance were investigated in this work. The results demonstrate that the tailings have much better filterability after appropriate flocculation, and filtration becomes a viable method of producing stackable tailings. Another method for treating fine tailings and MFT is centrifugation. Centrifugal force enhances the separation of fine clays from water and speeds up further consolidation. Pilot-scale test results show that centrifugation of the fine tailings produces consolidated cake that can be disposed of without the need for pond containment. These methods provide potential alternatives for eliminating or reducing the use of tailings ponds and for the recovery of water and the associated thermal energy.

See more of this Session: Solids-Liquid Separation Processes for the Petroleum Industry
See more of this Group/Topical: Separations Division

(305e) Effect of Biomass Feedstock On the Sustainability Metrics of Fischer-Tropsch Fuels

Ignasi Palou-Rivera and M.Q. Wang, Argonne National Laboratory, Argonne, IL

The use of biomass as feedstock for the manufacture of Fischer-Tropsch fuels has significant effects on their sustainability metrics. A number of studies have looked at this biomass use [1,2], with an emphasis on techno-economic results. Some of the studies incorporate limited life-cycle analyses [1,2,3], especially considering green-house gas emissions. Our study is a complete Well-To-Wheels Life Cycle Analysis of biomass-fed Fischer-Tropsch liquid fuels (gasoline, diesel, and jet-fuel) performed using our GREET(Greenhouse gases, Regulated Emissions, and Energy use in Transportation) Model developed at Argonne national laboratory under DOE sponsorship. The presented LCA has a particular focus on the effects of several design and operational parameters:

- ***Biomass characteristics***: forestry and agricultural residues, humidity level, relative amount (when not used as only feedstock), etc.
- ***Gasification characteristics***: biomass co-gasification with coal, mix of biomass-derived syngas with other (e.g. coal- and natural gas-derived syngas), gasifier technologies, etc.
- ***Process characteristics***: CO_2 capture and sequestration (CCS) degree, process integration, FT synthesis characteristics and catalyst type, etc.

The LCA includes an encompassing sensitivity analysis of the effects of all these parameters on major sustainability metrics. The considered metrics include energy use (total, fossil, etc), green-house gas emissions (CO_2e GHG), and emissions of other criteria pollutants (NO_x, PM10, PM2.5, SO_x, VOC, CO)

[1] Liu, G.; Larson, E.D.; Williams, R.H.; Kreutz, T.G.; Guo, X. *Making Fischer-Tropsch Fuels and Electricity from Coal and Biomass: Performance and Cost Analysis*. Energy Fuels, 2011, 25, 415-437

[2] NETL/DOE (National Energy Technology Laboratory/U.S. Department of Energy), *Affordable, Low-Carbon Diesel Fuel from Domestic Coal and Biomass*, January 2009.

[3] Xie, X.; Wang, M.; Han, J. *Assessment of Fuel-Cycle Energy Use and Greenhouse Gas Emissions for Fischer-Tropsch Diesel from Coal and Cellulosic Biomass*. Environ. Sci. Technol., 2011, 45, 3047-3053.

See more of this Session: Life Cycle Analysis of Renewable Feedstock-Based Processes and Products I
See more of this Group/Topical: Sustainable Engineering Forum

(388a) Metabolic Flux Analysis of *Escherichia Coli* MG1655 Under Octanoic Acid Stress

Yanfen Fu[1], Jong M. Yoon[1], Liam A. Royce[1], Maria Rodriguez-Moya[2], Ramon Gonzalez[2], Laura Jarboe[1] and Jacqueline V. Shanks[1], (1)Department of Chemical and Biological Engineering, Iowa State University, Ames, IA, (2)Departments of Chemical & Biomolecular Engineering and Bioengineering, Rice University, Houston, TX

Metabolic engineering has evolved to the point of fulfilling the dream of having our industrial chemicals produced renewably. Carboxylic acids (e.g., short chain fatty acids (SCFAs) such as octanoic acid (C8)) are one such chemical intermediate that can be produced with *Escherichia coli* engineered with a short chain thioesterase. However, C8 is toxic to *E. coli* growth at concentrations greater than 10 mM. To design a better host strain for the production of large amount of those SCFAs, metabolic flux analysis of *E. coli* grown under C8 stress was performed. When *E. coli* MG1655 was grown in M9 minimal media supplemented with 35 mM C8 under aerobic conditions, cell growth was inhibited by 55.1% ± 10.0% (mean ± standard deviation, n=4) compared to control (without C8). To study fluxes in central carbon metabolism, 10 g/L of a mixture of labeled and unlabeled glucose (25% uniform label glucose, 25% 1-label glucose and 50% natural glucose) was used as the sole carbon source for bacterial growth. Proteinogenic amino acid isotopomers were measured by 2D (^{13}C, ^{1}H) NMR. Notable differences of several amino acid isotopomer enrichments were observed between control condition and C8 stress condition, suggesting the precursor nodes of these amino acids in metabolic pathways are responding to the stress. More specifically, the difference suggested that the flux distribution among tricarboxylic acid (TCA) cycle, pyruvate node and α-ketoglutarate node and oxaloacetate node was changed. Under C8 stress, acetate production flux increased by ~50% and the flux from phosphoenolpyruvate to pyruvate increased by ~50%, while the TCA cycle pathway, the malic enzyme pathway, and the pentose phosphate pathway decreased by 25%, 90%, and 20% respectively. Downregulation of the TCA cycle can occur if an excess of NADH occurs, due to feedback inhibition of isocitrate synthase. Since our flux data shows that less NADH is produced via central carbon metabolism under C8 stress, an excess of NADH could occur if C8 stress inhibits regeneration of NAD^+ from NADH, via membrane disruption. Our transcriptome and proteomic data suggest that carboxylic acids may negatively impact the function and/or integrity of the cell membrane. Flux data combined with transcriptomic and proteomic analysis allow the development of hypotheses in context with the mechanism of C8 toxicity, which in turn helps to engineer strains with increased tolerance of C8.

See more of this Session: Microbial Engineering for the Synthesis of Bulk and High-Value Products
See more of this Group/Topical: Sustainable Engineering Forum

(349e) Recovery of Large Amount of Levoglucosan Produced by the Fast Pyrolysis of Acid Infused Biomass From the Reactor

Najeeb Kuzhiyil, Dustin Dalluge and Robert Brown, Center for Sustainable Environmental Technologies, Iowa State University, Ames, IA

We have previously shown that infusion of mineral acids converts alkali and alkaline earth metals in biomass into thermally-stable salts, reducing their catalytic activity during fast pyrolysis. The suppression of metal-catalyzed ring breaking reactions allows depolymerization to dominate pyrolysis reactions, resulting in enhanced yields of levoglucosan (LG). In a bench scale fluid bed pyrolyzer, however, the large amount of LG formed by the fast pyrolysis of acid infused switchgrass at 500°C led to bed agglomeration. It was inferred that the agglomeration was caused by LG polymerization and charring taking place at about 270°C and 280°C respectively. Analytical pyrolysis studies using micropyrolyzer/GC/FID showed that temperatures between 350 and 400°C are more suitable for collecting LG from acid-infused biomass. The present work explored various techniques such as low temperature pyrolysis and higher sweep gas flow rates to improve the recovery of LG using a bench scale fluid bed and auger reactors.

See more of this Session: Biomass Pyrolysis I
See more of this Group/Topical: Fuels and Petrochemicals Division

(522f) Synthesis of Green Diesel Fuel From Fatty Acid Feedstocks Via Electrochemical Hofer-Moest Decarboxylation

Cody Wagner, Charles Coronella and Rainer Busch, Chemical Engineering, University of Nevada, Reno, Reno, NV

Production of biodiesel by esterification or transesterification of long-chain fatty acids is a well known process, but the fuel product has characteristics making it not fully compatible with petroleum diesel engines. A better option is to convert fatty acid molecules to hydrocarbons by removing their carboxyl group. Decarboxylation of fatty acid feedstocks was performed by electrolysis using graphite electrodes, which selectively causes a Hofer-Moest (or non-Kolbe)-type reaction. Methanol was used as a solvent for a robust selection of fatty acid feedstocks.

The liquid product created by the electrochemical reaction was composed of mainly alkenes with at least one double bond, ethers, and methyl esters. This fuel product has advantageous cold-flow properties, no sulfur content, heating values close to those of diesel fuels, and is capable of running in current diesel engines as a drop-in replacement for petroleum diesel, unlike biodiesel. The temperature, pH, concentrations of ions, and type of base used were all varied to optimized the current density, which limits the production rate. A separation method was tested for removal of the fuel product from the electrolyte. An efficiency was then calculated for several varieties of fatty acid feedstocks, comparing the electrical energy input to the product generated from the electrolysis. Overall first-law thermodynamics and economics were also evaluated, examining production rates and the large-scale viability of the process.

See more of this Session: Developments In Biobased Alternative Fuels I
See more of this Group/Topical: Sustainable Engineering Forum

(647e) A Life Cycle-Based Analysis and Optimization Framework for Designing Sustainable, Multi-Product Biomass-to-Bioproducts Value Chains

Paritosh K. Sharma, Chemical Engineering, Louisiana State University, Baton Rouge, LA and Jose A. Romagnoli, Chemical Eng, Louisiana State University, Baton Rouge, LA

A life cycle-based analysis and optimization framework for designing sustainable, multi-product biomass-to-bioproducts value chains

In recent years there has been a marked surge in the search for alternative sources of energy that wean the world off of dependence on fossil fuels and reduce our carbon foot-print on this world. After a boom in U.S. corn-based ethanol in the early part of the 21st century, the interest has gradually shifted towards more viable sources of biofuels and biochemicals. Cellulosic ethanol, biodiesel, and syngas are examples of such fuels that are extremely attractive owing to the fact that the raw materials can be composed completely of "left-over" wastes of food crops and forest harvests that don't interfere with the human food chain and the natural ecosystem.

Even with increased research emphasis on biofuel and biochemical production technologies, large scale production is still hampered by many factors. Some of these factors that have received considerable interest in recent times include, cultivating more robust feedstock, genetically engineering enzymes and microbes that are more efficient and resistant to poisoning by by-products and improving process technologies that can render large scale production of biofuels a more profitable option for private entities. One area that has not received enough research interest in recent times is enterprise wide modeling and optimization in order to make the entire enterprise more competitive in the energy sector.

Modeling an enterprise that produces renewable-based fuels and chemicals can provide valuable insight into the inter-play of the bioproduct supply-chain. This idea had been motivated by a recent surge in the research areas concerning supply-chain optimization and how cost-cutting measures throughout a supply-chain can render an otherwise sluggish enterprise, profitable.

In the current formulation, we apply a distributed, systematic approach to model, optimize, and analyze three aspects of sustainability for a biomass-to-bioproducts value chain; business, environmental, and social sustainability. Life cycle analyses are conducted to generate forecasts for environmental parameters, infrastructural and employment data, input costs and supplies, product yields, and output market conditions, which are integrated with integer-based stochastic optimization models that yield strategic decisions such as feedstock and supplier selection, transportation logistics design, technology and product selection, waste mitigation design, and capacity design under uncertainty. A two-stage optimization strategy is used to increase computational efficiency while extracting the biggest value drivers for the successful construction and operation of a biorefinery.

The first stage in optimization is termed *portfolio screening*; a deterministic integer programming model is formulated to screen out economically and environmentally unfavorable choices of feedstock, conversion technologies, and products. Stakeholder value of the entire value chain, including suppliers and customers, is maximized in the model with a rolling-time horizon of 12 years. The model equations and constraints included are material, energy, capacity, cash, and debt balances, transformation equations, supply and demand constraints, and capacity and resource constraints. Carbon taxes and mitigation credits are used to represent environmental implications of biorefinery activities. The outputs of the model include integer decisions regarding feedstock, suppliers, technologies, and product selection. Furthermore, sensitivity analysis is carried out on the final supply chain and process configuration to determine process, supply chain, and market parameters that are the greatest economic, environmental, and social value drivers for the biorefinery.

The second stage of optimization is formulated as a *Real-options* stochastic optimization model. The purpose of the model is to design a timing strategy for feedstock production and pricing, incremental technology acquisition, and capacity design given uncertainty in input costs and supplies, technology costs, and product demands and prices. The model is similar in form and structure to the deterministic MIP from the first stage with stochastic forecasts as inputs. Two types of options are modeled; Growth options to model capacity expansion, and switching options to represent flexible biorefining systems with the ability to switch product volumes between different products. A binomial decision tree is used to discretize continuous stochastic forecasts for inputs. The final objective function is a probability weighted stakeholder value whose form is derived from the deterministic model. The screened portfolio choices that are obtained from the first-stage optimization are used as inputs. Integer decisions for the model include capacity establishment and expansion, investment in new technologies, and product switching by re-allocation of feedstock to different product groups.

The resultant framework is applied to a prospective biomass refinery in Southeastern United States. The feedstock choices include energy crops, namely, switchgrass, miscanthus, and energy cane. The feedstock decisions are affected by capital investment for land acquisition, fixed and operating costs for crop growth and harvest, transportation expenses, seasonal availability, environmental life cycle impact, and biomass and sugar yields of the crop. The initial product portfolio includes ethanol and butanol as fuels, and succinic and lactic acid and value-added chemicals. The conversion and separation technologies are demarcated into three groups; pretreatment, hydrolysis (dilute acid (DA), ammonia-fiber explosion (AFEX)), and fermentation (simultaneous saccharification and fermentation (SSF) and separate hydrolysis and fermentation (SHF)) which are common for all products, and separation and product recovery (distillation, molecular sieves, pervaporation, membrane separation) which are specific to each product. The product-technology set choices are affected by capital costs, operating costs, process yields, environmental life cycle impact, and market potential.

The screening model results yield miscanthus as the feedstock choice owing to high biomass yields and ease of processing, and ethanol and succinic acid as the products owing to great market potential and favorable profit margins. Dilute acid pretreatment, enzymatic hydrolysis, and SSF are selected as the conversion technologies owing to favorable capital cost structure and process economics including the energy and environmental load. Butanol and lactic acid not chosen primarily due to the high cost of product recovery while distillation and molecular sieves (ethanol) and membrane separation (succinic acid) are chosen for the products that are selected.

The results are then fed to the second stage of optimization with uncertainty being represented in the marginal returns that are generated by capacity and production decisions. The uncertainty in the return implies uncertain revenue and cost streams. The individual components of the uncertain revenue and cost streams include uncertain product prices, uncertain input costs and supplies, and uncertain process yields. The uncertain returns are assumed to capture all uncertainty during optimization while the distribution of returns is generated separately using the stochastic forecasts for the individual components. The results indicate an incremental capacity design to mitigate and manage supply, capital, operational, and market risks with the overall expected stakeholder value at $MM 3.4 indicating that positive value is being created. The incremental capacity design is compared to economies of scale based capacity design where all capacity is established during the first planning year. The stakeholder value for economies of scale based design is $MM -30.3 indicating a $MM 33.7 advantage to using incremental capacity design. Product switching capabilities are recognized in the final two time periods (4 years) where a price shock to the upside in succinic acid prices is simulated. Capacity is re-allocated to succinic acid production to exploit the favorable market conditions with a 2% increase in the stakeholder value thus depicting the advantages of designing flexible production systems for biomass refineries.

See more of this Session: Energy Systems Design and Alternative Energy Sources I
See more of this Group/Topical: Computing and Systems Technology Division

(106c) High-Performance Electrochemical Capacitors Based On Hybrid Graphene/Metal Oxide Nanostructured Textiles

Guihua Yu[1], Liangbing Hu[2], Yi Cui[2] and Zhenan Bao[1], (1)Chemical Engineering, Stanford University, Stanford, CA, (2)Materials Science and Engineering, Stanford University, Stanford, CA

Large scale energy storage system with low cost, high power and long cycle life is crucial for addressing the energy problem when connected with renewable energy production. To realize large-scale applications of the energy storage devices, there remain several key issues including the development of low-cost, high-performance materials that are environmentally friendly and compatible with low-temperature and large-scale processing. In this talk, we will present a novel structure based on graphene, transition metal oxide and textiles where solution-exfoliated graphene nanosheets conformably coated on highly porous textiles function as three-dimensional conductors and electrodeposited metal oxide nanostructures function as active electrode materials. Electrochemical capacitors based on hybrid graphene/metal oxide based textiles in aqueous electrolytes exhibit very promising characteristics with a power density of ~110 kW/kg, and excellent cycling performance of >95% capacitance retention over 5000 cycles, which are substantially better than those previously reported for this oxide-based system. Such low-cost, high-performance energy textiles based on all-solution-processed graphene/metal oxide hierarchical nanostructures offer great promise in grid-scale energy storage device applications.

See more of this Session: Nanomaterials for Energy Storage II
See more of this Group/Topical: Topical 5: Nanomaterials for Energy Applications

(39e) Atomic Layer Deposition of Nanoscale Materials for Energy Conversion

Marja N. Mullings, Jonathan R. Bakke, Carl Hgglund and Stacey F. Bent, Chemical Engineering, Stanford University, Stanford, CA

Fora number of technologies, including microelectronics and energy conversiondevices, feature sizes are becoming increasingly small, and the need forprecise control over costly materials continues to grow. In addition to decreasinglength scales, it is important that limited natural resources are carefully used.In fact, many photovoltaic and fuel cell components are comprised of a varietyof materials that may be expensive, toxic and/or rare. Placing such materialsonly where needed to derive the most desirable properties will be essential inreducing the costs of energy conversion devices. Atomic layer deposition (ALD)is uniquely poised to accomplish these goals.

Atomiclayer deposition is a growth technique consisting of self-limiting surfacereactions, which allow for sub-nanometer control of growth. The ability tofinely tune growth allows for careful control over material properties. Furthermore,ALD provides excellent capabilities for depositing a variety of materials onhigh surface area substrates including those that are highly structured.

Thiswork focuses on the ALD and materials characterization of transparentconducting oxides (TCOs) and surface passivation techniques that direct thegrowth of functional materials. Transparent conducting oxides (TCOs) are highlyconductive and transparent materials used in thin film solar cells, serving aselectrical contacts, chemical barriers and antireflection coatings. One of themost common and robust TCO materials, indium-doped tin oxide (ITO), has manyattractive properties; however, alternatives must be found to address thediminishing supplies, increasing demands and subsequently escalating costs ofindium. Replacing indium with zinc to create zinc tin oxide (ZTO) provides acost-effective and earth-abundant alternative. Various material properties ofZTOs are not well understood and there is little control over stoichiometryacross growth systems. At present, it is known that ZTO takes on an unstabletrigonal ilmenite structure ($ZnSO_3$) and stabilizes into the cubic spinel(Zn_2SnO_4) and tin dioxide (SnO_2) under hightemperature conditions.

Bydeveloping and carefully characterizing a ZTO ALD system, these structures willbe better understood and eventually optimized for use as TCOs. Atomic layerdeposition reaction conditions for tin oxide (SnO_x) and zinc oxide(ZnO), which are alternative TCOs, and ZTO will be described, including growthrates and temperature windows in our custom built reactors. Material propertiesof the ZTO films will be characterized with x-ray photoelectron spectroscopy(XPS), scanning Auger electron spectroscopy (AES), spectroscopic ellipsometry,scanning electron microscopy (SEM), x-ray diffractometry (XRD) and four-pointprobe measurements providing insight into surface, structural and electronicproperties.

(496f) Visualization and Temperature Analysis of Alternative Fuel Particle Combustion In a Drop Tube Furnace

Jingran Duan[1], Vignesh Venkatasubramanian[1], Brandon Norris[1], Steve R. Duke[1] and Anton Schindler[2], (1)Chemical Engineering, Auburn University, Auburn, AL, (2)Civil Engineering, Auburn University, Auburn University, AL

A lab-scale particle drop tube furnace with visualization windows was operated at conditions that simulate energy generation in cement manufacturing processes. A high resolution color camera obtained high speed and high magnification images of individual particles undergoing combustion in the drop tube. Image processing and analysis are described that use the three color components of the images to determine the spatial distribution of temperature for the particle surfaces. Temperature fields and temperature profiles will be presented for particles of pulverized coal, wood, and switch grass undergoing combustion in various combustion gas conditions. Combustion phenomena and characteristics observed in the visualization studies as well as char particle and effluent gas analysis are shown to elucidate effects of fuel type, particle size and shape, and operating conditions of the furnace. Development of a combustion and transport model based on the measured temperature profiles is described for prediction of performance of alternative fuels in cement manufacturing and energy generation.

See more of this Session: Alternative Fuels and Enabling Technologies II
See more of this Group/Topical: Fuels and Petrochemicals Division

(364c) Novel Lignocellulosic Biomass Pretreatment Using a Modified Taylor-Couette Mixer for Biofuels and Biomaterials

Wei Wang and Ilsoon Lee, Department of Chemical Engineering and Materials Science, Michigan State University, East Lansing, MI

Bioethanol derived from lignocellulosic biomass can be a good candidate for energy alternative, not only because lignocellulosic biomass is abundant and sustainable, but also bioethanol production forms carbon cycle in a more timely fashion compared to fossil fuel, which leads to reduction of green gas emission. Lignocellulosic biomass pretreatment, which is one of the essential steps in bioethanol production, accounts for the majority portion of the overall bioethanol production cost. Hence an effective and efficient pretreatment method is highly demanded. Although this research area has been deeply investigated several decades ago, it has not attracted tremendous attention until recently. In our research, cost-effective and breakthrough biomass pretreatment was conducted with a novel modified Taylor-Couette mixing system, aiming to overcome the natural recalcitrance of lignocellulosic biomass in a short processing time and facilitate a better accessibility of enzymes to polysaccharides. Compositional analysis proved significant removals in both lignin and hemicellulose and rich glucan content left in the residue only after 2 minutes or less. Parameters like mixing speed, retention time, biomass loading and corresponding enzyme digestibility will be presented.

See more of this Session: Developments In the Pretreatment of Lignocellulosics for Bioconversion I
See more of this Group/Topical: Sustainable Engineering Forum

(173c) Sn/SnOx /Graphene Nanocomposites: Synthesis and Anode Performance In Lithium-Ion Batteries

Mahbuba Ara[1], Dr. Steven Salley[2] and Dr.Simon Ng[2], (1)Chem. Eng., Wayne State University, Detroit, MI, (2)Chemical Engg., Wayne State University, Detroit, MI

The utilization of a Sn/SnO_x/graphene nanocomposite as anode material for lithium-ion batteries could potentially provide high reversible capacity as well as a reduction of capacity fade upon cycling, by increasing the Sn:Li_2O ratio in the anode matrix. A two step chemical synthesis route has been developed to produce a Sn/SnO_x/Graphene nanocomposite. In the first step, SnO_2 nanoparticles were prepared by a hydrolysis method and loaded on graphene nanosheets using an ultrasonic treatment to make SnO_2/graphene nanocomposites. In the second step, Sn/SnOx/graphene was prepared by the partial reduction of SnO_2/graphene in a 10% H_2 in Ar mixture at 700 °C. To get nanosized Sn/SnOx particles with size in the range of 5-10 nm, reaction time and temperature were varied to control the nucleation and to obtain a homogeneous distribution of Sn nanoparticles on the graphene sheets, while pH of the solution was optimized, as at a specific pH the surface charge of the particles affects particle organization on the surface. The small size and uniform distribution of Sn/SnOx nanoparticles are designed to reduce the volume expansion of Sn during cycling. Hence, the anode of lithium-ion battery endures lower strain and is less likely to be cracked upon cycling. Sn:C ratios of 1.5, 2.5 and 3.5 in the Sn/SnOx/graphene nanocomposites will be evaluated for a superior Li-storage performance with good cycle life and high capacity.

See more of this Session: Nanomaterials for Energy Storage III
See more of this Group/Topical: Topical 5: Nanomaterials for Energy Applications

(106b) Investigation of Charge Storage Mechanism of Nanostructured Metal Nitrides and Carbides Electrodes for Electrochemical Capacitors

Priyanka Pande, Paul G Rasmussen and Levi T. Thompson, Department of Chemical Engineering, University of Michigan, Ann Arbor, MI

Electrochemical capacitors (ECs) offer aunique combination of high power and energy densities filling the gap betweenconventional capacitors and batteries. These devices store energy via doublelayer (i.e. electrostatic mechanism) and/or pseudo-capacitance mechanisms thatinvolve charge transfer processes [1]. Early transition metal nitrides andcarbides are promising candidates for use as electrode materials due to theirhigh electronic conductivities, surface areas (up to 200 m^2/g) andelectrochemical stabilities [2,3]. For example, vanadium nitride has beenreported to have the highest capacitance up to 1340 Fg^{-1} [4].However the charge storage mechanism for these materials is not wellunderstood. In this paper we will report results from ion isolation experiments,impedance spectroscopy, charge-discharge and chronotentiometry and suggest chargestorage mechanisms that reconcile the results.

Nanostructured V, Mo, W nitrides andcarbides were synthesized via temperature-programmed-reaction of their oxideprecursors with anhydrous NH_3 or 15% CH_4/H_2followed by passivation in 1% O_2/He at room temperature to form aoxygen-rich passivation layer preventing bulk oxidation on exposure to air[2,3]. Physical characterization was performed using BET surface area analysis,scanning electron microscopy and X-ray diffraction. The stability of these materials in aqueous KOH and $H_2SO_4$was determined by cyclic voltammetry. The species contributing to chargestorage were identified by isolation of the electrolyte ions and pairing themwith inactive counter ions. H^+ and SO_4^{2-}, wereisolated as $H^+BF_4^-$ (tetrafluoroboric acid) and$[(C_2H_5)_4N^+]_2\ SO_4^{2-}$ (tetraethylammoniumsulfate, $(TEA)_2SO_4$), while K^+ and OH^-ions were isolated as $K^+(CF_3SO_3)^-$(potassium triflate, K-triflate) and $(C_2H_5)_4N^+OH^-$(tetraethylammonium hydroxide, TEA-OH). Cyclic voltammetry experiments werecarried out in each of these electrolytes. In order to establish the chargetransfer reaction chronopotentiometry experiments were carried out in varyingelectrolyte concentrations.

The redox behaviorconfirmed that most of the charge storage was via the pseudocapacitancemechanism. The ion isolation experiments indicated that OH^- and H^+were the primary species contributing to charge storage in VN in KOH (figure 1)and Mo_2N in H_2SO_4 systems, respectively. Thechronopotentiometry results (figure 2) suggest that the charge storage reactionfor VN was:

$$VN_xO_y + 2OH^- \leftrightarrow [VN_xO_y(OH)_2]^- + e^-$$

Results from impedancespectroscopy and three-electrode charge-discharge experiments will also bepresented

REFERENCES

[1] SimonP, Burke A, The ElectrochemicalSociety: Interface, Spring(2008) 38.

[2] Cladridge J B, York A P E,Brungs A J, Green Malcolm L H, Chem. Mater.12(2000) 132.

[3] WixomM R, Tarnowski D J, Parker J M, Lee J Q, Chen P -L, Song I, Thompson L T, Mat.Res. Soc. Symp. Proc. 496 (1998) 643.

[4] Choi D, Kumta P N, Electrochem. Solid-StateLett. 8 8 (2005) A418.

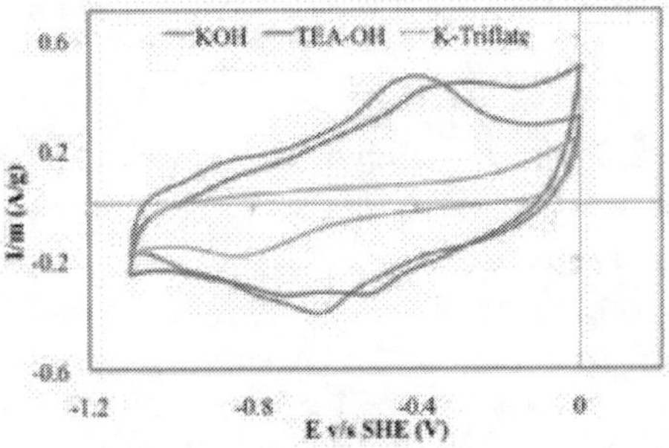

Figure 1: Cyclic voltammogram for VN in various electrolyte solutions at a scan rate of2mVs^{-1}.

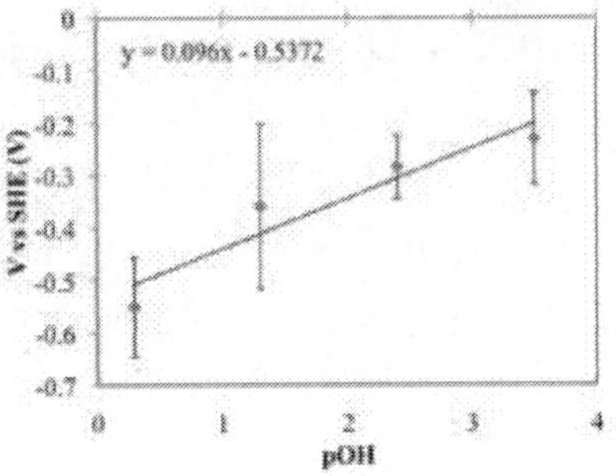

Figure 2: pOH *vs*Voltage for VN in KOH. Slope of 0.096 indicates that there are 2 OH^- ions reacting per electron transferred.

(305d) Pennycress-Derived Jet Fuel and Diesel LCA: Sustainable Advanced Biofuels

Jiqing Fan, Chemical Engineering, Michigan Technological University, Houghton, MI, David Shonnard, Michigan Technological University, Houghton, MI, Tom N. Kalnes, Research and Development, UOP LLC, Des Plaines, IL and Peter Johnsen, Arvens Technology Inc, Peoria, IL

Pennycress-derived jet fuel and diesel show promise as emerging biofuels that can help reduce dependence on petroleum imports, lower the trade deficit, and most importantly, lower greenhouse gas (GHG) emissions. Pennycress is a prolific seed producer with yields of more than 2000 pounds of per acre. The seeds contain 36% oil by weight, nearly twice the amount as soybeans. In the envisaged biofuel scenario, Pennycress would be grown between segments of a typical corn/soybean rotation, thus avoiding the concerns of a fuel crop competing with food crop production. As useful applications for pennycress meal develop, the oil is expected to become a sustainable alternative for advanced biofuels production.

The goal of this study is to determine the life cycle GHG emissions, cumulative energy demand, and fossil energy demand of pennycress-derived Hydroprocessed Renewable Jet (HRJ) fuel and Renewable Diesel (RD) as produced by the UOP Ecofining^{TM} process, and compute the GHG savings per MJ of fuel compared to petroleum-based jet fuels and diesel. The scope of this study encompasses the entire life cycle from pennycress cultivation and raw materials acquisition through the production and use of the fuels in vehicles and aircraft operations. The pennycress is grown in the Midwestern United States as a winter annual. After harvesting, the pennycress seeds are transported to a centralized processing facility where the oil is recovered and a de-oiled meal co-produced for use as a secondary energy source or animal feed. The oil is then transported to a hydroprocessing plant as a source of renewable feedstock for HRJ and RD production. Inventory data for pennycress cultivation, transportation, and oil recovery were provided by Arvens Technology Inc. Data for conversion of the pennycress oil to HRJ and RD were obtained from engineering design data, and was supplied by UOP. All the inventory data were assembled based on energy content of 1 MJ of final fuel product, which was the functional unit of this life cycle assessment (LCA) study. In the base case, energy allocation was applied to distribute environmental impacts among various products and co-products along the life cycle of each biofuel. The life cycle GHG emissions (excluding land use change) for RD and HRJ are 26.5 and 28.5 g CO_2 eq./MJ fuel, which represent savings relative to petroleum fuel counterparts of 72% and 69% respectively. Total energy consumption for each pennycress biofuels product is comparable to petroleum fuels with the exception that most of the energy is derived from renewable biomass as opposed to non renewable fossil. Scenario analyses were also conducted to determine response to model assumptions and data uncertainty, including allocation methodology, fertilizer application rate, source of H_2, and land use change. The results show that pennycress derived biofuels could qualify as advanced biofuels as defined by the Renewable Fuels Standard (RFS2).

(173f) An In Situ Method of Creating Metal Oxide-Carbon Nanocomposite As Anode Materials for Lithium-Ion Batteries

Zichao Yang, Jingguo Shen and Lynden A. Archer, Chemical and Biomolecular Engineering, Cornell University, Ithaca, NY

Transition metal oxides are attractive as anode materials for lithium-ion batteries (LIBs) due to their high lithium intercalation capacity and generally high natural abundances and their use on the nanoscale is being actively pursued. However, there are still challenges associated with the use of these metal oxides, including the pulverization problem and limited electrical conductivity of many metal oxides. It has been shown that these problems may be alleviated by creating metal oxide-carbon composites, which improve upon the mechanical flexibility and the electrical conductivity of the material.

There are different approaches for creating the carbon composites such as carbon nanopainting and attaching the metal oxide nanoparticles to carbon substrates. The current work describes an alternative method of creating metal oxide-carbon nanocomposites, in which the metal oxide nanoparticles are created *in situ* and embedded in a porous carbon matrix, which is made through calcination of a polyacrylonitrile-based polymer. Fe3O4 is used as a model compound.

Characterization techniques such as X-Ray diffraction, transmission electron microscopy and Raman spectroscopy were used to investigate the physical and chemical properties of the composite. Cyclic voltammetry of the composite shows lithium intercalation at 1.2V and 0.8V, which may be attributed to the reduction of Fe^{3+} to Fe^{2+} and Fe^{2+} to Fe^{0}, respectively. Very stable capacity is observed over >110 cycles at charging rate of 1C. Performance at other charging rates up to 5C is also shown and is seen to be stable. The contribution of factors including electrical conductivity, mechanical properties and porous nature of the carbon matrix in enhancing the electrochemical performance is investigated. The application of the method to other related materials such as MnO, Sn and Si is discussed.

See more of this Session: Nanomaterials for Energy Storage III
See more of this Group/Topical: Topical 5: Nanomaterials for Energy Applications

(690d) Active Species Concentration Effects On Performance for a Non-Aqueous All-Vanadium Redox Flow Battery

Aaron A. Shinkle, Alice E. S. Sleightholme, Levi T. Thompson and Charles W. Monroe, Department of Chemical Engineering, University of Michigan, Ann Arbor, MI

Redox flow batteries(RFBs) are being developed for large-scale energy storage, uninterruptablegrid-independent power supplies and load-leveling systems for solar or windpower. Aqueous chemistries are used for all current commercial RFBs, therefore, cell potentials are limited to thestability range of water (01.23 V vs. SHE). Since a battery's cellpotential determines its energy and power density, aqueous chemistries limitthe possibilities for the performance of RFBs.

Several non-aqueous systems have beenreported recently, including systems based on ruthenium, uranium, chromium,manganese, and vanadium coordination complexes [1-5]. In the case ofvanadium acetylacetonate, the use of a non-aqueous electrolyte results in a stablecell potential window of 2.2 V [5], more than 50% wider than the aqueoussingle-metal vanadium chemistry (1.26 V [6]). However, very few resultsfrom battery charge/discharge experiments have been shown for this non-aqueousvanadium system.

This talk will examine thecharge/discharge profile for the vanadium acetylacetonate ($V(acac)_3$)RFB system in acetonitrile, with tetraethylammonium tetrafluoroborate ($TEABF_4$)support. The effects of active-species concentration on batteryperformance were examined. Linear sweep voltammetry will be used todetermine the maximum charge/discharge rate of the battery cell as a functionof electrolyte composition. Figure 1 shows the first five cycles of acharge/discharge result for 0.4 M $V(acac)_3$,0.5 M $TEABF_4$ in acetonitrile. Coulombic, energy, and powerefficiency for charge/discharge experiments were examined as a function ofconcentration to illustrate the effects of active species crossover. Efficiency and energy density were balanced to determine an optimal activespecies concentration range for operation of the battery. Lastly, thenon-aqueous system was compared to the optimized aqueous all-vanadium system toidentify areas of future work.

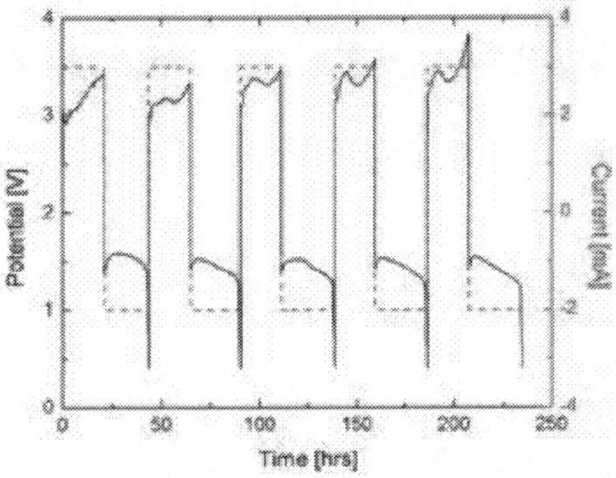

Figure1: First five cycles of a charge/discharge experiment for an all-vanadiumnon-aqueous redox flow battery chemistry. The system was charged at 3 mA until 50% theoretical stateof charge and discharged at 2 mA until 0.4 V. The system shown is 0.4 M vanadium acetylacetonate supported by 0.5 Mtetraethylammonium tetrafluoroborate in acetonitrile solvent. An H-cellconfiguration was used with graphite plate electrodes and a NeoseptaAHA anion exchange membrane.

REFERENCES

1. Matsuda Y, Takasu Y, Morita M, Tanaka K, Okada M, Matsumura-Inoue T(1985) Denki Kagaku 53:632

2. YamamuraT, Shiokawa Y, Yamana H,Moriyama H (2002) Electrochim Acta48:43

3. LiuQ, Shinkle AA, Li Y, Monroe CW, Thompson LT, Sleightholme AES (2010) Electrochem Comm 12:1634

4. SleightholmeAES, Shinkle AA, Liu Q, Li Y, Monroe CW, Thompson LT (2011) J Power Sources196:5742-5745

5. LiuQ, Sleightholme AES, Shinkle AA, Li Y, Thompson LT (2009) ElectrochemComm 11:2312

6. Poncede Leon C, Frias-Ferrer A, Gonzalez-Garcia J, Szanto DA, Walsh FC (2006) J Power Sources 160:716-732

See more of this Session: Energy Storage and Electrochemical Processes I
See more of this Group/Topical: International Congress on Energy 2011

(294d) Determination of Electroosmotic Drag and Proton Conduction Mechanism In Proton Exchange Membranes for Use In Low Temperature PEMFCs

Brian P. Setzler, Aurore Dabonot and Thomas F. Fuller, Chemical and Biomolecular Engineering, Georgia Institute of Technology, Atlanta, GA

One of the major barriers to the commercialization of proton exchange membrane fuel cells is durability. During operation, the electrochemically active surface area (ECA) of the cathode catalyst layer decreases primarily due to platinum dissolution and diffusion on both the nanometer and micrometer scale [1]. Nanoscale diffusion results in Ostwald ripening and loss of ECA due to an increase in the average particle size. Microscale diffusion results in precipitation of platinum particles in the membrane as platinum ions are reduced by crossover hydrogen [2]. Models of the catalyst degradation process have shown that the rate of ECA loss is dependent on the mobility of platinum ions in the membrane [3]. Therefore, a membrane with low platinum ion mobility that maintains high proton mobility is desired.

There are two mechanisms of proton conduction in proton exchange membranes: the vehicle mechanism and the Grotthuss (proton-hopping) mechanism. Because the Grotthuss mechanism is exclusive to protons, a membrane that conducts protons primarily by the Grotthuss mechanism is expected to have a lower platinum ion mobility. The relative importance of both proton conduction mechanisms can be determined by measuring the electroosmotic drag coefficient. The vehicle mechanism drags some water molecules with each proton, while the Grotthuss mechanism results in very little water transport.

The electroosmotic drag coefficient was measured as a function of water activity in Nafion 112, poly (phenylene sulfone) (PS) (IEC=1.0 meq/g) [4], and mixed H^+/Co^{2+} form Nafion 117 membranes by the concentration cell method described by Fuller and Newman [5]. Both membranes were found to have a drag coefficient dependent on the hydration state of the membrane. The drag coefficient was constant when the water vapor activity (a_w) was below 0.75. In the range $0.75<a_w<1.0$, the drag coefficient was also constant, but approximately twice as high as in the low humidity regime. In Nafion, $a_w=0.75$ corresponds to a water content of 6 moles H_2O per mole SO_3^- [6]. The drag coefficient of Nafion 112 was 1.1 for $a_w<0.75$ and 2.4 for $a_w>0.75$, while the drag coefficient of the PS membrane was 0.55 for $a_w<0.75$ and 0.95 for $a_w>0.75$. The Nafion 117 membrane in a mixed Co^{2+}/H^+ form had a drag coefficient of 2.0 for $a_w<0.75$ and 4.8 for $a_w>0.75$. When alloy catalysts are used, unstable metals like cobalt may have a high initial rate of dissolution, causing a change in the transport properties of the membrane. The results indicate that the Grotthuss mechanism makes a larger contribution to proton conduction in PS membranes than in Nafion membranes. This is expected to lead to a higher ratio of proton to platinum mobility in PS compared to Nafion.

References

1. Ferreira, P.J., et al., *Instability of Pt/C electrocatalysts in proton exchange membrane fuel cells - A mechanistic investigation.* Journal of The Electrochemical Society, 2005. **152**(11): p. A2256-A2271.

2. Bi, W., G.E. Gray, and T.F. Fuller, *PEM fuel cell Pt/C dissolution and deposition in nafion electrolyte.* Electrochemical and Solid State Letters, 2007. **10**(5): p. B101-B104.

3. Bi, W., et al., *The effect of humidity and oxygen partial pressure on degradation of Pt/C catalyst in PEM fuel cell.* Electrochimica Acta, 2009. **54**(6): p. 1826-1833.

4. Schuster, M., et al., *Highly Sulfonated Poly(phenylene sulfone): Preparation and Stability Issues.* Macromolecules, 2009. **42**(8): p. 3129-3137.

5. Fuller, T.F. and J. Newman, *Experimental-Determination of the Transport Number of Water in Nafion-117 Membrane.* Journal of The Electrochemical Society, 1992. **139**(5): p. 1332-1337.

6. Springer, T.E., T.A. Zawodzinski, and S. Gottesfeld, *POLYMER ELECTROLYTE FUEL-CELL MODEL.* Journal of The Electrochemical Society, 1991. **138**(8): p. 2334-2342.

See more of this Session: Fuel Cell Technology I
See more of this Group/Topical: Fuels and Petrochemicals Division

(522d) Co-Utilization of Methane In Steam-Biomass Gasification Using Concentrated Solar Energy

Aaron W. Palumbo, Erica L. Jorgensen and Alan W. Weimer, Chemical & Biological Engineering, University of Colorado at Boulder, Boulder, CO

Presented is a novel gasification application where biomass is steam-gasified and natural gas is steam-reformed simultaneously in a solar reactor. Solar-driven gasification is a proven technology, and can augment the biofuel industry as a domestic fuel source. Gasification performed in an indirectly heated gasifier can achieve temperatures greater than 1150 °C [1], mitigating or eliminating the production of CO_2, tar, char, and light hydrocarbon gases in the product gas. This reduces the need for gas clean-up technology and will not require a downstream water-gas shift reactor to adjust H_2:CO. Thermodynamics favor the formation of only hydrogen and carbon monoxide in the product synthesis gas. Therefore, a predictable H_2:CO can be controlled and tuned for downstream fuel conversion. Since the majority of biomass carbon is converted to CO, more than twice as much biomass is converted to a useable liquid fuel than in a traditional gasification process. Preliminary experiments show that reactor temperatures must exceed 1350 °C in order to reduce the effects of the water-gas shift reaction and reform practically all light hydrocarbon gases, including CH_4. This operating temperature is consistent with realizable concentrated solar energy systems.

1. Piatkowski, N., et al., *Solar-driven gasification of carbonaceous feedstock-a review.* Energy & Environmental Science, 2011. **4**(1): p. 73-82.

See more of this Session: Developments In Biobased Alternative Fuels I
See more of this Group/Topical: Sustainable Engineering Forum

(13e) A Dual-Stage Laminar Entrained Flow Reactor (LEFR) and Plug Flow Reactor (PFR) System for Studying Biomass Pyrolysis and Gasification

Mark W. Jarvis, Calvin Mukarakate, David Robichaud and Mark Nimlos, National Renewable Energy Laboratory, Golden, CO

An understanding of pyrolysis vapor cracking kinetics and mechanisms are important for developing approaches for reducing tars during gasification and ultimately reducing the cost of producing transportation fuels from gasification and fuel synthesis. In order to reduce tars using predictive design approaches such as Computation Fluid Dynamics (CFD), it is necessary to have robust chemical reaction models containing mechanism that lead to the formation of these species. Existing reaction models only contain limited, highly generalized mechanisms for the formation of tars and are not robust enough to be predictive. In particular, the initial steps of gasification are inadequately characterized. These steps are primarily cracking reactions that break down the pyrolysis vapor species into smaller molecules ultimately leading to syngas and tars. In this study, we have undertaken an experimental investigation of these types of cracking reactions. The solid phase vaporization reactions were separated from the gas phase cracking reactions by conducting experiments in a two-stage reactor. In the first stage, particles of biomass were pyrolyzed in a Laminar Entrained Flow Reactor (LEFR) at a low temperature (450°C) and in the second phase the vapors were cracked at variable residence times (0.2 to 1.5 s) and temperatures (430° to 680°C) in a Plug Flow Reactor (PFR). The tar species and other product gases were measured using a Molecular Beam Mass Spectrometer (MBMS). The data from the MBMS showed trends in the tar species concentrations as a function of residence time and temperature, which will allow the development of more accurate and robust chemical kinetics models of gasification. Eventually, these kinetic models will be used in CFD models of gasification in order to optimize it for reduced tar formation and increased syngas yield.

See more of this Session: Catalytic Conversion of Renewable Resources to Synthesis Gases and Pyrolysis Oils
See more of this Group/Topical: Sustainable Engineering Forum

(508b) Tin-Containing Beta Zeolite Catalyst for the Isomerization of Glucose: Insights On the Mechanism and the Active Site

Eranda Nikolla[1], Yuriy Roman[1], Manuel Moliner[2], Sonjong Hwang[1] and Mark E. Davis[1], (1)Chemical Engineering, California Institute of Technology, Pasadena, CA, (2)Chemistry, Instituto de Tecnologia Quimica (UPV-CSIC), Valencia, Spain

Biomass is being investigated as feedstock for the production of chemicals and transportation fuels of the future. Carbohydrates represent 75% of the annual renewable biomass. An approach to convert biomass-based carbohydrates to fuels and chemicals involves the production of chemical intermediates such as 5-hydroxymethylfurfural (HMF) that can serve as a gateway chemical for the production of high value products. We have recently reported that a tin containing zeolite beta catalyst (Sn-Beta) combined with a Brønsted acid catalyst such as HCl can efficiently convert sugars such as glucose to HMF under aqueous conditions with high selectivities (>70%). In this system, Sn-Beta zeolite is responsible for the isomerization of glucose to fructose, a process that has been a challenge to catalyze using heterogeneous catalysts under aqueous conditions. We have reported that Sn-Beta zeolite can efficiently catalyze the isomerization of glucose to fructose in aqueous media at low pH in the presence of HCl and an inorganic salt (Proc. Natl. Acad. Sci., 107 (2010) 6164, ACS Catal, 1 (4), (2011) 408).

Here, we will discuss efforts focused on obtaining a fundamental understanding of the nature of the active site in the Sn-Beta zeolite and the mechanism of the isomerization reaction of glucose to fructose. We have utilized solid-state NMR experiments to probe the environment of tin in the zeolite. Sn-Beta zeolite is synthesized with labeled Sn-119 in order to provide a solid that can be effectively analyzed by solid-state NMR. We have investigated the state of the tin when the zeolite is exposed to various conditions including glucose isomerization conditions, and the results will be discussed. We have also performed NMR experiments with deuterated and C-13 labeled glucose to study the interactions of the reactant with the tin active site in the zeolite. Detailed kinetic studies including isotope-labeling experiments have been conducted in order to obtain kinetic parameters such as activation energies and rate limiting steps for the isomerization of glucose to fructose. An overall picture of the state of the tin active center and the reaction pathway is emerging and will be presented.

See more of this Session: Catalytic Biomass Conversion to Chemicals I
See more of this Group/Topical: Fuels and Petrochemicals Division

(62c) Effects of Total Inorganic Carbon and Nitrogen Concentrations On Lipid Formation In Chlorella Vulgaris

Jinsoo Kim, Department of Chemical & Materials Engineering, University of Cincinnati, Cincinnati, OH, Joo-Youp Lee, Department of Chemical and Materials Engineering, University of Cincinnati, Cincinnati, OH and Kaniz F. Siddiqui, The Metropolitan Sewer District of Greater Cincinnati, Cincinnati, OH

Effects of totalinorganic carbon and nitrogen concentrations on lipid formation in *Chlorella vulgaris*

Jinsoo Kim[1], Joo-YoupLee[1], Kaniz F. Siddiqui[2]

[1] Chemical EngineeringProgram, University of Cincinnati, OH 45221-0012

[2] The Metropolitan SewerDistrict of Greater Cincinnati, Cincinnati, OH 45204-2002

Microalgae have recently received greatattention as a promising alternative bioenergy source because they can produce asubstantial amount of triacylglycerols (e.g. 20-50 % dry cell weight) and canbe readily converted into biodiesel. *Chlorella vulgaris* has been selected forthis growth kinetic study since it is one of the fastest growing microalgae andcontains a reasonable amount of triacylglycerols (e.g. 14-22 % dry cell weight). Dissolved carbon dioxide (CO_2(aq)) and bicarbonate ion (HCO_3^-) areessential inorganic carbon forms required for autotrophic algal growth, but fewstudies have been systematically conducted on the impact of inorganic carbon onalgal growth. In this respect, thegrowth kinetics of *Chlorella vulgaris*was previously investigated dependent on the total inorganic carbonconcentration (i. e., CO_2(aq) and HCO_3^-) As a result, it was observed that the growthkinetics was increased when the total inorganic carbon concentration wasincreased in the culture medium.

Under many nutrient limiting conditions,lipids are accumulated as a means of storage when energy (i.e. light) andcarbon (i.e. CO_2 for autotrophic growth) sources are available. However, many of these nutrients areessential for photosynthesis responsible for metabolic flux generation intolipids formation. Thesenutrient-deficient conditions decrease biomass growth rate, and offset thebenefit of lipid accumulation. Thegrowth of algal biomass such as *Chlorella vulgaris* is dependent on bicarbonate ion concentrationsavailable in the culture medium. Bicarbonateion also plays a critical role as a catalyst in the first step of fatty acidsbiosynthesis for the synthesis of malonyl-CoA.In this study, a change in total lipids content has been investigatedusing the Nile red stain method when the total inorganic carbon concentrationin culture medium is increased. This isexpected to provide fundamental understanding of the utilization of bicarbonateions for the design and operation of bioreactors.

Keywords: *Chlorella vulgaris*, total inorganic carbon, nitrogen deficiency,lipid content.

See more of this Session: Advances In Algal Biorefineries I
See more of this Group/Topical: Sustainable Engineering Forum

(650f) Characterization of Asphaltenes In Stock-Tank Oil and Prediction of Its Phase Behavior At Reservoir Conditions

Francisco M. Vargas[1], Senthilmurugan Balasubramanian[2], Sameer Punnapala[1], Alfredo Halique[1], Sama Al Hallaq[1], Salama Al Marzouqi[1], Sara Al Hashmi[1] and Ismail Alkhatib[1], (1)Chemical Engineering, The Petroleum Institute, Abu Dhabi, United Arab Emirates, (2)Petroleum Engineering, The Petroleum Institute, Abu Dhabi, United Arab Emirates

Asphaltene deposition is one of the most widely found flow assurance problems in upstream and downstream operations. It is well known that asphaltene precipitation is a necessary but not a sufficient condition for deposition. Although substantial amount of work has been done on this area, the mechanisms by which asphaltene precipitate and deposit are still being actively investigated.

Current methods used to determine the stability of asphaltenes in crude oil systems are based on near-infrared spectrometry and high pressure microscopy. Although these methods are widely used and accepted in the industry, they require time and can be expensive. This is the case particularly when a comprehensive study requiring different gas injections and temperatures needs to be performed.

In this work we propose as an alternative method the determination of the asphaltene behavior in stock-tank oil at room conditions by determining the onset of precipitation upon addition of different n-alkanes, which are known to be asphaltene precipitants. Determination of onset of asphaltene precipitation is done by using NIR spectrometry at room conditions. These data can then be used to tune asphaltene parameters of a model. This model will then be used to extrapolate the behavior at reservoir conditions including the associated gas. Comparison with data at high pressure and temperature is also presented and discussed.

Because the precipitation and flocculation of asphaltenes is a process greatly affected by kinetic factors, systematic experimentation was done to determine the effect of time on determining the asphaltene precipitation onset. Including this information, a detailed experimentation procedure was developed that matches the experiments done at high pressure and temperature. Different crude oils were studied to validate the proposed methodology.

Finally, the spectrometric methodology was also applied to the study of the effect of asphaltene inhibitors. Advantages of the proposed methodology over current experimental methods will be highlighted in this presentation. Also, important insight into the mechanism of action of asphaltene inhibitors will be presented and discussed.

(206d) Aqueous Phase Hydrogenation and Hydrodeoxygenation of Bio-Oil Model Compounds Under Mild Temperature and Pressure

Zhenglong Li, Biosystems & Agricultural Engineering, Chemical Engineering & Materials Science, Michigan State University, East Lansing, MI, Christopher M. Saffron, Biosystems & Agricultural Engineering, Department of Forestry, Michigan State University, East Lansing, MI, James E. Jackson, Chemistry Department, Michigan State University, East Lansing, MI and Dennis J. Miller, Chemical Engineering and Materials Science, Michigan State University, East Lansing, MI

Fast pyrolysis converts biomass into a liquid fuel intermediate known as bio-oil. Due to oxygen containing functional groups, bio-oil species are reactively unstable, tending to polymerize during long term storage and upgrading. The high temperatures and pressures used for conventional stabilizing and upgrading of bio-oil lead to coke production and catalyst deactivation. Milder treatments are needed to reduce coke formation and increase catalyst lifetime while lowering costs.

We have recently developed a mild strategy for bio-oil upgrading using electrocatalytic hydrogenation (ECH). During ECH, atomic hydrogen adsorbed on the catalyst surface reacts with adsorbed organic compounds. The availability of such surface hydrogen is controlled by the applied current and potential at ambient pressure, rather than by external hydrogen pressure, as in the classical hydrogenation. Useful temperature lowering is known as well since most ECH reactions proceed below the boiling point of water. We have now demonstrated the ECH of furfural and guaiacol, two model compounds derived respectively from pyrolysis of cellulose and lignin.

Cost is a key issue for any intended biomass upgrading process. Screening of catalysts for ECH of furfural therefore focused on inexpensive metals: nickel, iron, aluminum, copper and stainless steel. With nickel catalyst at room temperature and ambient pressure, 78% of furfural underwent ECH to furfuryl alcohol along with small amounts of hydrogenolysis product 2-methylfuran (selectivities 95:5). Further trials revealed that acid favored formation of 2-methylfuran, raising its selectivity from 5% to 14% as pH was decreased from 5.0 to 1.0. Additional factors affecting the product selectivity, such as temperature and initial furfural concentration, were also investigated.

Unlike the reactive aldehyde furfural, the electron rich aromatic guaiacol (2-methoxyphenol) resisted ECH with simple nickel catalysts. However, development of a carbon-supported ruthenium electrocatalyst enabled complete conversion of guaiacol to cyclohexanol and 2-methoxycyclohexanol at 80°C and ambient pressure. Effects of metal loading, electrolyte, temperature and substrate concentrations were explored. These model compounds studies highlight the promise of ECH as a strategy for stabilizing and upgrading bio-oils from biomass fast pyrolysis.

See more of this Session: Catalytic Biofuels Refining I
See more of this Group/Topical: Fuels and Petrochemicals Division

(650e) Study of Asphaltene Precipitation From Brazilian Petroleum

Mara T. Fukunaga[1], Eduardo T. Iamazaki[1], Felipe M.R. Cardoso[2], Watson Loh[1] and **Paulo T.V. Rosa**[1], (1)Institute of Chemistry, State University of Campinas, Campinas, Brazil, (2)State University of Campinas, Campinas, Brazil

Petroleum is a complex fluid containing several classes of compounds dispersed or solubilized in a liquid phase. During the petroleum production and processing, changes in pressure, temperature, and composition can decrease the stability of the dispersed systems, causing precipitation of a solid phase, composed mainly by asphaltenes. Asphaltenes are recognized as a mixture of species of high molecular weight, with a reasonable polarity, and different functional classes[1]. The aggregative nature of asphaltenes is related to the presence of heteroatoms in its structure that results in a superficial activity for this compounds[2,3]. The precipitation and deposition of asphatenes is a serious problem that can affects the production and refining process. Some methods are used to prevent deposit or its removal, but the use of precipitation inhibitors and/or dispersants is a feasible solution[4]. This work presents a kinetic study of asphaltene precipitation in two different types of Brazilian crude oil (P1 and P2). N-heptane was used as precipitant and three different types of precipitation inhibitor or dispersant (I1, I2 and I3) were tested. The efficiency of the additive was accessed by the sedimentation curves of the precipitated asphaltenes followed using a Hewlett Packard spectrophotometer, model 8453. Samples were prepared by weighing 0.1 g of oil, followed by the addition of different concentrations of additive, and approximately 10 g of n-heptane. After homogenizing the solution, 3 mL was placed into a quartz cuvette and the absorbance was monitored at 850 nm for a period of 3600 seconds. All tested additives presented some efficiency on the precipitation of asphaltenes but I2 needed lower concentration to stabilize the system.

[1] Loh, W., Mohamed, R.S., Ramos, A.C., "*Aggregation of asphaltenes obtained from a brazilian crude oil in aromatic solvents*", Petroleum Science and Technology, 17, 147-163, 1999.

[2] Mohamed, R.S., Loh, W., Ramos, A.C., "*Aggregation behavior of two asphaltenic fractions in aromatic solvents*", Energy & Fuels, 13, 2, 323-327, 1999.

[3] Mansoori, G.A., "*Modeling of asphaltene and other heavy organics deposition*", J. Pet, Sci. Eng., 17, 101-111, 1997.

[4] Stephenson, W.K., "*Producing asphaltenic crude oils: problems and solutions*", Petroleum Eng. International, 24-31, 1990.

See more of this Session: Heavy Oil and Flow Assurance I

(110d) Graphene Supported Bifunctional Catalysts for Oxygen Electrodes In Rechargeable Lithium-Air Batteries

Kapila Wadumesthrige, Chemical Engineering, Wayne State University, Detroit, MI, Dr. Steven Salley, Chemical Engg., Wayne State University, Detroit, MI and K. Y. Simon Ng, Department of Chemical Engineering and Material Science, Wayne State University, Detroit, MI

Rechargeable lithium-air batteries offer great promise for transportation applications due to their high specific energy and energy density compared to all other battery chemistries. Therefore such batteries could be developed for powering electric vehicles, enabling driving ranges comparable to gasoline powered automobiles. Although the theoretical discharge capacity of Li-Air cell is extremely high, the practical capacity is much lower and is always cathode limited. On charging, the cathode acts catalytically to evolve oxygen (the oxygen evolution reaction, OER) and on discharge acts to reduce oxygen (the oxygen reduction reaction, ORR). However, due to the slow kinetics and higher overpotential especially in OER together with clogging of Li_2O_2 in the pores of air electrode limit the re chargeability of Li-Air batteries. A key for rechargeable systems is the development of an air electrode with bifunctional catalyst on an electrochemically stable carbon matrix. The use of Graphene as a stable catalyst matrix for Air cathode has been studied in this work. A Li-Air cell constructed using an air cathode consists of nano Pt on graphene nanosheets (GNS) has showed promising improvement of 85% energy efficiency with average capacity of 1800 mAh/g and more than 30 cycles without significant capacity fading. Replacement of Pt with nano structured bifunctional catalysts of the general formula $La_{1-x}A_xCo_{1-y}B_y$ (A = Ca, Ce; B= Mn, Fe, Ni, Cu) is under investigation. The effects of microstructure and loading of several GNS (with different sizes and number of layers) on the pore size distribution pore volume and surface area, as well as the relationship of particle size and the synthesis route of catalyst to the electrochemical performance will be discussed. These developments will help to exceed the specific energies of 1000 Wh/kg for practical purposes.

See more of this Session: Novel Battery Chemistry and Technology
See more of this Group/Topical: Fuels and Petrochemicals Division

(564f) Conversion of Fischer-Tropsch Hydrocarbons Over H-ZSM-5 and Ga-ZSM-5 for the Production of Fuels

Benjamin D. Harris, Chemical Engineering, University of Michigan, Transportation Energy Center, Ann Arbor, MI and Johannes W. Schwank, Chemical Engineering, University of Michigan, Ann Arbor, MI

Fischer-Tropsch (FT) catalysts have long been used for hydrocarbon production from synthesis gas. The product spectrum primarily consists of straight-chained alkane and alkene compounds beginning with methane and ending with the production of heavy waxes. The average carbon number of the products is determined by the operating conditions and catalyst chosen. For certain applications, the absence of aromatic compounds is preferred. However, for certain fuels to meet required standards, the aromatic volume percentage must exceed that which can be produced by an FT reactor.

H-ZSM-5 and Ga-ZSM-5 catalysts have the ability to increase the branched hydrocarbon and aromatic composition of a hydrocarbon stream produced from an FT reactor system. Hydrocarbons too heavy to be used in fuels can be cracked over H-ZSM-5 located within or downstream of an FT reactor. Ga-ZSM-5 catalysts can be used to upgrade undesired C2-C4 alkanes and alkenes into aromatic compounds in a downstream reactor. This work explores the combination of multiple reactor beds used in series for the potential production of fuels from synthesis gas, and the effect of the inorganic gases present in this stream on the aromatization reactions.

See more of this Session: Alternative Fuels and Enabling Technologies III
See more of this Group/Topical: Fuels and Petrochemicals Division

(294f) A Two-Phase Two-Dimensional Steady State Model of a Cylindrical PEM Fuel Cell Cathode

Debangsu Bhattacharyya, Department of Chemical Engg., West Virginia University, Morgantown, WV, **M. Srinivasarao**, Chemical Engineering, Indian Institute of Technology, Chennai, India, Brian Bullecks, Chemical Engg. Dept., Clarkson University, Potsdam, NY and Raghunathan Rengasamy, Department of Chemical Engineering, Texas Tech University, Lubbock, TX

Traditional polymer electrolyte membrane fuel cells (PEMFCs) are planar. Cost and water management are two major issues with this design. To improve the gravimetric and volumetric power densities of the PEMFC and to reduce the cost, a novel cylindrical PEMFC design has been developed. The performance of the air-breathing cylindrical PEMFC is found to be superior to a state-of-the-art planar cell in the high current density region. To understand the effect of various design parameters and operating conditions on the performance of the cylindrical PEMFC and to optimize its design, a two-dimensional, two-phase, steady-state model of the air-breathing cylindrical cell is developed. For comparison, a similar model of the planar cell is also developed. The model of the planar cell is validated with the data from a state-of-the-art commercial cell while the model of the air-breathing cylindrical cell is validated with the experimental data from an in-house cell. For comparison, same membrane electrode assembly (MEA) is used in both the cells. The ohmic resistances used in both the models are determined by electrochemical impedance spectroscopy (EIS). The developed models are utilized for analyzing the reasons for superior performance of the cylindrical PEMFC. The model of the air-breathing cylindrical PEMFC is extended to a pressurized cell by modeling the gas flow channel. The performance of the pressurized cell is found to be considerably higher than the air-breathing cell. A number of parametric studies are done with the model of the pressurized cell. The temperature is found to have a modest effect on the performance in the temperature range studied. As the air flow rate is increased from a low flow rate, the cell performance improves significantly. Further increase in the flow rate has negligible effect. The study suggests that there is an optimum platinum loading because of an interplay between the change in the activation losses and concentration losses as the platinum loading is changed. The study also shows that an optimum loading of ionomer content can improve the cell performance especially at higher current densities. The parametric studies show that a multivariable optimization study can significantly improve the cell performance in the entire polarization range.

See more of this Session: Fuel Cell Technology I
See more of this Group/Topical: Fuels and Petrochemicals Division

(339f) Improving the Quality of Spent Mushroom Substrate (SMS) As Feedstock for Biorefineries

Maria Nydia Ruiz-Felix, Chemical Engineering, Villanova University, Villanova, PA, Justin M. Yeash, Department of Chemical Engineering, Villanova University, Villanova, PA and Justinus A. Satrio, Chemical Engineering Department, Villanova University, Villanova, PA

In the current ecological climate there is a strong push to find a renewable energy resource to reduce the usage of fossil fuels which are depleting and are causing greenhouse gas emissions. Biorenewable materials, particularly lignocellulosic biomass, have received increased attention recently as the only carbon-containing resources to produce renewable energy, particularly liquid transportation fuels. Many biomass resources that have been explored in the past include wood chips, tree bark, corn stover, switch grass, bagasse, and other organic wastes. Some of these sources are byproducts of commercial agriculture and some provide opportunities for biomass to energy agribusiness, where the crop is grown just for energy harvesting. At Villanova University, we are interested in research on biomass to energy conversions because the school is situated in the proximity of some major agricultural areas in the Mid-Atlantic region, like Lancaster County in Pennsylvania, which produces numerous biorenewable materials that can be used for energy production. Another important region near Villanova University is Chester County, which is known to be "the mushroom growing capital of the world". Currently, Chester County accounts for around 37% of mushroom production in the U.S.

Mushroom growing produces a large amount of waste product called spent mushroom substrate (SMS). After the mushroom fungi grow and fruit the nutrients in the substrate are no longer favorable for mushroom growth and the substrate is considered waste. SMS poses a unique opportunity as a renewable energy resource since it is available in very large amounts, especially in Chester County and the surrounding areas in Pennsylvania. SMS is typically land applied in nearby agricultural fields or sold as compost. These avenues provide little benefit to the mushroom growing companies and are unsustainable due to the methane released through the decomposition of the substrate.

As is, SMS is not the most ideal feedstock for biorefining due to its high moisture and ash/inorganic materials content. Our research is to explore how to improve the quality of SMS to make it an ideal feedstock for biorefining by using torrefaction as a pretreatment method. Torrefaction is the process by which a biomass source is heated in the absence of oxygen to between 200 and 300°C for residence times ranging from 15 to 60 minutes. In this process the biomass will lose 30% of its mass, but only lose 10% of its initial energy which results in significant energy densification. This energy densification is important for biomass feedstocks which may require transportation from the source to further processing facilities. Additionally, torrefaction helps to make biomass properties uniform which enables processing facilities to expend less energy adjusting to various feedstock properties. One of the crucial steps in the conversion of biomass to energy is making sure that the biomass characteristics are within optimal specifications for pyrolysis and further thermal degradation.

To be reported are the results of the torrefaction of SMS at various process temperatures and lengths of processing time. Properties being evaluated are heating values, ash content, and elemental chemical composition. These properties will be analyzed to determine the effects that torrefaction and cleaning by mechanical separation have on the quality of the SMS. Results from fast pyrolysis of torrefied/cleaned SMS compared to untreated SMS by using pyro-GC/MS will be reported as well. Finally, assessment on the sustainability of the torrefaction and cleaning by mechanical separation process for improving the quality of SMS will be presented.

See more of this Session: Sustainable Biomass Feedstock Production and Supply for the Emerging Biorefinery Industry
See more of this Group/Topical: Sustainable Engineering Forum

Kinetics of Self-Hydrolysis of Concentrated Sodium Borohydride Solutions at High Temperatures

Rajasree Retnamma[1], Lin Yu[2], Carmen M. Rangel[3], Augusto Q. Novais[1], Karl Johnson[4], Michael A. Matthews[2]

[1]Energy Systems Modeling and Optimization Unit, National Laboratory of Energy and Geology, Lisbon, Portugal
[2]Department of Chemical Engineering, University of South Carolina, USA
[3]Fuel Cells and Hydrogen Unit, National Laboratory of Energy and Geology, Lisbon, Portugal
[4]Department of Chemical and Petroleum Engineering, University of Pittsburgh, USA

1. Introduction

Hydrolysis of sodium borohydride ($NaBH_4$) solution over metal catalyst is accepted as a potential technology for delivering H_2 to portable fuel cells. However, unresolved issues such as minimizing the amount of water and the nature of hydration of metaborate by-products limit its hydrogen storage capacities. An alternative method, steam/water vapor hydrolysis, can enhance the hydrogen storage capacities without catalyst, if operating conditions can be optimized [1]. In this approach, solid $NaBH_4$ absorbs water and deliquesces, forming a highly concentrated viscous solution at near boiling point of water, followed by hydrogen evolution from the concentrated viscous solution. It is also well understood that self-hydrolysis occurs even at room temperature when $NaBH_4$ is mixed with water and the rate becomes significant at high temperatures. Self-hydrolysis must be arrested to increase the shelf-life of the solution.

The knowledge of the kinetics of self-hydrolysis of concentrated solutions at elevated temperature is thus crucial for: (1) developing steam/water vapor hydrolysis technology; and (2) handling and storage of $NaBH_4$ solutions. Concentrated solutions have not been studied extensively so far and the reported studies, except two [2, 3], were for buffered dilute solutions of $NaBH_4$ (< 0.4 wt%) at low temperatures (< 25 ^{0}C). The present study reports the kinetics of self-hydrolysis of unbuffered, unstabilized concentrated $NaBH_4$ solutions (10-20 wt%) at high temperatures (25-80 ^{0}C) by a differential isothermal isoconversional method.

2. Theoretical

The $NaBH_4$ hydrolysis (self or catalytic) reaction occurs according to [4]:

$$NaBH_4 \text{ (aq or s)} + (2+x)\, H_2O \text{ (l)} \longrightarrow NaBO_2.\ x\, H_2O \text{ (aq or s)} + \text{heat} \qquad (1)$$

where x is the excess hydration factor.

The mechanism of this process is highly complex and too complicated to be described by a kinetic model within the practical operating conditions [5]. The isoconversional method is often used to describe such kinetics, where, assuming a single step kinetic approximation, the rate expression is [6].

$$\frac{d\alpha}{dt} = k'_{ap}(T, C_0)\, f(\alpha) \qquad (2)$$

$k'_{ap}(T, C_0)$ is the apparent rate constant, which depends on temperature T and initial reactant concentration C_0, α is the extent of conversion and $f(\alpha)$ the conversion function representing the mechanism of the process.

The $NaBH_4$ hydrolysis process exhibits maximum rate at the beginning of the reaction, which decreases continuously with the extent of conversion. The most appropriate form of $f(\alpha)$ for such decelerating type process is the 'reaction order model' [7]:

$$f(\alpha) = (1-\alpha)^m \qquad (3)$$

where m is the reaction order.

Combining Eqs. (2)- (3), and considering the definition of extent of reaction α gives:

$$\frac{d\alpha}{dt} = k_{ap}(T)\, C_0^{m-1}\, (1-\alpha)^m \qquad (4)$$

where,

$$k'_{ap}(T, C_0) = k_{ap}(T) * C_0^{\,m-1} \qquad (5)$$

For a fixed conversion (α), Eq. (4) reduces to

$$\left(\frac{d\alpha}{dt}\right)_\alpha = k''_{ap}(T, \alpha)\, C_0^{m-1} \qquad (6)$$

where,

$$k''_{ap}(T, \alpha) = k_{ap}(T) * (1-\alpha)^m \qquad (7)$$

Equation (6), after a logarithmic transformation:

$$\ln\left(\frac{d\alpha}{dt}\right)_\alpha = \ln k''_{ap}(T, \alpha) + (m-1)\ln C_0 \qquad (8)$$

For an isothermal process, the values of $\ln k''_{ap}(T, \alpha)$ and m are determined by Eq. (8) for each temperature and α, for different initial borohydride concentrations.

The temperature dependence of the rate constant $k_{ap}(T)$ is then expressed by Arrhenius relationship:

$$k_{ap}(T) = A_{ap} \exp\left(\frac{-E_{ap}}{RT}\right) \quad (9)$$

The apparent activation parameters, i.e. apparent activation energy (E_{ap}) and pre-exponential factor (A_{ap}) are estimated by linear regression of ln $k_{ap}(T)$ vs. 1/T data.

3. Results and discussion

3.1. Kinetic study

Effect of $NaBH_4$ conversion and temperature on reaction rate

The experimental kinetic data used were reported by Yu and Matthews [8], in which the evolution of metaborate formation and $NaBH_4$ consumption were generated *in-situ* by ^{11}B NMR technique for unbuffered, unstabilized 10, 15 and 20 wt% $NaBH_4$ solutions at 25, 60 and 80 ^{0}C. The estimated reaction rates from the experimental data for various $NaBH_4$ conversion, initial $NaBH_4$ concentration and temperature are given in Table 1. As evident, the reaction rate increases with both temperature and initial $NaBH_4$ concentration for a fixed $NaBH_4$ conversion and decreases with increase in $NaBH_4$ conversion for a fixed temperature and initial $NaBH_4$ concentration. Table 1 is then exploited to determine the reaction order and apparent activation parameters with respect to $NaBH_4$ conversion and temperature.

Table 1. Reaction rates for different $NaBH_4$ conversions at various temperatures and initial $NaBH_4$ concentrations.

$NaBH_4$ (wt %)	T(^{0}C)	Reaction rates (r, mol L^{-1} hr^{-1}) at various $NaBH_4$ conversions									
		5 %	10 %	15 %	20 %	25%	30 %	35 %	40 %	45%	50%
10		0.086	0.048	0.025	0.015						
15	25	0.136	0.080	0.037	0.027						
20		0.241	0.108	0.057	0.036						
10		0.637	0.559	0.485	0.415	0.345	0.273	0.210	0.158	0.110	0.100
15	60	0.740	0.654	0.569	0.483	0.398	0.317	0.242	0.181	0.144	0.130
20		0.897	0.797	0.697	0.598	0.499	0.402	0.310	0.227	0.160	0.143
10		2.209	2.089	1.970	1.848	1.723	1.597	1.468	1.340	1.201	1.062
15	80	2.384	2.241	2.130	1.981	1.862	1.720	1.687	1.470	1.352	1.232
20		2.698	2.555	2.410	2.264	2.115	1.964	1.810	1.653	1.493	1.326

Effect of $NaBH_4$ conversion and temperature on reaction order

Fig. 1 shows the reaction order as estimated by linear regression of rate versus $NaBH_4$ concentration for various $NaBH_4$ conversions and. As seen, the reaction order remains constant at a given temperature irrespective of $NaBH_4$ conversion, while it decreases with increase in temperature irrespective of $NaBH_4$ conversion. The order decreases from first- to 0.25 with increase in temperature from 25 to 80 ^{0}C. The estimated rate constant, $k_{ap}(T)$ (not given), decreases with increase in $NaBH_4$ conversion at constant temperature and increases with temperature at constant $NaBH_4$ conversion. It is noteworthy that the variation of reaction order with temperature has been reported by Vyazovkin and Sbirrazzuolui [9] for complex epoxy-anhydride cure kinetics.

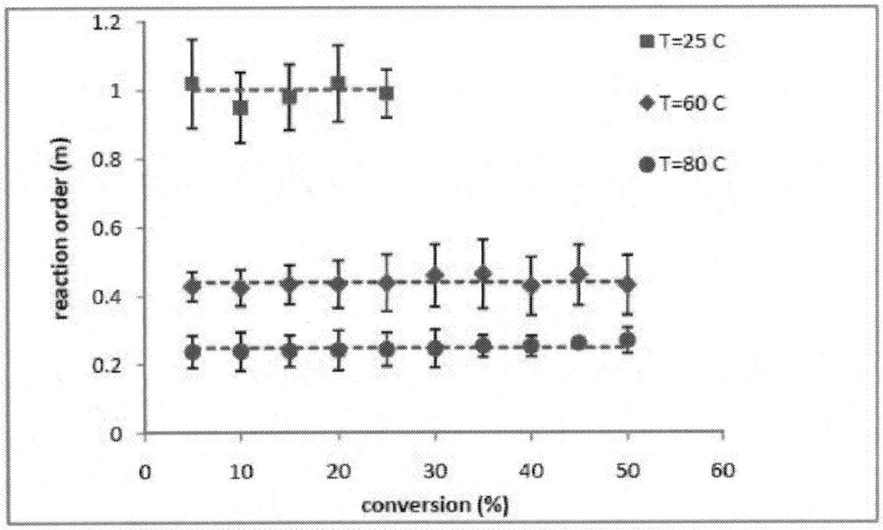

Figure 1. Effect of $NaBH_4$ conversion (%) on reaction order

It is well known that the solution viscosity increases during the course of hydrolysis reaction and this has been considered the reason for the observed decrease in rate with reaction time or with $NaBH_4$ conversion [5]. The effect of temperature on solution viscosity can be described in terms of Arrhenius-type relationships [10]:

$$\eta = \eta_0 \exp\left(\frac{E_{af}}{RT}\right) \qquad (10)$$

where η is the viscosity, η_0 the pre-exponential factor and E_{af} the activation energy of flow. This implies that increase in temperature decreases the solution viscosity. Viscosity is an indication of the solution resistance to flow, and reduced flow resistance at high temperature enhances the interaction between reactant molecules. This in turn lessens the relative contribution of the effect of the concentration on the reaction rate. The deviation from first-order kinetics at low temperatures to fractional-order kinetics at higher temperatures observed in the present study, is attributed to this difference in interaction between reactant molecules with respect to temperature due to change in solution viscosity. It is worthwhile noting that the earlier studies also reported first-order kinetics for self-hydrolysis of $NaBH_4$ for lower conversions and temperatures and for buffered solutions [11-15]

Effect of $NaBH_4$ conversion on activation energy

The classical kinetic theory defines reaction orders to be constant with temperature. In the present case the observed difference in the reaction order with temperature (Fig. 1) suggests a $NaBH_4$ conversion effect on the activation energy of the hydrolysis reaction (E_a), as given in Table 2. As seen, activation energy increases with $NaBH_4$ conversion/with formation of metaborate by-products.

It is well documented in literature that the formation of highly basic metaborate by-products in the course of the hydrolysis reaction results in highly viscous solutions. Investigation on the influence of soluble solids content on viscosity, by various researchers, showed that depending on the type of soluble solids content the activation energy of flow (E_{af}) might increase or decrease. The effects of soluble solid contents are described by the following expressions [10]:

$$\eta = \eta_1 C^{b_1} \qquad (11)$$

$$\eta = \eta_2 \exp(b_2 C) \qquad (12)$$

where η_i and b_i are constants and C is the concentration. Hence one may assume that the activation energy of the flow increases with an increase in $NaBH_4$ conversion due to metaborate by-product formation, which in turn increases the activation energy of the hydrolysis reaction.

Table 2. Apparent activation energy determined for various $NaBH_4$ conversions

$NaBH_4$ Conversion (%)	$^aE_{ap}$ (kJ mol^{-1})	R^2	Standard error	F	P-value
5	64.43 ± 14.5	0.9944	1.14	3914.89	0.025
10	71.03 ± 11.5	0.9998	0.91	6721.39	0.038
15	82.50 ± 13.5	0.9998	1.03	6725.19	0.008
20	89.62 ± 12.5	0.9994	1.96	2152.88	0.013

[a]with $\pm$ 95% confidence interval

3.2. Modeling

The dependence of apparent activation energy (E_{ap}) on α (Table 2) is described by [16]:

$$E_{ap} = \varepsilon_0 + \varepsilon_1 \ln(1-\alpha) \qquad (13)$$

with $\varepsilon_0 = 54.2 \pm 1.7$ kJ mol^{-1} and $\varepsilon_1 = -114 \pm 12.2$ kJ mol^{-1} (within a 95% confidence interval)

The change in apparent reaction order with temperature (Fig. 1) can be evaluated if the parameters of the kinetic compensation effect (KCE) are known [16]. The present study shows that $NaBH_4$ hydrolysis follows the following KCE relationship:

$$\ln A_{ap} = \alpha^* + \beta^* E_{ap} \qquad (14)$$

with α^* =0.84 ± 1.2 (mol $L^{-1})^{1-(m+n)}$ hr^{-1} and β^*= 0.36 ± 0.04 kJ mol^{-1}(within a 95% confidence interval)

Combining Equations (4), (9), (13) and (14) results in:

$$\frac{d\alpha}{dt} = A_0(1-\alpha)^{(m+n)} \qquad (15)$$

where:

$$n = \beta^*\varepsilon_1 - \frac{\varepsilon_1}{RT} \qquad (16)$$

$$A_0 = \exp\left(\alpha^* + \beta^*\varepsilon_0 - \frac{\varepsilon_0}{RT}\right)C_0^{m+n-1} \qquad (17)$$

Equation (15) shows that the overall process is apparently described with a reaction order n+m as illustrated in Fig 1. If the values of n are estimated from equation (16) and n+m from Fig 1, the values for m obtained are given in Table 3. As expected, m tends to a constant value, i.e. 0.22 ± 0.01.

Table 3. Values of n, n+m and m at different temperatures

T(^{0}C)	$n^{(1)}$ (Eq. 16)	$n+m^{(2)}$ (Fig. 1)	$m^{(3,*)}$
25	0.820	1.04	0.214
60	0.222	0.44	0.218
80	0.024	0.245	0.220

[1]dependent on temperature, [2]apparent reaction order
[3] independent on temperature, [*] m = 0.22± 0.01 (mean ± standard deviation)

4. Conclusions

The complex $NaBH_4$ self-hydrolysis process is analyzed by a differential isothermal isoconversional method, based on a single-step kinetic approximation. The overall process is apparently described by a 'reaction order' model with temperature dependent reaction order in a wide practical operating window (10-20 wt% $NaBH_4$ solution and 25-80 ^{0}C).

The apparent activation parameters, i.e. apparent activation energy and pre-exponential factor are conversion dependent and interrelated through the kinetic compensation effect, which explains the observed effect of the temperature on the reaction order.

The apparent activation parameters obtained provide some insight into the theoretical kinetic mechanism, which however remains a matter of heated debates.

5. References

[1] Beaird AM, Davis TA, Matthews MA. Deliquescence in the hydrolysis of sodium borohydride by water vapor, Ind Eng Chem Res 2010;49(20):9596-9599.

[2] Gonçalves, A., Castro, P., Novais, A.Q., Fernandes, V., Rangel, C. and Matos, H., Dynamic modeling of hydrogen generation via hydrolysis of sodium borohydride, In Proceedings of PRES'07, 24-28 June, Ischia, Italy, (In Chemical Engineering Transactions, Vol. 12, 2007. Editor: Jiri Klemes, AIDIC, Italy, 243.)

[3] Andrieux J, Demirci UB, Hannauer J, Gervais C, Goutaudier C, Miele P. Spontaneous hydrolysis of sodium borohydride in harsh conditions. Int J Hydrogen Energy 2011;36:224-233.

[4] Cavaliere S, Hannauer J, Demirci UB, Akdim O, Miele P. Ex situ characterization of N_2H_4-, $NaBH_4$- and NH_3BH_3- reduced cobalt catalysts in $NaBH_4$ hydrolysis. Catal Today 2011;170:3-12.

[5] Retnamma R, Novais AQ, Rangel C.M. Kinetics of hydrolysis of sodium borohydride for hydrogen production in fuel cell applications: A review. Int J Hydrogen Energy 2011; 36: 9772-9790 (and references thereof).

[6] Simon P. Isoconversional methods: Fundamentals, meaning and application. J. Therm. Anal. Calorimetry 2004;76:123-132.

[7] Vyazovkin S, Burnham AK, Criado JM, Perez-Maqueda AA, Popescu C, Sbirrazzuoli N. ICTAC Kinetics committee recommendations for performing kinetic computations on thermal analysis data. Therm Chim Acta 2011;520:1-9.

[8] Yu Lin, Matthews MA. Hydrolysis of sodium borohydride in concentrated aqueous solution. Int J Hydrogen Energy 2011;36:7416-7422.

[9] Vyazovkin S, Sbirrazzuoli N. Kinetic methods to study isothermal and nonisothermal epoxy-anhydride cure. Macromol Chem Phys 1991;200:2294-2303.

[10] Ibarz A, Garvin A, Costa J. Rheological behavior of sole (prunus spinosa) fruit juices. J Food Engng 1996;27:423-430.

[11] Pecsok R. Polarographic studies on the oxidation and hydrolysis of sodium borohydride. J Ame Chem Soc 1953;75:2862-2864.

[12] Stockmayer WH, Miller RR, Zeto RJ. Kinetics of borohydride hydrolysis. J. Phys. Chem 1961;65:1076-1077.

[13] Davis RE, Swain CG. General acid catalysis of the hydrolysis of sodium borohydride. J Ame Chem Soc 1960;82(22):5949-5950.

[14] Davis RE. Bromels E, Kibby CL. Borohydrides III. Hydrolysis of sodium borohydride in aqueous solution. J Ame Chem Soc 1962;84:885-889.

[15] Mesmer RE, Jolly WL. The hydrolysis of aqueous hydroborate. Inorg Chem 1962;1:608-612.
[16] Budrugeac P, Popescu C, Segal E. Change of the apparent reaction order with temperature as a consequence of the reaction mechanism complexity. J. Therm. Anal. Calorimetry 2001;64:821-827.

(583a) Production of Concentrated Monosaccharride Solutions From Biomass Using CO2-H2O Mixtures At Varying Temperatures

Jeremy S. Luterbacher[1], Jefferson W. Tester[1] and Larry P. Walker[2], (1)Department of Chemical and Biomolecular Engineering, Cornell University, Ithaca, NY, (2)Department of Biological and Environmental Engineering, Cornell University, Ithaca, NY

Sustainably producing concentrated solutions of monosaccharrides from biomass is a key bottleneck in the conversion of lignocellulosic biomass to biofuels or bioproducts. Most pretreatment and enzymatic hydrolysis processes are run at low-solid concentration (<10 wt%) and use chemical catalysts, while high-solids enzymatic hydrolysis experiments are almost always performed with air-dried pretreatment mixtures.

This study uses a stirred 1 L reactor to study high solids (40 wt%) and high-pressure (200 bar) H_2O-CO_2 pretreatment of mixed hardwood and switchgrass. In addition, two-temperature stage pretreatment was introduced and optimized to take advantage of the temperature dependent biomass depolymerization reaction sequence (a short high-temperature stage at 200-210°C for 0 to 24 min was followed by a long low-temperature stage at 160-170°C for 60 min). Optimal glucan to glucose yields around 80% were obtained after 72 hrs of dilute hydrolysis (1 wt% glucan) for both substrates at similar pretreatment conditions (210°C, 16 min for hardwood and 210°C, 1 min for switchgrass, both followed by 60 min at 160°C). These optimally pretreated substrates were used as feedstock in high-solids enzymatic hydrolysis reactions, which gave glucose yields above 80% for both hardwood and switchgrass after 48 hrs of hydrolysis. Therefore, without additional chemical catalysts or any drying, two-temperature stage H_2O-CO_2 pretreatment coupled with high-solids enzymatic hydrolysis can produce monosaccharide solutions of 185 gr/L (170 gr/L of glucose and cellobiose 15gr/L of hemicellulose sugars) for mixed hardwood and 148 gr/L (115 gr/L of glucose and cellobiose 33gr/L of hemicellulose sugars) for switchgrass. This suggests that H_2O-CO_2 pretreatment is an attractive alternative to chemically catalyzed processes such as dilute acid pretreatment.

See more of this Session: Developments In the Pretreatment of Lignocellulosics for Bioconversion III
See more of this Group/Topical: Sustainable Engineering Forum

(609a) Exciton Engineering with Carbon Nanotubes and Graphene for Solar Energy Conversion: From Exciton Antennae to Nano-Heterojunctions

Michael Strano[1], Geraldine LC Paulus[2], Moon-Ho Ham[2], Chang Young Lee[2], Changsik Song[2], Kourosh Kalantar-Zadeh[3], Wonjoon Choi[4], Jae-Hee Han[1], Ryuichiro Maruyama[5], Esther S. Jeng[6], Daniel A. Heller[2], Woo-Jae Kim[1], Paul W. Barone[7] and Cristiano Fantini[8], (1)Chemical Engineering, Massachusetts Institute of Technology, Cambridge, MA, (2)Department of Chemical Engineering, Massachusetts Institute of Technology, Cambridge, MA, (3)School of Electrical and Computer Engineering, RMIT University, Melbourne, Australia, (4)Department of Mechanical Engineering, Massachusetts Institute of Technology, Cambridge, MA, (5)Materials Laboratories, Sony Corporation, Kanagawa, Japan, (6)Department of Chemistry, University of Illinois at Urbana-Champaign, Urbana, IL, (7)Chemical Engineereing, Massachusetts Institute of Technology, Cambridge, MA, (8)Departamento de Fisica, Universidade Federal de Minas Gerais, Belo Horizonte-MG, Brazil

Exciton engineering with carbon nanotubes and graphene for solar energy conversion: from exciton antennae to nano-heterojunctions

Michael S. Strano

Charles and Hilda Roddey Associate Professor of Chemical Engineering

66-566 Department of Chemical Engineering

77 Massachusetts Avenue

Cambridge, MA 02139-4307

Email: strano@MIT.EDU

phone: (617) 324-4323

fax: (617) 258-8224

http://web.mit.edu/stranogroup/

Our laboratory has been interested in how low dimensional materials, such as single walled carbon nanotubes or graphene, can augment and enhance solar conversion efficiencies and demonstrate new photonic concepts. We will discuss two recent examples of our work in this space.

In the first example, there is significant interest in combining carbon nanotubes with semiconducting polymers for photovoltaic applications because of potential advantages from smaller exciton transport lengths and enhanced charge separation. However, to date, bulk heterojunction (BHJ) devices have demonstrated relatively poor efficiencies, and little is understood about the polymer/nanotube junction. To investigate this interface, we fabricate a planar nano-heterojunction comprising well-isolated millimeter-long single-walled carbon nanotubes underneath a poly(3-hexylthiophene) (P3HT) layer (1). The resulting junctions display photovoltaic efficiencies per nanotube ranging from 3% to 3.82%, which exceed those of polymer/nanotube BHJs by a factor of 50-100. The increase is attributed to the absence of aggregate formation in this planar device geometry. It is shown that the polymer/nanotube interface itself is responsible for exciton dissociation. Typical open-circuit voltages are near 0.5 V with fill factors of 0.25-0.3, which are largely invariant with the number of nanotubes per device and P3HT thickness. A maximum efficiency is obtained for a 60 nm-thick P3HT layer, which is predicted by a Monte Carlo simulation that takes into account exciton generation, transport, recombination, and dissociation. We combine for the first time both optical T-matrix and kinetic Monte Carlo models to investigate the photocurrent generation in two state-of-the-art PHJ photovoltaics. The combined model takes into account the rates of exciton generation, transport, recombination and dissociation using literature values. By including the optical, electronic and structural properties of the different materials, we are able to predict the short-circuit current of recently reported P3HT/SWNT PHJ and also P3HT/PCBM PHJ solar cells from the literature. The experimental data for each of these devices show a maximum photocurrent output at a P3HT thickness of 60-65 nm, in contradiction to the expected value equal to the diffusion length of excitons in P3HT (8.5nm). The model demonstrates how a bulk exciton sink can explain this shifted maximum in the P3HT/SWNT case, whereas the maximum is mainly determined by PCBM interdiffusing in P3HT in the P3HT/PCBM case. This platform is promising for further understanding the potential role of polymer/nanotube interfaces for photovoltaic applications.

In the second example, there has been renewed interest in solar concentrators and optical antennas for improvements in photovoltaic energy harvesting and new optoelectronic devices. We dielectrophoretically assemble single-walled carbon nanotubes (SWNTs) of homogeneous composition into aligned filaments that can exchange excitation energy, concentrating it to the centre of core–shell structures with radial gradients in the optical bandgap (2). We find an unusually sharp, reversible decay in photoemission that occurs as such filaments are cycled from ambient temperature to only 357 K, attributed to the strongly temperature-dependent second-orderAuger process. Core–shell structures consisting of annular shells of mostly (6, 5) SWNTs (E_g = 1.21 eV) and cores with bandgaps smaller than those of the shell (E_g = 1.17 eV (7,5)–0.98 eV (8,7)) demonstrate the concentration concept: broadband absorption in the ultraviolet–near-infrared wavelength regime provides quasi-singular photoemission at the (8, 7) SWNTs. This approach demonstrates the potential of specifically designed collections of nanotubes to manipulate and concentrate excitons in unique ways.

1. Ham MH, Paulus GLC, Lee CY, Song C, Kalantar-zadeh K, Choi W, Han JH and **Strano MS**: Evidence for High-Efficiency Exciton Dissociation at Polymer/Single-Walled Carbon Nanotube Interfaces in Planar Nano-heterojunction Photovoltaics. ACS NANO, 4 (2010) 6251-6259

2. Han JH, Paulus GLC, Maruyama R, Jeng ES, Heller DA, Kim WJ, Barone PW, Lee CY, Choi JH, Ham MH, Song C, Fantini C, **Strano MS**: Exciton Antennas and Concentrators from Core-Shell and Corrugated Carbon Nanotube Filaments of Homogeneous Composition. NATURE MATERIALS, 9, 833 - 839 (2010).

See more of this Session: Semiconductor Nanowires and Nanotubes for Next Generation Photovoltaics From Fundamentals to Devices
See more of this Group/Topical: Topical 5: Nanomaterials for Energy Applications

(231f) Integrations of Biosurfactant Production Into Advanced Biorefineries

William Colonna[1], **Mustafa E. Marti**[1], Michelle Pynn[2], Gabriel Reznik[2], Kevin Jarrell[2], Buddhi Lamsal[3] and Charles E. Glatz[1], (1)Chemical and Biological Engineering, Iowa State University, Ames, IA, (2)Modular Genetics, Inc., Woburn, MA, (3)Food Science and Human Nutrition, Iowa State University, Ames, IA

Several classes of biosurfactants are produced by microorganisms. One such surfactant is FA-Glu, which is related to the natural biosurfactant, surfactin. FA-Glu consists of a long-chain β-hydroxy fatty acid coupled to glutamic acid. It is produced by a strain of *Bacillus subtilis* genetically-engineered by Modular, Inc. It has several potential applications including use as a dispersant for oil spills, where biodegradability is desirable.

Economic production and eventual application will be favored if substrate costs can be reduced and low-cost recovery methods developed. Several such opportunities arise in advanced biorefineries. Product value would be further enhanced if such production could achieve favored "green" rating by avoiding solvents other than water and simple alcohols and effect emulsification and low toxicity could be demonstrated. This report focus on production considerations.

Production of FA-Glu by fermentation on glucose is compared to results using lower cost substrates available in advanced biorefineries, including fiber components of soy and glycerol. FA-Glu yields from soybean hulls (SBH) ranged from 29-51% of that from glucose on a substrate weight basis.. When SBH polysaccharides were hydrolyzed (~65%) by pre-treatment with carbohydrases, FA-Glu levels were only ~21-37% of those from glucose. The cells consumed the enzymatically-released sugars; however, very little of this was converted into FA-Glu. Addition of glucose to the enzymatic SBH hydrolysate improved FA-Glu yields, but only slightly. The low FA-Glu yields from SBH could be due to inhibitors released from the hulls by the enzymes or inhibitory substances in the enzyme preparations.

Glycerol, a by-product of biodiesel production, proved to be a suitable carbon source for FA-Glu production. Cells grown on 2% w/v glycerol produced ~25% more FA-Glu than cells grown on 2% glucose. Glycerol is abundant and low in cost. Its use as a feedstock for production of microbial biosurfactants would provide a means for efficient utilization of soy by-products and improving the ag economy.

The FA-Glu was purified by collecting the foam from the aerated broth, clarification by centrifugation or filtration, and adsorption and elution from an anion exchange column to a purity of 97.9%. Final product color can be improved by liquid-liquid extraction, but this may compromixe "green" status. A precipitation based alternative process is being examined as a means of eliminating solvent use.

(737b) Effects of Annealing On $Zr_8Ni_{19}X_2$ (X = Ni, Mg, Al, Sc, V, Mn, Co, Sn, La, Hf) for Ni/MH Battery Operation

Jean Nei[1], Kwo Young[2], Simon K. Y. Ng[1] and Steven O. Salley[1], (1)Chemical Engineering and Materials Science, Wayne State University, Detroit, MI, (2)Energy Conversion Devices Inc./Ovonic Battery Company, Rochester Hills, MI

Zr, Ti-based AB_2 metal hydride (MH) alloys are considered to be one of the most promising candidates to replace the misch metal based AB_5 alloys presently used as the negative electrode in nickel/metal hydride (Ni/MH) batteries. Since most of the studies on AB_2 MH alloys focus on multi-element and highly disordered families, the understanding of basic Ti-Ni and Zr-Ni intermetallic alloys is crucial for fully utilizing the synergetic effects between the main AB_2 Laves phases and the non-Laves secondary phases.

For the best balance between large storage capacity (lower hydride heat of formation) and high discharge capability (higher hydride heat of formation), a stoichiometry of A:B close to 1:2 is desirable. However, there is no AB_2 intermetallic alloy for either Zr-Ni or Ti-Ni binary systems. The Zr-Ni phase diagram shows several intermetallic alloys, and the closest intermetallic compounds to AB_2 are Zr_7Ni_{10} and Zr_8Ni_{21}. Zr_7Ni_{10} has slightly stronger metal-hydrogen bond strength and solidifies congruently from liquid. Zr was replaced partially by Ti to reduce the metal-hydrogen bond strength. Other substituting elements were also studied in order to improve the electrochemical properties. Zr_8Ni_{21} was chosen as the focus of this report. Zr_8Ni_{21} does not solidify directly from the liquid as in the case of Zr_7Ni_{10}. Instead, the Zr_2Ni_7 phase is first solidified from the liquid with a Zr-to-Ni ratio of 8:21 and later reacts with the remaining liquid to form Zr_8Ni_{21} alloy peritectically. If the condition of equilibrium is not reached during cooling, one or more secondary phases will precipitate out. Therefore, an annealing process was adopted for this study.

The effects of annealing at 960 °C for 8 hours on the phase structures of a series of partially substituted Zr_8Ni_{21} alloys were studied. The substituting elements included Mg, Al, Sc, V, Mn, Co, Sn, La, and Hf. In general, the abundance of dominant phase increased and the number of smaller secondary phases diminished after annealing. Similar to the choice of C14/C15 main phase in AB_2 alloys, the number of outer-shell electrons plays an important role in determining the main phase of the

A_8B_{21} samples. As number of outer shell electrons of the substituting element increases, the main phase evolves from tetragonal Zr_7Ni_{10}, to orthogonal Zr_7Ni_{10}, Zr_2Ni_7 and then Zr_8Ni_{21}.

The annealing effects on the gas phase and electrochemical hydrogen storage properties were studied. It's observed that as the Zr-content in the main phase of the alloy increases (orthogonal Zr_7Ni_{10} > tetragonal Zr_7Ni_{10} > Zr_8Ni_{21} > Zr_2Ni_7), the gas phase storage capacity increases. After annealing, due to the increase in the abundance of main phase, the capacities of alloys with higher capacities before annealing increase while others decrease. The sample with Hf-supplement shows the highest electrochemical discharge capacity of 200 mAh/g. After annealing, all samples with the same main phase as the as-cast ones show degradation in electrochemical capacity due to the reduction in number and abundance of secondary phases. All supplements show improvement in surface exchange current from the base alloy. Except for La and Hf supplements, annealing reduces the exchange current density. The bulk hydrogen diffusion constant decreases for most of the supplements except V and Sn. All supplements, excluding Sn, show improvement in bulk diffusion after annealing. The gas phase full capacity has strong correlation to the full electrochemical discharge capacity. Among all samples, as-cast La-supplement shows the best overall electrochemical properties.

See more of this Session: Energy Storage and Electrochemical Processes II
See more of this Group/Topical: International Congress on Energy 2011

(609d) Photoelectrochemical Complexes for Solar Energy Conversion That Dynamically, Reversibly and Autonomously Self-Assemble

Ardemis A. Boghossian[1], Jong Hyun Choi[2], Moon-Ho Ham[1], Esther S. Jeng[3], Rachel A. Graff[4], Daniel A. Heller[1], Alice C. Chang[3], Aidas Mattis[5], Timothy H. Bayburt[6], Yelena V. Grinkova[6], Adam Scott Zeiger[7], Krystyn J. Van Vliet[8], Erik K. Hobbie[9], Stephen G. Sligar[6], Colin A. Wraight[6] and Michael Strano[3], (1)Department of Chemical Engineering, Massachusetts Institute of Technology, Cambridge, MA, (2)Mechanical Engineering, Purdue University, West Lafayette, IN, (3)Chemical Engineering, Massachusetts Institute of Technology, Cambridge, MA, (4)Chemical and Biomolecular Engineering, University of Illinois at Urbana-Champaign, Urbana, IL, (5)Department of Biochemistry, University of Illinois at Urbana-Champaign, Urbana, IL, (6)Biochemistry, University of Illinois Urbana-Champaign, Urbana, IL, (7)Materials Science and Engineering, Massachusetts Institute of Technology, Cambridge, MA, (8)Materials Science and Engineering, MIT, Cambridge, MA, (9)Polymers Division, National Institute of Standards and Technology, Gaithersburg, MD

Photoelectrochemical complexes for solar energy conversion that dynamically, reversibly and autonomously self-assemble

Naturally occurring photosynthetic systems use elaborate pathways of self-repair to limit the impact of photo-damage. Here, we demonstrate a complex consisting of an aqueous solution containing photosynthetic reaction centers (RCs), membrane scaffold proteins (MSPs), phospholipids, and single-walled carbon nanotubes (SWCNTs) solubilized with the surfactant sodium cholate (SC) that mimics this process.(1) The components reversibly self-assemble into a highly ordered structure upon dialysis of the surfactant. The resulting assembled structure is photoelectrochemically active and consists of 4-nm-thick lipid bilayer disks (nanodisks, NDs) arranged parallel to the surface of the SWCNT with the RC housed within the bilayer such that its hole injecting site faces the nanotube surface. The structure can be assembled and disassembled autonomously with the addition or removal of surfactant. We model the kinetic and thermodynamic forces that drive the dynamics of this reversible self-assembly process.(2) The assembly is monitored using spectrofluorimetry during dialysis and subsequent surfactant addition and used to fit a kinetic model to determine the forward and reverse rate constants of ND and ND-SWCNT formation. The calculated ND and ND-SWCNT forward rate constants are 79 $mM^{-1}\ s^{-1}$ and $5.4 \times 10^2\ mM^{-1}\ s^{-1}$, respectively, and the reverse rate constants are negligible over the dialysis time scale. We find that the reaction is not diffusion-controlled since the ND-SWCNT reaction, which consists of entities with smaller diffusion coefficients, has a larger reaction rate constant. Using these rate parameters, we were able to develop a kinetic phase diagram for the formation of ND-SWCNT complexes, which indicates an optimal dialysis rate of approximately $8 \times 10^{-4}\ s^{-1}$.

The assembly is thermodynamically metastable and can only transition reversibly if the rate of surfactant removal exceeds a threshold value. Only in the assembled state do the complexes exhibit photoelectrochemical activity. We experimentally demonstrate a regeneration cycle that uses surfactant to switch between assembled and disassembled states, resulting in an increased photoconversion efficiency of more than 300% over 168 hours and an indefinite extension of the system lifetime. The system can disassemble and reassemble over an indefinite number of cycles. We also fit the model to these cyclic ND-SWCNT assembly and disassembly experiments and hence mimic the thermodynamic forces used in regeneration processes detailed previously. Such forces may form the basis of both synthetic and natural photoelectrochemical complexes capable of dynamic component replacement and repair.

1. Ham MH, Choi JH, Boghossian AA, Jeng ES, Graff R.A, Heller DA, Chang AC, Mattis A, Bayburt TH, Grinkova YV, Zeiger AS, Van Vliet KJ, Hobbie EK, Sligar SG, Wraight CA and **Strano MS**: Photoelectrochemical complexes for solar energy conversion that chemically and autonomously regenerate. NATURE CHEMISTRY, 2 (2010) 929-936.

2. Boghossian AA, Choi JH, Ham MH and **Strano MS**: Dynamical and Reversible Self-Assembly of Photoelectrochemical Complexes Based on Lipid Bilayer Disks, Photosynthetic Reaction Centers, and Single-Walled Carbon Nanotubes. Langmuir, 27 (2011) 1599-160

See more of this Session: Semiconductor Nanowires and Nanotubes for Next Generation Photovoltaics From Fundamentals to Devices
See more of this Group/Topical: Topical 5: Nanomaterials for Energy Applications

(641e) Effects of Reaction Conditions On the Acid-Catalyzed Hydrolysis of Miscanthus Dissolved In An Ionic Liquid

Sean J. Dee and Alexis T. Bell, Chemical and Biomolecular Engineering, University of California - Berkeley, Berkeley, CA

Miscanthus, a perennial grass, is an attractive biofuel feedstock compared with corn and poplar due to its high energy yield per hectare and low agricultural inputs such as water and soil nutrients. Recent studies have shown the three primary components of miscanthus, cellulose, hemicellulose, and lignin, can be dissolved in ionic liquids (ILs) at levels approaching 20 wt%. An interesting question is then whether the cellulosic and hemicellulosic components of the dissolved miscanthus can be hydrolyzed to sugars in high yield. While acid-catalyzed hydrolysis of cellulose and hemicellulose has been studied extensively in ILs, the conversion of biomass is not as well understood. Recent literature has claimed that effective chemical conversions of biomass in ILs requires either thermochemical delignification which can destroy both hemicellulose and cellulose, or multistage processing that is inefficient and makes IL recycling energy intensive. This study investigated the acid-catalyzed hydrolysis of miscanthus dissolved in the ionic liquid 1-ethyl-3-methylimidazolium chloride ([Emim][Cl]). The effects of varying temperature, acid concentration, miscanthus concentration, and the gradual addition of water were undertaken. The similarities and differences between the rate of hydrolysis of microcrystalline cellulose (Avicel) and amorphous hemicellulose (Xylan from birchwood) compared to the cellulosic and hemicellulosic components of miscanthus were determined. The influence of miscanthus delignification on the hydrolysis rate was also explored. The present work has demonstrated that 90% biopolymer conversion from miscanthus could be achieved, with 93% selectivity to saccharine products in a single stage hydrolysis without pretreatment.

See more of this Session: Chemical and Catalytic Conversions and Processes for Renewable Feedstocks
See more of this Group/Topical: Sustainable Engineering Forum

(294e) Development of Novel Electrocatalyst Support In Proton Exchange Membrane Fuel Cells for Automotive Applications

Sehkyu Park, Yuyan Shao, Haiying Wan, Vilayanur Viswanathan, Jun Liu and Yong Wang, Pacific Northwest National Laboratory, Richland, WA

Proton exchange membrane (PEM) fuel cells have received great attention as an alternative power source for transportation and stationary applications. A membrane-electrode-assembly (MEA) that generates electricity by electrochemical reactions consists of proton exchange membrane as an electrolyte and two electrodes for the anode and the cathode, respectively. Pt or Pt alloy catalyst is used to oxidize hydrogen molecule at the anode and to reduce gaseous oxygen at the cathode. Pt or Pt alloy nanoparticles is typically dispersed on high surface area carbon to increase its surface area available for electrochemical reactions. However, carbon oxidation can be thermodynamically favored even at low potentials (> 0.2 V *vs.* standard hydrogen electrode (SHE)). The corrosion rate of carbon in the presence of Pt nanoparticles greatly increases at positive potentials (> 1.0 V *vs.* SHE) of the cathode during start-stop cycles.

In this study, durable hybrid electrocatalyst supports for the cathode were developed using graphitized carbons as scaffolds and metal oxides to disperse Pt nanoparticles onto carbon surface protected by metal oxides. The electrochemical property for new electrocatalysts with different hybrid supports was examined using thin film rotating disk electrode (TFRDE). Performance and durability of the cathodes loaded with various electrocatalysts were evaluated under the catalyst support accelerated stress test. A metal oxide modified graphitized carbon resulted in better durability (> 3X) than commercially available catalyst support, thereby showing great promise in becoming alternative and durable electrocatalyst support in PEM fuel cells for transportation applications.

See more of this Session: Fuel Cell Technology I
See more of this Group/Topical: Fuels and Petrochemicals Division

(373d) Development and Testing of a Novel Cu-Based ORR Catalyst within An Acidic Fuel Cell

Matthew S. Naughton[1], Matt A. Thorseth[2], Andrew A. Gewirth[2] and Paul J. A. Kenis[3], (1)Department of Chemical & Biomolecular Engineering, University of Illinois at Urbana-Champaign, Urbana, IL, (2) Chemistry, University of Illinois at Urbana-Champaign, Urbana, IL, (3)Chemical & Biological Engineering, University of Illinois at Urbana-Champaign, Urbana, IL

Development andtesting of a novel Cu-based ORR catalyst within an acidic fuel cell

Matt S. Naughton,Matt A. Thorseth, Andrew A. Gewirth, and Paul J. A. Kenis

Departmentof Chemical & Biomolecular Engineering,

University of Illinois at Urbana-Champaign,

600 S. Mathews Avenue, Urbana, IL61801, USA.

Section: Alternate Fuels and NewTechnology (16D)

Session: ***16D03 Fuel Cell Technology***

(Funding from DOE science)

Fuel cells hold promise as alternativepower sources due to their ability to bypass Carnot efficiency limitations by directlyconverting chemical energy into electrical energy. However, the high costs of Ptcatalysts and membranes, as well as component durability issues, have barredwidespread implementation [1]. A key area for cost reduction is thedevelopment of novel cathode catalysts, which can greatly reduce fuel cellcosts [2]. In addition to these cost advantages, non-Pt catalysts are oftenmore tolerant to contaminants such as methanol at the cathode. Copper-basedORR catalysts have shown promise in alkaline media, but a high-performance catalysthas not yet been successfully demonstrated in acidic media [3].

Previously, we developed a pH-flexibleflowing electrolyte microfluidic fuel cell, which uses an external referenceelectrode to individually analyze cathode and anode performance [1]. Thismicrofluidic configuration combines the versatility of a traditional threeelectrode cell with the conditions found in an operating fuel cell, allowingfor *in-situ* studies of catalyst and electrode performance. Externalcontrol over the flowing electrolyte stream allows for controlled introductionof contaminants and maintains their concentrations over the course ofexperimentation [4].

Here, we present our work on thedevelopment and testing of a novel Cu-based ORR catalyst operating in an acidicfuel cell. The effects of varied loading and catalyst forms are demonstrated.The catalyst performance in the presence of contaminants such as methanol andethanol is quantified. Development of this catalyst has the potential toreduce the cost of acidic fuel cell systems and improve the understanding ofCu-based ORR catalysis.

[1] Brushett *et al.*, Journal of the ElectrochemicalSociety, 2009, 156, B565

[2] Gewirth *et al.*, Inorganic Chemistry, 2010, 49,3557-3566
[3] Thorum *et al.*, Angewandte Chemie, 2009, 121, 171-173

[4] Naughton *et al.*, Journal of Power Sources, 2011,196, 1762-1768

See more of this Session: Fuel Cell Technology II
See more of this Group/Topical: Fuels and Petrochemicals Division

(436a) The Role of Oxygen In the Stability of Biomass Pyrolysis Oils

Foster Agblevor, Biological Engineering, Utah State University, Logan, UT and Ofei Mante, Biological Systsems Engineering, Virginia Polytechnic Institute and State University, Blacksburg, VA

One of the major challenges in the conversion of biomass pyrolysis to high value products and fuels is the poor stability of the oils. Biomass pyrolysis oils have shown increases in viscosity during storage at ambient conditions and when heated above ambient temperature. The poor stability of the biomass pyrolysis oils have been attributed to the high intrinsic oxygen content of the oils. Researchers have been working on stabilizing the oils for various applications. In order to demonstrate the role of oxygen in the stability of biomass pyrolysis oils, we prepared three pyrolysis oil samples using catalysts and no catalysts. Two of the oils prepared with different catalysts were stored at room temperature for several months and were also subjected to distillation. The two catalyst oils (catalyst #1 oil and catalyst #2 oil) had oxygen content of 21 wt% whereas the non-catalytic oil had oxygen content of 43 wt%. The viscosities of the two oils were measured when fresh and after storage. The viscosity of catalyst #1 oil remained almost constant after 18 months of storage whereas the catalyst #2 oil sample almost solidified after 5 months of storage at ambient conditions. A sample the catalyst #1 oil whose viscosity did not change after 18 months of was stored without lid for six months and its viscosity measured. After six months storage without lid, the viscosity of catalyst #1 oil also increased but at a less rate compared to the non-catalyst oil and the catalyst #2 oil. The catalyst #1 oil sample could be completely distilled both in vacuum and atmospheric pressure whereas catalyst #2 oil and the non-catalyst oil formed solid after 45wt% of the material was distilled. The results clearly showed the intrinsic oxygen content of biomass pyrolysis oil is not a good measure of the stability of the oil instead, the distillation of the oil may be a better measure of the oil stability. Further, atmospheric oxygen also contributes significantly towards the destabilization of the biomass pyrolysis oils. Specific oxygenated functional d groups in the biooils are responsible for the for poor stability which is not measured by the total oxygen content of the biooils.

See more of this Session: Biomass Pyrolysis II
See more of this Group/Topical: Fuels and Petrochemicals Division

(523d) Deconstruction of Wheat Straw Lignin by Streptomyces Viridosporus As Insight Into Biological Degradation Mechanism

Jijiao Zeng[1], Department of Biological Systems Engineering, Washington State University, Pullman, WA, Deepak Singh, Edeniq Inc, Visalia, CA, Dhrubojyoti D. Laskar, Washington State University, Pullman, WA and Shulin Chen, Biological Systems Engineering, Washington State University, Pullman, WA

Deconstruction of wheat straw lignin by *Streptomyces viridosporus* asinsight into biological degradation mechanism

JijiaoZeng[1], Deepak Singh[2],Dhrubojyoti Dey Laskar[1]and Shulin Chen[1*]

1. Department of Biological Systems Engineering, Washington State University,Pullman, Washington 99164-6120, USA

2. Edeniq Inc, 2505 N Shirk Road, Visalia, 93291 CA

*Corresponding author. Tel: +1 509 335 3743; fax: +1 509 335 2722.

E-mail address: chens@wsu.edu

Abstract:

Lignins are secondary natural organic compositions of plant cell wall besides cellulose and hemicelluse. Structurally lignins are linked with hemicelluloses and cellulose to form recalcitrant complex that resist the environmental stress. Biological systems created by nature over millions of years of evolution can be capitalized for targeted deconstruction of plant cell walls through selective cleavage and structural modification of lignin. Such modification causes changes in cell wall structure for facilitating enzymatic attack to cellulose for sugar release. The study of biological degradation mechanism on lignocellulosic materials could provide the new knowledge to convert plant biomass to biofuel and high value co-products. In this study, we have investigated the effect of wheat straw lignin structural changes by *Streptomyces viridosporus* on downstream enzymatic hydrolysis. ^{13}C CPMAS NMR revealed that the rising of carbonhydrate/lignin ratio accompany with enhanced intensity of methoxyl group in bio-treated wheat straw. The degraded lignin were isolated and analyzed by Fourier Transmission Infrared (FTIR), Pyrolysis Gas Chromatography/Mass Spectrometry (Py-GC/MS) and two dimenssional heteronuclear multiple quantum coherence nuclear magnatic resonance (2D HMQC NMR) to elucidate their structural change. The results indicate that the biological pretreated lignin were associated with critical change and/or modification on their structure implying selective removal of guaiacyl (G) lignin unit which is more condense than syringyl (S) unit. Moreover, the treated lignin also dramatically reduced the inhibition of cellulose hydrolysis. This information provides new insights for understanding the biological degradation mechanism on lignin for cellulose utilization and exploring a possible route to convert lignin derived co-products.

See more of this Session: Developments In the Pretreatment of Lignocellulosics for Bioconversion II
See more of this Group/Topical: Sustainable Engineering Forum

(106d) MnO2 Doped Graphene As Electrodes in Supercapacitor

Lixin Wang[1], Kapila Wadumesthrige[1], Steven O. Salley[2] and K. Y. Simon Ng[3], (1)Chemical Engineering, Wayne State University, Detroit, MI, (2)Chemical Engineering and Material Science, Wayne State University, Detroit, MI, (3)Wayne State University, Detroit, MI

Abstract:

Graphene doped carbon nanotubes for high-energy supercapacitors

A supercapacitor is an electrochemical capacitor with a relatively high energy density and is characterized by fast discharge rates and low maintenance. These characteristics have already been explored in various applications such as consumer electronics, wind energy system, smart grid applications and electric vehicles. The primary purpose behind the burgeoning effort in supercapacitor research is to achieve greater energy densities at a lower cost. Vertical carbon nanotube (CNT) is a promising electrode material for supercapacitor production with a superior conductivity, large voltage window of stability, high surface area, steady mechanical strength and low cost. Capacitance values up to 135 F/g were reported with multi-walled carbon nanotubes. However, the relatively small inherent capacitance of CNTs confines their use to supercapcitor applications. In this study, using different methods, graphite will be doped into CNT walls to improve its capacitance. The doped CNTs will then be used in supercapacitor electrode applications and their performance tested in order to find a suitable method of improving overall capacitance. 200 F/g capacitance is expected after the optimization of doping materials and doping methods.

See more of this Session: Nanomaterials for Energy Storage II
See more of this Group/Topical: Topical 5: Nanomaterials for Energy Applications

(436e) A Model-Assisted Tool for the Characterization of Char In Biomass Pyrolysis and Biomass Catalytic Pyrolysis

Shoucheng Du, Julia Valla and George M. Bollas, Department of Chemical, Materials & Biomolecular Engineering, University of Connecticut, Storrs, CT

Coke formationappears as a reason for catalyst deactivation and results in undesirableproduct selectivity in biomass pyrolysis. However, there is no effective way toseparate coke and char in the majority of current approaches, which preventsstudying coke formation and therefore the real reason for catalyst deactivation.The purpose of this study is to develop a model-assisted tool for quantitativeanalysis of coke and char composition in the residual after biomass pyrolysisand biomass catalytic pyrolysis.

In thisstudy, alpha-d-glucose was used as a biomass representative molecule and ZSM-5 fromW.R. Grace & Co. was used for catalytic pyrolysis experiments. Char sampleswere prepared with biomass (catalytic) pyrolysis by devolatilizing glucose atdifferent temperatures in nitrogen atmosphere. The gasification of biomass charswas carried out in a thermogravimetric analyzer (TGA) in carbon dioxideatmosphere at different heating rates.

In order tocharacterize the biomass pyrolysis char, three representative models,volumetric model (VM), grain model (GM) and random pore model (RPM) were developedand applied to fit the experimental data from TGA.

The following equation is the randompore model for fluid-solid reactions developed by Bhatia and Perlmutter [1].

$$\frac{dX}{dt} = k_0 e^{-\frac{E}{RT}}(1-X)\sqrt{1-\psi \ln(1-X)}$$

In the above equation, the pre-exponentialfactor k_0, activation energy E, the structuralparameter can be obtained, by fitting the model to experimental dataof char/coke gasification or oxidation. Similar approaches have been used byMiura et al. [2] and Fermoso et al. [3] in the characterization of coal/coal-biomassblend derived chars. On the basis of these works, a method is being developed todifferentiate coke from char, using temperature programmed gasification(reaction of char and coke with CO_2) to exploit the fact that the morereactive char is gasified at lower temperatures than the less reactive coke. Independentapplication of RPM for coke and char followed by integration for a linear rampin temperature yields a correlation for the total mass loss X as afunction of the composition in coke and char. The concept applies similarly onthe other two models, VM and GM. Moreover, a pore-scale effectiveness factor isintroduced to modify RPM, allowing different pore sizes to grow at differentrates [4].

The final correlations for each model arethen fitted to experimental measurements of TGA gasification at differentheating rates, resulting in a model-assisted tool for char and cokeidentification. Parameter estimation is performed using GAMS global solvers.The results of the models are also compared and verified with scanning electronmicroscopy to validate the approach and generate a standard method for char andcoke characterization. Experimental results, statistical analysis of theliterature and model fitting parameters will be presented. Results illustratethe applicability of the method for the characterization of biomass char andcoke.

References:

(1) Bhatia, S. K.;Perlmutter, D. D. *AIChE Journal.* **1980**, *26*, 379-386.

(2) Miura, K.; Nakagawa,H.; Nakai, S.; Kajitani, S. *Chemical Engineering Science.* **2004**, *59*,5261-5268.

(3) Fermoso, J.; Gil, M.V.; Pevida, C.; Pis, J. J.; Rubiera, F. *Chemical Engineering Journal.* **2010**,*161*, 276-284.

(4) Singer, S. L.;Ghoniem, A. F. *Energy & Fuels.* **2011**, *25*, 1423-1437.

See more of this Session: Biomass Pyrolysis II
See more of this Group/Topical: Fuels and Petrochemicals Division

(349h) Studying Effects of Inorganic Salts On Biomass Pyrolysis Using a Captive Sample Reactor

Calvin Mukarakate[1], David Robichaud[1], Bryon S. Donohoe[2], Mark W. Jarvis[1], Kristiina Iisa[3] and Mark R. Nimlos[1], (1)National Renewable Energy Laboratory, Golden, CO, (2)Biosciences Center, National Renewable Energy Laboratory, Golden, CO, (3)National Renewable Energy Laboratory, Boulder, CO

Pyrolysis and gasification are two promising thermochemical technologies for the conversion of biomass to bio-oil and synthesis gas (syngas) respectively. Understanding biomass pyrolysis behavior is important for designing economical pyrolysis and gasification systems. Inorganic compounds present in biomass have been shown to cause selective formation of pyrolysis products, including catalytic degradation of biomass. In this study, 300 mm poplar sections were soaked in 0.1 M inorganic salt solutions and dried in a vacuum oven overnight at 60 °C. Poplar sections were impregnated in 100 mM solutions of various inorganic salts (NaCl, KCl, $CaCl_2$, NaOH, KOH, $Ca(OH)_2$, $CuCl_2$, $FeCl_3$, $Cu(SO_4)_2$ and $Fe(SO_4)_3$. The dried samples were then pyrolyzed in a captive sample reactor (CSR) at 450 °C and the poplar samples, both before and after pyrolysis, were analyzed by multiple microscopy modalities. Samples soaked in $CuCl_2$, $FeCl_3$, and $Fe(SO_4)_3$ displayed intriguing morphological and cell wall structural changes before pyrolysis whereas those soaked in alkali and alkaline earth metal salts retained the same morphology as untreated poplar. The morphological changes of post-pyrolysis chars from poplar samples treated with transition metal salts resemble changes observed for non-treated samples at significantly higher temperatures (>900 C). The pyrolysis products produced from these samples using CSR/molecular beam mass spectrometry (MBMS) system will be discussed.

See more of this Session: Biomass Pyrolysis I
See more of this Group/Topical: Fuels and Petrochemicals Division

(206f) Kinetic Model for Hydrodeoxygenation Reaction of M- Cresol On Alumina- Supported Pt Catalyst

Phuong T. Do, Andrew J. Foster, Jingguang G. Chen and Raul F. Lobo, Chemical Engineering, University of Delaware, Newark, DE

Variety of oxygenated compounds are detected in the pyrolysis oils, which leads to a number of associated problems such as low stability and reduced heating values. As a consequence, complete deoxygenation of bio oils with minimum hydrogen uptake is one of the proposed strategies toward improving the bio- oil quality. The oxygenated compounds come with various functionalities and consequently the deoxygenation chemistry is highly sensitive to these functional groups. In order to design effective catalysts to eliminate oxygen from all oxygenated functionalities, the first step is to gain complete understanding of the deoxygenation mechanisms of individual functional groups. This study has focused on the phenolic functional groups, which abundantly appear in bio oils. 3-methylphenol or m-cresol was chosen as a representative molecule for methylated phenol compounds. The reactions were carried out under plug-flow mode and in gas-phase condition, which will allow more effective reaction kinetic modeling compared to liquid-phase reactions. Conversions of m-cresol occurred over 1.7 wt% Pt/ Al_2O_3 catalyst at atmospheric pressure and temperatures ranging from 513 K to 563 K. In presence of hydrogen, the deoxygenation of m- cresol into hydrocarbons proceeds through two consecutive steps: hydrogenation followed by dehydration reactions. On Pt, 3- methylphenol is first hydrogenated to 3-methylcyclohexanol, which is consequently deoxygenated to unsaturated cyclic methylcyclohexenes via dehydration reaction on the support acid site. Once methylcyclohexenes were formed, they were quickly transformed into methylcyclohexane and toluene via hydrogenation and dehydrogenation reactions on metal sites. From the hydrogen consumption standpoint, toluene is a more favorable hydrocarbon product than methylcyclohexane.

In order to achieve a thorough understanding of effects of the bifunctional catalyst on the overall conversion and product selectivity of m- cresol reactions, Langmuir- Hinshelwood kinetic model involving dual sites (i.e. metal and support) was employed to fit the reaction data. Besides four surface reactions listed above, desorption of toluene from the Pt metal was considered as one of the rate-limiting steps. In order to achieve a reasonable regression, equilibirium adsorption constants (K_i) of adsorbed species on metal and support estimated from density functional theory (DFT) calculations and equilibrium reaction constants (K_M) were constrained during the fitting. The important kinetic parameters such as elementary rate constants (ki) and activation barriers (Δ Ea) were extracted from the model. Turn-over rate (TOR) were normalized by the fitted rate constants with the number of metal and acid active sites provided by CO chemisorption and temperature-programmed desorption (TPD) of isopropylamine experiments. Identification of rate-limiting step and acknowledgement of individual roles of metal and acid sites in overall conversion and product selectivity of m-cresol reactions will assist us in design the most effective catalysts to convert phenolic compounds into desirable aromatic hydrocarbons.

See more of this Session: Catalytic Biofuels Refining I
See more of this Group/Topical: Fuels and Petrochemicals Division

(429d) Sub- and Supercritical Water Based Processes for Conversion of Algae Biomass to Fuels and Green Chemicals A Review

Philip A. Marrone, SAIC, Newton, MA and Bryan V. Yeh, SAIC, Oakland, CA

Sub- and Supercritical Water Based Processes for Conversion of Algae Biomass to Fuels and Green Chemicals – A Review

Philip A. Marrone and Bryan V. Yeh

Science Applications International Corporation (SAIC)

60 Wells Ave, Suite 103

Newton, MA 02459

In recent years, there has been increasing motivation to develop alternative liquid fuel and petroleum feedstock options due to concerns over increasing instability in the price and supply of traditional petroleum fuels. Biomass is an ideal feedstock for the production of liquid fuels and green chemicals because it is carbon neutral, is available or can be produced domestically in large quantities, and can spur economic benefits to local industries that produce and/or process it. One type of biomass in particular that has attracted attention is algae. Because of algae's rapid growth and ability to produce substantial amounts of lipids under the right conditions, there has been significant on-going research and development work on understanding and improving algae growth and processing. On the commercial side, numerous companies have been established to commercialize the process of converting algae to fungible liquid fuels.

Most attempts to process algae into fuel have focused on one of two broad approaches: 1) extract the oil (lipid) content of algae cells after harvesting and convert the oil chemically to biodiesel, or 2) convert the entire algal biomass to fuel via a thermochemical approach and biocrude intermediate. Most processes to date have proven to be economically nonviable. One of the key costs in most processes is drying of the algae after harvesting. A way to get around this bottleneck is to use a hydrothermal medium for algae processing. This not only avoids the high cost of drying but allows one to take advantage of the significant benefits afforded by supercritical water with regards to solubility, density, and mass transport. Although much early work associated with supercritical water centered upon its use as a medium for oxidative waste destruction, more recent interest has been on using it as a medium for non-oxidative thermochemical processing of biomass such as algae. Depending on the conditions utilized, one can generate solid, liquid, or gaseous products which can then undergo further refinement to the desired final product.

This presentation will review and highlight some of the important work performed to date on the use of sub- and supercritical water to process algae into liquid fuels. The focus will include developments both at the research level and commercial scale.

See more of this Session: Alternative Fuels and Enabling Technologies I
See more of this Group/Topical: Fuels and Petrochemicals Division

(496d) Thermophlic Lypolytic Bacteria From Yellowstone National Park with Potential for Biodiesel Production

Lew P. Christopher[1], Vasudeo Zambare[2] and Archana Zambare[2], (1)Center for Bioprocessing Research and Development and Department of Chemical and Biological Engineering, South Dakota School of Mines & Technology, Rapid City, SD, (2)Center for Bioprocessing Research & Development, South Dakota School of Mines & Technology, Rapid City, SD

Lipases (EC 3.1.1.3) are hydrolytic enzymes with a broad use in biocatalysis due to their versatility for industrial applications, their thermostability, relatively low cost and need for added cofactors. In the last two decades, lipases have gained interest because of their potential use among others in the food, detergent, and biodiesel industries. To date, bacterial isolates from the Yellowstone National Park have been relatively poorly studied for lipase production. In this work, soil samples from mushroom, bath and octopus springs were collected and separately enriched at 65^0C in a basal medium (pH 7) supplemented with 1% (v/v) of tributyrine oil, tributyrine oil emulsion, soybean oil or waste cooking oil. The primary screening for lipase production was carried out using olive oil emulsion agar plates followed by rhodamine B agar plate techniques. Selected isolates were studied for extracellulase lipase production under submerged fermentative conditions using various nutritional and environmental parameters. The present investigation identifies lipolytic bacterial cultures producing highly thermostable extracellular lipases with properties potentially useful in biodiesel production.

See more of this Session: Alternative Fuels and Enabling Technologies II
See more of this Group/Topical: Fuels and Petrochemicals Division

(13g) Low Temperature Gasification of Impregnated Biomass Feedstock

Foster Agblevor, Biological Engineering, Utah State University, Logan, UT, Ofei Mante, Biological Systsems Engineering, Virginia Polytechnic Institute and State University, Blacksburg, VA, Allen Aradi, Afton Chemical Inc, Richmond, VA and Tze-Chi Jao, Afton Chemical, Richmond, VA

Hybrid poplar biomass feedstocks were impregnated with catalysts and gasified at three different temperatures in a bubbling fluidized bed reactor. The impregnated biomass were gasified in a nitrogen atmosphere at 400 °C, 600° C, and 800 °C. The gas yield increased with increasing temperature as expected and the char and liquid yields decreased with increasing temperature. Te catalytic effect was observed at 400 C when the impregnated gasification was compared with the control that did not contain any catalyst. At 400 C, the impregnated sample had a gas yield of 51wt% compared to 41 wt% for the non-catalytic gasification at 450 C. The char yield was 11 wt% compared to 14 wt% for the non-catalytic run at 450 C. The char yield decreased from 11 wt% to 6.5 wt% when the gasification temperature was increased from 400 to 800 C. Although there was 88 wt% gas yield at 800 C, the thermal effect was very strong and the catalyst was not effective in cracking the lignin degradation products. Thus, this catalyst needs considerable improvement to effectively crack the lignin fraction of the biomass pyrolysis products.

See more of this Session: Catalytic Conversion of Renewable Resources to Synthesis Gases and Pyrolysis Oils
See more of this Group/Topical: Sustainable Engineering Forum

(339e) Coupled Optimization and Simulation for Multi-Biomass Source-to-Biorefinery Supply Chain Modeling and Analysis

William Faulkner[1], Joseph Amundson[1], Dr. Fazleena Badurdeen[2] and Dr. Jeffrey Seay[3], (1)University of Kentucky, Lexington, KY, (2)Institute for Sustainable Manufacturing, University of Kentucky, Lexington, KY, (3)Chemical and Materials Engineering, University of Kentucky, Lexington, KY

As cost associated with fossil fuels rise due to various environmental, social, and economic factors, it has become prudent to examine the possibility of exploiting alternative sources of raw materials for the production of energy and chemicals. Biomass, of various forms, has been proposed as an option to fill this need for alternative raw materials. Several problems with biomass as an alternative to fossil fuel derived materials instantly arise: low bulk energy density, the resulting high transport costs, seasonal variability of biomass supply, and the risks inherent to relying on a largely agriculturally based system are several examples. To begin to realize energy and chemical production with biomass as the feedstock on an economic scale, supply chain optimization and a robust means of simulating variation over long periods of time is critical.

Due to a relatively low energy density of the given biomass feedstocks, the majority of the biorefinery supply chain costs occur in the process of raw material transportation. By optimizing this aspect of the biorefinery supply chain, given the plant and potential feedstock source locations, the cost associated with raw material transportation can be minimized. In order to examine the potential longevity and general robustness of the output from the optimization model, simulation is a valuable tool. In this way, the effects of perturbations on the various feed stock supplies due to weather variability, crop yield variability, or other factors can be examined and analyzed.

This presentation focuses on the development of a coupled system comprised of an optimization model for the upstream portion of biorefinery supply chains in Western Kentucky and a discrete-event simulation program. The optimization model minimizes the transportation costs associated with moving biomass feedstock from centralized storage facilities to the biorefinery location. Multiple feedstocks available in the region under consideration including chicken litter, forest residue, and corn stover were taken into consideration. Various alternative shipping methods as well as a centralized versus a distributed production facility layout are considered as well. The discrete-event simulation model allows for the examination of the robustness of the determined optimal supply chain design. This combination of optimization and simulation for the examination of multi-biomass source-to-biorefinery supply chains could potentially serve as a template for future work in biorefinery supply chain model development and decision-making.

See more of this Session: Sustainable Biomass Feedstock Production and Supply for the Emerging

(388f) Building a Better Biofuel Production Platform: Engineering Metabolic Control Techniques to Control the Photosynthetic Cyanobacterium Synechocystis Sp. PCC 6803

Stevan Albers, Department of Cell and Molecular Biology, Colorado State University, Fort Collins, CO and Christie A.M. Peebles, Chemical Engineering, Colorodo State University, Fort Collins, CO

Building a Better Biofuel Production Platform:

Engineering metabolic control techniques to control the photosynthetic cyanobacterium *Synechocystis sp.* PCC 6803

Stevan C. Albers[1], Christie A.M. Peebles[1,2]

[1]Cell and Molecular Biology Graduate Program, [2]Department of Chemical and Biological Engineering

Colorado State University, Fort Collins, Colorado

Biofuel production in microorganisms has great potential to fulfill this nation's need for fungible "green" transportation fuels. Because of the simplicity of bacterial cells, photosynthetic cyanobacteria make good candidates as a platform for biofuel production. Many metabolic engineering techniques and procedures have been optimized for various heterotrophic bacteria, but few techniques have been designed with autotrophic cyanobacteria in mind. Designing rational control systems in cyanobacteria is the first step in controlling pathways that utilize carbon dioxide to produce long chain hydrocarbon like molecules. Understanding and controlling the metabolic pathways that produce these molecules is critical to demonstrate the true functionality of these photosynthetic organisms.

Synechocystis sp. PCC 6803 is a very versatile organism for a variety of reasons. It is a photoautotrophic bacterium for which it's haploid genomic DNA has been sequenced. Because of the simplicity of cyanobacterial cells, higher system efficiencies can be reached when compared to other photosynthetic algal cells. Manipulation techniques like natural uptake, electroporation, and homologous recombination are well annotated in this species. The organism is capable of rapid doubling times while harboring such metabolic pathways as flavonoid, clavulanic acid, and fatty acid biosynthesis while also harboring pathways capable of toluene, atrizine, DDT degradation. Our group plans to focus on understanding specific pathways endemic to *Synechocystis* and utilizing these pathways to further the molecular engineering toolset available for this versatile organism. Specifically, we will be discussing the design, construction, and expression of various promoter constructs capable of rational gene control in *Synechocystis sp.* PCC 6803. This talk will focus on our progress to date.

See more of this Session: Microbial Engineering for the Synthesis of Bulk and High-Value Products
See more of this Group/Topical: Sustainable Engineering Forum

(641d) Catalytic Oxidative Dehydration of Butanol Isomers: 1-Butanol, 2-Butanol and Isobutanol

Ivan C. Lee, US Army Research Laboratory, Adelphi, MD

The catalytic production of economically important four-carbon olefins was achieved through catalytic oxidative dehydration of 1-butanol, 2-butanol, and isobutanol using a millisecond contact time reactor. Both alumina foam and rhodium-alumina foam catalysts convert these four-carbon alcohols into four-carbon olefins, with contact time equal to 25 ms over a wide range of equivalence ratios (f) from 0.176 to 6.338 (or C/O_{air} ratio from 0.0659 to 2.11) . The cracking of the carbon backbone was found to be minimal. A mixture of butene isomers was obtained due to the isomerization of the carbocation intermediates. The degree of this isomerization was determined by analyzing the ratios of butenes/(total olefins), 1-butene/2-butene, cis-2-butene/trans-2-butene, and isobutene/(linear butenes). In addition, a reaction mechanism of the catalytic oxidative dehydration of butanol in short contact times was proposed. In this mechanism, the C-O bond of the majority of butanol was broken to form olefinic species on the alumina surface. These surface olefinic species would undergo successive oxidation to form CO and/or CO_2, depending on the availability of oxygen.

See more of this Session: Chemical and Catalytic Conversions and Processes for Renewable Feedstocks
See more of this Group/Topical: Sustainable Engineering Forum

(218e) Gas Evolution In a Flow-Assisted Zinc-Nickel Oxide Battery

Yasumasa Ito, Michael Nyce, Robert Plevilech, Martin Klein, Daniel Steingart and Sanjoy Banerjee, Chemical Engineering, Energy Institute, City College of New York, New York, NY

Zinc-based rechargeable batteries are oneof the most attractive electrical energy storage systems due to its advantagessuch as low cost and green. On the other hand, short cycle life due to internalshort circuit is the most critical disadvantage. We have carried outexperiments in flow-assisted zinc-nickel oxide batteries, and concluded thatflowing electrolyte contributes to the modification of zinc morphology andextend the battery cycle life significantly. More than 1500 cycles have beenobtained while keeping the Coulombic efficiency at or above 90% by utilizing adeep-discharge reconditioning step every 15 cycles [1].

Theloss of Coulombic efficiency is attributed to remaining zinc on anodes ondischarge, hydrogen and oxygen evolution occurring as parasitic reaction, andreasons such as corrosion in the sintered nickel electrodes. The extent of gasevolution is determined by the balance of thermodynamics and kinetics of thehydrogen and oxygen, and those of the main reduction-oxidation reactions in thebattery. We have conducted experiments on gas evolution in a sealed zinc-nickeloxide battery for with and without flow assisted cases. The effects of flowingelectrolyte on gas evolution were quantitatively evaluated and the impacts onbattery performances were discussed.

Metalzinc as negative (anode) electrode was deposited on polished nickel coatedcopper sheets as substrate. Sintered nickel oxide electrode was used as thecathode. The test cell was a prismatic cell which has a capacity of 2.2 Ah.Zinc oxide at a concentration of 60 g L^{-1} was initially dissolvedinto 45 wt % potassium hydroxide (KOH) solution as electrolyte. The electrolyteflows from the bottom to the top of the cell at an average velocity of 3 mm s^{-1}in the flowing case. Evolved gases were accumulated in the head space of thecell. The concentration of the oxygen in the head space was measured by anoxygen sensor. The internal pressure of the cell was measured by a pressuretransducer which is also attached to the head space of the cell. Galvanostaticbattery cycling experiments were carried out for a charge and discharge ratesof 1C and 0.5C, respectively. Charge was terminated when a battery cell wasfully charged to the full rated capacity of the sintered nickel oxideelectrodes, and discharge was terminated when the cell voltage dropped to 1.0V.

Theresults show that, on charge, hydrogen is evolved from zinc (anode) side.Because of the formation of concentration boundary layer and the lowerconcentration of zinc (zincate) ion near the zinc anodes in the non-flowingcase, hydrogen evolution tends to occur more instead of zinc electrodeposition. On the other hand, on discharge, hydrogen can theoretically be evolved from thecathode (nickel oxide). However, from a thermodynamic point of view, hydrogenevolution should not start until the cell voltage drops much lower, near 0 V.Other possibilities are zinc corrosion and that hydrogen evolved on charging isreleased from voids within the zinc electrodeposit duringdischarging. Time elapsed images of the zinc anode in a cycle showed that theevolved bubbles on charging, which are presumed to be hydrogen, do not detachfrom the surface on charging and detach on discharging. Also, there are quite afew bubbles that appear as zinc dissolves on discharging. From a thermodynamicpoint of view, these bubbles should not be oxygen and, in fact, oxygen in thecell doesn't increase on discharging. Therefore, these are expected to behydrogen which is generated on charging.

Onthe other hand, xygen is drastically evolved towardthe end of charge for both the cases. This oxygen evolution is from thepositive (nickel oxide) electrodes and this phenomenon in zinc–nickel oxidebatteries has been reported in many previous works. The results also show thatoxygen is evolved more in the non-flowing case than the flowing case. Though wedid not compare the differences of crystal structure of the nickel oxideelectrodes between the flowing and non-flowing cases, considering that oxygenevolution basically occurs on overcharging, and that electrolyte (KOH)concentrations affect the performance of sintered nickel oxide electrodes,oxygen evolution tends to occur more than the reaction between nickel oxide andhydroxide ion in the non-flowing case, because of the concentration gradient ofthe hydroxide ion at the electrodes surfaces. On discharging, oxygen is notevolved for both the cases, and the curves rather show a tendency of slightdecrease of oxygen in the cell. This indicates that oxygen is slightlyrecombined on discharging, probably by the reduction on the positive plates.

[1] Y. Ito, M. Nyce, R. Plivelich, M. Klein, D. Steingart, S. Banerjee, J PowerSources, 196, 2340-2345 (2011).

See more of this Session: Flow Batteries for Energy Storage (In Memory of Bernard Lee)
See more of this Group/Topical: Fuels and Petrochemicals Division

Density Functional Theory Study of the Effect of *3d* and *4d* Transition Metals on $NaMgH_3$ Complex Metal Hydride for Hydrogen Storage Applications

Authors: Fernando Soto[1], Daniela S. Mainardi[2]

[1]Institute for Micro-manufacturing, Louisiana Tech University
[2]Institute for Micro-manufacturing, Chemical Engineering, Louisiana Tech University

In order to achieve the Hydrogen Economy, it is necessary to use hydrogen as an energy carrier. Hydrogen as an energy carrier is a possible solution to neutralize the climate changes occurred due to carbon emission. It is also highly attractive to ensure a smooth and efficient provision of energy without depending on other nations. Vehicles powered by fuel cells are one of the applications in which hydrogen can be used as an energy carrier. However, the absence of a commercially viable hydrogen storage technology is delaying the large-scale use of fuel cell powered vehicles. In order to promote research in this area, the U.S. Department of Energy (DoE) has set targets for the year 2010 and 2015 for a system of on board hydrogen storage. Previous results suggest that the addition of transition metals enhances the desorption kinetics of complex metal hydride and lowers the hydrogen dissociation temperature and activation energies.

In the present work, study of catalytic additives with improved performance over Ti allows us to generalize the work on the procedure for 3d/4d doping in NaMgH3 and use it in systems with higher gravimetric densities and more favorable thermodynamics. For this purpose, plane wave density functional theory (PW-DFT) calculations helps us examine bulk and surface models of NaMgH3, Mg2FeH6 and KMgH3 complex metal hydride (CMHs) with impurities from the 3d element block. In this respect, first- principles calculations determined the cohesive energies of pure and doped bulk/surface CMHs as well as the adsorption/substitution energies. Furthermore, DFT coupled Molecular Dynamics (DFT-MD) calculations at elevated temperatures elucidate the effect of Ti placed on different sites on surface models. The results indicate that 5 out of 7 additives at the bulk model are thermodynamically stable relative to the pure model; therefore, possible. The most favorable elements in this regard came out to be; Ti, V, and Co in that order. Afterwards, from dynamics calculations performed at surface slab models, we conclude that Ti additives replacing sodium lattice sites and Ti additives placed at a four-fold vacant position are the stable arrangement in this system with potential formation of TiMgxHx complexes. Meanwhile, alloyed systems with 4d transition metals may be extremely stable for on-board hydrogen storage applications. Also, dynamics calculations performed at the surface slab models suggests that Ti additives replacing sodium lattice sites and Ti additives placed at a four-fold hollow site are the preferred, stable arrangements in NaMgH3 with possible formation of TiMgxHx complexes. This conclusion may serve as a guideline for the future design of nano-structured complex hydride for on-board hydrogen storage utilization.

Keywords: density functional theory, hydrogen storage.

1. Introduction

Finding a feasible solution to the problem of hydrogen storage for vehicular applications is essential for the transportation sector because it accounts for two-thirds of energy consumption in the U.S. [1]. Also, it is necessary to ensure an energy supply and avoid energy dependence from regions such as the Middle East which are politically unstable [2]. It is imperative to reduce carbon emissions in order to improve the air quality and avoid harsh climatic changes [3]. In order to promote research in this area, the U.S. Department of Energy (DoE) has set targets for the year 2010 and 2015 for a system of on board hydrogen storage. The use of hydrogen within H2/O2 fuel cell systems requires an adequate and accessible hydrogen storage medium. The production of liquid or highly compressed hydrogen requires an tremendous amount of energy and may be dangerous in certain situations. Therefore, solid-state hydrogen storage is currently the safest and most effective way of storing hydrogen. As explained by Grochala et al. an ideal hydrogen storage material (HSM) will have a low-molar weight, be inexpensive, have rapid kinetics for absorbing and desorbing H2 in the 25-120 °C temperature region, and store large quantities of hydrogen reversibly [4]. As noted by Grochala et al., at the moment, no single material fulfills all of these requirements. High stability and poor kinetics are significant challenges, which have to be solved in order for any material to be considered for on-board automobile applications [5].Therefore, the identification of material that can meet this criterion is a grand challenge that has attracted the interest of the global scientific community.

Light chemical elements are essential to establish a reversible hydrogen storage system [6]. In order to achieve hydrogen gravimetric and volumetric densities required for mobile applications, host elements in the hydrogen storage system should be lighter than Al metal. Typical elements considered among the light elements are: Li, Be, B, C, N, O, F, Na, Mg, and Al. As explained by Wu et al., metal hydrides are either too stable for efficient hydrogenation cycling or too unstable that desorption occurs at or below room temperatures but adsorption requires high pressures of hydrogen [7]. Nonetheless, CMH have been identified as promising candidates for on board hydrogen storage applications. CMHs may enhance kinetics and reversibility and inhibit toxic-gas side products. Since the pioneer work of Bogdanovic and Schwickardi in 1997, the focus has shifted to CMHs such as alanates, borohydrides, amides/imides, and destabilized CMH [8, 9].

Since the work of Bogdanovic and Schwickardi in 1997, it has been established that certain element addition can help CMH meet the criteria for HSM and are capable of showing hydrogen reversibility at moderate conditions [8, 10, 11]. Thus, enhanced CMH are a promising class of material for hydrogen storage that needs further study because high stability and poor kinetics are important challenges which have to be solved for the CMH to be considered for on-board automobile applications [12]. Recently several types of CMHs have attracted the attention as promising enhanced CMHs candidates for HSM. These CMHs are: Ternary hydrides composed of transition metals and different alkaline earth metals [13], alkaline transition-metal borohydrides [14], and ABH_3perovskite structures [15]. Ti-doped CMHs have shown to improve the kinetics of complex metal hydride decomposition and the thermodynamic reactions relative to the pristine material. Thus, the role of Ti and other *3d*/*4d* transition elements in different types of CMHs need further studying in order to achieve the goals established by the DoE [9].

$NaMgH_3$

Recently, Mg-based hydrides have attracted interest as a hydrogen storage medium due to their light weight and low cost [10].Several Mg-based hydrides show the perovskite structure expressed as ABH_3, where A is an alkali element and B is magnesium. $NaMgH_3$ is the only magnesium based ternary hydride with Na and it has attracted the attention as a possible candidate for hydrogen storage material because it has high gravimetric hydrogen density (6 %) and high volumetric hydrogen density (88 kg/m^3) [11]. This magnesanate has not been widely investigated as a candidate for hydrogen storage. The crystal structure of this material is known and it derives from cubic $KMgH_3$ by an orthorhombic distortion and has a perovskite structure similar to $GdFeO_3$ [10]. It belongs to the *Pnma* space group with orthorhombic symmetry in the temperature range of 4 to 370 K [11,12].The primitive unit cell contains four formula units for a total of 20 atoms.The $NaMgH_3$ unit cell is shown in figure 1.

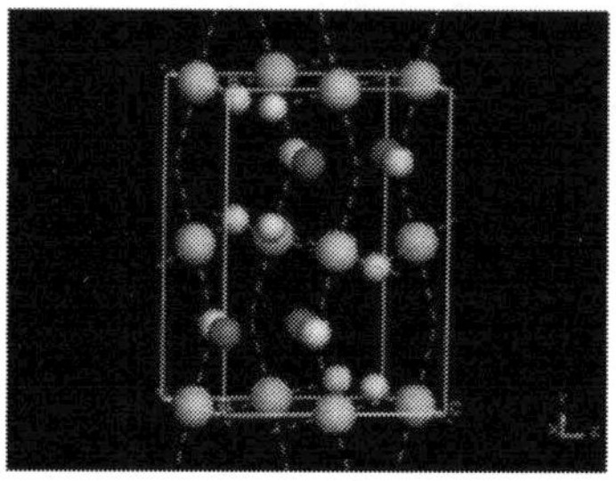

Figure 1. Unit Cell of Sodium Magnesium Trihydride.

Hydrogen disassociation in this material occurs at atmospheric pressure and approximately 670 K, which is well above the operable range for on board hydrogen storage applications. Thus, $NaMgH_3$ is more thermally stable than binary MgH_2, which has a hydrogen dissociation temperature ~550 K. The dehydrogenation reaction scheme takes place in two steps [12]:

$$NaMgH_3 \rightarrow NaH + Mg + H_2 \qquad (1)$$

$$NaH \rightarrow Na + \frac{1}{2} H_2 \qquad (2)$$

Numerous studies have been carried out to develop suitable Mg-based systems that show fast sorption kinetics at moderate temperatures. These studies achieved partial success in improving the kinetics by addition of dopants. Ma and co-workers demonstrated high affinity of transition metal cations toward hydrogen. Catalysts are usually mixed with the hydride by mechanical ball milling, which introduces defects and impurities. However, much research still needs to be done in order to identify appropriate catalysts, study the existence of stable intermediate phases and their crystal structure and understand the kinetics. Much insight can be gained from first principles simulations for the purpose of designing alloy catalysts.

For instance, Bouhadda et al. used ab initio calculations to report on the crystal, electronic and thermodynamic properties of $NaMgH_3$. Bouhadda et al. reported relaxed structures in agreement with experimental structures (Table 1).

Perovskite Structure	Experimental Lattice Constants (Å)	Calculated Lattice Constants (Å)
$NaMgH_3$	a=5.463 b=7.703 c=5.411	a=5.410 b=7.628 c=5.358

Table 1. Experimental and Calculated Cell Parameters

In addition, this author also reported that this structure is a wide band gap semiconductor with a calculated band gap of 3.4 eV. Bouhadda et al. explains that the interaction between Na and MgH_3 is ionic and that the interaction between Mg and H is both ionic and covalent.
To study the thermodynamic properties of the material, Bouhadda et al. calculated the standard enthalpy of formation H_f with the following reaction equation.

$$Na + Mg + 3/2\ H_2 \rightarrow NaMgH_3 \quad (3)$$

The result of 151.83 KJ/mol is well in agreement with the results obtained from Ikeda et al. (145.05 KJ/mol, 147 KJ/mol) [15]. Thus, this formation enthalpy is considerably larger than that of metal hydride NaH (-114 KJ/mol) [15]. Klaveness et al. also calculated the enthalpy of formation for this material obtaining similar results as those presented by Bouhadda et al. and Ikeda et al. Thus, the first principles approach tends to reproduce experimental results very well, and this is true also for ionic hydrides such as $NaMgH_3$ [16,17].

Furthermore, Shane et al. reported that the high rates of H motion in $NaMgH_3$ are due to regions with poorly organized crystal structure [17]. The author concludes that if the disorder could be sustained, it could provide a route towards improved reaction kinetics. Hao et al. presented a theory-guided computational study explaining the existence of intrinsic defects in $NaMgH_3$. The author reported that schottky defects are the dominant defects in this material and that hydrogen diffusion is determined by mobility of schottky defects. Hao et al. concluded that the diffusion induced by schottky defects can be tweaked by adding suitable dopants such as Cu and Co that can adjust the material's Fermi level [18]. Komiya et al. studied the substitution energies in $NaMgH_3$ by replacing the Na ion with an element from the same 1A group element from the periodic table (Li, K, Rb, Cs) and replacing the Mg ion with an element from the 2A group elements (Be, Ca, Sr, Ba) [19]. Komiya et al. reported positive substitution energies on the Na site, indicating that doping the structure makes the material unstable, except when doping with the K ion. Also, the addition of Be, Sr, and Ba destabilizes the structure and doping with Ca increases the stability. These results show that addition of impurities has a positive correlation on the destabilization of the material.

Moreover, ABH_3 structures such as $NaMgH_3$ may enhance kinetics and reversibility and inhibit toxic gas side products compared to the parents' borohydride, amide, and alanate compounds. Also, higher H mobility in this material due to their perovskite structure cause improved hydrogenation rates compared to pure MgH_2. It is clear from the literature review presented that further investigation is needed to investigate the positive correlation between hydrogen mobility, addition of dopants and hydrogenation/dehydrogenation reversibility. Therefore, the aim of this paper is to study the stability and dynamics of pure and doped $NaMgH_3$ for on-board hydrogen storage applications.

2. Computational Methods

2.1 Bulk Calculations

In this study, we use the first principles pseudopotential total energy method through the generalized gradient approximation (GGA) calculations with the Perdew and Wang (PW91) functional as implemented in the CASTEP® module [5] of the Materials Studio® software by Accelrys, Inc. [6] to study bulk and surface models of $NaMgH_3$. Spin-polarized calculations on Fe-, Ni-, and Co-doped systems show no significant effect to be created energetically, with respect to the non-spin polarized calculations. Hence, further conducted geometry optimization calculations were spin-unpolarized, and the cut-off energy was 400 eV with a fine k-point mesh in all cases.

We created hydride models alloyed with 3*d* and 4*d* block transition metals to investigate their effects on the energetics of bulk/surface of $NaMgH_3$ systems. The alloying elements substitute one Na atom at a center lattice location of each structure, resulting in a concentration between 5.65 and 7.57 at.wt. % in the case of 3*d* transition metals. Afterwards, bulk hydrides and surface slabs relaxed entirely in terms of all atomic degrees of freedom to obtain the ground-state energy system. After complete relaxation of the systems was obtained, single-point energies were calculated for the optimized structures. Since the main aim of this work is to illustrate the effect of transition metals on the prototype $NaMgH_3$ system, calculation of cohesive energies (ΔE_{coh})of the doped systems were doneaccording to equation (4).

$$\Delta E_{coh} = E_{TOT}(Na(X)MgH_3) - [NE(Na) + NE(Mg) + NE(H) + E(X)] \qquad (4)$$

Where E_{TOT} is the total energy of the system, N is the number of atoms, E(Na) is the energy of a free neutral atom of Na, E(Mg) is the energy of a free neutral atom of Mg, E(H) is the energy of a single hydrogen atom and E (X) is the energy of a free neutral transition metal X.

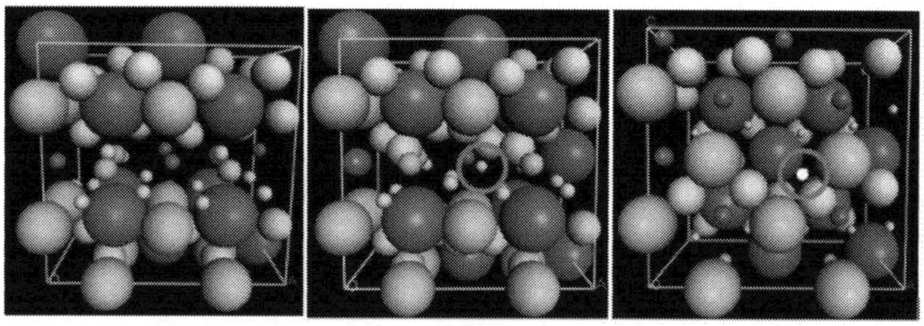

(a) (b) (c)

Figure 2. $NaMgH_3$ 2x1x2 Supercell consisting of five f.u. **(a)** pure model, **(b)**Ti_{sub}@Na in model, and **(c)**Vac_{sub}@Na in model.

2.2 Surface Calculations

Study of surface properties is vital to understand the catalytic behavior of impurities. An explanation for this is that the arrangement of atoms, distribution of electrons, and their response to external behavior is different at the surface when compared to the bulk.

Thus, besides the studies regarding bulk doped $NaMgH_3$ systems, we considered the adsorption/substitution behavior of Ti on the $NaMgH_3$ (001) surface. Accordingly, the calculations were performed for surface structures with a slab width of 2x1x2, and a gap of approximately 5 Å at the top of the $NaMgH_3$ surface. Moreover, k-point sampling is performed at the gamma point only. Doping on the surface structure proceeded in two distinct steps: first, one is the substitution adsorption, in which the dopant replaces a Na atom at the center lattice location of the uppermost layer of $NaMgH_3$ (001) surface. The second form of adsorption in the on-top of surface adsorption.

The substitutional/adsorption energy calculation for the (001) surface proceeds via the following equation:

——— ——— (5)

Where N is the total number of atoms in the model and X is the transition metal added to the system.

3. Results and Discussion

The equilibrium cell parameters a, b and c of pure $NaMgH_3$ are 5.48, 7.68 and 5.40 Å, respectively. A brief comparison with experimental values showed that; these calculated values are of 0.20 % different than the initially set unit cell dimensions, which is a strong indication of the good agreement between the computational and experimental results. The calculated results for the lattice parameters are shown in Table 2.

Lattice Parameter (Å)	
experimental	**DFT (GGA/PW91)**
a = 5.463	5.47
b = 7.703	7.68
c = 5.411	5.39

Table 2. Experimental and Calculated Lattice Parameter

The calculated energy gap is 3.308 eV which is similar to the values found in literature [15]. Calculated band gap figure is shown in figure 3.

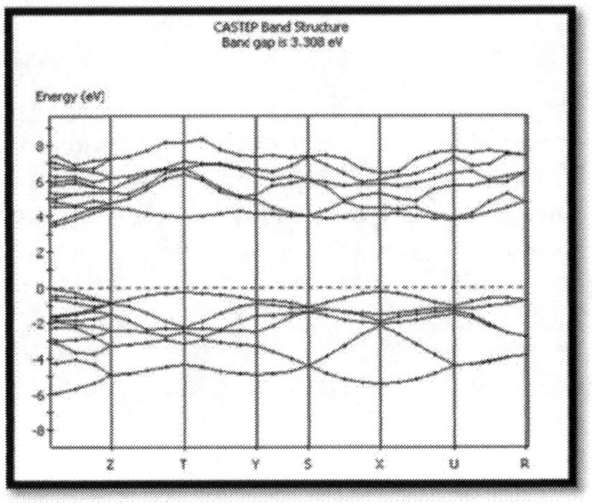

Figure 3. Calculated Band Gap as obtained from CASTEP®

Furthermore, electron density maps corroborate the results reported by Bouhadda et al., the interaction between Na and MgH_3 is ionic and that between Mg and H involves both ionic and covalent bonding schemes. In this respect, the theory level used is appropriate and the results encourage us to continue with the same theory level for subsequent ab initio calculations (Figure 4).

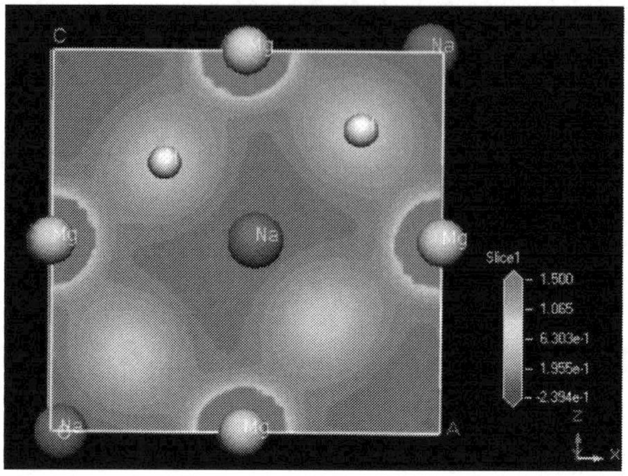

Figure 4. Total Charge Density of $NaMgH_3$ parallel to the x and z axis. Where the green balls are Mg atoms, purple balls are Na atoms and white balls are hydrogen atoms.

3.1.1 Bulk

Next, the primary goal in calculating the cohesive energy energies for each alloyed system was to investigate the effect of *3d* and *4d* block elements in the stability of the system. Modification in $NaMgH_3$ models took place via addition of selected impurity elements X (where X can be any element from *3d* and *4d* block: Sc, Ti, V, Nb, Y, Mo, etc) to the desired $NaMgH_3$ model. The cohesive energy of the pure model served as the reference value; thus, any energy above 0 eV may be considered as stable. It is interesting to note that several elements added as dopants in the $NaMgH_3$ structure are thermodynamically stable and; therefore, appealing for further consideration.

3d Transition Metals

In this respect, for the *3d* block, Ti, V, Sc, Co, Ni and Cu in the decreasing order, are the most favorable elements with respect to the cohesive energy of the pure system for *3d* transition metals (Figure 5). For instance, the cohesive energy due to the presence of the Ti dopant is 2.89 eV more favorable than the pure case where no dopants are available. Meanwhile, the removal of a Na ion from the pure model is 3.95 eV less preferred than the pure model. Also, V is 2.64 eV more energetically favorable, Sc has a value of +2.16 eV, and for Co and Cu those values are +1.62 eV and +0.62 eV respectively. On the other hand, Zn and Mn where found to be unfavorable in this system with cohesive energies of -2.58 eV and -0.68 eV, respectively. These results and their corresponding bond lengths with Mg and H are tabulated below.

Model	X in Bulk Site	System	*E_{coh} (eV)	E_{sub} (eV/atom)
Bulk	Na	$Na_{16}Mg_{16}H_{48}$	0.00	
	0	$Na_{15}Mg_{16}H_{48}$	-3.95	
	Sc	$(Sc)Na_{15}Mg_{16}H_{48}$	2.16	0.027
	Ti	$(Ti)Na_{15}Mg_{16}H_{48}$	2.89	0.036
	V	$(V)Na_{15}Mg_{16}H_{48}$	2.64	0.033
	Mn	$(Mn)Na_{15}Mg_{16}H_{48}$	-0.68	-0.009
	Co	$(Co)Na_{15}Mg_{16}H_{48}$	1.62	0.02
	Cu	$(Cu)Na_{15}Mg_{16}H_{48}$	0.62	0.008
	Zn	$(Zn)Na_{15}Mg_{16}H_{48}$	-2.58	-0.032

Table 3. Cohesive Energies of the different Bulks Models

X @Na site Bond Length (Å)	Na	Sc	Ti	V	Mn	Fe	Co	Ni	Cu	Zn
X-Mg	3.24	3.09	3.06	2.99	3.02	2.72	2.67	3.18	3.23	3.26
X-H	2.51	2.11	1.93	1.83	1.74	1.67	1.63	1.71	1.75	1.92
Mg-H	1.99	1.98	2.00	2.00	2.04	1.99	2.00	1.98	2.00	1.99

Table 4. Inter Bond Lengths of the different 3*d*Bulk Models

It should be pointed out that; that the cohesive energies of these 3*d* block elements in the bulk follow a strong correlation with the cohesive energies of their metallic elements counterpart. Also, the cohesive energies of the 3*d* block elements show a negative correlation with the ionization energies of the pure metallic elements and they also show a positive correlation with the melting point in Kelvin of the pure crystalline elements.

Thus, a high cohesive energy value of the metallic element helps to predict a high cohesive energy value in the bulk model, and transition elements with a high ionization potential energy appear to be responsible for the low thermal stability of the bulk model and the melting point of the metallic elements follow a strong positive correlation with the cohesive energy of the bulk suggesting that all these physical quantities have to be taken into account when destabilizing $NaMgH_3$ with partial substitution of the cation Na. Such prediction helps to identify additional elements that could improve on the effect of 3*d* block transition elements on CMHs.

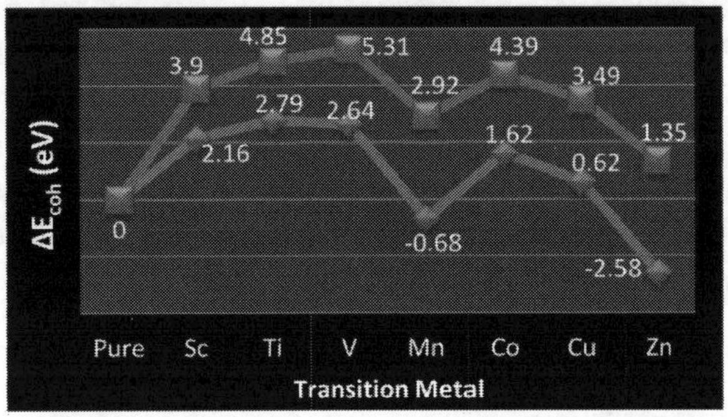

Figure 5. Cohesive Energy Results. Bulk substitution of transition metal in $NaMgH_3$(blue), pure crystalline transition metal (red).

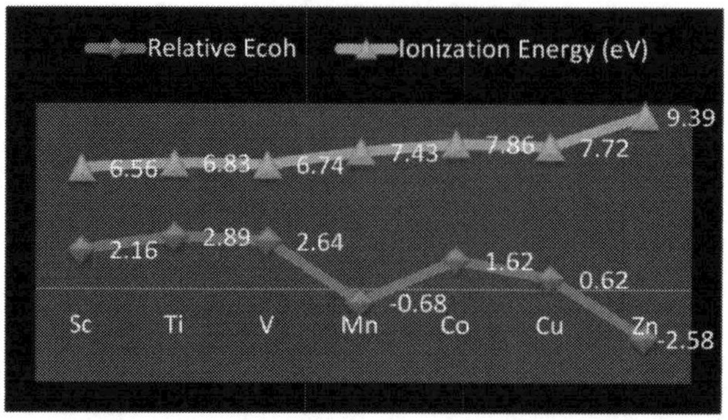

Figure 6. Cohesive Energy Results. Bulk substitution of transition metal in $NaMgH_3$ (blue), Ionization Energy (green)

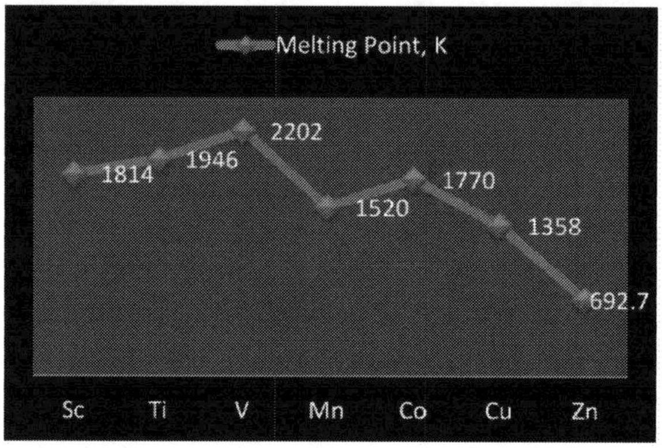

Figure 7.Melting point in Kelvin for transition metals in their pure crystalline state.

3d Total Charge Density

To further understand the effect of transition metals on bulk $NaMgH_3$ the total charge density of pure and Ti-doped models are shown below. The total charge density can be used to provide quantitative information at the atomic scale surrounding any atom or fragment of interest. The color corresponds to the total charge density, with the red regions corresponding to the highest density, which surrounds the atoms. The majority region of the model displays a very low charge density, colored blue. The plots suggest that the interaction between Na and Mg-H complex is mainly ionic in both models. However, the Ti-doped model shows an increased association of Ti with surrounding H and Mg atoms, as reported in table 4 above. Thus, a shortening of the X-H and X-Mg bond lengths implies an elongation and possible weakening of Mg-H bond lengths, making the desorption of H atoms more feasible.

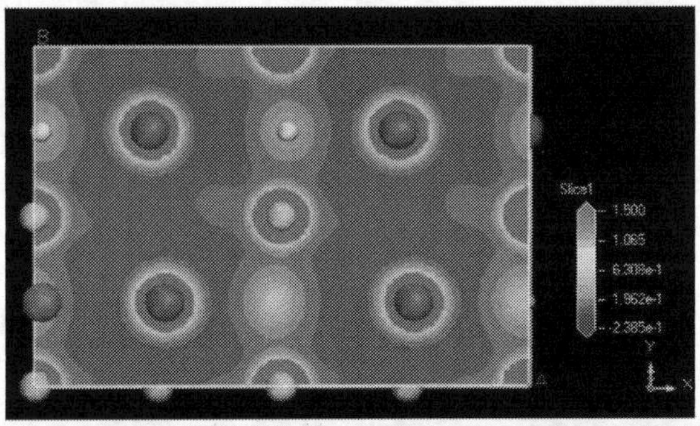

Figure 8. Total Charge Density of pure bulk $NaMgH_3$ parallel to the x and y axis. Where the green balls are Mg atoms, purple balls are Na atoms and white balls are hydrogen atoms.

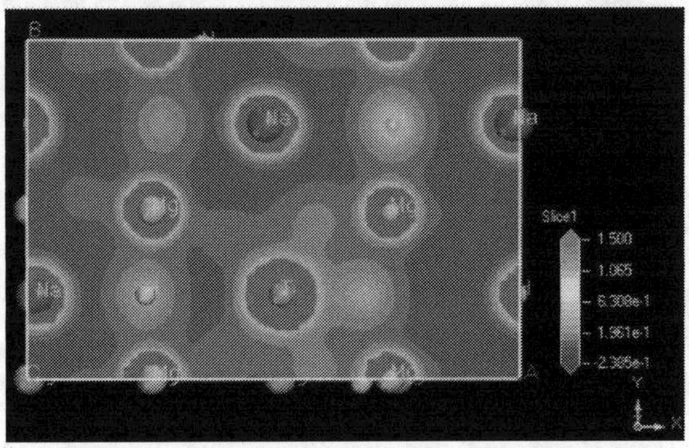

Figure 9. Total Charge Density of Ti-doped bulk $NaMgH_3$ parallel to the x and y axis. Where the green balls are Mg atoms, purple balls are Na atoms, white balls are hydrogen atoms and gray ball is Ti.

4d Transition Metals

In a similar fashion, cohesive energies for *4d* block were calculated, Tc, Mo, Nb, Zr, and Y in the decreasing order, are the most favorable elements with respect to the cohesive energy of the pure . However, for the *4d* block, all the models tested show a high positive energy with respect to the pure model. Thus, the alloyed systems are too stable for hydrogen storage applications. For instance, the cohesive energy due to the presence of the Tc dopant is 7.36 eV more favorable than the pure case where no dopants are available. This cohesive energy is 2.5 times greater than the cohesive energy of the system due to the presence of Ti. However, the cohesive energy of the system due to the presence of *4d* block elements shows a strong positive correlation with the ionization energy of the crystalline element. Meanwhile, the correlation with the different melting points of the *4d* element block and the cohesive energy of the crystalline elements is not as positive when compared to the *3d* transition metal results.These results and their relationships are shown below.

Model	X in Bulk Site	System	*E_{coh} (eV)	E_{sub} (eV/atom)
Bulk	Na	$Na_{16}Mg_{16}H_{48}$	0.00	
	Y	$(Y)Na_{15}Mg_{16}H_{48}$	3.17	0.04
	Zr	$(Zr)Na_{15}Mg_{16}H_{48}$	5.15	0.064
	Nb	$(Nb)Na_{15}Mg_{16}H_{48}$	5.46	0.068
	Mo	$(Mo)Na_{15}Mg_{16}H_{48}$	6.21	0.077
	Tc	$(Tc)Na_{15}Mg_{16}H_{48}$	7.36	0.92

Table 5. Cohesive Energies of the different 4*d* Bulks Models

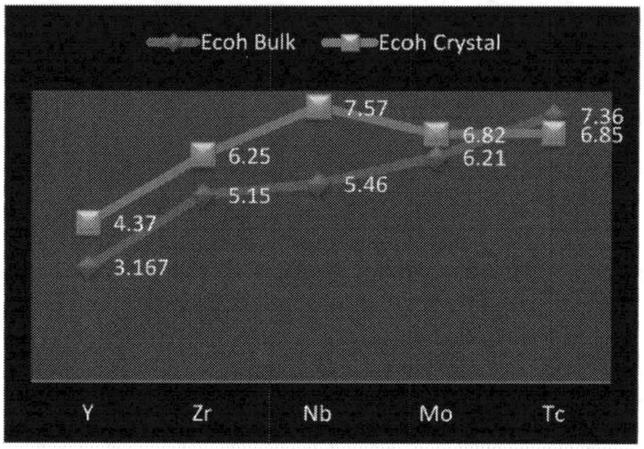

Figure 10. Cohesive Energy Results. Bulk substitution of 4*d* transition metal in $NaMgH_3$ (red), pure crystalline transition metal (blue).

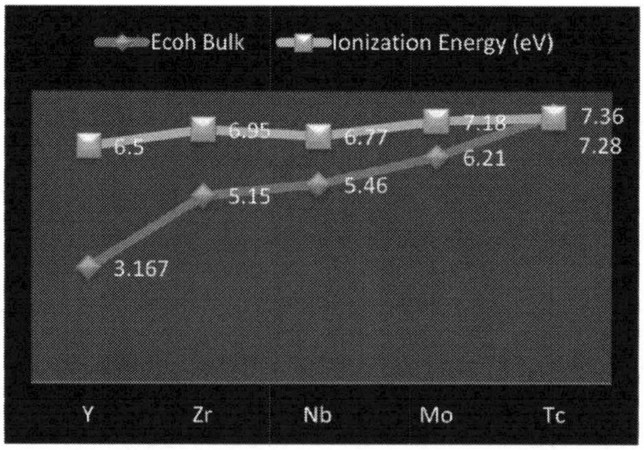

Figure 11. Cohesive Energy Results. Bulk substitution of 4*d* transition metal in NaMgH3 (blue), Ionization Energy (green)

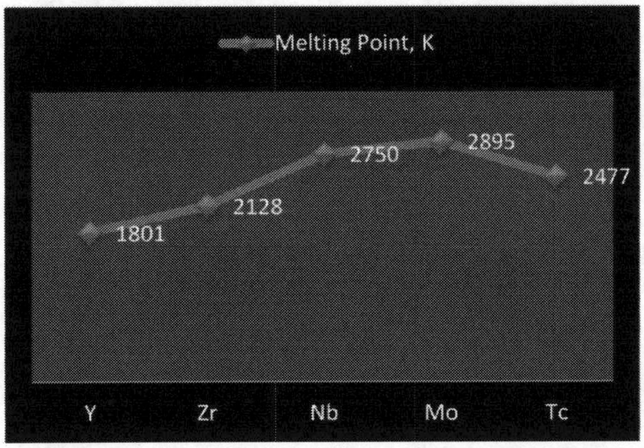

Figure 12. Melting point in Kelvins for *4d* transition metals in their pure crystalline state.

4d Total Charge Density

The total density charge of the Nb-doped bulk model is taken as a sample from the *4d* transition metal block. This plot also suggests that the addition of a dopant affects the local electron density of the model. This local modification of the *4d* doped bulk model leads to an increase association of the dopants with the surrounding Mg and H atoms, thus; leading to a possible destabilization of the Mg-H complex as well.

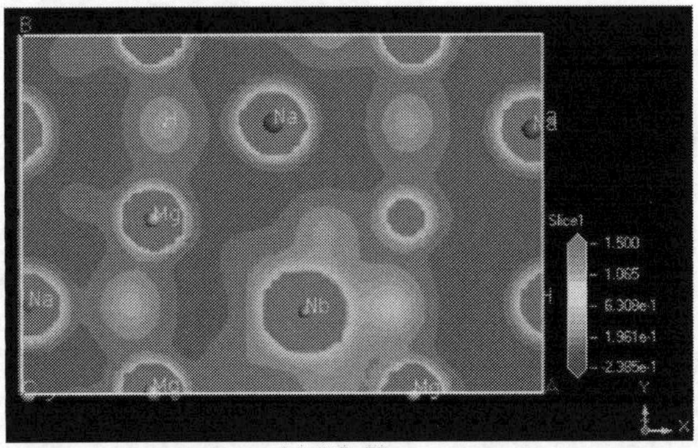

Figure 13. Total Charge Density of Nb-doped bulk $NaMgH_3$ parallel to the x and y axis. Where the green balls are Mg atoms, purple balls are Na atoms, white balls are hydrogen atoms and gray ball is Nb

3.2 Surfaces

Next, DFT calculations were conducted on five (001) surface models : Ti @ Na lattice site, Ti @ hollow site, Ti @ top of Na lattice site, Ti @ top of hollow site, and Ti @ Mg site. It is possible to observe the preferred Ti dopant site in terms of the substitutional/addition energy relative to the pure surface model. In this regard, possible formation of $TiMg_xH_x$ complexes was systematically investigated. The models used are showed below:

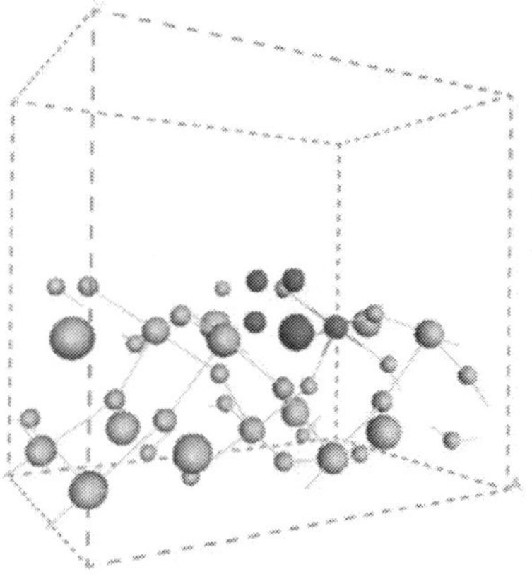

Figure 14 Model with Dopant (red) Sites

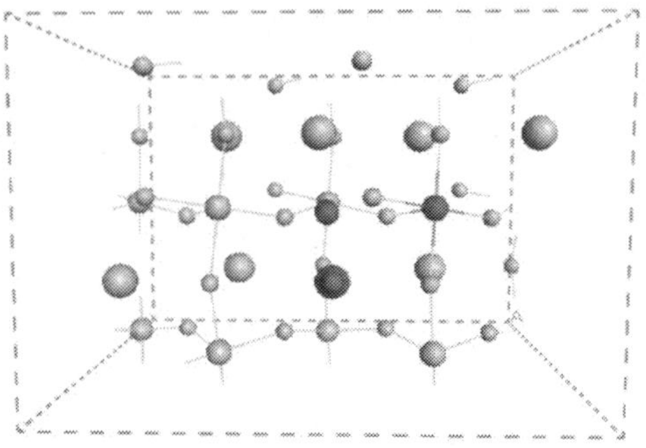

Figure 15 Model with Dopant (red) sites (Top View)

From our calculations, Ti @ Hollow site and Ti @ Na site are the preferred sites for the dopant in this system. Ti @ Hollow site has a cohesive energy of +5.47 eV and Ti @ Na site has a cohesive energy value of +4.60 eV (Table 5). Furthermore, Ti @ Mg shows a more favorable cohesive energy value than Ti @ Top sites with values of +4.23 eV and +3.22 eV, respectively. Simulation results also showed that it is more favorable to remove a hydrogen atom from a doped model than from the pure model, with values of +3.05 eV and –3.46 eV, respectively. Thus, for subsequent calculations only the most favorable models will be tested.

Site	System	E_{coh} (eV)	ΔE_{add}(eV/ atom)
Ti@ Na site	$TiNa_7Mg_8H_{24}$	4.60	N/A
Ti @ Hollow Site	$TiNa_8Mg_8H_{24}$	5.47	0.13
Ti@ Top Na site	$TiNa_8Mg_8H_{24}$	3.22	0.08
Ti@ Top Hollow Site	$TiNa_8Mg_8H_{24}$	3.22	0.08
Ti@ Mg Site	$TiNa_8Mg_7H_{24}$	4.23	0.11
H removed from Ti@ Na site	$TiNa_7Mg_8H_{23}$	3.05	N/A
H removed from Pure model	$Na_8Mg_8H_{23}$	-3.46	N/A

Table 5. Surface Energy Results

Formation of Ti-Mg-H complexes at the surface as shown in Figure 17 and Figure 18 may improve the reaction mechanism, in order to observe this phenomena surface models are investigated using DFT-MD at different temperatures (423 K and 448 K).

Figure 17 Optimized Ti@ Na site surface model

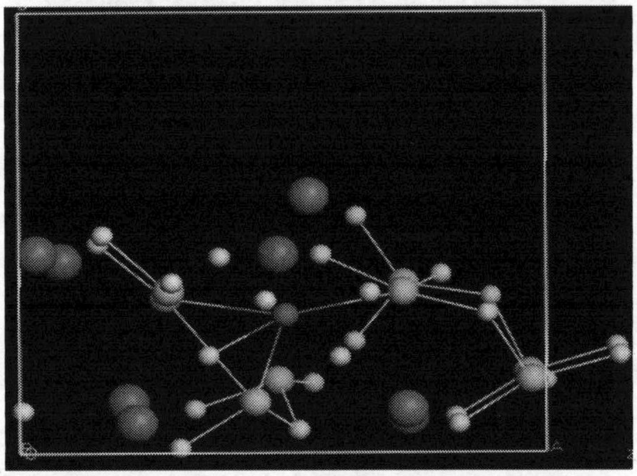

Figure 18 Optimized Ti@ Hollow site surface model

4. Dynamics

After completing the calculations regarding surface doped models and identifying the preferred sites of Ti dopants, the effect of high temperature on the Ti-Mg_x–H_x complexes is studied with ab initio molecular dynamics at 423K and 448 K in the canonical and micro-canonical regimes as mentioned earlier. First, via DFT coupled MD simulations, the effect of Ti impurity on surface models was studied to determine the stability of the complexes and to aid us in determining the role they play in hydrogen behavior at the surface. The calculations were spanned over an 8 ps time frame. The systems were considered to reach equilibrium once the drift in the constant of motion was calculated to be less than 8 meV/ atom. Also, dynamic equilibrium was analyzed through the use of radial distribution functions at the 1 ps interval. Electron density and electron density difference maps were generated to verify the distinguished effects of the Ti additive on the electronic structure of the system.

First, dynamic simulations were carried out for Ti@ Na site at 423 K for a total simulation time of 4 ps. At the end of the 4 ps the Ti-Mg_x–$H_x$was still present with a Ti-Mg_5 –H_7

formation. The calculated bond lengths as confirmed with radial distribution function plots were as follow: Ti-Mg 2.97 Å, Ti-H 1.87 Å. Then, the temperature is increased to 448 K to further study the temperature effect for a total simulation time of 8 ps. In a similar fashion, at the end of the 4 ps the Ti-Mg_x–$H_x$was still present with a Ti-Mg_4 –H_7 formation. The calculated bond lengths as were as follow: Ti-Mg 3.10 Å, Ti-H 1.93Å.

Next, the electron density difference in the surface model due to the Ti substitution at the Na lattice site is investigated. The electron density difference illustrates how the charge density changes during a chemical reaction or on docking of a molecule to a surface.

The electron density difference with respect to atoms can be represented in the following manner:

$\Delta\rho = \rho Ti@$ Na site- $\Sigma(\rho i)$ (6)

Where the subscript i runs over all atoms. This equation shows the changes in the electron distribution due to the formation of all the bonds. Furthermore, Figure 19 shows the formation of chemical bonds between Ti and the surrounding atoms by delocalization of the atomic electron density. In these plots, a loss of electron is indicated in blue, while regions rich in electrons are indicated in red. White colors indicate regions with small changes in the electron density. Based on this plots, it is clearly shown that Ti lost electrons; thus, confirming that electron affinity and ionization energy are important parameters in hydrogen storage research.

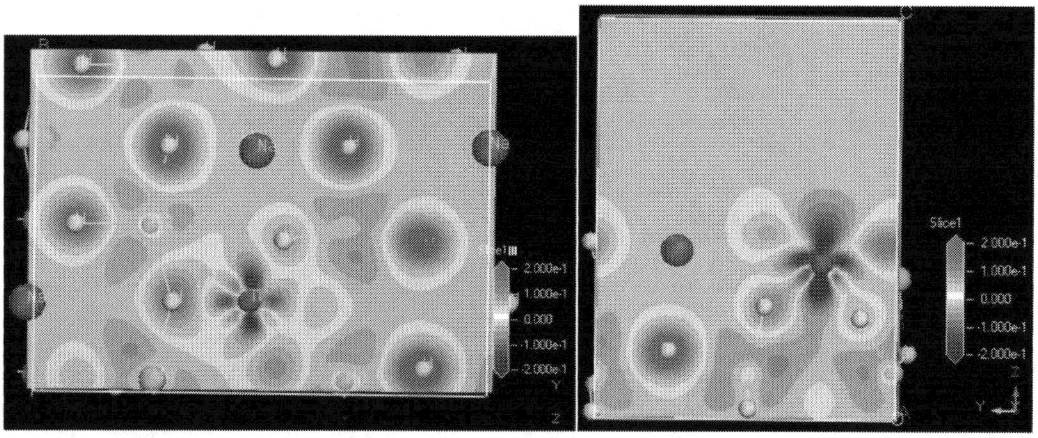

Figure 19 Electron Density Difference Maps of Ti@ Na lattice surface parallel to the x/y surface and y/z surface. Where red ball is Ti, green ball is Mg, purple ball is Na and white ball is H.

Then, following the same procedure DFT-MD simulations were carried out for Ti@ hollow site at 423 K with a canonical ensemble equilibrium run of 2 ps followed by a micro-canonical ensemble production run of 2 ps for a total simulation time of 4 ps. At the end of the 4 ps the Ti-Mg_x–$H_x$was still present with a Ti-Mg_6 –H_7 formation. The calculated bond lengths were as follow: Ti-Mg 2.92 Å, Ti-H 1.91 Å. Then, the temperature is increased to 448 K to further study the temperature effect for a total simulation time of 8 ps. Similarly, at the end of the 4 ps the Ti-Mg_x–H_x was still present with a Ti-Mg_5 –H_7 formation. The calculated bond lengths were as follow: Ti-Mg 2.94 Å, Ti-H 1.87Å.

Comparing the dynamic results of both models, it seems that the complex formed at Ti @ Hollow site shows increased association as predicted by cohesive energy values. However, it is apparently necessary to increase the number of accessible neighbors to the dopant to accurately elucidate the complex formation and exact role it plays in absorption/desorption mechanisms in CMHs.

In light of the MD simulation results, bulk/surface geometry optimization results several observations are reached. To summarize, Ti-Mg_x–$H_x$are observed after geometry optimization of doped surface models, DFT-MD show the existence of these complexes even at high temperatures, cohesive energy calculations at the surface models show that Ti@Na lattice site and Ti@ hollow site are the preferred Ti locations in this system, and the mechanism in which Ti @ hollow site ejects Na ion from lattice site requires further study with an extended slab model.This phenomenon will be further studied to determine the role that this plays in creation of nucleation sites or possible Na or Mg rich sites that ultimately leads to saturation and desorption of H from the system.

4. Conclusion and Future Work

The cohesive energies of the 3*d* block elements in the bulk follow a strong correlation with the cohesive energies of their metallic elements counterpart. For the 3*d* block elements, Ti, V, Sc, Co, Ni and Cu in the decreasing order, are the most favorable elements with respect to the cohesive energy of the pure system. On the other hand, Zn and Mn doped models are unstable in comparison to the pure model. Also, 4*d* block elements are too stable in comparison to the pure model to be used as viable dopants for CMHs. Furthermore, 4*d* block elements show a positive correlation with crystalline elements ionization energy, suggesting that dopants replacing Na ions should have high ionization energies to further weaken the Mg-H covalent bond, and ultimately reducing the thermodynamic stability of the system.

From the cohesive energy calculation on surface models, we conclude that dopants at the hollow and Na lattice sites are the most favorable sites for dopant locations in the $NaMgH_3$ system. Thus, future studies will have more emphasis on those models. Ti @ Na lattice site shows a possible Ti-Mg_4H_6 complex formation upon completion of geometry optimization. In the same manner, Ti @ Hollow site shows a possible $TiMg_5H_7$ complex formation; thus, showing an increased association of the bond lengths when compared to the Ti@ Na lattice site model. The increased association of the dopants in the bulk and surface model leads to elongation of Mg-H bond lengths, thus; possible weakening that facilitates hydrogen desorption is feasible.

It is worth mentioning that the Ti placed @ the hollow site of $NaMgH_3$ ejects a Na ion from its lattice site. This phenomenon will be further investigated to determine the role that it plays in desorption of H from the system. Thus, two destabilization methods can be rationalize based on bulk and surface models: 1) additives with a mix of low ionization energy (high ionization energy for 4*d* transition metals), low melting point and low crystalline cohesive energy should be considered to locally destabilize the CMHs by increasing the association with Mg-H complexes which ultimately results in regions rich in H and Mg atoms that initiate nucleation/growth sites that facilitate H kinetics and 2) additives with high ionization energy or low electron-negativity affinity should be considered to lower the overall thermal stability of the system which may lead to a decrease in H desorption temperature.

Furthermore, as a second step in the research program, the role of 3*d* transition metals such as Ti and Mn in the improved dehydrogenation kinetics of $NaMgH_3$ needs to be understood. Reactants and products will be optimized using DFT codes within the GGA approximation to determine the reaction barrier height for the first reaction. Reaction pathways will be determined through transition state search using LST/QST methods.

Acknowledgements

Financial support provided by NSF through the "Louisiana EPSCoR Research Infrastructure Improvement" grant (EPS 1003897) is acknowledged.

References

1) Hirst, E; Moyers, J.C.; "Efficiency of Energy Use in the United States." Science. 1973. 179. 1299-1304.

2) Salameh, M.G; "Quest for Middle East oil: the US versus the Asia-Pacific region." Energy Policy. 2003.31.1085-1091.

3) Kone, A.C; Buke, T.; "Forecasting of CO2 emissions from fuel combustion using trend analysis." Renewable and Sustainable Energy Reviews. 2010. 14. 2906-2915.

4) Grochala, W; Edwards, P. P.; "Thermal Decomposition of the Non-Interstitial Hydrides for the Storage and Production of Hydrogen." Chem. Rev. 2004. 104. 1283-1315.

5) Alapati, S. V; Johnson, K. J.; Scholl, D. S. "Identification of Destabilized Metal Hydrides for Hydrogen Storage Using First Principles Calculations." J. Phys. Chem. B. 2006. 110. 8769-8776.

6) Choudhury, P.; Bhetanabotla, V.R.; Stefanakos, E.; "Manganese Borohydride as a Hydrogen –Storage Candidate: First-Principles Crystal Structure and Thermodynamic Properties." J. Phys. Chem. C. 2009. 113. 13416-13424.

7) Wu, H; Zhou, W.; Udovic, T.J.; Rush, J.J.;Yildirim, T.; " Crystal Chemistry of Perovskite-Type Hydride NaMgH3 : Implications for Hydrogen Storage." Chem. Mater. 2008. 20. 2335-2342.

8) Bogdanovic, B; Schwickardi, M.; "Ti-doped alkali metal aluminum hydrides as potential novel reversible hydrogen storage materials." J. Alloys. Compd. 1997.253. 1-9.

9) Ozolins, V; Majzoub, E.H.; Wolverton, C.; "First-Principles Prediction of Thermodynamically Reversible Hydrogen Storage Reactions in the Li-Mg-Ca-B-H System." J. Am. Chem. Soc. 2009. 131. 230-237.

10) Lovvik, O.M; Swang, O.; "Crystal structures and electronic structures of alkali aluminohexahydrides from density functional calculations."J. Alloys. Compd. 2005. 404. 757-761.

11) Aguayo, A; Singh, D.J.; "Electronic Structure of the complex hydride NaAlH4."Physical Review B. 2004. 69.103-108.

12) Alapati, S. V; Johnson, K. J.; Scholl, D. S. "Identification of Destabilized Metal Hydrides for Hydrogen Storage Using First Principles Calculations." J . Phys. Chem. B. 2006. 110. 8769-8776.

13) Lovvik, O.M; Opalka, S. M.; "Density Functional calculations of Ti-enhanced NaAlH4." Physical Review B. 2005. 71. 1-9.

14) Yvon, K; Bertheville, B.; "Magnesium based ternary metal hydrides containing alkali and alkaline-earth metals." J. Alloys Compd. 2006. 38. 1-8.

15) Bouhadda, Y.; Boudouma, Y.; Fennineche, N.; Bentabet, A.; " Ab initio calculations study of the electronic, optical and thermodynamic properties of $NaMgH_3$, for hydrogen storage." J. Phys. Chem. Solids. 2010. 71. 1264-1268.

16) Klaveness, A.; Swang, O.; Fjellvag, H.; "Formation enthalpies of $NaMgH_3$ and $KMgH_3$: A computational study." Europhys. Lett. 2006. 76. 285-290.

17) Shane, D.T.; Corey, R.L.; Bowman, R.C.; Zidan, R.; Stowe, A. C.; Hwang, S.; Kim, C.; Conradi, M. S.; " NMR Studies of the Hydrogen Storage Compound $NaMgH_3$." J. Phys. Chem. C. 2009.113. 18414-18419.

18) Hao, S.; Sholl, D.S.; "Role of Schottky Defects in Hydrogen and Metal Diffusion in NaH, MgH_2, and $NaMgH_3$." J. Phys. Chem. Lett. 2010. 1. 2968-2973.

19) Komiya, K.; Morisaku, N.; Rong, R.; Takahashi, Y.; Shinzato, Y.; Yukawa, H.; Morinaga, M.; " Electronic Structure Calculations of Perovskite-type Hydrides for Hydrogen Storage." J. Alloys. Compd.2008. 453. 157-160.

20) Accelrys, "Materials Studio®," 4.0 ed. San Diego: Accelrys Software, Inc., 2006.

21) M. D. Segall, P. J. D. Lindan, M. J. Probert, C. J. Pickard, P. J. Hasnip, S. J. Clark, M. C. Payne;"First-principles simulation: ideas, illustrations and the CASTEP code." Journal of Physics: Condensed Matter. 2002.14.2717-2743

(38b) Nanoscale Organic Hybrid Materials (NOHMs): Opportunities In Electrical Energy Storage

Lynden A. Archer, Chemical and Biomolecular Engineering, Cornell University, Ithaca, NY

This talk focuses on two new classes of nanoparticle organic hybrid materials (NOHMs), which manifest liquid-like ion transport properties at high particle loadings. In one group (Nanoparticle Ionic Materials (NIMs)), electrostatic and steric forces dominate the interparticle potential and appear to be responsible for the system's ability to relax to an equilibrium state. In the other group, only steric forces are present, yet the system is able to reach equilibrium in the absence of a suspending solvent. I will discuss equilibrium structure, rheology, and ionic transport properties of these hybrids. I will also study them for applications as anodes, cathodes, and electrolytes for advanced secondary lithium batteries.

See more of this Session: Nanomaterials for Energy Storage I
See more of this Group/Topical: Topical 5: Nanomaterials for Energy Applications

Contents lists available at ScienceDirect

Metabolic Engineering

journal homepage: www.elsevier.com/locate/ymben

Styrene biosynthesis from glucose by engineered *E. coli*

Rebekah McKenna, David R. Nielsen *

Chemical Engineering, Arizona State University, 501 E. Tyler Mall, ECG 301, Tempe, AZ 85287-6106, USA

ARTICLE INFO

Article history:
Received 18 March 2011
Received in revised form
2 June 2011
Accepted 14 June 2011

Keywords:
Styrene
Phenylalanine ammonia lyase
E. coli
Cinnamic acid
L-Phenylalanine
Aromatic

ABSTRACT

Styrene is a large volume, commodity petrochemical with diverse commercial applications, including as a monomer building-block for the synthesis of many useful polymers. Here we demonstrate how, through the *de novo* design and development of a novel metabolic pathway, styrene can alternatively be synthesized from renewable substrates such as glucose. The conversion of endogenously synthesized L-phenylalanine to styrene was achieved by the co-expression of phenylalanine ammonia lyase and *trans*-cinnamate decarboxylase. Candidate isoenzymes for each step were screened from bacterial, yeast, and plant genetic sources. Finally, over-expression of *PAL2* from *Arabidopsis thaliana* and *FDC1* from *Saccharomyces cerevisiae* (originally classified as ferulate decarboxylase) in an L-phenylalanine over-producing *Escherichia coli* host led to the accumulation of up to 260 mg/L in shake flask cultures. Achievable titers already approach the styrene toxicity threshold (determined as ~300 mg/L). To the best of our knowledge, this is the first report of microbial styrene production from sustainable feedstocks.

1. Introduction

Styrene is a versatile, large commodity chemical for which 60% of its global annual consumption supports the production of numerous, industrially important polymers, and co-polymers (SRI, 2010). In 2006, over 6 million metric tons of styrene were produced by U.S. manufacturers, a market that was valued at nearly $28 billion and projected to grow by 4.3% per year through 2010 (SRI, 2010). Today, all commercially available styrene is derived from the world's dwindling petroleum resources. Conventional styrene synthesis is achieved through the chemocatalytic dehydrogenation of petroleum-derived ethylbenzene (Wu et al., 1981) which requires over 3 metric tons of steam per metric ton of styrene produced. This exorbitant requirement renders styrene production as the most energy-intensive among commodity chemical production routes, consuming nearly 200 trillion BTU of steam for its domestic annual production alone (DoE, 2002). With that being said, the goal of this study was to engineer a biocatalyst capable of synthesizing styrene from renewable resources as a more sustainable and greener source of styrene and styrene-derived polymers.

Beyond its petrochemical origins, styrene has been observed as a trace metabolite in foods, in particular cheeses, where it acts as an aroma defect. For instance, the yeast *Penicillium camemberti* has been reported to be capable of synthesizing low levels of styrene from excess L-phenylalanine, but neither a defined pathway nor the requisite genes have thus far been elucidated (Pagot et al., 2007). Styrene is also known to be naturally synthesized by select plant species, including several trees in the *Styracaceae* family (including several *Styrax* sp.). Here, styrene is also synthesized from excess L-phenylalanine where it then subsequently accumulates as a minor constituent (< 0.55% of total dry weight) within benzoin resins (which are predominantly composed of benzoic acid) (Fernandez et al., 2005). Again, however, neither the enzymes nor genes associated with said pathway have been identified to date. Although it is possible to purify styrene from plant resins via distillation or liquid–liquid extraction (Clark, 1990), considering the extremely low productivity, poor net yields, and low inherent value of styrene, its potential, large-scale biological production by such a mechanism is rendered as completely uneconomical and unsustainable. A more sustainable and inexpensive approach, however, would involve the engineering of microorganisms that possess the unique ability to synthesize styrene directly from renewable resources.

In recent years, a variety of additional, novel synthetic routes have been proposed and engineered in microorganisms for the production (from renewable substrates such as glucose) of a number of other useful, functionalized monoaromatic compounds with structural similarity to styrene. For example, a biosynthetic pathway for the production of *p*-hydroxystyrene (a monomer used in synthesis of photo-resist polymers) from renewable sugars has been constructed using both *Escherichia coli* (Qi et al., 2007) and *Pseudomonas putida* (Verhoef et al., 2009) as host platforms. Meanwhile, both phenol (a precursor and monomer

* Corresponding author. Fax: +1 480 727 9321.
E-mail address: David.R.Nielsen@asu.edu (D.R. Nielsen).

1096-7176/$ - see front matter © 2011 Elsevier Inc. All rights reserved.
doi:10.1016/j.ymben.2011.06.005

Please cite this article as: McKenna, R., Nielsen, D.R., Styrene biosynthesis from glucose by engineered *E. coli*. Metab. Eng. (2011), doi:10.1016/j.ymben.2011.06.005

for phenolic resins) (Wierckx et al., 2005) and *p*-hydroxybenzoate (a precursor to parabens, which are used as preservatives) (Verhoef et al., 2007) have also been synthesized as individual products from glucose by engineered strains of *P. putida*. Interestingly, each of the above non-natural metabolites were derived using L-tyrosine (or its immediate precursor, 4-hydroxyphenylpyruvate) as a pathway precursor, thereby making them each phenolics (Boudet, 2007). To date, there remain few examples of engineered biosynthetic pathways for the production of non-phenolic, monoaromatic compounds using microbial biocatalysts. Moreover, there exist no previous reports regarding the development of a styrene biosynthetic pathway or the engineering of microbes capable of synthesizing styrene from renewable resources. With this in mind, the present study describes the *de novo* design and development of a functional styrene biosynthetic pathway and the engineering of *E. coli* strains capable of styrene biosynthesis from glucose.

The proposed styrene biosynthesis pathway utilizes endogenously synthesized (from glucose) L-phenylalanine as an intermediate precursor which is converted to styrene by a series of two enzymatic steps, as in Fig. 1. First, endogenously occurring L-phenylalanine is converted to *trans*-cinnamic acid (tCA) through its deamination, as catalyzed by phenylalanine ammonia lyase (PAL). Said activity and substrate specificity has been previously reported for a number of PAL isoenzymes that have been identified and characterized in yeast (Gilbert and Tully, 1982; Qi et al., 2007; Vannelli et al., 2007b), plants (Cochrane et al., 2004; Young et al., 1966), and (although less prevalently) bacteria (Moffitt et al., 2007a; Xiang and Moore, 2005; Xiang and Moore, 2006; Young et al., 1966). Among previously characterized PAL isoenzymes, considerable variability with respect to both substrate specificity and activity has been reported. The most commonly studied PAL isoenzyme in recombinant systems, including *E. coli*, remains the bifunctional PAL/TAL (TAL: tyrosine ammonia lyase, which also catalyzes the deamination of L-tyrosine to *p*-coumaric acid) of the yeast *Rhodotorula* sp. (Cui et al., 2008; Gilbert et al., 1985, 1983; Gilbert and Tully, 1982; Santos et al., 2011; Vannelli et al., 2007a). Meanwhile, a number of PAL isoenzymes have been also been studied from plant sources (Jones, 1984) where, for example, *Arabidopsis thaliana* has been characterized as possessing four distinct PALs (encoded by *PAL1*, *PAL2*, *PAL3*, and *PAL4*) (Cochrane et al., 2004). More recently, a number of prokaryotic PALs have been isolated and characterized, beginning with that which is encoded by *encP* from *Streptomyces maritimus* (Xiang and Moore, 2002, 2005). Soon after, two additional PAL isoenzymes were discovered in the cyanobacteria *Nostoc puntiformes* and *Anabaena variabilis* (Moffitt et al., 2007a; Xiang and Moore, 2005, 2006). Interestingly, each of these prokaryotic PALs was also found to be highly specific for L-phenylalanine, and thus do not also display TAL activity that is so common among yeast PALs.

The second step in the proposed styrene biosynthesis pathway involves the subsequent decarboxylation of tCA by a phenylacrylate decarboxylase (PADC) displaying *trans*-cinnamate decarboxylase activity to yield styrene as the final product. Several genes encoding PADC activity have been characterized and reported in the literature, including *pdc* from *Lactobacillus plantarum* and *padC* from *Bacillus subtilis* (Tran et al., 2008). It currently remains unknown, however, if their enzyme products could specifically decarboxylate tCA to produce styrene. Meanwhile, the yeast *Saccharomyces cerevisiae* has demonstrated the ability to synthesize styrene when supplied with exogenous tCA (Clausen et al., 1994); however, the underlying mechanism and genetic basis for said activity presently remains unclear. For instance, previous characterizations of *PAD1* have shown that its over-expression in *S. cerevisiae* results in increased resistance to tCA (as achieved by its conversion to styrene) (Larsson et al., 2001). However, it has also been demonstrated that the expression of *PAD1* alone is insufficient for achieving PADC activity (Clausen et al., 1994), and it was later speculated that the expression of a second enzyme was necessary to convert tCA to styrene (Jiang et al., 2005). Most recently, it was reported that tCA decarboxylase activity in *S. cerevisiae* maintains an essential dependence on the co-expression of both *PAD1* and *FDC1*, the latter a gene previously characterized as encoding ferulic acid decarboxylase (note that ferulic acid is also a phenylacrylic acid) (Mukai et al., 2010).

The present study describes the *de novo* design of a styrene biosynthetic pathway, as supported through the comprehensive screening of composite pathways enzymes from various genetic sources. The synthesis of styrene from glucose was ultimately achieved through the co-expression of PAL and tCA decarboxylase enzymes in an L-phenylalanine over-producing *E. coli* host platform.

2. Materials and methods

2.1. Chemicals

Chemicals used in this study include L-phenylalanine (98.5%, VWR, Westchester, PA), *trans*-cinnamic acid (99%, MP Biomedicals, Solon, OH), styrene (99%, Alfa Aesar, Ward Hill, MA), methanol (99.8%, VWR, Westchester, PA), and trifluoroacetic acid (99.5%, EMD, Darmstadt, Germany). All other chemicals used in this study are from Sigma-Aldrich (St. Louis, MO).

2.2. Strains and media

All strains, plasmids, and oligonucleotide primers used in this study are listed in Table 1. Custom oligonucleotide primers were synthesized by Integrated DNA Technologies (Coralville, IA). cDNA of *N. puntiformes* and *A. variabilis* were gifts from Prof. Bradley Moore (UCSD). Strains of *B. subtilis*, *L. plantarum*, and *S. cerevisiae*, as well as the plasmids pSTV28 and pTrc99A were all gifts from Prof. Kristala Prather (MIT). Genomic DNA was prepared from whole cells using the ZR Fungal/Bacterial DNA MiniPrep (Zymo Research, Irvine, CA) according to vendor protocols. Strains were

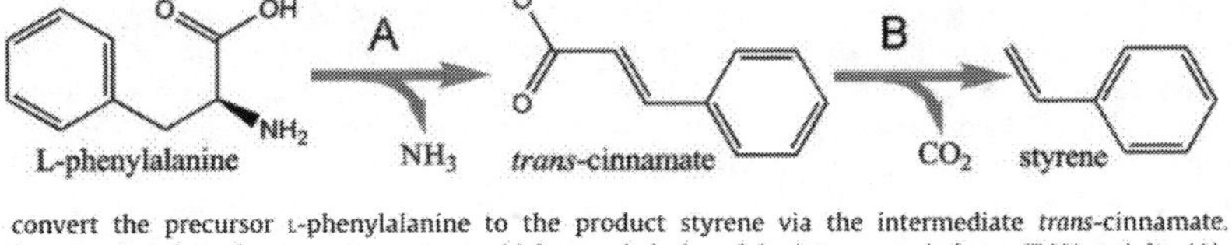

Fig. 1. Enzymatic pathway to convert the precursor L-phenylalanine to the product styrene via the intermediate *trans*-cinnamate. The two-step pathway from L-phenylalanine is achieved by the co-expression of one or more genes which encoded phenylalanine ammonia lyase (PAL) activity (A), and one or more genes which encoded *trans*-cinnamic acid decarboxylase (CADC) activity (B).

Please cite this article as: McKenna, R., Nielsen, D.R., Styrene biosynthesis from glucose by engineered E. coli. Metab. Eng. (2011), doi:10.1016/j.ymben.2011.06.005

Table 1
Strains, plasmids, and oligonucleotide primers used in the present study.

	Description	Source
Strains		
E. coli NST74	*aroH367, tyrR366, tna-2, lacY5, aroF394(fbr), malT384, pheA101(fbr), pheO352, aroG397(fbr)*	ATCC 31884
E. coli BL21Star(DE3)	*F-ompT hsdS$_B$ (r$_B$-m$_B$-) gal dcm rne131 (DE3)*	Invitrogen
E. coli NEB-10 beta	*araD139 Δ(ara,leu)7697 fhuA lacX74 galK16 galE15 mcrA f80d(lacZΔM15)recA1 relA1 endA1 nupG rpsL rph spoT1Δ(mrr-hsdRMS-mcrBC)*	New England Biolabs
S. cerevisiae W303	Source of *pad1* and *fdc1*	Prather Lab, MIT
L. plantarum	Source of *pdc*	Prather Lab, MIT
B. subtilis 3610	Source of *padC*	Prather Lab, MIT
Plasmids		
pTrc99A	P_{trc}, pBR322 ori, *lacI^q*, *Amp^r*	Prather Lab, MIT
pSTV28	P_{lac}, pACYC184 ori, *Cm^r*	Prather Lab, MIT
pSpal1At	*PAL1* of *A. thaliana* inserted into the *EcoRI* and *SphI* sites of pSTV28	This study
pSpal2At	*PAL2* of *A. thaliana* inserted into the *EcoRI* and *SphI* sites of pSTV28	This study
pSpalAv	*pal* of *A. variabilis* inserted into the *EcoRI* and *BamHI* sites of pSTV28	This study
pSpalNp	*pal* of *N. punctiforme* inserted into the *EcoRI* and *BamHI* sites of pSTV28	This study
pSencPSm	*encP* of *S. maritimus* inserted into the *EcoRI* and *BamHI* sites of pSTV28	This study
pTpad1Sc	*PAD1* of *S. cerevisiae* inserted into the *NcoI* and *XbaI* sites of pTrc99A	This study
pTpad1Sc-fdc1Sc	*FDC1* of *S. cerevisiae* inserted into pTpad1Sc	This study
pTfdc1Sc	*FDC1* of *S. cerevisiae* inserted into the *SalI* and *HindIII* sites of pTrc99A	This study
pTpadCBs	*padC* of *B. subtilis* inserted into the *BamHI* and *SbfI* sites of pTrc99A	This study
pTpdcLp	*pdc* of *L. plantarum* inserted into the *BamHI* and *SbfI* sites of pTrc99A	This study
pUN15-pal1	Clone U10120 containing AT2G37040 (*PAL1*) from *A. thaliana*	ABRC
pUN15-pal2	Clone U12256 containing AT3G53260 (*PAL2*) from *A. thaliana*	ABRC
Primers (5′→3′)		
F_pal_AV_EcoRI	TAAGAATTCAAGGGGATAAATAATGAAGACACTATCTCAAGC	
R_pal_AV_BamHI	ATTGGATCCTTAATGCAAGCAGGGT	
F_pal_NP_EcoRI	TAAGAATTCAAGGGGATAAATAATGAATATAACATCTCTACAAC	
R_pal_NP_BamHI	ATTGGATCCTTACGTTGACTTTAAGCT	
F_pal1_AT_EcoRI	TAAGAATTCAAGGGGATAAATAATGGAGATTAACGGGGCAC	
R_pal1_AT_SphI	ATTGCATGCTTAACATATTGGAATGGGAGCTC	
F_pal2_AT_EcoRI	TAAGAATTCAAGGGGATAAATAATGGATCAAATCGAAGCAATG	
R_pal2_AT_SphI	ATTGCATGCTTAGCAAATCGGAATCGGAG	
F_encP_SM_EcoRI	TAAGAATTCAAGGGGATAAATAATGACCTTCGTCATAGAGCT	
R_encP_SM_BamHI	ATTGGATCCTTAGTGCGCCGCCACG	
F_padC_BS_BamHI	AAAGGATCCCGACTAAGGGAGGATAAGATGGAAAACTTTATCGGAAG	
R_padC_BS_SbfI	ATACCTGCAGGATGTTTATTATAATCTTCCCGCG	
F_pdc_LP_BamHI	ATAGGATCCCTCTGGAGGCAGTTCTAATGACAAAAACTTTTAAAACACT	
R_pdc_LP_SbfI	ATACCTGCAGGCCAGAATGTTTCACGTGAA	
F_pad1_SC_NcoI	TTACCATGGAGGAACCTAGGCACACAATGGTCCTATTTCCAAGAAGAA	
R_pad1_SC_XbaI	ATTTCTAGATTACTTGCTTTTTATTCCTTCCC	
F_fdc1_SC_SalI	ATAGTCGACAGACATCAAAGGACGGTTCATGAGGAAGCTAAATCCAGCT	
R_fdc1_SC_HindIII	ATTAAGCTTTTATTTATATCCGTACCTTTTCCAAT	

routinely cultured in Luria-Bertani (LB) broth (supplemented with antibiotics as necessary). Cultures were assayed for their ability to synthesize L-phenylalanine, *trans*-cinnamate, and styrene by cultivation in phosphate-limited minimal media (herein referred to as "MM1") with glucose. MM1 was adapted from Qi et al. (2007), and is composed of glucose (nominally 15 g/L), $MgSO_4 \cdot 7H_2O$ (0.5 g/L), $(NH_4)_2SO_4$ (4.0 g/L), MOPS (24.7 g/L), KH_2PO_4 (0.3 g/L), K_2HPO_4 (0.7 g/L), and 5 mL/L ATCC Trace Mineral Supplement (Catalog No. MD-TMS) (EDTA (0.5 g/L), $MgSO_4 \cdot 7H_2O$ (3 g/L), $MnSO_4 \cdot 7H_2O$ (0.5 g/L), NaCl (1 g/L), $FeSO_4 \cdot 7H_2O$ (0.1 g/L), $Co(NO_3)_2 \cdot 6H_2O$ (0.1 g/L), $CaCl_2$ (0.1 g/L), $ZnSO4 \cdot 7H_2O$ (0.1 g/L), $CuSO_4 \cdot 5H_2O$ (0.01 g/L), $AlK(SO_4)_2$ (0.01 g/L), H_3BO_3 (0.01 g/L), $Na_2MoO_4 \cdot 2H_2O$ (0.01 g/L), Na_2SeO_3 (0.001 g/L), $Na_2WO_4 \cdot 2\,H_2O$ (0.10 g/L), and $NiCl_2 \cdot 6H_2O$ (0.02 g/L)).

2.3. *Toxicity assays*

To determine extent and effects of metabolite toxicity on *E. coli*, the impacts of the exogenous addition of tCA and styrene (at increasing final concentrations) to growing cultures was investigated. Seed cultures of *E. coli* NST74 were prepared in 5 mL of LB broth and grown at 32 °C overnight while shaking at 250 rpm. The seed culture (1 ml) was then used to inoculate 50 ml of LB broth in a 250 mL shake flask. Cultures were grown to an optical density (OD_{600}) of either ~0.6 or ~2.0, at which time either tCA or styrene was added to the flasks at an array of final concentrations between 0 to 1 g/L. Two different initial OD_{600} levels were used to investigate the potential effects of product toxicity as a function of cell density and/or growth stage. In all cases, culturing then resumed at 32 °C for another 6–8 h while cell growth, as determined by OD_{600} measurements, was periodically monitored using a UV/vis spectrophotometer (Beckman Colter DU800, Brea, CA).

2.4. *Cloning of candidate genes encoding PAL activity from A. variabilis, N. punctiforme, and A. thaliana*

All genes used in this study were PCR amplified using a BioRad iCycler system with Phusion DNA Polymerase (Finnzymes, Espoo, Finland) using custom oligonucleotide primers. PCR cycling and reaction conditions were standardized according to manufacturer instructions. Candidate PAL encoding genes were amplified from the cDNA of *A. variabilis* and *N. punctiforme*. Candidate PAL encoding genes from *A. thaliana* were derived from cDNA library plasmids containing the specific loci of interest (Table 1) obtained from the ABRC (Ohio State University, Columbus, OH). Whereas *A. thaliana* posses four distinct PALs, we focused on those encoded by *PAL1* and *PAL2* as these have displayed the greatest activity when expressed in recombinant *E. coli* (noting also that PAL3 was

Please cite this article as: McKenna, R., Nielsen, D.R., Styrene biosynthesis from glucose by engineered *E. coli*. Metab. Eng. (2011), doi:10.1016/j.ymben.2011.06.005

found to be of 'very low activity') (Cochrane et al., 2004). Amplified linear DNA fragments were purified using the Zyppy Clean and Concentrator kit (Zymo Research, Orange, CA). Purified fragments were treated by endonuclease digestion using appropriate restriction enzymes (all from New England Biolabs, Ipswich, MA). Amplified fragments containing the *pal* from both *N. punctiforme* and *A. variabilis* were digested with *BamHI* and *EcoRI* whereas fragments containing *PAL1* and *PAL2* from *A. thaliana* were digested with *EcoRI* and *SphI*. All digestions were performed at 37 °C according to manufacturer's protocols. The expression vector pSTV28 was similarly digested with either *BamHI* and *EcoRI* or *EcoRI* and *SphI*. Digested fragments were gel purified using the Zyppy Gel DNA recovery kit (Zymo Research, Orange, CA). Linearized fragments were ligated using T4 DNA ligase (New England Biolabs, Ipswich, MA) at 4 °C overnight. Chemically competent *E. coli* NEB10-Beta (New England Biolabs, Ipswich, MA) cells were used for all transformations. Transformants were selected by plating on LB solid agar containing 34 mg/L chloramphenicol and culturing at 37 °C overnight. The transformant pool was subsequently screened according to both colony PCR (employing the same primers as used in the initial amplification) and restriction digest mapping of the resultant plasmids to identify those clones harboring the successful construct. This approach resulted in construction of plasmids pSpalAv, pSpalNp, pSpal1At, and pSpal2At, as listed in Table 1.

2.5. *Cloning of candidate genes encoding PADC activity from L. plantarum, B. subtilis, and S. cerevisiae*

Candidate PADC encoding genes, including *pdc*, *padC*, *PAD1*, and *FDC1*, were amplified via PCR using genomic DNA templates derived from *L. plantarum*, *B. subtilis*, and *S. cerevisiae*, respectively. PCR amplified DNA fragments were purified before treatment by endonuclease digestion. Fragments containing *padC* and *pdc* were each digested with *BamHI* and *SbfI*. The *E. coli* expression vector pTrc99A was similarly digested with *BamHI* and *SbfI*. Meanwhile, the amplified fragment containing *PAD1* was digested with *NcoI* and *XbaI* whereas the *FDC1* containing fragment was digested with *SalI* and *HindIII*. The *E. coli* expression vector pTrc99A was similarly digested with either *NcoI* and *XbaI* (for the insertion of *PAD1*) or *SalI* and *HindIII* (for the insertion of *FDC1*). All digested fragments were gel purified then ligated at 4 °C overnight before their transformation into chemically competent *E. coli* NEB10-Beta. Selection was then achieved by plating transformed cells on LB solid agar containing 100 mg/L ampicillin and culturing at 37 °C overnight. After confirmation of the correct transformant, these works resulted in the generation of plasmids pTpadCBs, pTpdcLp, pTpad1Sc, and pTfdc1Sc, as listed in Table 1. The *SalI–HindIII* digested *FDC1* fragment was then also cloned into the same sites in the newly generated plasmid pTpad1Sc by an analogous protocol, resulting in the plasmid pTpad1Sc-fdc1Sc.

2.6. *Assaying PAL activity in crude lysates of recombinant E. coli*

Each of the newly created PAL harboring plasmids (namely pSpalAv, pSpalNp, pSpal1At, and pSpal2At) were individually transformed into *E. coli* BL21(DE3) (Invitrogen, Carlsbad, CA). Seed cultures of each of the resultant strains were prepared (in triplicate) in 5 mL LB broth supplemented with 34 mg/L chloramphenicol and cultured at 32 °C while shaking at 250 rpm overnight. 50 µl of each seed was then used to again inoculate 5 ml LB broth. These cultures were grown until reaching an OD_{600} of ~0.6, at which point each was induced by the addition of isopropyl β-D-1-thiogalactopyranoside (IPTG) at a final concentration of 0.2 mM. Induced cultures were incubated for an additional 6 h after which an equal number of cells (determined by OD_{600} measurement) were collected by centrifugation at 1400 × *g* for 4 min. The cell pellet was re-suspended in 900 µL distilled water. Cell lysis was achieved using the FastBreak Cell Lysis Reagent kit (Promega, Madison, WI) and the supernatant collected after centrifugation at 11,000 × *g* for 2 min. PAL activity was analyzed at room temperature in pH 7.5 50 mM Tris-HCl buffer containing 250 mM L-phenylalanine. Activity assays were initiated by the addition of 5 µL of crude cell lysate. The production of tCA was followed at 290 nm on a Beckman Colter UV/vis spectrophotometer for a total of 5 min at 20 s intervals. A molar extinction coefficient of 9000 M^{-1} cm^{-1} and a 1 cm path length were used to establish enzyme activity in terms of U mg^{-1} protein. The PAL protein content in each crude lysate was determined via first separation by SDS-PAGE using Mini-PROTEAN TGX 4–20% precast gels (Bio-Rad, Hercules, CA) and standard protocols. Concentration was then analyzed using the ImageJ software package (NIH, Bethesda, MD) and calibrated vs. Precision Plus Unstained Protein Standards (Bio-Rad, Hercules, CA).

2.7. *Assaying PAL activity in recombinant E. coli whole cells*

Seed cultures of *E. coli* BL21(DE3) harboring one of pSpalAv, pSpalNp, pSpal1At, or pSpal2At, were prepared in 5 mL LB broth and grown overnight. Shake flasks (250 mL) containing 50 mL of LB were inoculated with 1 mL of each seed culture. Cultures were grown until an OD_{600} of ~0.6 was reached, at which point the cultures were induced by IPTG addition at a final concentration of 0.2 mM. Cultures were then incubated for an additional 6 h (resulting in an OD_{600} of ~2) before an equal number of cells were collected and centrifuged at 1400 × *g* for 5 min. The pellet was washed once with PBS (phosphate buffered saline) at pH 7 before being re-suspended in 12 ml PBS buffer. Finally, the appropriate substrate, L-phenylalanine or L-tyrosine at a final concentration of 1 g/L, was added to the suspension. The suspensions were then shaken at 32 °C for a total of 3 h. Samples (1 mL) were taken every hour, centrifuged, and 750 µL of supernatant was collected for HPLC analysis to monitor the production of either *tCA* or *p*-coumaric acid.

2.8. *Assaying PADC activity in recombinant E. coli whole cells*

The plasmids pTpad1Sc-fdc1Sc, pTpad1Sc, pTfdc1Sc, pTpdcLp, and pTpadcBs were each individually transformed into chemically competent *E. coli* BL21(DE3). A seed culture of each strain was then grown in LB broth overnight. Shake flasks (250 mL) containing 50 mL of LB were inoculated with 1 ml of seed culture. Cultures were grown at 32 °C until an OD_{600} of ~0.6, at which point they were induced by adding 0.2 mM IPTG and then incubated for an additional 6 h. Cells were then collected and re-suspended in 12 ml PBS buffer (as previously described) and the substrate (*tCA* or *p*-coumaric acid) was added at a final concentration of 1 g/L. Samples (1 mL) were removed from the culture at both the time of initiation as well as after 12 h of incubation at 32 °C and analyzed by HPLC using the methods described herein.

2.9. *Co-expression of PAL and CADC isoenzymes in E. coli NST74 to convert L-phenylalanine to styrene*

The L-phenylalanine over-producing strain *E. coli* NST74 (Table 1) was co-transformed with the plasmids pSpal2At and pTfdc1Sc and selected for on LB agar supplemented with 100 mg/L ampicillin and 34 mg/L chloramphenicol. The resultant transformant was then grown overnight at 32 °C in 5 mL LB broth. Shake flasks (250 mL) containing 50 mL LB were inoculated with 1 mL of seed culture. The culture was then grown at 32 °C until

Please cite this article as: McKenna, R., Nielsen, D.R., Styrene biosynthesis from glucose by engineered *E. coli*. Metab. Eng. (2011), doi:10.1016/j.ymben.2011.06.005

an OD_{600} of ~0.6, at which time it was induced with 0.2 mM IPTG and then incubated for an additional 8 h. Cells were then collected and re-suspended in 12 ml PBS buffer (as previously described) and L-phenylalanine added at a final concentration of either 400 or 900 mg/L. Samples were taken periodically and analyzed by HPLC to determine the content of L-phenylalanine, tCA, and styrene, according to the methods presented below.

2.10. *Co-expression of PAL and CADC isoenzymes in E. coli NST74 to convert glucose to styrene in shake flask cultures*

The L-phenylalanine over-producing strain *E. coli* NST74 was co-transformed with each of the following combinations of plasmids: pSpalAv and pTfdc1Sc, pSpalNp and pTfdc1Sc, pSpal1At and pTfdc1Sc, and pSpal2At and pTfdc1Sc.Single colonies were then selected from the resulting transformants and grown in 5 mL LB broth for 12 h at 32 °C while shaking at 250 rpm to prepare a seed culture. Each seed (1 mL) was then used to inoculate 50 mL MM1. These cultures were performed in 100 mL serum bottles outfitted with septa caps that were tightly sealed upon inoculation. A closed system was used to avoid volatile product (i.e., styrene) losses. A large headspace was used to preclude the depletion of oxygen from the bottle. Cultures were grown for 10 h prior to being induced by the addition of IPTG at a final concentration of 0.2 mM. Culturing continued for 29–48 h post-induction while 1 ml samples were periodically taken from each culture and analyzed for relevant metabolite contents via HPLC, using the methods described herein.

2.11. *Metabolite analysis by HPLC*

Samples were prepared by removing 1 mL of culture from a shake flask culture and pelleting the cells at 11,000 × *g* for 2 min. The supernatant (0.75 mL) was then transferred to a glass HPLC vial and sealed with a Teflon-lined cap. HPLC analysis was carried out using a Hewlett Packard 1100 series HPLC system equipped with an auto sampler, diode array (UV/vis) detector, and reverse-phase Hypersil Gold SBC18 column (4.6 mm × 150 mm; Thermo Fisher, USA). Samples (5 μL) were injected for analysis at a total constant flow rate of 1.0 ml/min and constant column temperature of 45 °C. The column was eluted with 'solvent A' (consisting of double-distilled water) and 'solvent B' (consisting of methanol (99.8% grade) plus 0.1% trifluoroacetic acid (TFA)). The eluent began as a mixture of 95% solvent A and 5% solvent B before a linear gradient was applied over 8 min to then reach a mixture of 20% solvent A and 80% solvent B. This eluent composition was then held constant for 2 min before a second linear gradient was then applied over the course of 4 min to achieve a final mixture of 95% solvent A and 5% solvent B. The eluent was monitored at each of 215 nm for L-phenylalanine and 258 nm for *tCA*, *p*-coumaric acid, hydroxystyrene, and styrene. Under these conditions L-phenylalanine, *p*-coumaric acid, *tCA*, *p*-hydroxystyrene, and styrene were eluted at 4.5, 6.7, 8.67, 8.78, and 10.4 min, respectively.

2.12. *Confirmation of styrene biosynthesis by GC–MS*

Culture supernatant (1 ml) was added to hexane (1 ml) and vortexed for 20 min at maximum speed. The biphasic mixture was centrifuged for 1 min at 11,000 × *g* to settle. 750 μl of the hexane layer was removed for analysis by GC–MS. GC–MS analysis was performed on a Hewlett Packard 5890 Series II gas chromatograph with a flame ionizing detector and Supelco MDN-5S (30 m × 0.25 mm id) fused-silica capillary column using helium as the carrier gas. The injector, column, and detector temperatures were initial set at 280, 240, and 220 °C, respectively. The column temperature was then increased from 40 to 320 °C at 14 °C/min.

3. Results and discussion

3.1. *Assaying styrene toxicity*

Under the studied conditions, maximal titers of L-phenylalanine produced by *E. coli* NST74 after 48 h of culture in MM1 (with 1.5% glucose) reached about 700–1000 mg/L (results not shown). Whereas the theoretical yield of L-phenylalanine on glucose is 0.55 g/g (Baez-Viveros et al., 2004), achievable yields by *E. coli* NST74 under the present conditions in shake flask cultures were only 0.052–0.074 g/g (or about 10% of theoretical). Assuming complete conversion of all endogenously produced L-phenylalanine to styrene were possible (for example, if our engineered pathway could achieve a particularly high flux), we would expect to be able to synthesize up to 440–630 mg/L of styrene under the same culture conditions (corresponding to a glucose yield of 0.051–0.072 g/g), barring any physiological limitations such as product toxicity. Through a preliminary screening study, however, we have found that the products *trans*-cinnamate ($\log K_{ow}$=2.13) and styrene ($\log K_{ow}$=3.05) can each inhibit *E. coli* growth when added to cultures at final concentrations above ~800 and ~300 mg/L, respectively, as demonstrated in Fig. 2. For reference, *p*-hydroxystyrene ($\log K_{ow}$=2.28) has previously been shown to inhibit *P. putida* at concentrations of ~540 mg/L (Verhoef et al., 2009). Solvent hydrophobicity is typically a good

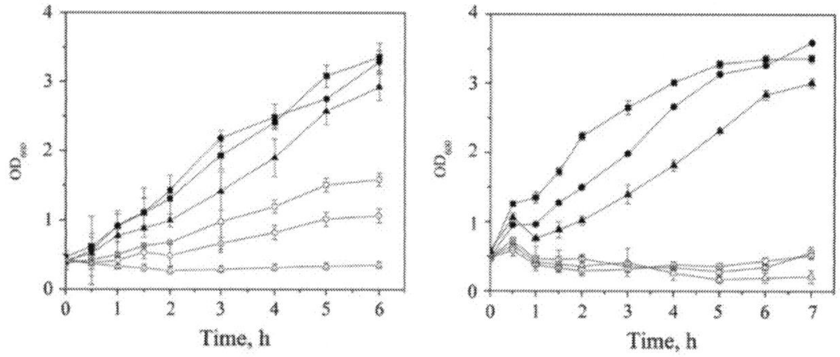

Fig. 2. Left: growth response of *E. coli* NST74 to tCA at concentrations of 0 mg/L (solid square), 200 mg/L (solid circle), 400 mg/L (solid triangle), 600 mg/L (open square), 800 mg/L (open circle), and 1000 mg/L (open triangle). Right: growth response of *E. coli* NST74 to styrene at concentrations of 0 mg/L (solid square), 100 mg/L (solid circle), 200 mg/L (solid triangle), 300 mg/L (open square), 400 mg/L (open circle), and 500 mg/L (open triangle). Error bars reported at one standard deviation from triplicate experiments.

Please cite this article as: McKenna, R., Nielsen, D.R., Styrene biosynthesis from glucose by engineered *E. coli*. Metab. Eng. (2011), doi:10.1016/j.ymben.2011.06.005

indicator of its toxicity towards microorganisms and may be quantified by log K_{ow}, the log of the octanol–water partition coefficient (Ramos et al., 2002). Even at low concentrations, compounds with a log K_{ow} between 1.0 to 5.0 tend to be very toxic (Heipieper et al., 1994). Accumulation of hydrophobic aromatics within the cytoplasmic membrane has been shown to disrupt its integrity, resulting in the leakage of ions, metabolites, lipids, and proteins, as well as affecting the cells ability to maintain its internal pH and an appropriate transmembrane proton gradient (Ramos et al., 2002; Weber et al., 1993). The measured toxicity thresholds were found to be consistent at both low (OD_{600} 0.6) and high (OD_{600} 2, data not shown) initial cell densities (reflecting early and late exponential growth, respectively), indicating growth inhibition was independent of cell density and growth stage. As a result of the inherent toxicity of styrene we anticipated that its biosynthesis using an NST74 host platform would be limited to below its maximum potential under our culture conditions. While no prior studies have specifically explored either the effects of styrene toxicity on *E. coli* or strategies to improve its tolerance, several *Pseudomonas* sp. have been shown to display enhanced styrene tolerance characteristics (Weber et al., 1993). Whereas the present study is solely focused on the prototyping of a novel pathway for styrene biosynthesis, it is clear from our preliminary assays that styrene toxicity must eventually be overcome or effectively circumvented if renewable styrene production is ever to become viable or sustainable.

Table 2
Specific activity of PAL isoenzymes from *A. variabilis*, *N. punctiforme*, and *A. thaliana* on L-phenylalanine and L-tyrosine when expressed in recombinant *E. coli* BL21(DE3). Errors reported at one standard deviation from triplicate experiments.

Strain	Substrate	Activity (U mg^{-1} protein)
BL21(DE3)	L-Phenylalanine	n.d.
	L-Tyrosine	n.d.
pSencPSm	L-Phenylalanine	n.d.
	L-Tyrosine	n.d.
pSpalAv	L-Phenylalanine	2.38 ± 0.64
	L-Tyrosine	n.d.
pSpalNp	L-Phenylalanine	0.91 ± 0.32
	L-Tyrosine	n.d.
pSpal1At	L-Phenylalanine	2.42 ± 1.07
	L-Tyrosine	n.d.
pSpal2At	L-Phenylalanine	4.08 ± 0.11
	L-Tyrosine	n.d.

n.d.—not detected.

3.2. Screening candidate PAL isoenzymes for activity in recombinant *E. coli*

The activities of recombinant PALs from various genetic sources were analyzed according to both *in vitro* (crude lysate) and *in vivo* (whole, resting cell) assays. All recombinant PALs, expressed from plasmids pSencPSm, pSpalAv, pSpalNp, pSpal1At, and pSpal2At (Table 1) in an *E. coli* BL21(DE3) background were recovered as crude lysates. According to our *in vitro* assay results, all of the tested PALs showed comparable levels of activity on L-phenylalanine as substrate (Table 2), with the exception of EncP from *S. maritimus* whose activity was non-measureable. It has previously been shown through kinetic studies on prokaryotic PALs that those derived from *N. puntiformes* and *A. variabilis* possessed 500–1000 times greater activity than EncP from *S. maritimus*; furthermore, the k_{cat}/K_m value of PAL from *A. variabilis* was found to be greater than that of *N. puntiformes* (72.2 and 43.8 $mM^{-1}\,s^{-1}$, respectively) (Moffitt et al., 2007b). Thus, it is possible that EncP was in fact functionally expressed in our study; however, its activity was too low to measure according to the protocols employed. The relative activities of candidate PALs were then further explored through the use of whole-cell assays in recombinant *E. coli*. Resting cells suspended in PBS buffer (pH 7) were supplemented with L-phenylalanine or L-tyrosine and product (tCA or *p*-coumaric acid, respectively) formation was monitored over the course of 3 h. Although the PALs from *A. thaliana* were found to possess the greatest specific activity in crude lysates (Table 2), assays of *in vivo* function provided a more stark contrast into the relative activities among all candidate PALs, as shown in Fig. 3. Under the studied conditions, the PALs of *A. thaliana* were found to display the greatest activity (being significantly higher that of either prokaryotic PAL) although being comparable for both isoenzymes. This result is more consistent with previous studies wherein k_{cat}/K_m values of PAL1 and PAL2 were estimated as 25,500 and 51,200 $M^{-1}\,s^{-1}$ (or six orders of magnitude greater than those reported for *N. puntiformes* and *A. variabilis*) when expressed in recombinant *E. coli* (Cochrane et al., 2004). The elevated relative activity of PAL2 is also consistent with the observation of higher initial rates of tCA accumulation when expressed from BL21(DE3) pSpal2At, as seen in Fig. 3.

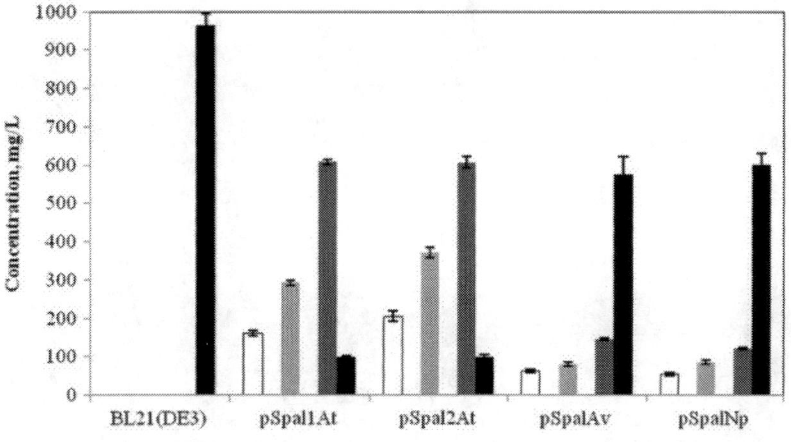

Fig. 3. Phenylalanine ammonia lyase activity from candidate genes cloned from *A. thaliana*, *A. variabilis*, and *N. punctiformes* in recombinant *E. coli* BL21(DE3) whole cells. 50 ml cultures were grown for 8 h (induced with 0.2 mM IPTG after 1.5 h), spun down and re-suspended in 10 ml PBS buffer. The conversion of 1 g/L L-phenylalanine (black) to *trans*-cinnamic acid after 1 h (no color), 2 h (light gray), and 3 h (dark gray). Error bars reported at one standard deviation from triplicate experiments.

Please cite this article as: McKenna, R., Nielsen, D.R., Styrene biosynthesis from glucose by engineered *E. coli*. Metab. Eng. (2011), doi:10.1016/j.ymben.2011.06.005

Again, however, none of the tested PAL isoenzymes were found to display any activity on L-tyrosine as the substrate (data not shown), which is consistent with the results of our *in vitro* assays (Table 2). Taken together, these analyses demonstrate that all of the studied PALs display rigid substrate specificity for L-phenylalanine. This important result contrasts numerous prior works on aromatic pathway development in recombinant *E. coli*, which have solely relied upon the use of bifunctional yeast PAL/TALs (Cui et al., 2008; Gilbert et al., 1985; Gilbert et al., 1983; Gilbert and Tully, 1982) and which was specifically used to synthesize *p*-hydroxystyrene (Qi et al., 2007). As these results illustrate that the first committed step in our pathway is highly specific for the intended substrate (L-phenylalanine) alone, we anticipate that this advantageous outcome will ultimately help to control product purity (specifically styrene over *p*-hydroxystyrene or a mixture of products) while also improving the activity and flux of our desired pathway. Meanwhile, the same cannot be assured had the first committed step of our pathway been catalyzed by PAL/TAL, as was the case for all engineered *p*-hydroxystyrene pathways reported to date (Qi et al., 2007; Verhoef et al., 2009).

3.3. *Screening candidate PADC isoenzymes for tCA decarboxylase activity in recombinant E. coli*

Candidate PADC isoenzymes from *L. plantarum*, *B. subtilis*, and *S. cerevisiae* were screened for their ability to decarboxylate tCA to produce styrene when expressed in *E. coli*. Plasmids harboring the candidate PADC-encoding genes (Table 1) were individually transformed into *E. coli* BL21(DE3), as described above. Whole, resting cells were prepared in PBS buffer supplemented with 1 g/L of tCA acid or *p*-coumaric acid. The production of styrene or *p*-hydroxystyrene, respectively, was then followed periodically, and the results after 12 h of culture are compared in Fig. 4. With the exception of the strain expressing *PAD1* from *S. cerevisiae* alone, all other strains displayed decarboxylase activity on *p*-coumaric acid, leading to *p*-hydroxystyrene biosynthesis. These results are consistent with previous reports on the functional expression of *pdc* and *padC* in recombinant *E. coli* (Qi et al., 2007) to support p-hydroxystyrene biosynthesis from glucose. However, our findings further and importantly show that the enzymes from *L. plantarum* and *B. subtilis* cannot catalyze the conversion of tCA to styrene (thus making them more specific for *p*-coumaric acid). In the present study it was instead found that the sole expression of *FDC1* from *S. cerevisiae* (which was previously characterized to encode a ferulic acid decarboxylase) was sufficient for achieving decarboxylase activity on either tCA or *p*-coumaric acid (with perhaps a slight preference towards tCA, based on overall conversion). Though it was previously reported that the co-expression of both *FDC1* and *PAD1* is necessary to achieve tCA decarboxylase activity in the native *S. cerevisiae* (Mukai et al., 2010), we now report that functional tCA decarboxylase activity in *E. coli* depends solely upon *FDC1* over-expression and is not dependent upon the co-expression of *PAD1*. Furthermore, as can be seen in Fig. 4, comparable styrene titers were achieved when either *FDC1* was expressed alone or together with *PAD1*, indicating that the co-expression of *PAD1* does not increase (or otherwise alter) tCA decarboxylase activity in *E. coli*, as has previously been suggested in the native *S. cerevisiae* (Larsson et al., 2001). It is, however, plausible that the expression of *ubix* (which has been shown to be 50% similar to *PAD1* (Mukai et al., 2010)) in our *E. coli* background could have served to compensate for the absence of *PAD1* expression, enabling tCA decarboxylase activity to be achieved in *E. coli* expressing *FDC1* alone. Future studies will explore the deletion of *ubix* from the background of *E. coli* to further investigate the recombinant function of *FDC1*. Most importantly, these results have demonstrated that *FDC1* over-expression in *E. coli* uniquely enables the decarboxylation of tCA to styrene.

Although the proposed styrene biosynthesis pathway is somewhat analogous to that which was previously developed for

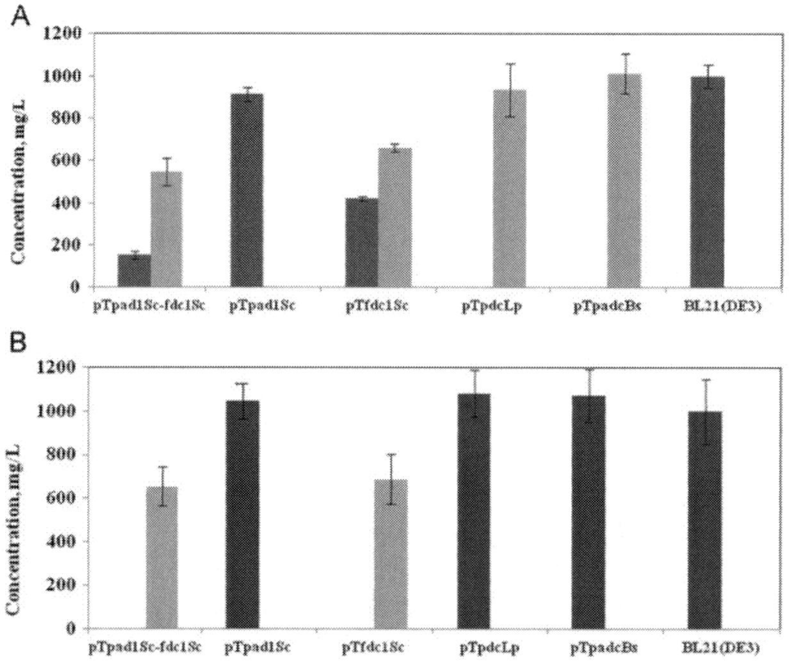

Fig. 4. Phenylacrylic acid decarboxylase activity from candidate genes cloned from *S. cerevisiae*, *L. plantarum*, and *B. subtilis* in recombinant *E. coli* BL21(DE3) whole cells. The conversion of (A) 1 g/L *p*-coumaric acid (dark gray) to *p*-hydroxystyrene (light gray) and (B) 1 g/L *trans*-cinnamic acid (dark gray) to styrene (light gray) after 12 h. Error bars reported at one standard deviation from triplicate experiments.

Please cite this article as: McKenna, R., Nielsen, D.R., Styrene biosynthesis from glucose by engineered *E. coli*. Metab. Eng. (2011), doi:10.1016/j.ymben.2011.06.005

p-hydroxystyrene biosynthesis (Qi et al., 2007; Verhoef et al., 2009), it is essential to realize that the specific enzyme 'parts' used in those two prior studies are wholly inadequate for achieving the present objective of styrene biosynthesis from glucose. In other words, said previously characterized enzymes were not simply transferrable among the two (styrene and *p*-hydroxystyrene) distinct pathways, but rather that styrene biosynthesis could only be supported through the identification and characterization of a specific tCA decarboxylase.

3.4. Biosynthesis of tCA and styrene from glucose in recombinant E. coli

The biosynthesis of tCA from glucose was first investigated using strains of *E. coli* NST74 that individually carried the plasmids pSpalAv, pSpalNp, pSpal1At, and pSpal2At. Cultures were grown in MM1 media with 1.5% glucose. 48 h after induction with 0.2 mM IPTG, the production of tCA was detected in each culture at final titers of 600, 473, 648, and 918 mg/L, respectively. Neither styrene nor *p*-coumaric acid were detected in any samples.

To test for styrene biosynthesis from glucose, *FDC1* from *S. cerevisiae* was then co-expressed with each of the PAL-encoding genes according to the following constructed strains: NST74 pSpalAv pTfdc1Sc, NST74 pSpalNp pTfdc1Sc, NST74 pSpal1At pTfdc1Sc, and NST74 pSpal2At pTfdc1Sc. Conversion of glucose to styrene was followed after IPTG induction and results for the strain NST74 pSpal2At pTfdc1Sc are shown in Table 3 (other results not shown as this dataset was representative of the results and trends observed with all other strains). Substantial L-phenylalanine accumulation was not observed until 17 h post-induction, after which time styrene titers then also rose considerably, reaching a final titer of 260 mg/L in the culture medium 29 h post-induction. The final styrene titers for the strains NST74 pSpalAv pTfdc1Sc, NST74 pSpalNp pTfdc1Sc, and NST74 pSpal1At pTfdc1Sc were found to be 210, 183, and 188 mg/L, respectively. As styrene-producing strains were cultured, a strong 'hydrocarbon' aroma was readily detected upon opening the sealed culture bottles. The high volatility of styrene necessitated the use of sealed jars for culturing. To ensure that sufficient oxygen remained available to the culture, a large headspace volume (200 mL headspace vs. 50 mL media) was used. Thus, cultures were maintained under aerobic conditions throughout the culture (as confirmed by the inclusion of the indicator dye resazurin). However, the use of such a large headspace volume also allows for the significant accumulation of styrene vapor, which should also be accounted for when assessing productivity. Under dilute and near-ambient conditions, the equilibrated headspace vapor composition can be estimated by application of Henry's Law. Using a dimensionless Henry's Law constant of 0.113 (Yang, 1992), it can be predicted that a headspace equilibrated with an aqueous phase containing 260 mg/L styrene will contain an additional 29 mg/L styrene, for a total of 18.9 mg in the 250 mL flask (note this would be equivalent to an aqueous titer of 376 mg/L if no volatilization had occurred). In an analogous manner, the final styrene production by strains NST74 pSpalAv pTfdc1Sc, NST74 pSpalNp pTfdc1Sc, and NST74 pSpal1At pTfdc1Sc would be estimated as 15.2, 13.4, and 13.6 mg, respectively.

Table 3
Biosynthesis of L-phenylalanine, *trans*-cinnamic acid, and styrene by *E. coli* NST74 pSpal2At pTfdc1Sc from glucose in MM1 media. Errors reported at one standard deviation from triplicate experiments.

Time (h)	L-Phenylalanine (mg/L)	*trans*-Cinnamic acid (mg/L)	Styrene (mg/L)
0	13.9 ± 0.03	0	0
13	56.4 ± 0.13	3.7 ± 0.01	7.3 ± 0.12
17	236.5 ± 0.57	6.3 ± 0.01	23 ± 0.38
21	167.4 ± 0.40	9.1 ± 0.02	205 ± 3.40
25	152.2 ± 0.36	13.5 ± 0.02	243 ± 4.03
29	179.5 ± 0.43	14.0 ± 0.03	260 ± 4.31

Following these fermentation studies, culture supernatants were extracted using hexane, and the extracts were analyzed by GC–MS to confirm that it was in fact styrene that was being synthesized by our cultures. Comparing the spectra of the dominant metabolite peak which was recovered from the extract to the NIST08 spectral database (Babushok et al., 2007), styrene was found to be the most probable compound. As can be seen from Fig. 5, the mass spectra of the extracted sample and the library reference provide an excellent match, providing confirmation that styrene was in fact synthesized by our engineered strains of *E. coli*.

Throughout the duration of culture, never was all of the L-phenylalanine observed to be fully assimilated into the styrene pathway (as seen in Table 3). Furthermore, tCA titers were observed to remain low throughout, indicating that almost all of the synthesized tCA could be quickly converted to styrene. Taken together, these observations suggest that low PAL activity presently remains as the flux limiting condition in the engineered styrene biosynthesis pathway. As the first committed step in the styrene pathway, high PAL activity is essential and must be

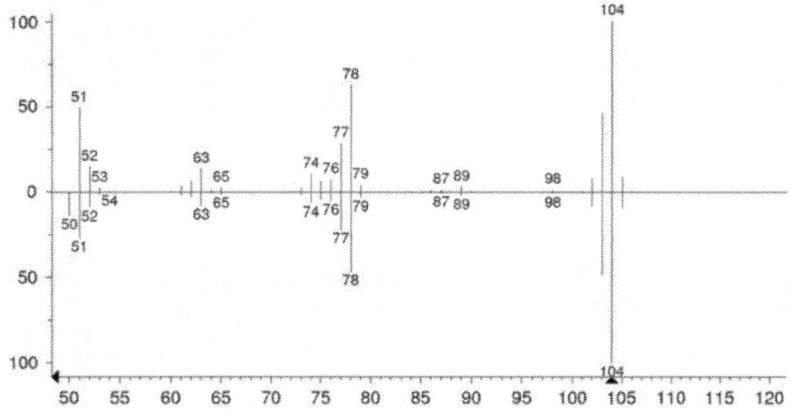

Fig. 5. Head to tail comparison of the standard mass spectra showing the relative abundance of the mass-to-charge ratio of styrene from the NIST08 library (lower) with that of the dominant metabolite peak obtained in hexane extractions of the culture broth (upper).

Please cite this article as: McKenna, R., Nielsen, D.R., Styrene biosynthesis from glucose by engineered *E. coli*. Metab. Eng. (2011), doi:10.1016/j.ymben.2011.06.005

improved in subsequent generations of our strains. To start, the expression of PAL can likely be improved through the use of codon optimized variants. This approach has shown to be useful, particularly in cases for which pathways involve the expression of plant-derived genes in *E. coli*. For example, codon optimization was applied to the amorphadiene oxidase from *Artemisia annua* (a plant species) as part as a synthetic pathway to produce artemisinic acid (Keasling, 2008). Whereas the native protein originally showed neither *in vivo* nor *in vitro* activity when expressed in *E. coli*, a codon optimized variant resulted in its functional expression in *E. coli*, contributing to successful pathway development.

Whereas improvements in PAL expression can lead to enhancements in the specific metabolite flux, the net flux can also be improved by increasing the pathway 'driving force'. That is, by promoting the increased availability of the pathway's immediate precursor, L-phenylalanine. This notion was tested in resting cells assays wherein *E. coli* NST74 pSpal2At pTfdc1Sc cell suspensions were supplemented with exogenous L-phenylalanine at initial concentrations of either 400 or 950 mg/L. As shown in Fig. 6, the conversion of 400 mg/L L-phenylalanine to styrene occurred rapidly and completely (i.e., no residual L-phenylalanine or tCA was detectable), yielding a final styrene titer of 250 mg/L after 30 h. Meanwhile, when 950 mg/L L-phenylalanine was added to the resting cell cultures, nearly 500 mg/L styrene could be produced as all of the L-phenylalanine was consumed. However, in this case, nearly an additional 250 mg/L tCA also remained in the culture. Noting that such a final styrene titer actually surpasses the toxicity threshold of growing cells, it is not altogether surprising that the reaction was unable to proceed to completion. It is speculated that the final styrene titer was only capable of surpassing the previously-determined toxicity threshold in resting cell studies because the biotransformation in this case was no longer growth dependent. That is to say that the stationary phase resting cells had previously assembled the necessary pathway enzymes and were not reliant upon endogenously synthesized L-phenylalanine as a precursor. Thus, the impediment of conversion observed at further elevated styrene titers most likely represents an inhibition imposed at the level of the pathway enzymes rather than an impairment of cell growth, as was observed in our previous characterization experiments. Further experiments will be required to characterize and

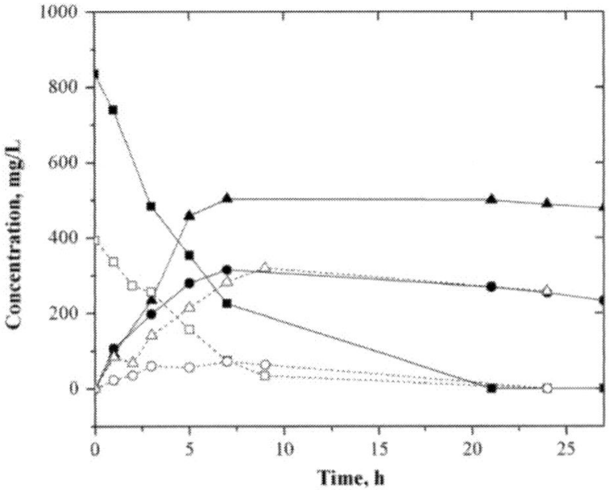

Fig. 6. Whole cell production of styrene (triangle) from *trans*-cinnamic acid (circle) after exogenous L-phenylalanine (square) addition by *E. coli* NST74 pSpal2At pTfdc1Sc. Open and filled shapes correspond to the initial addition of 400 and 950 mg/L L-phenylalanine.

understand this potential inhibitory mechanism. These results demonstrate that if the L-phenylalanine precursor pool can be enhanced the net production of styrene can also be increased. Clearly, however, the addition of exogenous L-phenylalanine is not a sustainable approach towards enhancing styrene biosynthesis, but rather endogenous L-phenylalanine production must be enhanced in our host platform. Subsequent studies will explore the incorporation of additional modifications into the genome of the *E. coli* NST74 host that have previously been shown by other groups to result in L-phenylalanine yield enhancements. For instance, it has been reported that the over-expression of endogenous transketolase I (encoded by *tktA*) supports elevated L-phenylalanine biosynthesis (Gosset et al., 1996). TktA over-expression enables the enhanced biosynthesis of erythrose 4-phosphate (E4P) which, when condensed with phosphoenolpyruvate (PEP), yields 3-deoxy-D-arabino-heptulosonate 7-phosphate (DAHP), the first committed intermediate in the aromatic amino acid biosynthesis pathway (Flores et al., 1996; Lütke-Eversloh and Stephanopoulos, 2008). Meanwhile, disruption of the global carbon storage regulatory system of *E. coli* through deletion of *csrA*, increases PEP biosynthesis, leading to up to a 2-fold enhancement in L-phenylalanine yield (Tatarko and Romeo, 2001).

For microbially derived, renewable styrene to become an economically viable and sustainable alternative to petroleum-derived predecessor, titers and productivity must ultimately be improved. Although the styrene titers achieved by our 1st generation strains were modest, in comparison to the toxicity assays performed above it can be seen that they are already approaching the inhibitory threshold. Thus, whereas increasing L-phenylalanine yields could eventually translate into elevated styrene production, product toxicity would soon become the subsequent limiting factor and must be addressed. As complex phenotypes such as solvent tolerance are simply not monogenic in nature (Alper et al., 2006), numerous specific mutations must applied in concert to achieve the specifically desired result. Since little is presently known of styrene toxicity or tolerance, rational approaches to engineering more tolerant strains may be less suitable than higher throughput, combinatorial strategies. Although commonly employed, combinatorial approaches towards enhancing desired phenotypes often rely upon the aggressive use of chemomutagenesis, such techniques are less desirable as they can also result an unforeseen and difficult to understand negative impacts on host fitness and/or productivity (Bonomo et al., 2006), in addition to requiring laborious screening and selection procedures. A more effective approach for the present application might involve the use of an alternative host platform which inherently boasts greater tolerance (Fischer et al., 2008), a strategy which has been applied for the production of biofuels, such as n-butanol (Nielsen et al., 2009). The bacterium *P. putida* S12 has been engineered, for example, as a solvent tolerant platform for the biosynthesis of both *p*-hydroxybenzoate and *p*-hydroxystyrene (Verhoef et al., 2007, 2009), and might also make excellent starting point for styrene production. Since our pathway is derived from the ubiquitous, proteinogenic amino acid L-phenylalanine, its transference to an alternative host platform remains wholly compatible. Alternatively, improved product tolerance can also be achieved through genome evolution, as accomplished through serial adaptations to increasing styrene concentrations. Such an approach has worked well for achieving tolerance to biofuels like n-butanol and could also be coupled with genomic library screens to provide the first comprehensive view of styrene inhibition and tolerance in *E. coli* (Reyes et al., 2011). In an interesting example, prior works have shown that *Pseudomonas* sp. may be adapted to styrene when also grown in the presence of acetate or similar carboxylic acids as the sole

Please cite this article as: McKenna, R., Nielsen, D.R., Styrene biosynthesis from glucose by engineered *E. coli*. Metab. Eng. (2011), doi:10.1016/j.ymben.2011.06.005

carbon source (Weber et al., 1993). It has been suggested that this evolved phenotype may be associated with genetic changes leading to reductions in membrane fluidity, as well as through the activation of genes believed to be specifically associated with enhanced tolerance to aromatic compounds. As an alternative strategy, combinatorial procedures such as the global Transcription Machinery Engineering (gTME) (Alper et al., 2006; Nicolaou et al., 2010) may be implemented to improve styrene tolerance in *E. coli* or alternative host platforms selected due to their elevated tolerance baseline. The principals of gTME have been successfully employed to enhance ethanol tolerance in both yeast (Alper et al., 2006) and *E. coli* (Alper and Stephanopoulos, 2007).

In addition to the development of styrene tolerant phenotypes, product toxicity can also be enhanced through the use of in situ product recovery (ISPR). Common ISPR approaches involve solvent extraction (Gyamerah and Glover, 1996; Malinowski, 2001; Weilnhammer and Blass, 1994), adsorption (Nielsen et al., 2010; Nielsen and Prather, 2009), gas and vacuum stripping (Loser et al., 2005), and membrane pervaporation (Vane, 2005). Each of these approaches has been successfully applied for the continuous recovery of biofuel compounds such as ethanol, for example, and even aromatics like L-phenylacetylcarbinol (Khan and Daugulis, 2010) and benzaldehyde (Jain et al., 2010). The latter three approaches may be particularly well-suited for styrene recovery given its volatile nature, as previously discussed.

4. Conclusion

For the first time, the present study has demonstrated the biosynthesis of styrene from renewable resources using an engineered microbial platform. Whereas low activity of pathway enzymes and product toxicity remain as challenges which limit the productivity of our styrene-producing strains, continued improvements will lead to the development of robust biocatalysts for the sustainable production of this important, large volume, commodity chemical.

Acknowledgments

We thank Prof. Bradley Moore (UCSD) and Prof. Kristala Prather (MIT) for their kind gifts. We also thank Tom Colella (ASU) for his skillful technical assistance in GC–MS analysis.

References

Alper, H., Moxley, J., Nevoigt, E., Fink, G.R., Stephanopoulos, G., 2006. Engineering yeast transcription machinery for improved ethanol tolerance and production. Science 314, 1565–1568.

Alper, H., Stephanopoulos, G., 2007. Global transcription machinery engineering: a new approach for improving cellular phenotype. Metabolic Engineering 9, 258–267.

Babushok, V.I., Linstrom, P.J., Reed, J.J., Zenkevich, I.G., Brown, R.L., Mallard, W.G., Stein, S.E., 2007. Development of a database of gas chromatographic retention properties of organic compounds. Journal of Chromatography A. 1157, 414–421.

Baez-Viveros, J.L., Osuna, J., Hernandez-Chavez, G., Soberon, X., Bolivar, F., Gosset, G., 2004. Metabolic engineering and protein directed evolution increase the yield of L-phenylalanine synthesized from glucose in *Escherichia coli*. Biotechnology and Bioengineering 87, 516–524.

Bonomo, J., Warnecke, T., Hume, P., Marizcurrena, A., Gill, R.T., 2006. A comparative study of metabolic engineering anti-metabolite tolerance in *Escherichia coli*. Metabolic Engineering. 8, 227–239.

Boudet, A.M., 2007. Evolution and current status of research in phenolic compounds. Phytochemistry 68, 2722–2735.

Clark, G., 1990. Phenethyl alcohol. Perfumer Flavor 15, 37–44.

Clausen, M., Lamb, C.J., Megnet, R., Doerner, P.W., 1994. PAD1 encodes phenylacrylic acid decarboxylase which confers resistance to cinnamic acid in *Saccharomyces cerevisiae*. Gene 142, 107–112.

Cochrane, F.C., Davin, L.B., Lewis, N.G., 2004. The Arabidopsis phenylalanine ammonia lyase gene family: kinetic characterization of the four PAL isoforms. Phytochemistry 65, 1557–1564.

Cui, J.D., Jia, S.R., Sun, A.Y., 2008. Influence of amino acids, organic solvents and surfactants for phenylalanine ammonia lyase activity in recombinant *Escherichia coli*. Letters in Applied Microbiology 46, 631–635.

DoE, Steam system opportunity assessment for the pulp and paper, chemical manufacturing, and petroleum refining industries. In: Energy, O. o. E. E. a. R. E. U. S. D. o. (Ed.), 2002.

Fernandez, X., Lizzani-Cuvelier, L., Loiseau, A.M., Perichet, C., Delbecque, C., Arnaudo, J.F., 2005. Chemical composition of the essential oils from Turkish and Honduras Styrax. Flavour and Fragrance Journal 20, 70–73.

Fischer, C.R., Klein-Marcuschamer, D., Stephanopoulos, G., 2008. Selection and optimization of microbial hosts for biofuels production. Metabolic Engineering 10, 295–304.

Flores, N., Xiao, J., Berry, A., Bolivar, F., Valle, F., 1996. Pathway engineering for the production of aromatic compounds in *Escherichia coli*. Nature Biotechnology 14, 620–623.

Gilbert, H.J., Clarke, I.N., Gibson, R.K., Stephenson, J.R., Tully, M., 1985. Molecular-cloning of the phenylalanine ammonia-lyase gene from rhodosporidium-toruloides in *Escherichia-coli* K-12. Journal of Bacteriology 161, 314–320.

Gilbert, H.J., Stephenson, J.R., Tully, M., 1983. Control of synthesis of functional messenger-rna coding for phenylalanine ammonia-lyase from rhodosporidium-toruloides. Journal of Bacteriology 153, 1147–1154.

Gilbert, H.J., Tully, M., 1982. Synthesis and degradation of phenylalanine ammonia-lyase of rhodosporidium-toruloides. Journal of Bacteriology 150, 498–505.

Gosset, G., Yong-Xiao, J., Berry, A., 1996. A direct comparison of approaches for increasing carbon flow to aromatic biosynthesis in *Escherichia coli*. Journal of Industrial Microbiology 17, 47–52.

Gyamerah, M., Glover, J., 1996. Production of ethanol by continuous fermentation and liquid-liquid extraction. Journal of Chemical Technology and Biotechnology 66, 145–152.

Heipieper, H.J., Weber, F.J., Sikkema, J., Keweloh, H., de Bont, J.A.M., 1994. Mechanisms of resistance of whole cells to toxic organic solvents. Trends in Biotechnology 12, 409–415.

Jain, A.N., Khan, T.R., Daugulis, A.J., 2010. Bioproduction of benzaldehyde in a solid–liquid two-phase partitioning bioreactor using *Pichia pastoris*. Biotechnology Letters 32, 1649–1654.

Jiang, H., Wood, K.V., Morgan, J.A., 2005. Metabolic engineering of the phenylpropanoid pathway in *Saccharomyces cerevisiae*. Applied and Environmental Microbiology 71, 2962–2969.

Jones, D.H., 1984. Phenylalanine ammonia-lyase: regulation of its induction, and its role in plant development. Phytochemistry 23, 1349–1359.

Keasling, J.D., 2008. Synthetic biology for synthetic chemistry. ACS Chemical Biology 3, 64–76.

Khan, T.R., Daugulis, A.J., 2010. Application of solid–liquid TPPBs to the production of L-phenylacetylcarbinol from benzaldehyde using *Candida utilis*. Biotechnology and Bioengineering 107, 633–641.

Larsson, S., Nilvebrant, N.O., Jonsson, L.J., 2001. Effect of overexpression of *Saccharomyces cerevisiae* Pad1p on the resistance to phenylacrylic acids and lignocellulose hydrolysates under aerobic and oxygen-limited conditions. Applied Microbiology and Biotechnology 57, 167–174.

Loser, C., Schroder, A., Deponte, S., Bley, T., 2005. Balancing the ethanol formation in continuous bioreactors with ethanol stripping. Engineering in Life Sciences 5, 325–332.

Lütke-Eversloh, T., Stephanopoulos, G., 2008. Combinatorial pathway analysis for improved L-tyrosine production in *Escherichia coli*: identification of enzymatic bottlenecks by systematic gene overexpression. Metabolic Engineering 10, 69–77.

Malinowski, J.J., 2001. Two-phase partitioning bioreactors in fermentation technology. Biotechnology Advances 19, 525–538.

Moffitt, M.C., Louie, G.V., Bowman, M.E., Pence, J., Noel, J.P., Moore, B.S., 2007a. Discovery of two cyanobacterial phenylalanine ammonia lyases: kinetic and structural characterization. Biochemistry 46, 1004–1012.

Moffitt, M.C., Louie, G.V., Bowman, M.E., Pence, J., Noel, J.P., Moore, B.S., 2007b. Discovery of two cyanobacterial phenylalanine ammonia lyases: kinetic and structural characterization. Biochemistry 46, 1004–1012.

Mukai, N., Masaki, K., Fujii, T., Kawamukai, M., Iefuji, H., 2010. PAD1 and FDC1 are essential for the decarboxylation of phenylacrylic acids in *Saccharomyces cerevisiae*. Journal of Bioscience and Bioengineering 109, 564–569.

Nicolaou, S.A., Gaida, S.M., Papoutsakis, E.T., 2010. A comparative view of metabolite and substrate stress and tolerance in microbial bioprocessing: from biofuels and chemicals, to biocatalysis and bioremediation. Metabolic Engineering 12, 307–331.

Nielsen, D.R., Amarasiriwardena, G.S., Prather, K.L.J., 2010. Predicting the adsorption of second generation biofuels by polymeric resins with applications for in situ product recovery (ISPR). Bioresource Technology 101, 2762–2769.

Nielsen, D.R., Leonard, E., Yoon, S.H., Tseng, H.C., Yuan, C., Prather, K.L., 2009. Engineering alternative butanol production platforms in heterologous bacteria. Metabolic Engineering.

Nielsen, D.R., Prather, K.J., 2009. In situ product recovery of n-butanol using polymeric resins. Biotechnology and Bioengineering 102, 811–821.

Pagot, Y., Belin, J.M., Husson, F., Spinnler, H.E., 2007. Metabolism of phenylalanine and biosynthesis of styrene in *Penicillium camemberti*. Journal of Dairy Research 74, 180–185.

Please cite this article as: McKenna, R., Nielsen, D.R., Styrene biosynthesis from glucose by engineered *E. coli*. Metab. Eng. (2011), doi:10.1016/j.ymben.2011.06.005

Qi, W.W., Vannelli, T., Breinig, S., Ben-Bassat, A., Gatenby, A.A., Haynie, S.L., Sariaslani, F.S., 2007. Functional expression of prokaryotic and eukaryotic genes in *Escherichia coli* for conversion of glucose to p-hydroxystyrene. Metabolic Engineering 9, 268–276.
Ramos, J.L., Duque, E., Gallegos, M.-T., Godoy, P., Ramos-González, M.I., Rojas, A., Terán, W., Segura, A., 2002. Mechanisms of solvent tolerance in gram-negative bacteria. Annual Review of Microbiology 56, 743–768.
Reyes, L.H., Almario, M.P., Kao, K.C., 2011. Genomic library screens for genes involved in n-butanol tolerance in *Escherichia coli*. PLoS ONE 6, e17678.
Santos, C.N.S., Koffas, M., Stephanopoulos, G., 2011. Optimization of a heterologous pathway for the production of flavonoids from glucose. Metabolic Engineering 13, 392–400.
SRI, 2010. Styrene. Access Intelligence LLC Inc.
Tatarko, M., Romeo, T., 2001. Disruption of a global regulatory gene to enhance central carbon flux into phenylalanine biosynthesis in *Escherichia coli*. Current Microbiology 43, 26–32.
Tran, N.P., Gury, J., Dartois, V., Nguyen, T.K.C., Seraut, H., Barthelmebs, L., Gervais, P., Cavin, J.-F., 2008. Phenolic acid-mediated regulation of the padC gene, encoding the phenolic acid decarboxylase of *Bacillus subtilis*. Journal of Bacteriology 190, 3213–3224.
Vane, L.M., 2005. A review of pervaporation for product recovery from biomass fermentation processes. Journal of Chemical Technology and Biotechnology 80, 603–629.
Vannelli, T., Wei, Qi, W., Sweigard, J., Gatenby, A.A., Sariaslani, F.S., 2007a. Production of p-hydroxycinnamic acid from glucose in *Saccharomyces cerevisiae* and *Escherichia coli* by expression of heterologous genes from plants and fungi. Metabolic Engineering 9, 142–151.
Vannelli, T., Xue, Z.X., Breinig, S., Qi, W.W., Sariaslani, F.S., 2007b. Functional expression in *Escherichia coli* of the tyrosine-inducible tyrosine ammonia-lyase enzyme from yeast Trichosporon cutaneum for production of p-hydroxycinnamic acid. Enzyme and Microbial Technology 41, 413–422.
Verhoef, S., Ruijssenaars, H.J., de Bont, J.A., Wery, J., 2007. Bioproduction of p-hydroxybenzoate from renewable feedstock by solvent-tolerant *Pseudomonas putida* S12. Journal of Biotechnology 132, 49–56.
Verhoef, S., Wierckx, N., Westerhof, R.G., de Winde, J.H., Ruijssenaars, H.J., 2009. Bioproduction of p-hydroxystyrene from glucose by the solvent-tolerant bacterium *Pseudomonas putida* S12 in a two-phase water-decanol fermentation. Applied and Environmental Microbiology 75, 931–936.
Weber, F.J., Ooijkaas, L.P., Schemen, R.M., Hartmans, S., de Bont, J.A., 1993. Adaptation of *Pseudomonas putida* S12 to high concentrations of styrene and other organic solvents. Applied and Environmental Microbiology 59, 3502–3504.
Weilnhammer, C., Blass, E., 1994. Continuous fermentation with product recovery by in-situ extraction. Chemical Engineering & Technology 17, 365–373.
Wierckx, N.J., Ballerstedt, H., de Bont, J.A., Wery, J., 2005. Engineering of solvent-tolerant *Pseudomonas putida* S12 for bioproduction of phenol from glucose. Applied and Environmental Microbiology 71, 8221–8227.
Wu, C., Koylinski, T., Bozik, J., 1981. Preparation of styrene from ethylbenzene US Patent 4,255,599.
Xiang, L.K., Moore, B.S., 2002. Inactivation, complementation, and heterologous expression of encP, a novel bacterial phenylalanine ammonia-lyase gene. Journal of Biological Chemistry 277, 32505–32509.
Xiang, L.K., Moore, B.S., 2005. Biochemical characterization of a prokaryotic phenylalanine ammonia lyase. Journal of Bacteriology 187, 4286–4289.
Xiang, L.K., Moore, B.S., 2006. Biochemical characterization of a prokaryotic phenylalanine ammonia lyase (187, pg4286, 2005). Journal of Bacteriology 188, 5331.
Yang, C.L.Y. a.H.C., 1992. Henry's Law Constant for Compound in Water. Gulf Publishing Company, Houston.
Young, M.R., Towers, G.H.N., Neish, A.C., 1966. Taxonomic distribution of ammonia-lyases for L-phenylalanine and L-tyrosine in relation to lignification. Canadian Journal of Botany 44, 341.

Please cite this article as: McKenna, R., Nielsen, D.R., Styrene biosynthesis from glucose by engineered *E. coli*. Metab. Eng. (2011), doi:10.1016/j.ymben.2011.06.005

(349g) Reaction-Transport Model for the Pyrolysis of Shrinking Cellulose Particles

Yenhan Lin, Joungmo Cho, George W. Huber and Jeffrey M. Davis, Department of Chemical Engineering, University of Massachusetts Amherst, Amherst, MA

A reaction-transport model is developed for the pyrolysis of non-porous cellulose particles. The model includes recently measured kinetic parameters, intra- and extra-particle heat transfer, and particle shrinkage as gaseous products are formed. The intra-particle temperature gradient, which gives rise to spatially-varying reaction rates, is mainly influenced by the initial particle size, the external temperature, and the heat transfer coefficient. Numerical predictions for measurable properties, including the temporal evolution of the residual mass and the final char yield, are validated through comparison to data from experiments for various particle sizes, and excellent agreement is found. Predicted char yields and effectiveness factors are presented for a wide range of parameters, and operating conditions are identified under which the process is transport vs. reaction controlled due to the limitations of intra-particle heat transport.

See more of this Session: Biomass Pyrolysis I
See more of this Group/Topical: Fuels and Petrochemicals Division

(583e) Hydrolysate De-Acidification Using Resin-Wafer Electrodeionization

Yupo J. Lin, Saurav Datta, Michael Henry, Anthony Fracaro and Seth Snyder, Argonne National Laboratory, Argonne, IL

Dilute acid pretreatment of lignocellulosic feedstock offers an economic route for the extraction of fermentable sugars from non-food biomass for biofuel production by bioconversion. However, removal of acids from the hydrolysate is a critical step to ensure optimal conditions for the subsequent bioprocessing/conversion steps, i.e., enzyme hydrolysis of the cellulose to release C6 sugars and for the final fermentation to produce biofuel. The removal (conditioning step) of inorganic and organic acids (released together with C5 sugar molecules from the hemicellulose) after the acid treatment stage poses a great economic challenge to the overall biofuel production cost. To address this challenge, we used a proprietary resin wafer electrodeionization (RW-EDI) technology to remove both the inorganic and organic acids in a single step. Preliminary process economic analysis indicates that the easily scalable and small foot-print unit operation of RW-EDI can offer an economic and viable means of addressing the hydrolysate conditioning challenge. Selective removal of inorganic acids and organic acids was demonstrated when appropriate operation conditions were employed for the RW-EDI operation. This unique feature makes possible the separate recovery of both inorganic acid (e.g., sulfuric acid) and organic acids, which act as fermentation inhibitors, so that the inorganic acid can be recycled to the dilute acid pretreatment stage. The RW-EDI also provides an option to remove other fermentation inhibitor such the furfural and hydroxy methyl furfural (HMF) during the hydrolysate conditioning. In this presentation, process design and performance of hydrolysate de-acidification using RW-EDI will be discussed. Pilot-scale demonstration of technology viability using synthetic hydrolysate will be also presented along with preliminary assessments of economic and process benefits.

See more of this Session: Developments In the Pretreatment of Lignocellulosics for Bioconversion III
See more of this Group/Topical: Sustainable Engineering Forum

(349f) High Pressure Fast-Pyrolysis and Fast-Hydropyrolysis for Conversion of Biomass to Liquid Fuels

Vinod Kumar Venkatakrishnan, Andrew D. Smeltz, Sergey V. Semikolenov, Fabio H. Ribeiro, W. Nicholas Delgass and Rakesh Agrawal, School of Chemical Engineering, Purdue University, West Lafayette, IN

The transportation sector has been heavily dependent on the use of liquid fuels produced from fossil-based petroleum sources. In the context of a petroleum-deprived future, it is imperative to look for sustainable carbon sources in conjunction with more efficient process pathways for producing high-energy-density liquid fuels. Sustainably available (SA) biomass comprised of crop and forest residues, agriculture and municipal waste, etc. is one such carbon source for producing liquid fuels to meet the large demand for transportation. Among the traditional routes for biomass conversion to liquid fuel, the fast-pyrolysis process, involving rapid heating of biomass at near atmospheric pressure to temperatures of ~500-550 °C, is known to produce a high oxygen content, acidic liquid product with a similar energy content as the feed biomass and requiring further processing under severe conditions to be upgraded to transportation grade fuel. As an alternative to this two step approach, we look at a novel *H_2Bioil* [1] process based on continuous fast-hydropyrolysis, which includes fast-pyrolysis in a moderate to high-pressure (upto 50 bar) hydrogen environment to produce hydropyrolysis vapors which are upgraded in the vapor phase by catalytic hydrodeoxygenation and quenched to form a high-energy-density, deoxygenated liquid product that can supplement petroleum-based liquid fuels or potentially be used directly as a fuel.

In this presentation we summarize our reactor design and on-going process development to achieve high-pressure fast-pyrolysis and fast-hydropyrolysis of cellulose and lignocellulosic materials. Furthermore, the role played by hydrogen in fast-hydropyrolysis will be briefly discussed, by comparison with high-pressure inert gas fast-pyrolysis experiments with model and real biomass feedstocks. In addition to basic oil analysis techniques such as elemental analysis, Karl-Fischer, and TAN, we have utilized atmospheric-pressure, chemical-ionization mass spectrometry (APCI-MS) to identify and quantitatively compare major product species present in the liquid product. Investigation of the effect of catalysts on the hydrodeoxygenation process will also be presented.

[1]Agrawal R, Singh NR. Synergistic Routes to Liquid Fuel for a Petroleum-Deprived Future, *AIChE J.* 2009; 55: 1898-1905

See more of this Session: Biomass Pyrolysis I
See more of this Group/Topical: Fuels and Petrochemicals Division

Phase behavior of asphaltenes + organic diluents

Kasra Nikooyeh and John M. Shaw
Department of Chemical and Materials Engineering
University of Alberta
Corresponding Author: John M. Shaw (jmshaw@ualberta.ca)

Asphaltenes, defined operationally on the basis of ASTM standards that include mixing and macroscopic filtration steps, are typically asserted to be soluble in toluene [1]. A similar definition is used to classify organic diluents as solvents and non-solvent for asphaltenes. In the analysis of phase separation mechanisms of asphaltenes from hydrocarbon media, the assumption of solubility in the original medium is a key one. From a solution thermodynamics standpoint, dissolution of a solid into a liquid includes the enthalpy of transition to liquid for the solute plus the enthalpy associated with homogeneous mixing at a molecular level [2]. The present work challenges the application of the concept of solubility to mixtures of asphaltenes with organic diluents and suggests an alternative explanation for the behavior of asphaltenes in organic liquids based on thermodynamic measurements.

Solution calorimetry and differential scanning calorimetry (DSC) were performed on mixtures of asphaltenes with a variety of diluents ranging from n-alkanes to bicyclic aromatics. Athabasca n-pentane asphaltenes were mixed with diluents at compositions ranging from 0.1 to 10 wt%. In each case 25 mg of asphaltenes were added to 25 ml of a mixture with an underlying composition. Solution calorimetry measurements performed using a TA Instruments TAM Precision solution calorimeter, in the range 15 to 80 °C provided information on phase change with composition at fixed temperature. DSC measurements with a TG-DSC 111 (Setaram, France) calorimeter, provided information on phase change with temperature at fixed composition and on the reversibility of such phase behavior.

These solution calorimetry measurements reflect the net enthalpy change resulting from addition of asphaltenes to diluents. The value of enthalpy of solution is a weak function of temperature and composition if dissolution is complete, i.e.: the solute undergoes a solid-liquid phase change. This behavior was demonstrated for the benchmark case pyrene (solute) + toluene (diluent). The enthalpies measured for asphaltenes in all diluents used in this study, both integral [ΔH^{sol}] and differential, [d ΔH^{sol}/dx] measurements, however, change significantly as temperature and composition are varied.

The enthalpies trend to higher values with increasing temperature and to lower values at increasing asphaltene concentration at fixed temperature on a per unit mass of asphaltene basis as shown in Figure 1. These results are in agreement with the hypothesis that only a fraction of asphaltenes undergo a solid-liquid phase change upon mixing with diluents and that fraction is a function of diluent properties, global composition and temperature.

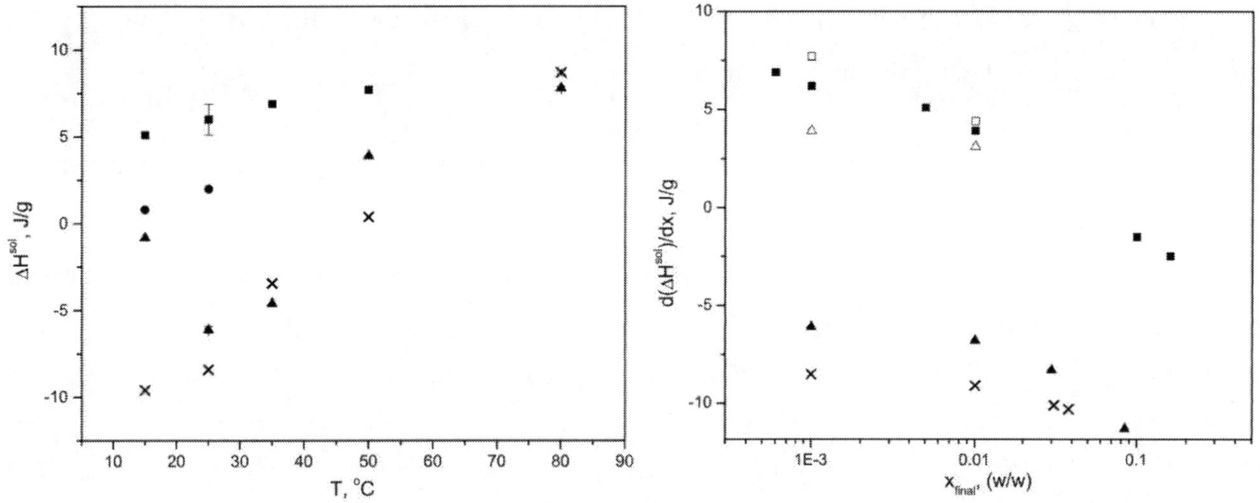

Figure 1 a) Enthalpy of dispersion of Athabasca C_5 Asphaltenes at 0.1 wt % in toluene (■), tetrahydrofuran (•), 1-methylnaphthalene (▲) and quinoline (×). b) Enthalpy of dispersion of Athabasca C_5 Asphaltenes in toluene at 25 °C(■) and 50 °C (□) and 1-mthylnaphtalene at 25 °C(▲) and 50 °C (Δ) quinoline (×) as a function of concentration

Differential scanning calorimetry measurements for asphaltene + toluene and 1-methylnaphthalene mixtures allow for determination of the enthalpy of solution at higher concentrations. The results with toluene, an example of which is plotted in figure 2, show that mixing asphaltenes with toluene at high concentrations (5 wt% and above) is largely exothermic, a result consistent with the solution calorimetry data.

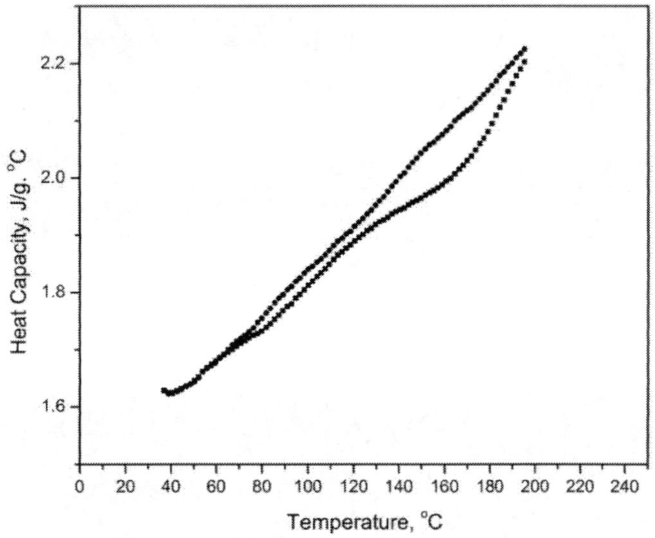

Figure 2. Apparent heat capacity of C_7 asphaltenes added to toluene, derived from differential scanning calorimetry measurements. The heating rate was set to 5 °C/min and the composition range was 5 to 20 wt%. First run (■), second run, (•)

The apparent heat capacities for asphaltenes derived from the DSC thermograms for toluene + asphaltene mixtures also illustrate that the impact of mixing on phase behavior is irreversible, in the temperature interval ~120 to ~200 °C as shown in Figure 2 where the values differ significantly between repeated runs with the same sample.

Over all, these results are inconsistent with the application of dissolution thermodynamics concepts, as classically defined, to asphaltene + diluent mixtures. The ranges of values, and their trends with temperature and composition are consistent with overlapping exothermic and endothermic processes occurring simultaneously that involve only a fraction of the asphaltenes.

References

1. ASTM-D6560, *Standard Test Method for Determination of Asphaltenes (Heptane Insolubles) in Crude Petroleum and Petroleum Products*. 2000, ASTM International: West Conshohocken, PA.
2. Zhang, Y., et al., *Estimating the Interaction Energy of Asphaltene Aggregates with Aromatic Solvents*. Energy Fuels, 2005. **19**(3): p. 1023-1028.

(573c) Ab-Initio Study of Glycerol Dehydration Mechanisms with Explicit Solvent Treatment

Tim Courtney[1], Giannis Mpourmpakis[2], Jingguang G. Chen[1], Dion G. Vlachos[1] and Stavros Caratzoulas[1], (1)Catalysis Center for Energy Innovation, University of Delaware, Newark, DE, (2) Institute of Electronic Structure and Laser, FORTH, Heraklion 71110, Crete, Greece, Crete, Greece

Glycerol produced as a waste product in biodiesel production has interesting potential as a low-cost biomass-derived feedstock for chemical processing. A process of current academic and commercial interest is the acid-catalyzed dehydration of glycerol to acrolein, a valuable platform chemical. Such a process would improve the economic viability of biodiesel production while also producing acrolein from renewable resources rather than from propylene – the current commercial process. Despite a breadth of study on catalysts for this chemistry, relatively little has been done to investigate the reaction mechanism at a fundamental level.

We investigated the acid-catalyzed dehydration mechanism of glycerol to acrolein using *ab-initio* methods. Solvent effects were accounted for using an implicit solvent model and a number of explicit water molecules. We show that the reaction proceeds first through a hydride transfer (dehydration) inhibited by the solvent (water molecules) and forms an aldol intermediate, which tautomerizes to the unsaturated diol via a solvent mediated proton transfer. Though the latter is easily reversible to the lower energy aldol, it can also form acrolein as a final product. We shall argue that the mechanism favors the formation of acrolein through the diol, in stark contrast to the mechanisms previously proposed in the literature, which form acrolein through the aldol. Gas phase calculations in the literature have suggested that production of acetaldehyde and formaldehyde are kinetically favored over production of acrolein. However, our study clearly demonstrates that systems with explicit water molecules correctly predict acrolein selectivity, in agreement with experiments. These insights provide leverage for the rational design of more effective catalysts to promote this chemistry.

See more of this Session: Catalytic Biomass Conversion to Chemicals II
See more of this Group/Topical: Fuels and Petrochemicals Division

(364d) Modeling and Scale up of CO2-Water Pretreatment of Guayule Biomass

Ehsan Moharreri and J Richard Elliott, Chemical and Biomolecular Engineering, The University of Akron, Akron, OH

A carbon dioxide – water pretreatment process was developed for guayule biomass. The guayule biomass was mixed with water and then loaded to a reactor where it was exposed to a biphasic environment consisting of CO2 rich vapor phase and water rich liquid phase. Then, the biomass was enzymatically hydrolyzed and the sugar content was measured at the end. The effects of temperature, pressure, moisture content, accessible volume and holding time in the pretreatment step were investigated. One hundred and twenty (120) experimental runs were carried out in the temperature range of 145 C to 210 C, pressure was between 3.4 MPa to 34 MPa, the moisture content ratio was between 0.6 to 6, the accessible volume was between 0.2 to 0.1 L and the holding time was between 20 to 50 minutes. A model was fit to the experimental data based on dissolved CO2 in the liquid phase using CO2-H2O phase envelop and a temperature dependant constant. The model fits to one hundred and twenty (120) data points with an average absolute percent deviation(AAD%) of 13. Then, the same process was implemented to a pilot scale vessel with a 1 liter volume. The process was carried out in batch and semi-continuous methods and twenty (20) data points were produced to confirm the model. The model predicted the pilot plant runs with a 15% average absolute percent deviation (AAD%). In the lab scale of 190 C, 26 MPa, moisture ratio of 1.5, accessible volume of 0.22L, biomass load of 15 grams and 35 minutes in holding time resulted in 78% glucose yield and 66% hemicelluloses yield. In the pilot scale of 200 C, 10MPa, moisture ratio of 2, accessible volume of 0.88, biomass load of 60 grams and 35 minutes in holding time resulted in 78% total sugar yield.

See more of this Session: Developments In the Pretreatment of Lignocellulosics for Bioconversion I
See more of this Group/Topical: Sustainable Engineering Forum

HOSE: PRESSURE-ASSISTED EXTRACTION OF BITUMEN FROM OIL SANDS FOR HIGH YIELD AND ENVIRONMETALLY ACCEPTABLE BYPRODUCTS

Xinyue Zhao, Civil & Environmental Engineering, University of Utah, Salt Lake City, UT,
Zhixiong Cha, Civil and Environmental Engineering, University of Utah, Salt Lake City, UT and
Chia-Jung Cheng, Civil & Environ. Eng., University of Utah, Salt Lake City, UT
Andy Hong, Civil & Environ. Eng., University of Utah, Salt Lake City, UT
Willem Duyvesteyn, TaRDI, Technology and Resource Development, Inc., Reno, NV

Abstract

The hot water extraction process has long been used to extract bitumen from oil sands. Chemical additives to the oil sand slurry increase bitumen recovery while making the process water and tailings very difficult to treat and dispose of. A heightened oil sand extraction (HOSE) process used in this study takes advantage of microbubbles formed during pressure cycles to release bitumen from the sand particles. Bitumen recoveries over 90% were achieved using no chemical additives and only a small amount of water with a solid-to-water-volume ratio of 2. Clean spent sands containing 1-3% residual organic components settle rapidly, resulting in >99.9% removal of suspend solids from the water column within 30 minutes with a 90% reduction in turbidity within 2 h. The HOSE process appears promising as an environmentally friendly way to extract bitumen. It also shows promise for the economic processing of lower bitumen grade oil sands.

Introduction

Hot water extraction of bitumen from oil sands has been commercially applied for decades [1]. The extraction process is greatly aided by the addition of caustics and other reagents, which adversely impact handling and disposal of process water and tailings. Recent efforts of increasing the energy efficiency by reducing operation temperature might necessitate longer conditioning time, conditioining intensity and/or additional chemical additives that may require further treatment downstream. In particular, Utah oil sands with high bitumen viscosities are expected to require more intense agitation and caustic wetting agents such as sodium hydroxide. Process water treatment post extraction is another critical issue. A large amount of water is used in the extraction process. While it is desirable to improve water use efficiency by recycling process water, a concern is that the bitumen recovery rate will be adversely affected. The hot water extraction process also creates a large amount of coarse and fine tailings that are quite difficult to dispose of. Huge tailing ponds have been constructed for the fine tailings and they are used to store the process water and certain tailings on site. These tailings contain water, dissolved salts, organics, clays and other minerals, and residual bitumen. While the solids in the fine tailings settle slowly and water evaporates concurrently, it still takes years for certain fine tailings to consolidate [2].

With effective recovery of bitumen from oil sands and minimal adverse impacts of tailings to the environment as important objectives, we have developed a new heightened oil sands extraction (HOSE) process that greatly accelerates the release and collection of bitumen from oil sands, while requiring no intensive agitation, caustics, or other chemical additives that aggravate hydrocarbon contamination of the process water and subsequent treatment of the process water and disposal of the tailings. The HOSE technique involves compression and decompression cycles, delivering a gas and oversaturating with it in the sands/water mixture during compression; creating expanding microbubbles during decompression. This helps to dislodge bitumen from the sands. The new process aims at high bitumen yields without chemical additives that are prone to creating problematic end products.

Materials and Methods

Bitumen was extracted from oil sands in a Parr 5100 low pressure stainless steel reactor. Gas (CO_2, air or mixed of the two) was injected into the reactor charging the headspace pressure to 100 psi. The reactor content was heated by an electrical heater. An entrainment impeller was used to distribute injected gas and provide mechanical stirring up to 800 rpm. A factorial experimental design was used to identify the impact of experimental parameters on bitumen yield, process water quality, and spent sand settling characteristics. Studied parameters include process temperature, pressure, water-to-solid ratio, stirring speed, injecting gas type, and number of pressure cycle numbers with each parameter set at two levels.

Results and Discussion

Heightened Oil Sand Extraction was conducted on a Canadian ol sand sample at varying temperatures, pressures, gas types and numbers of pressure cycles. Figure 1 shows the bitumen recovery rate from this Canadian oil sands according to cycle numbers at various treatment conditions. Higher temperatures enhances the bitumen recovery as it reduces the viscosity of the oil sands. CO_2 gas works better than air in bitumen extraction, possibly due to increased solubility of the CO_2 in the bitumen and its increased compatibility with it.

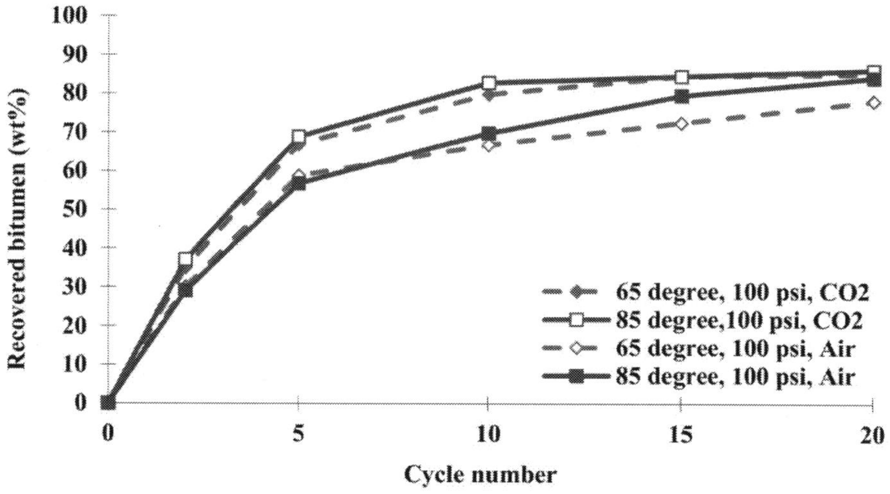

Figure 1. Bitumen recovery from a Canadian oil sand (weight %) as extracted by hot water showing the effects of temperature in the two gases used (other conditions: P = 6.9 atm;

water-to-solid volume ratio = 2).

HOSE also achieved good bitumen recovery from Utah oil sands, which due to its high bitumen viscosity was apparently more difficult to achieve by conventional hot water extraction. Figure 2 shows bitumen recovery from a Utah oil sand according to cycle numbers at varying temperature. Because of the high viscosity of Utah bitumen, lower operating temperatures (75°C and 65°C) resulted in lower bitumen recovery. At higher temperatures, it was still possible to achieve a bitumen recovery >91%

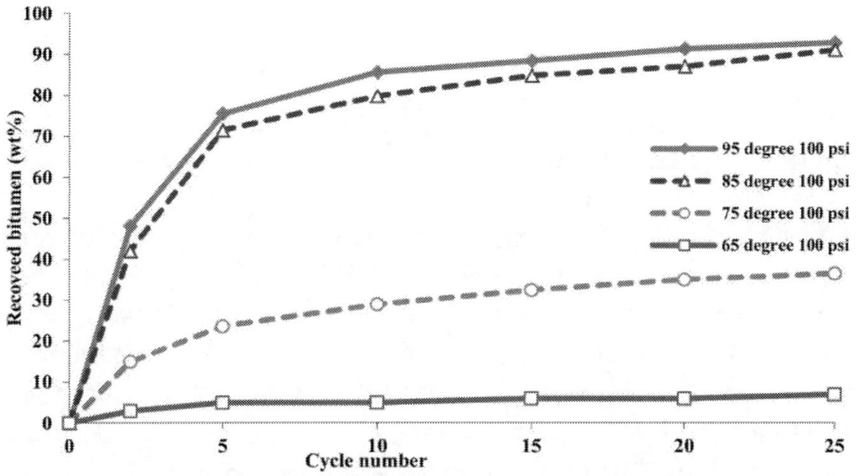

Figure 2. Bitumen recovery from Utah oil sands (weight %) as extracted by hot water according to the temperatures, pressures, and number of pressure cycles (other conditions: air used; water-to-solid volume ratio = 2).

High bitumen yields were achieved by the new HOSE process, leaving spent sands that could readily separate from the water column via gravity settling. The spent sands from HOSE treatment were removed by settling, resulting in 99.98% removal of the suspend solids from the process water in less than 30 minutes. While sands were settling, the turbidity of process water reduced rapidly (as shown in Figure 3); a reduction in the turbidity of more than 90% was obtained within 2 hours. The settled sands contained only 1-3% organics and its potential incorporation as construction fill is being explored.

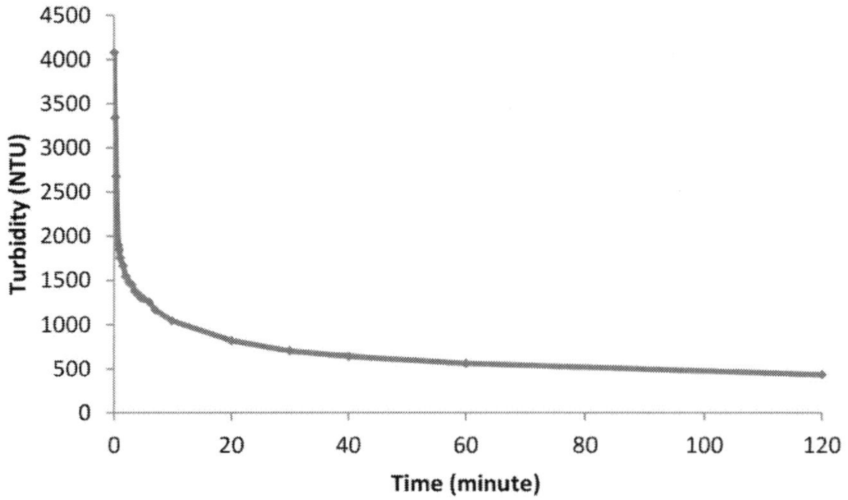

Figure 3. Change in turbidity of process water with settling time.

Water quality parameters were assessed following the HOSE process water and are listed in Table 1 . and are compared with (i) typical data from a convetional hot water extraction process tailings pond water and (ii) discharge limits. HOSE process water contained less solids, alkalinity, and COD than typical oil sand tailing pond water. The HOSE process water appeared to be amenable to treatment by sand filtration or ozonation before extraction reuse or discharge. Preliminary sand filtration results showed reductions of COD in the process water from 242 mg/L to 78 mg/L and of turbidity down to 0.45 NTU.

Table 1. Qualities of tailings pond water and HOSE process relative to discharge limits.

	Hot water extraction tailings pond water	HOSE process water	Discharge limit
TDS (mg/L)	1900-2221[2]	611	500[a]
TSS (mg/L)	<7000[3]	200	30[b]
pH	8.2-8.4[2]	7.0-7.8	6-9[c]
Alkalinity (meq/L)	12.7-15.6[2]	0.42-2.39	
Hardness (mg/L as $CaCO_3$)	91-112[2]	135-330	100[c]
COD (mg/L)	86-973[2]	156-232	

a. EPA secondary maximum contaminant level
b. U.S. Federal discharge limits
c. EPA national recommended water quality criteria

Reference

1. J. Masliyah, Z. Zhou, Z. Xu, J. Czarnecki, H. Hamza, Understanding water-based bitumen extraction from Athabasca oil sands, The Canadian Journal of Chemical Engineering 82 (2004), 628-654.

2. E.J. Allen, Process water treatment in Canada's oil sands industry: I. target pollutants and treatment objectives, Journal of Environmental Engineering and Science, Mar. 1st 2008.
3. Z.Guo, Integrated oil sands tailings pond water treatment, Saskatchewan Research Council, Dec. 8th 2010.

(231b) Concentration of a Solids-Free Lignocellulose Hydrolysate for Very High Gravity Fermentation: Inhibition Effects

Steven J. Schneiderman, Todd J. Menkhaus and Patrick C. Gilcrease, Chemical and Biological Engineering, South Dakota School of Mines and Technology, Rapid City, SD

A major problem in the efficient conversion of lignocellulose hydrolysates to ethanol is low ethanol concentrations in the fermentation broth due to low initial sugar concentrations. This is undesirable because low ethanol concentrations lead to increased separation costs downstream. Distillation is energy intensive, and constitutes more than 50% of the total energy consumption of a typical ethanol plant. An obvious solution is to start with a high initial sugars concentration, which is referred to as a very high gravity (VHG) fermentation. By increasing the glucose concentration in the hydrolysate from 40 to 100 g/l, the distillation energy consumption would be reduced by 60%. In the first step of our integrated process, polyelectrolyte flocculants facilitate the efficient removal of biomass solids from a pine wood hydrolysate. Once the hydrolysate is solids-free, an evaporation or reverse osmosis step can be readily used to remove water and concentrate the fermentable sugars. VHG fermentations present some challenges of their own, including higher energy costs associated with concentrating the sugars, simultaneous concentration of some or all of the fermentation inhibitors, and osmotic stresses on the fermentation organism at higher substrate concentrations. Because some of the inhibition effects are synergistic, this could become more problematic at VHG conditions.

The goal of this study was to investigate the potential problems associated with fermenting a concentrated lignocellulose hydrolysate, and to determine operating conditions that maximize productivity and minimize costs for the combined sugars concentration-fermentation-distillation process. A batch culture design of experiments was used to quantify the inhibition effects of five common inhibitors on the productivity of our *S. cerevisiae* D_5A strain. The inhibitors studied were glucose, ethanol, furfural, HMF, and acetic acid at concentration ranges of 25 – 300 g/l, 0 – 150 g/l, 0 – 5 g/l, 0 – 5 g/l and 0 – 10 g/l, respectively. Our experimental results were used to regress kinetic model parameters, and define an acceptable operating window for VHG fermentation. The hydrolysate concentration step (evaporation/reverse osmosis) was then tailored to minimize total separation costs (sugars concentration and ethanol distillation) while maintaining a high level of ethanol productivity. Finally, the fermentation of a concentrated pine hydrolysate at optimum conditions was demonstrated at the bench scale.

See more of this Session: Integrated Processes for Biochemical Conversion of Renewable Feedstocks to Fuels and Chemicals I
See more of this Group/Topical: Sustainable Engineering Forum

(609b) Semiconductor Nanowire Fabric As a New Photovoltaic Material

Chet Steinhagen[1], Vahid Akhavan[1], Vince Holmberg[1] and **Brian A. Korgel**[2], (1)Chemical Engineering, University of Texas at Austin, Austin, TX, (2)Department of Chemical Engineering, University of Texas at Austin, Austin, TX

Significant quantities of semiconductor nanowires can be synthesized using solution-based methods, such as supercritical fluid-liquid-solid (SFLS) and solution-liquid-solid (SLS) growth. These nanowires are mechanically flexible because of their narrow (<50 nm) diameter and high aspect ratio, with lengths greater than millimeters in some cases. Because of the absence of crystallographic defects like grain boundaries, the nanowires are extremely strong, exhibiting nearly ideal mechanical strength. The nanowires can be formed into non-woven fabric, or paper, that is flexible and mechanically robust. We are in the process of trying to develop these nanowire fabrics as a new class of flexible photovoltaic material. This presentation will highlight the nanowire fabric itself--its properties and how it is made--as well as data from prototype photovoltaic silicon (Si) and $CuInSe_2$ (CIS) nanowire fabric.

See more of this Session: Semiconductor Nanowires and Nanotubes for Next Generation Photovoltaics From Fundamentals to Devices
See more of this Group/Topical: Topical 5: Nanomaterials for Energy Applications

(231c) Separate Processing of Hemicellulose and Cellulose to Optimize Ethanol Production From Corn Stover

Alexandre Chapeaux[1], Nancy Dowe[1] and Daniel J. Schell[2], (1)National Bioenergy Center, National Renewable Energy Laboratory, Golden, CO, (2)National Renewable Energy Laboratory, Golden, CO

Separate processing ofhemicellulose and cellulose to optimize ethanol production from corn stover

Alexandre Chapeaux, NancyDowe, Daniel J. Schell

Using a series of thermochemical (pretreatment)and biochemical steps, lignocellulosic biomass is converted to sugars that subsequentlycan be made into fuels and chemicals. One such process uses heat and dilute sulfuric acid to hydrolyzes the hemicellulosic fractionof the biomass to monomeric and oligomeric sugars.The removal of hemicellulose and to a lesser extent lignin from the biomass significantlyimproves the performance of the next step in the process, enzymatic hydrolysisof cellulose to glucose. However, enzymatic cellulose hydrolysis can beperformed in the presence or absence of the hemicellulosic sugars producedduring pretreatment. It is well known that removing the hemicellulose sugarsand other compounds produced during pretreatment improves enzymatic celluloseconversion yields. In previous work, we assessed the potential of a processthat removes the hemicellulosic sugars prior to enzymatic cellulose hydrolysis byexploring the interaction between biomass solids loading, yields, and processcosts. However, this process is more expensive because of the additional costfor solid-liquid separation equipment required to remove th-richsugar liquor from the cellulosic solids.

In this study, the goal was to deepen ourunderstanding of the advantages and disadvantages of the separatehemicellulose/cellulose process. The feedstock was dilute-acid-pretreated cornstover produced in a 200 kg dry biomass/d continuous horizontal reactor. Thepretreated slurry was separated into a liquor stream (hemicellulosic sugars)and a washed cellulosic solid stream. Enzymatic cellulose hydrolysis was investigatedat four enzyme loadings (10, 20, 30 and 40 mg protein/g cellulose) and atinsoluble solids concentrations from 9% to 21% (w/w). The resulting glucose wasfermented to ethanol using several yeasts and a glucose-xylose co-fermentingbacterium, *Zymomonas mobilis* 8b. The sugars in the liquorstream were fermented to ethanol with three strains of co-fermenting *Z. mobilis* after dilution of the streamwith water to lower sugar concentrations. Finally, the overall performance ofthe process was assessed after partitioning water between the two phases in away that optimizes overall ethanol yields. We then compared these results to aprocess configuration in which the streams are not separated. In the latterprocess, enzymatic cellulose hydrolysis is performed in the presence of thebackground hemicellulosic sugars produced during pretreatment of the biomass.

See more of this Session: Integrated Processes for Biochemical Conversion of Renewable Feedstocks to

(173a) Silicon Nanostructures for Lithium Ion Battery Anodes

Justin Harris[1], Aaron Chockla[1], Matthew Panthani[1], Colin Hessel[1], Dariya Reid[1] and **Brian Korgel**[2], (1) Chemical Engineering, The University of Texas at Austin, Austin, TX, (2)Department of Chemical Engineering, The University of Texas at Austin, Austin, TX

Colloidal synthetic methods have been developed for a wide variety of nanocrystal and nanowire materials. Among these materials, silicon has been one of the most challenging to synthesize in in solution. The general approach is to feed molecular reactants into hot solvent environment to precipitate the nanomaterial of interest. In the case of silicon (Si), suitable reactants are needed that readily decompose in the solvent environment and the reaction must be controlled sufficiently to yield nanomaterials with the desired size and shape. Recently, we have been developing a variety of methods for the synthesis of Si nanomaterials, including amorphous and crystalline Si colloidal particles and Si nanorods and nanowires. These methods can yield significant quantities of Si nanomaterials and one potential application for them is as a replacement for carbon as the anode material in lithium ion batteries. Si has the ability to alloy with lithium ($Li_{4.4}Si$) to achieve very high charge storage capacity of as much as 4200 mAh/g, which is more than an order of magnitude higher than the storage capacity of carbon. Lithiation, however, leads to a massive volume expansion (~400%), which leads to pulverization. Si nanostructures have the ability to tolerate this volume expansion, while enabling significant lithiation. Recent data of the charge storage capacity of colloidal silicon nanoparticles and nanowires will be presented.

See more of this Session: Nanomaterials for Energy Storage III
See more of this Group/Topical: Topical 5: Nanomaterials for Energy Applications

(453d) Landfill Gas Utilization In Gas Turbines: Experimental Results for Combustion of Landfill Impurities In Premixed Flames

Aydin Jalali[1], Mirmohammadyousef Motamedhashemi[2], Fokion N. Egolfopoulos[1] and Theodore Tsotsis[2], (1)Department of Aerospace and Mechanical Engineering, University of Southern California, Los Angeles, CA, (2)Mork Family Department of Chemical Engineering and Material Science, University of Southern California, Los Angeles, CA

Gas turbines and other conventional internal combustion engines can combust landfill gas (LFG), which is an important renewable energy source, to generate electricity. However, the corrosion and damage that can be caused by its impurities may reduce the operating life of these combustion engines when operating on LFG. One class of major impurities present in LFG is Siloxanes. In our research, the decomposition of Hexamethyldisiloxane (also referred to as L_2) present in LFG is experimentally investigated in order to provide a better knowledge of the way it is degraded during the combustion of LFG. Several experimental techniques were utilized in order to study premixed flat flames, and to gain a better understanding about the durability of L_2 in the flame environment and the kinetics of its decomposition. In our experiments, we have utilized the counterflow technique, which allows one to accurately probe and analyze the mechanisms leading to the formation of silica microparticles resulting from the decomposition of L_2. We also applied laser extinction techniques for the *in situ* monitoring of particle formation, as well as GC-MS techniques for measuring the kinetics of siloxane decomposition.

It was found that microcrystalline silica particles were generated during LFG combustion, leading to a rapid coverage of the surface of Ni/Cr sheets placed in the flame environment, and forming a light white layer of solid particles, whose surface was analyzed by atomic force microscopy (AFM). The size of these particles was then estimated using SEM/EDAX. The effect of L_2 concentration on its conversion along the flame was found for a number of different feed concentrations. Volume fractions of particles within the flame were also measured *via* the laser extinction method indicating a linear relationship between the concentration in the fuel and the corresponding volume of particles in the flue-gas. The temperature profile has been measured experimentally in order to identify the kinetics of burning of the siloxane compounds. This fundamental insight is important in terms of being able to accurately determine the maximum allowable siloxane content for biogas that is safe to use without leading to deposits and microparticle formation.

See more of this Session: Fundamentals of Biomass Utilization
See more of this Group/Topical: Environmental Division

(110e) Kinetics of MnO2 Using a Rotating Disk Electrode

Abhinav Gaikwad, Joshua Gallaway and Dan Steingart, Chemical Engineering, Energy Institute, City College of New York, New York, NY

Electroactive materials are commonly characterized by cyclic voltammetery and chronopotentiometry experiments. The commonly used MnO_2 electrodes are a mixture of MnO_2 particle, graphite to improve the overall conductivity and a binder to hold the electrode together. These electrodes require very slow scan rate in range of uV/s in-order to compensate for the slow mass transfer thorough the porous electrode to study the true kinetics and thermodynamics of the electrode. Using a thin-film of MnO_2 electrode drop casted on a rotating disk electrode the limitation on mass transter effect can be eliminated and cyclic voltammetery study can be carried out at rapid speed. The rotating disk electrode study of various MnO_2 be used advantageously to study the cyclability of the electrode and benchmarking of various MnO_2 materials.

See more of this Session: Novel Battery Chemistry and Technology
See more of this Group/Topical: Fuels and Petrochemicals Division

(588e) Hydrogen Car Fill-up: Thermodynamic Modeling Using the Generic Cubic Equation of State

Fernando Olmos, Chemical and Biomolecular Engineering, UCLA, Los Angeles, CA and Vasilios Manousiouthakis, Chemical Engineering Department, University of California at Los Angeles, Los Angeles, CA

In order to provide hydrogen fueling stations with required information about hydrogen's pressure, temperature, and mass flowrate, hydrogen was modeled as a non-ideal gas employing the generic cubic equation of state (GC-EOS), along with its special four cases (VdW, RK, SRK, PR). With the purpose of validating the GC-EOS hydrogen thermodynamic property predictions, comparisons were carried out with NIST hydrogen's data, real-life hydrogen fueling station data, and with a mass balance equation in integrable form. Both the results of the GC-EOS and the mass balance model matched the NIST data and the hydrogen fueling station data describing the time evolution of the molar volume inside of the hydrogen vehicle tank. In addition, based on the concept of residual properties, an explicit expression for constant-pressure and constant-volume heat capacity was derived based on the GC-EOS as a function of molar volume and temperature. The results from the heat capacities, based on GC-EOS, matched NIST hydrogen's data at low pressures, and at pressures ranging from 100 to 1000 bar there is deviation less than 2%. Finally, the fueling station was modeled employing a total mass and energy balance, along with the GC-EOS, establishing that, in the absence of heat transfer, the maximum temperature achieved by hydrogen inside of the vehicle's tank does not depend on how high or low the mass flowrate is, but rather it is a function of the amount of mass accumulated in the tank.

See more of this Session: Fundamentals of Hydrogen Production and Utilization
See more of this Group/Topical: Environmental Division

(231e) Fischer Tropsch Synthesis Via Biomass Derived Synthesis Gas

Syed Ali Gardezi, Chemical Engineering, University of South Florida, Tampa, FL, Babu Joseph, Chemical and Biomedical Engineering, Clean Energy Research Center, University of South Florida, Tampa, FL and John T. Wolan, University of South Florida, Tampa, FL

Due to the recent surge in energy costs and uncertainties in fuel prices, there is significant interest in seeking alternative sources of energy particularly from renewable such as biomass. Indirect liquefaction of biomass (via the thermo-chemical conversion to syngas followed by liquefaction via Fischer-Tropsch Synthesis or FTS) offers a commercially viable route for meeting the challenge of producing renewable fungible liquid fuels. Apart from biomass, FTS is also the critical step for converting natural gas and coal to liquid fuels.

A FTS technology has been developed for producing middle distillates (diesel and aviation fuel) by using pine chips as the main feedstock. In this "two step" process, pine chips were subjected to an entrained flow gasification process developed by Pearson et al. [1]. The resulting tar free gas is comprised mainly of carbon monoxide, hydrogen, carbon dioxide, methane and some organic contaminants such as benzene, toluene, and naphthalene along with substantial amount of water. For further liquefaction this gas was cleansed using multiple adsorbents. An inline "Shaw Moisture Meter" was used to continuously monitor the moisture content of the feed to the liquefaction process.

For the production of liquid fuels, eggshell cobalt catalysts were used. Cobalt catalysts have several advantages e.g. they produce linear paraffinic hydrocarbons and minimize undesirable water gas shift reaction. This eggshell design is a well known strategy for producing a narrow hydrocarbon distribution. In this regard previous research work conducted by Iglesia et al. was also consulted [2]. A bench scale fixed bed reactor was used for the conversion of syngas. Modeling of the reactor startup and of adjustment of the reactor packing ensured successful operation resulting in a high conversion and narrow distribution of hydrocarbons in the range of diesel and aviation fuel.

References :

[1] Pearson et al., U.S. Patent 4,857,076, issued Aug 15, 1989.

[2] E. Iglesia, S. L. Soled, J. E. Baumgartner, S. C. Reyes, *J. Catal.* **153**, 108 (1995).

See more of this Session: Integrated Processes for Biochemical Conversion of Renewable Feedstocks to Fuels and Chemicals I
See more of this Group/Topical: Sustainable Engineering Forum

(411a) Control Emulsions, Suspensions, and Solids Separation with Inline Particle Size Measurements In Dark Crude Oil

Anjan P. Pandey and Des O'Grady, METTLER TOLEDO AutoChem, Columbia, MD

Hydrocarbon pipeline flow is typically a complex multi-phase system comprised of bubbles, droplets, and particles including asphaltenes, waxes, gas hydrates, emulsions, and inorganic particles. To control separations, flow, and process efficiency, analytical measurements are often applied by petroleum engineers and chemists. Yet extreme temperature and pressure conditions make offline sampling and analysis difficult and time consuming. Offline measurements can also alter or destroy multiphase components and cannot be applied to make real-time process optimization and control decisions. With the implementation of inline particle characterization technology, particle or droplet phase behavior can be measured quickly, in real time, without pulling samples.

This paper focuses on the application of real time, in-process particle and droplet measurement tools to avoid processing problems in the following R&D and full scale production applications:

- Controlling asphaltene and wax precipitation in dark crude oil
- Improving oil-water and water-oil separation through in-process droplet measurements
- Visualizing and inhibit gas hydrate formation to ensure consistent flow rates
- Improving oil/sand tailings settling rates and optimizing flocculent and chemical surfactant effectiveness

See more of this Session: Solids-Liquid Separation Processes for the Petroleum Industry
See more of this Group/Topical: Separations Division

(13c) Multi-Scale Study On the Pyrolysis of Sustainable Biomass Feedstock

Jessica D. Murillo, Chemistry, Environmental Sciences, Tennessee Technological University, Cookeville, TN, Joseph J. Biernacki, Chemical Engineering, Tennessee Technological University, Cookeville, TN and C. Pat Bagley, College of Agricultural and Human Sciences, Tennessee Technological University, Cookeville, TN

Multi-Scale Study on the Pyrolysis of Sustainable Biomass Feedstock

Jessica D. Murillo

Department of Chemistry, Environmental Sciences, Tennessee Technological University, Cookeville, TN 38505

Joseph J. Biernacki

Department of Chemical Engineering, Tennessee Technological University, Cookeville, TN 38505

C. Pat Bagley

College of Agricultural and Human Sciences, Tennessee Technological University, Cookeville, TN 38505

Pyrolysis of lignocellulosic material for the production of biofuels and value-added products offers a short- to long-term renewable energy option to help curb fossil fuel use, gain energy independence, and mitigate global climate change. The main components of lignocellulsic material are cellulose, hemicellulose and lignin; however, there are many issues with using such a complex chemical feedstock for the production of renewable fuels. During pyrolysis the main components degrade via primary and secondary reactions, producing a spectrum of products, including volatiles, char, and bio-oil. To gain a more complete understanding of the pyrolytic behavior of lignocellulosic biomass, a deeper knowledge of the morphology and chemical structure is required.

Preliminary meso-scale experimental results from the pyrolysis of tall fescue hay, sorghum-sudangrass, Bermuda grass, switchgrass, and alfalfa hay, have paved the way for study at microscopic lengths. The physical and chemical characteristics of each grass hay was studied using X-ray diffraction, mid-infrared spectroscopy (MIR) and electron scanning microscopy to determine the microstructure and to quantify the content of lignin, cellulose, and hemicelluloses. Comparison studies on the reaction kinetics of grass samples and isolated pseudo-components were performed using thermogravimetric analysis (TGA).

Improving the sustainability of biofuels through modification of the biomass production subsystem: Application to corn and lignocellulosic ethanol

Concerns about dwindling fossil fuel reserves and climate change, energy independence, and depletion of ecosystem goods and services have driven the development of alternative technologies such as biomass-based fuels (biofuels). Although biofuels may be a step towards alleviating the aforementioned concerns due to the renewable nature of the biomass feedstock, there are several drawbacks associated with them. Food crop-based fuels such as corn ethanol, which comprise the majority of the current biofuel mix in the US, require large amounts of fossil-based inputs purchased from the economy in order to grow the biomass. These inputs, such as fertilizers, pesticides, and fossil fuels greatly diminish the renewable nature of the plant itself and force the production of biofuels to be reliant on non-renewable fossil-based inputs. In addition to reliance on non-renewable inputs, biofuels are heavily reliant on ecosystem goods and services. Current large-scale farming practices result in the runoff of nutrients from fertilizers into rivers and streams, which can overwhelm the carrying capacity of the nitrogen (N) and phosphorous (P) cycles of these aquatic ecosystems with devastating results. Freshwater for irrigation is needed if natural precipitation is not sufficient. Topsoil, a vital local non-renewable resource, is also depleted through erosion in traditional monocultural row cropping. Ecosystem goods and services are also indirectly consumed through the life cycle of the inputs used in farming.

This research aims to improve the sustainability of biofuels, specifically corn ethanol, by modifying the corn farming system to use less purchased inputs and consume less ecosystem goods and services both directly and indirectly through the reduction of fossil-based inputs needed from the economy. The following design changes will be considered:

1) No-till/conservation tillage. This will reduce soil erosion and require less labor and fuel for preparing soil for seeding. Application of more herbicides may be necessary as tilling helps remove weeds.
2) Planting winter cover crops in the growing offseason, also known as "double cropping". This reduces bare soil exposure in the offseason, reducing erosion. Also, offseason weed growth can be reduced requiring less herbicide when the corn is planted. Depending on the species planted, soil nutrients can also be rebuilt in the offseason. However, additional labor, energy, and other inputs will be needed for planting and harvesting the cover crop.
3) Intercropping other species amounts corn during the growing season. This is often referred to as a polyculture. This can reduce soil exposure and hence reduce erosion and reduce fertilizer consumption due to fixation of nutrients by the intercropped species. However, separation of corn from the intercropped species may require additional inputs.
4) Constructed wetlands and/or riparian buffers for nutrient runoff capture. This can prevent nutrients from entering rivers and streams and reduce eutrophication.

5) Using treated sludge from wastewater treatment facilities (biosolids) for fertilizer, displacing the need for fossil-based artificial fertilizers. However, there is a limited supply of biosolids and may not be enough for large-scale application. Also, biosolids may be more labor and energy intensive to apply.
6) Fossil fuel substitution, i.e., replacing partially or fully the fossil fuels used in the farming system with biofuels to reduce the reliance on non-renewable inputs. Any disadvantage of biofuels will be compounded when fed back into the production of more biofuels.

The impact of these design changes will be quantified both directly and indirectly in the life cycle through the use of Ecologically Based Life Cycle Assessment (Eco-LCA) to determine the overall effect of the changes on the sustainability of corn ethanol. This LCA method will quantify traditional LCA impacts, such as emissions and energy consumption, but will also quantify the multitude of ecological goods and services consumed throughout the life cycle. This ecological accounting is vital to ensure the sustainability of any system since everything relies directly or indirectly on ecosystem services. Additionally, the resulting corn farming design is an example of the design of a technological-ecological network, where both technological and ecological variables are considered in the design of the overall system. Such systems are more sustainable than systems that use technological variables only, since technology alone cannot lead to sustainability due to second law constraints.

Although these design changes may improve the sustainability of corn ethanol, the scaling up of biofuel production can compete with food and lead to indirect land use change where native ecosystems such as forests are converted to farmland for biomass production, which may create a net greenhouse gas increase. Therefore, it is worth considering other biomass sources that can be produced on marginal lands. The case of producing ethanol from a low input, high diversity (LIHD) lignocellulosic grassland on degraded farmland will also be presented. LIHD biomass requires much fewer inputs than most monocultural biomasses and may be a sustainable source of liquid fuel. LIHD ethanol will also be compared with food crop-based ethanol as well as other cellulosic ethanol sources.

(453a) Removal of Selenite with Microbial Fuel Cells Utilizing Shewanella Oneidensis MR-1

Prithiviraj Chellamuthu, Department of Biological Sciences, University of Southern California, Los Angeles, CA, Mandy Ward, Department of Earth sciences, University of Southern California, Los Angeles, CA, Kenneth Nealson, Department of Earth Sciences, University of Southern California, Los Angeles, CA and Miriam Kronen, Department of Microbiology, University of Heidelberg, Heidelberg, Germany

Selenium is a trace element with important physiological roles in immune function and oxidative stress; however, at elevated levels it is known to be toxic to all forms of living organisms. Apart from naturally occurring selenium in rocks, selenium compounds are widely used in electronics, oil production and agricultural applications, which has resulted in the contamination of ground water. Selenite (+4 oxidation state) and selenate (+6 oxidation state) are the oxidized forms of selenium. Both are soluble, mobile and toxic in soil and water systems. Even at low levels of contamination these elements have deleterious effects on fauna, where bio-magnification plays a significant role in inducing toxicity. The EPA has set a limit of .05 parts per million (ppm) in drinking water. Hence remediation is important and a variety of methods have been proposed, including activated alumina, coagulation/filtration, reverse osmosis and bioremediation. Recent investigations have shown that bioremediation, which utilizes indigenous or engineered bacteria for the removal of selenium compounds from the soil and/or water systems could be a viable option.

Microbial fuel cell (MFC) systems utilize the catalytic ability of electrogenic bacteria, unique organisms that can convert energy from organic compounds to electricity through the respiration of solid electrodes. There has been much research into the applications of MFCs, including recent investigations that suggest that the cathode of a MFC can be used for bioremediation and the extraction of valuable metals. Bioremediation results from the ability of the bacteria to convert toxic compounds into more manageable, nontoxic forms. In a MFC, transformation of toxic compounds occurs when used as electron acceptors or oxidants in the cathode compartment. Researchers have shown the applicability of MFC bioremediation technology for the removal of various metals including copper, chromium, selenium and arsenic. In order to fully exploit the potential of this technology it is important to understand the electron transfer mechanism involved in this reduction/ remediation process.

Shewanella oneidensis MR-1 is a facultative anaerobic bacterium known for its versatility in using a variety of soluble and insoluble electron acceptors. This versatility stems from an array of multi-heme cytochromes which makes them an interesting candidate for bioremediation of toxic heavy metals such as uranium, selenium and chromium. Understanding respiratory electron transfer pathways will help in the design of engineered organisms for better removal of toxic contaminants from water. Using a MFC, we have identified the key cytochromes involved in the reduction of selenium under fuel cell conditions. We have also optimized the maximum concentrations for optimal removal rates without affecting the performance of the MFC. Using various electrochemical techniques, the operating parameters, such as internal resistance and maximum power output, have been determined. Our results indicate this technology could satisfy the stringent EPA water quality standards for removal of selenium.

See more of this Session: Fundamentals of Biomass Utilization
See more of this Group/Topical: Environmental Division

(411e) Technology Development and Demonstration of a Low-Temperature Fluidized-Bed Biomass Dryer

Charles J. Coronella, Mike P Matheus and Victor R. Vasquez, Chemical Engineering, University of Nevada, Reno, Reno, NV

At the University of Nevada, Reno, we initiated a research project to study mechanisms of drying wastewater solids (sludge) in 2007. Preliminary studies led to the development of a new low-temperature fluidized-bed sludge dryer. The sludge dryer has been tested in batch and in continuous studies, and operates at very low temperatures. We report results of a parametric study based on air flow rate and dryer temperature. The dryer produced dried solids at unusually low temperatures, between 125 ° F and 175 ° F, with heat provided at 210 ° F. Collected sludge has been characterized in terms of its moisture content, particle size, and ash content. The moisture of the collected product ranges between 4% and 15%, and can be controlled by proper selection of operating conditions.

The dryer is a novel configuration, and consists of a bed of inert solids operating in the bubbling bed regime. Heat is provided to the bed by internal heat transfer tubes. Wet sludge falls into the warm bed, which is bubbling vigorously. The rapid particle movement causes the wet sludge to break apart, increasing surface area, and thereby enhancing heat transfer. As the sludge dries and fragments, it becomes relatively light dust particles. The small dry particles have specific density 1.2, and are entrained in the fluidizing air. Solid particles are removed from the air in a filter outside the bed, and collected and stored.

The capacity of the dryer is limited by the rate of air flow. Air leaving the bed can remove a finite amount of moisture, up to its saturation of humidity. Increasing too high has a deleterious effect on the performance of the fluidized bed.

We have successfully completed a demonstration of our dryer with a continuous feed of up to 20 lb/hour of wet (85% moisture) sludge onsite at the Truckee Meadows Water Reclamation Facility (TMWRF). We are currently seeking partners for a 200 lb/hour demonstration. Trials and lessons from commercialization with the support of the UNR Tech Transfer Office will be discussed.

See more of this Session: Solids-Liquid Separation Processes for the Petroleum Industry
See more of this Group/Topical: Separations Division

(231d) Ethanol and Furfural Production From Corn Stover Using Hybrid Fractionation Process with Zinc Chloride and Simultaneous Saccharification and Fermentation (SSF)

Chang Geun Yoo[1], **Tae Hyun Kim**[1] and Monlin Kuo[2], (1)Agricultural and Biosystems Engineering, Iowa State University, Ames, IA, (2)Department of Natural Resources Ecology and Management, Iowa State University, Ames, IA

Two-stage hybrid fractionation process was investigated to produce cellulosic ethanol and furfural from corn stover. Zinc chloride ($ZnCl_2$) was used to selectively solubilize hemicellulose in the first stage. During the second stage, the remaining treated solids were converted into ethanol using commercial cellulase and *Saccharomyces cerevisiae* or recombinant *Escherichia coli*, KO11.

Hybrid fractionation process recovered >90% of glucan, ~90% of xylan, and ~75% of lignin at optimal reaction conditions: 1st stage: 5% acidified $ZnCl_2$, 7.5 ml/min, 150 °C (10 min) and 170 °C (10 min); 2nd stage: Simultaneous saccharification and fermentation (SSF). Furfural yield from hemicellulose hydrolysates was 58 %. SSF of $ZnCl_2$ treated solid with *Saccharomyces cerevisiae* resulted in 69-98 % of theoretical maximum ethanol yields based on glucan present in the treated solids. After fermentation, solid residues contained mostly lignin. Based on total lignin in untreated corn stover, the lignin recovery yield was ~75%.

See more of this Session: Integrated Processes for Biochemical Conversion of Renewable Feedstocks to Fuels and Chemicals I
See more of this Group/Topical: Sustainable Engineering Forum

(484e) Comparison of Membrane Performance: Analysis with Fouling Models

Mingyan Zhou[1], James E. Kilduff[1] and **Georges Belfort**[2], (1)Department of Civil and Environmental Engineering, Rensselaer Polytechnic Institute, Troy, NY, (2)Department of Chemical and Biological Engineering, Rensselaer Polytechnic Institute, Troy, NY

Membrane research and engineering applications often involve optimizing flux and rejection for a particular application. This process frequently presents the need to compare the performance of membranes having different inherent permeabilities; an important performance criterion is resistance to fouling. To identify the most appropriate (least biased) approaches to compare membrane performance, an analysis using fouling models, with an emphasis on the combined pore blockage and cake filtration model, was performed. We illustrate how the models can be used to provide a general framework for assessing fouling and developing ways to compare membrane performance. Comparisons at constant transmembrane pressure using the same initial flux and the same initial pressure were emphasized. This analysis demonstrates that membrane resistance has a significant effect on flux decline - flux decline is smaller for membranes having greater resistance, especially when the feed solution and membrane-foulant interactions are similar. This effect cannot be normalized regardless of whether flux is plotted as a function of filtration time and volume throughput. A better approach is to compare the effects of fouling by evaluating fouling resistance as a function of volume throughput. Alternatively, comparing membranes based on total resistance provides some advantages because it can be related to permeate production and energy cost. When comparing membranes having different resistance and rejection, the effects of fouling should be compared by evaluating fouling resistance as a function of mass accumulation on the membrane surface. The effects of cake compressibility are also discussed.

See more of this Session: Theories and Applications of Cross Flow Filtration
See more of this Group/Topical: Separations Division

DESIGN CHALLENGES AND MODELING FOR AN EFFICIENT LIQUID HYDROGEN STORAGE TANK FOR AUTONOMOUS SYSTEMS

*Richard O. Stroman and Michael W. Schuette**
Naval Research Laboratory, Washington DC
**Global Strategies Group (North America) Inc., Washington DC*

Introduction

Small autonomous systems are rapidly adding capability and taking on new roles due to advances in navigation, controls, communications and sensing. Despite these improvements, however, the mission duration remains limited by the amount of useful energy available. There are two ways to provide more energy: raise the conversion efficiency or store more energy at the outset.

In recent years some vehicles have used fuel cells to convert fuel to electrical power for the propulsion plant and payload, because they tend to be more efficient than combustion engines in the 0.5 – 5 kW power range. One such vehicle is the Naval Research Laboratory's Ion Tiger, a 15.9 kg unmanned air vehicle (UAV) with a proton exchange membrane fuel cell (PEMFC) power system. PEMFCs convert H_2 to electrical power with about 50% efficiency, which in combination with the high specific energy of H_2 (33,400 Wh/kg) makes them ideal for long endurance systems.

The Ion Tiger demonstrated a 26 hr flight using 0.5 kg of gaseous H_2 (GH_2) stored at 5000 psi in a 22 L, 3.8 kg pressure vessel. [1] The pressure vessel and regulator together had to weigh less than 4 kg to satisfy the aircraft weight budget, which dictated a minimum storage efficiency of 12 wt% H_2 when combined with the goal of 24+ hr flight. Higher storage efficiency is necessary to increase the Ion Tiger flight endurance beyond 1 day, and our experience suggests that GH_2 is not a solution. Increasing the H_2 pressure would provide more H_2 storage in the same volume, but would require a disproportionately heavier pressure vessel and decrease storage efficiency. Furthermore, pressures above 5000 psi can require heavier materials not susceptible to H_2 embrittlement. Therefore, we opted to explore liquid H_2 (LH_2) as an alternative to GH_2.

The mass density of LH_2 is 62.6 g/L at 80 psi and 25 K, which is nearly three times larger than that of GH_2 at 5000 psi and 298 K (23.0 g/L). [2] The greater mass density of LH_2 could in principle provide nearly three times the storage efficiency of the present GH_2 system. LH_2 has been largely abandoned for most fuel cell applications because boil off cannot be eliminated entirely, so long term storage is problematic. For UAVs, however, LH_2 presents an interesting opportunity because they can be launched immediately after fueling and then continuously burn fuel in flight. The high mass efficiency of an LH_2 tank may enable unique small, long endurance electric UAVs. The major challenges when designing an LH_2 tank for an aircraft are to build a lightweight container with very low heat leak from ambient, and to accommodate transients in the fuel withdrawal rate.

Three aspects of the LH_2 storage tank design problem are described here: design of the insulation, design of the vent tube assembly (VTA), and the selection of an operating scheme. Validation of the insulation and VTA design models is also addressed.

Design Considerations

The tank design process was dominated by two (non-rigorous) optimization goals: maximize fuel storage and minimize venting. Fuel storage is directly related to the system weight – if the system weighs less, more of the weight budget can be allocated to fuel. Venting is related to the rate of heat leak from ambient – a lower rate of heat leak is less likely to raise tank pressure to the point at which it must be vented. Therefore the practical design goals were to minimize weight and minimize the heat leak rate.

There are additional reasons to minimize weight and heat leak rate. Lighter weight aircraft tend to require less propulsion power, be more maneuverable, and carry heavier payloads. With the rate of heat leak from ambient minimized, that the rate of boil off can be controlled over the widest possible range to match the withdrawal rate. Reducing the heat leak rate provides more control flexibility, "turn down" capability and endurance.

A dewar consisting of two nested vessels separated by a vacuum space is used in most high performance applications where exceptionally low heat leak is required. The vacuum eliminates conduction and convection heat transfer between the inner and outer vessels, and multi-layer insulation (MLI) is often added to reduce the rate of radiation heat transfer. Two heat conduction paths between the inner and outer vessels are largely unavoidable: the inner vessel supports and the structure(s) for filling and venting the inner vessel. The first design task is to determine the amount of MLI and the vacuum quality required by estimating the rate of radiation heat leak as a function of those parameters. The second is to design the vent tube by estimating the rates of conduction and radiation heat leak it introduces.

The fuel consumption rates for a typical Ion Tiger flight are shown in Table 1. Ideally the rate of LH_2 boil off would vary to match the rate of withdrawal, but because it is determined by the temperature difference between the LH_2 and ambient, it is nearly constant. To accommodate transients in the withdrawal rate one can allow the tank pressure to fluctuate, use a heater to "trim" the rate of boil off, or both. The third design task is to determine how large a heater is required, how often it should turn on, and how large an open space (filled with GH_2) above the LH_2 is required to damp transients in the withdrawal rate by allowing the tank pressure to vary.

Table 1. Fuel flow and heating rates required in a typical Ion Tiger flight.

Phase	H_2 Flow Rate [g/s]	Heat Rate, Vaporize LH_2 [W]	Heat Rate, Warm GH_2 [W]
Warm Up	8.33×10^{-4}	0.34	3.0
Climb Out	1.17×10^{-2}	4.78	41.4
Cruise	5.67×10^{-3}	2.32	20.1
Landing	8.33×10^{-4}	0.34	3.0

The total heat leak design goal was 90% of the heat rate required to boil off LH_2 at cruise power, shown as 2.32 W in Table 1. This limit kept the design goal feasible while providing some turn down capability. Early estimates of the relative rates of heat leak through each path led to the detailed heat leak budget in Table 2.

Table 2. Heat leak budget.

Path	Transfer Mode	Rate [W]
Supports	conduction	0.025
Insulation	radiation	1.889
VTA	conduction and radiation	0.175
Total	all	2.089

Design of the Insulation

A 1-D steady state heat transfer model was developed to determine how many radiation shields would be necessary to satisfy the heat leak budget. The model accounted for radiation heat transfer among layers of double aluminized Mylar insulation, using the Stefan-Boltzmann law [3-5] and absorbtivity data provided by the manufacturer. Conduction through the rarefied gas (assuming He at a pressure of 1×10^{-5} torr) [6] and through the Dacron net spacer were also included. Boundary conditions were the temperatures of the inner and outer vessels, assumed to be 25 K and 298 K, respectively. A nonlinear equation solver (the *fsolve* function in MATLAB) was used to determine which temperature profile through the stack of radiation shields would conserve energy, and the radiation flux through the stack. The number of MLI layers was increased until the desired radiation flux was reached.

Comparison to experimental data showed that the model consistently underpredicted the rate of heat transfer, and hence the number of MLI layers required. The discrepancy was attributed to insufficient fidelity in the spacer heat conduction model – the spacer reduces the insulation effectiveness by providing a conduction path, and the rate of conduction was underpredicted. The rate of heat conduction through the spacer layers depends on the contact area between the woven spacer and the radiation shields, and on the flux area, neither of which were well characterized. Attempts to fit these parameters were unsuccessful, so a published empirical model [7] was used instead. The empirical model predictions compared well with a test rig built for model validation (see Validation section) and confirmed that considerable effort is necessary to produce an accurate physics based MLI model. The empirical model should be sufficient for most designs.

Design of the Vent Tube Assembly

A 1-D steady state center differencing model was developed to estimate the rate of heat transfer through the VTA as a function of tube material, tube length, wall thickness and gas flow rate. The VTA heat leak model accounted for conduction through the tube wall, convection to the H_2 gas and radiation exchange between the tube and outer vessel (see Figure 1).

The tube and gas channel were broken into discrete volumes with the rates of heat transfer among them determined by Fourier's Law for conduction, Newton's Law of Cooling for convection and the Stefan-Boltzmann law for radiation. [3-5] Constant temperature boundary conditions were imposed at each end of the tube and at the gas entrance to the tube. A nonlinear equation solver (the *fsolve* function in MATLAB) was used to determine which tube and gas temperature profiles $T(x)$ conserved energy in each discretization. The resulting tube temperature profile was used to determine the heat flow rate into the inner vessel.

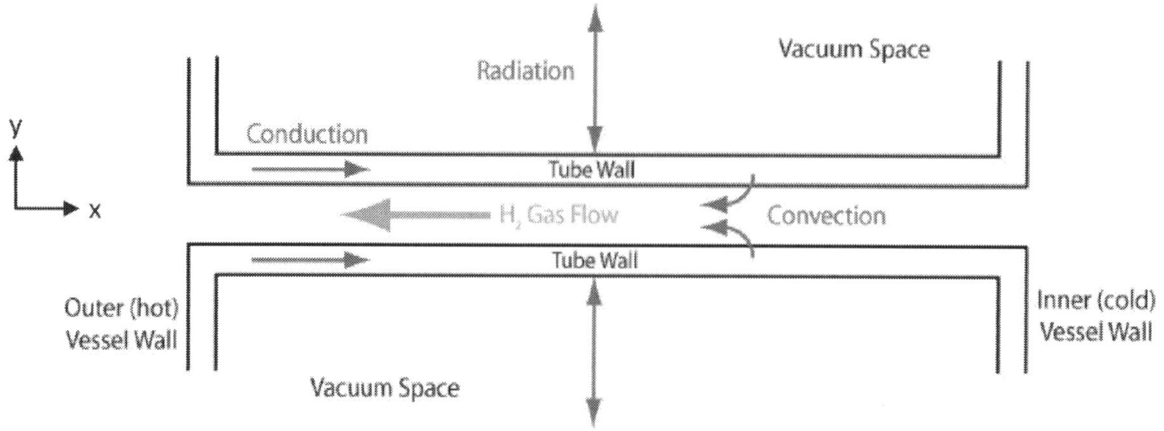

Figure 1. Heat transfer modes included in the VTA heat leak model.

The gas and tube thermal conductivities k varied considerably through the 25 – 298 K temperature range experienced by the tube, so the model interpolated among data from an online NIST database [8] to determine $k_{tube}(T)$ *and* $k_{H2}(T)$. The gas velocity (and hence the rate of convective heat transfer) was calculated in each discretization by accounting for the variation in gas density. The gas density was determined as a function of temperature and pressure by interpolating among data which functioned as an empirical equation of state. [2] It was assumed that the VTA was positioned in the vacuum space between the MLI and outer vessel, so it absorbed radiation from the outer vessel directly and via reflection from the MLI. This geometry was approximated by treating the VTA as if it were enclosed by a black body, i.e. assuming that the MLI was a perfect reflector. A typical radiation absorbtivity for each tube material was chosen from the literature, but the calculated temperature profiles showed only a small dependence on the value selected because the net rate of radiation heat transfer was small. The numbers of tube and gas discretizations were increased until the result ceased to vary with the number of discretizations.

An example result from the VTA heat leak model is shown in Figure 2, where the heat leak rate due to various tube lengths and gas flow rates have been plotted. Each curve in Figure 2 assumes the same tube material, inside diameter and outside diameter.

The importance of heat leak being intercepted by the gas through convection is clear – for each gas flow rate, there is a length beyond which the entire conduction and radiation heat leaks are intercepted. Given a maximum allowable rate of vent tube heat leak, expected gas flow rate and tube geometry, results such as in Figure 2 can be used to determine the minimum tube length necessary. It can also be used to explore different tube materials and combinations of inside diameter and gas flow rate.

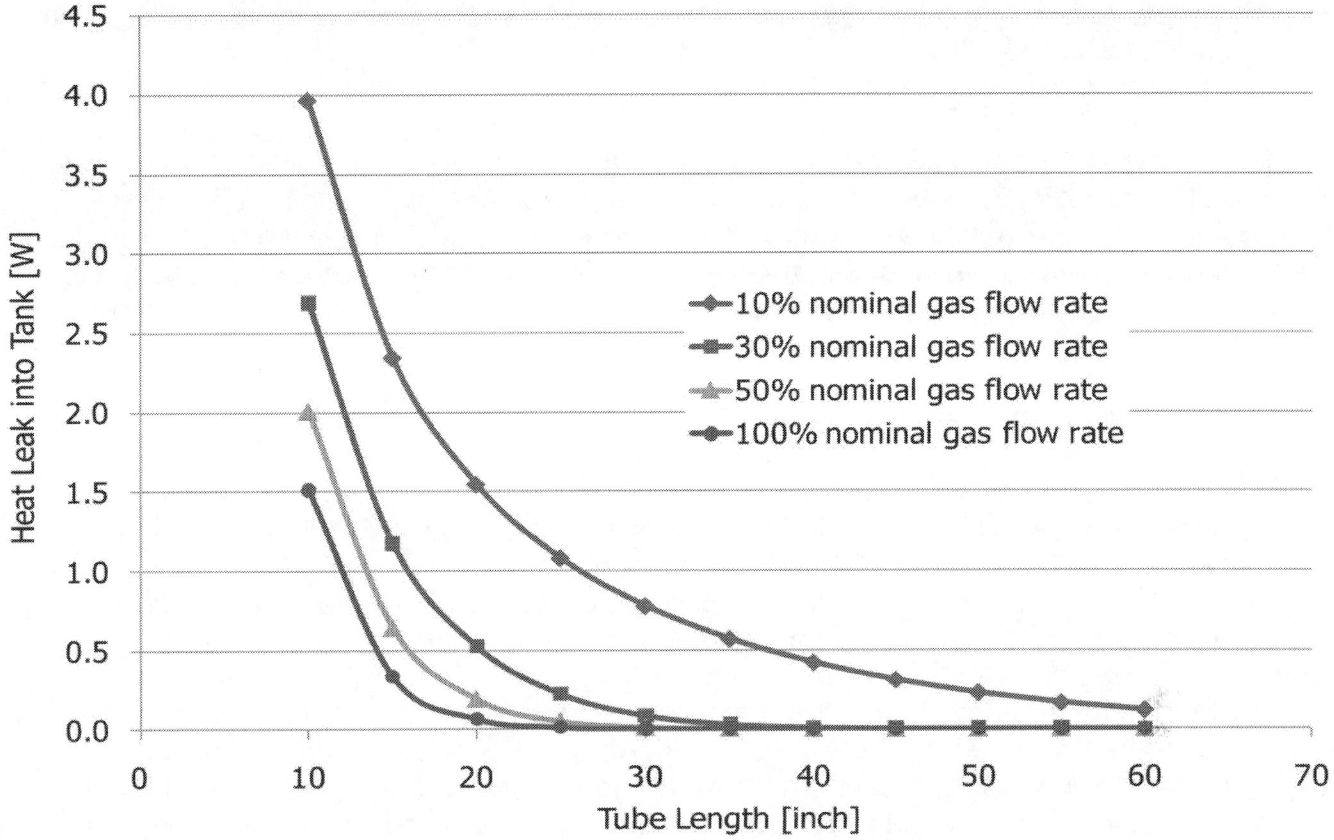

Figure 2. Predicted VTA net heat leak rate as a function of tube length and GH_2 flow rate.

A safety concern related to the VTA is the possibility of a catastrophic loss of vacuum which would cause rapid LH_2 vaporization, and if the gas cannot escape, rising tank pressure. In order to avoid a boiling liquid expanding vapor explosion (BLEVE) the VTA must vent GH_2 at the failure boil-off rate without exceeding the burst pressure of the inner vessel. The failure heat flux and boil off rate were estimated by assuming that gaseous N_2 from the air would condense on the outside of the inner vessel. This established temperature boundary conditions at the inside (25 K) and outside (77 K) surfaces of the inner vessel wall, which allowed an estimate of the heat flux to be calculated. Note that neglecting the thermal resistance associated with the LH_2 boiling made the heat flux estimate conservative (high).

The method outlined by [9] was used to estimate the maximum flow rate through tubes of various diameter and length. In each case the gas pressure in the inner vessel was assumed to be 75% of the inner vessel burst pressure. For the chosen tube length and inside diameter, the predicted gas flow rate was sufficient to avoid a BLEVE when compared to the predicted failure boil off rate.

Validation of Heat Transfer Models

A test rig was built to validate the model; the rig consisted of two concentric aluminum cylinders separated by a vacuum space. The inner cylinder was wrapped with MLI and separated from the outer cylinder by low thermal conductivity supports. The inner cylinder was filled with liquid N_2 (LN_2) and the total rate of heat transfer to the inner cylinder was estimated from the rate of LN_2 boil off.

The total rate of heat leak was measured for three different boundary temperatures: 207±1 K, 298±4 K and 387±3 K. The relative amounts of conduction and radiation heat transfer were estimated by assuming a dependence on $(T_h - T_c)$ for conduction and $(T_h^4 - T_c^4)$ for radiation. The rate of radiation heat transfer agreed with the empirical model to within 10% using a degradation factor of 7.4, which describes the deviation from ideal behavior introduced by seams and penetrations. In all of the measurements, the flow rate of LN_2 was sufficient to ensure that the heat leak rate was dominated by radiation heat transfer, confirming the expectation that MLI plays a major role in the design.

Load Based Transient Analysis

Fuel consumption rate data from GH_2 powered Ion Tiger flights were analyzed to estimate the transient performance required of the LH_2 tank. All aircraft have varying propulsion power requirements which depend on vehicle mass, drag and the way in which it maneuvers, but because the Ion Tiger is relatively small, the propulsion power also varies substantially with weather and thermal phenomena. Due to these variations, it is important to accurately estimate the range and typical rate of change in the fuel withdrawal rate.

To accommodate periods of low fuel consumption rate, a pressure relief valve (PRV) must be included in the tank design. Tank pressure increases when the boil off rate exceeds the consumption rate; when the pressure reaches the upper limit for the inner vessel, the PRV opens to release GH_2. Having a larger open space above the liquid allows more LH_2 to boil off before the PRV must open, thereby decreasing the amount of lost fuel, but this comes at the cost of storing less LH_2 at the outset. Having a lower overall heat leak rate and using a heater in periods of high fuel consumption is preferred, because then a smaller gas volume can be used to damp transients. The smaller gas volume leaves more space for LH_2 storage.

The design goal was to maintain tank pressure between 40 and 80 psig; to do this a transient performance model was developed. The model used propulsion load data and heat leak rate as inputs and the tank fill fraction and pressure as initial conditions. It integrated the difference between the LH_2 boil-off rate and the rate of fuel consumption, and then used an empirical equation of state [2] to estimate the tank pressure at each time step. When the pressure reached the upper limit, the tank vented, and when the pressure reached the lower limit, the heater turned on. Deadband of ½ the pressure range was assumed for the heater controller – the heater turned off at the nominal tank pressure. Model outputs were the predicted tank pressure, GH_2 blow-off rate and heater duty cycle as functions of time.

Figure 3 shows the predicted tank pressure as a function of time given an Ion Tiger fuel consumption curve, total heat leak rates of 2.0, 2.1 and 2.2 W, no PRV or heater, and a gas volume of 4 L. Note that changes in the heat leak rate as small as ±100 mW integrated over 26 hr meant the difference between exceeding the pressure limits of the tank and starving the fuel cell. The same is true to a greater extent when integrated over a 72 hr flight. Figure 3 shows the clear need for a low heat leak rate from ambient, a PRV, and a heater for adequate control of the LH_2 storage system. The high heat leak case exceeded the 80 psia inner vessel pressure limit and the low heat leak case could not support the fuel consumption rate over the entire flight.

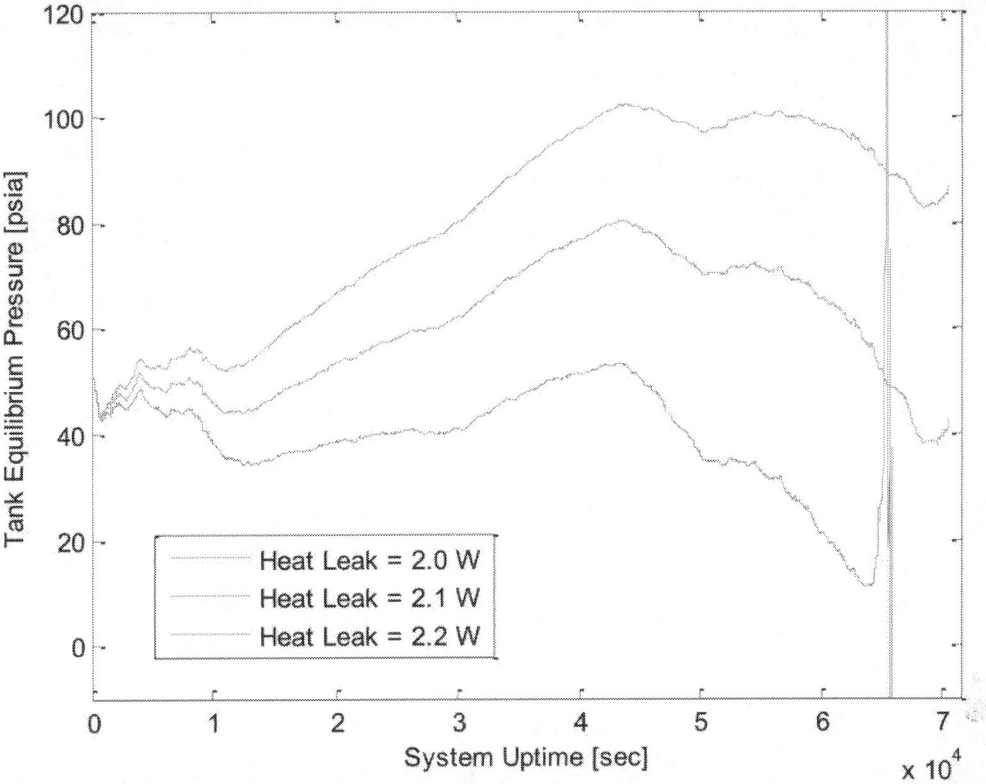

Figure 3. Predicted tank pressure assuming heat leak rates of 2.0, 2.1 or 2.2 W and a gas volume of 4 L. The large transient in the 2.0 W curve is an integration error due to the tank pressure falling below ambient pressure, which cause the equation of state to fail.

Assuming a total heat leak rate of 2.09 W, the transient performance model was used to compare two design scenarios: a large gas space and no heater vs. a small gas space with a heater. In the first case, the gas volume and PRV damp are the only means to damp transients, so the gas volume must be larger. In the second case, the heater supports the pressure during periods of high withdrawal rate, so the gas space can be smaller. The model predicted that adding a heater would improve tank pressure control and ensure it remained within the desired 60±20 psi range, as shown in Figure 4. The smaller gas space led to an increase in venting, but it was more than offset by larger volume of fuel stored at the outset. The tank with the heater was predicted to hold 18% more LH_2 at the end of the 26 hr simulated flight, as shown in Table 3.

Table 3. Tank performance in a simulated 26 hr flight, with and without a heater. * indicates predicted values.

Case	Initial fuel volume [L]	Total Heat Leak [W]	Fuel vented* [g]	Total Capacity* [g]
No heater, 4 L gas volume	18.0	2.09	0.0	1161
Heater, 0.5 L gas volume	21.5	2.09	19.4	1367

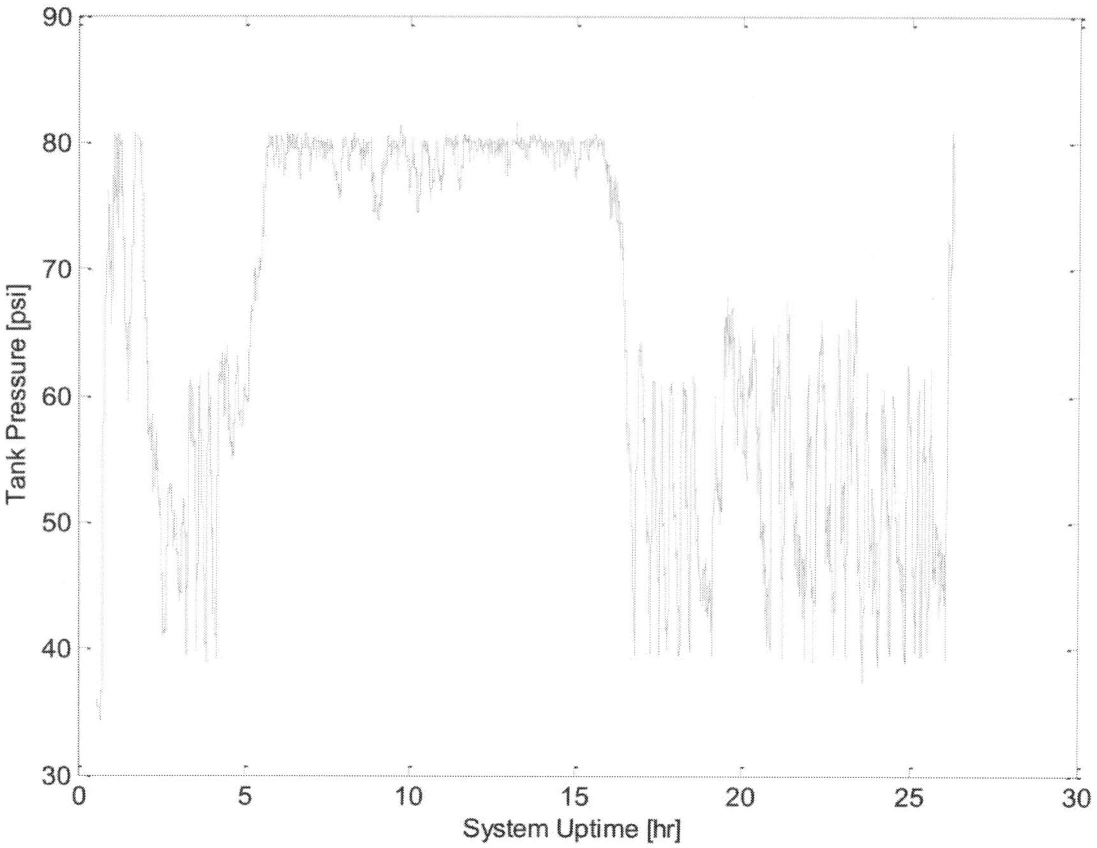

Figure 4. Predicted tank pressure with a heat leak rate of 2.09 W, 0.5 L gas volume, PRV set to 80 psia and heater activated at 40 psia.

Conclusions

The design of a LH_2 storage tank for small aircraft applications is challenging in large part because of the restrictions on weight and heat leak rate. For the Ion Tiger LH_2 storage tank design, these two restrictions drove most of the material selection, mechanical design and thermal design decisions. Steady state heat transfer and transient flow models can provide valuable insights into the performance of proposed designs, and guide the design process, once the basic design concepts have been established. As shown by the insulation modeling, however, validation of the models is essential and empirical models are sometimes preferred to physics based approaches.

Acknowledgements

The authors gratefully thank the Office of Naval Research for funding this work.

References

1. K. Swider-Lyons, R.O.Stroman, G. S. Page, J. F. Mackrell, J. A. Rodgers, M. W. Schuette. *The Ion Tiger Fuel Cell Unmanned Air Vehicle*. in *Proceedings of the 44th Power Sources Conference*. 2010.

2. *NIST Thermophysical Properties Database*. [cited 2011; Available from: http://webbook.nist.gov/chemistry/fluid/.
3. Incropera, F.P. and D.P. DeWitt, *Fundamentals of heat and mass transfer*. 5th ed. 2002, New York: J. Wiley. xix, 981 p.
4. Kays, W.M., M.E. Crawford, and B. Weigand, *Convective heat and mass transfer*. 4th ed. McGraw-Hill series in mechanical engineering. 2005, Boston: McGraw-Hill Higher Education. xxx, 546 p.
5. Brewster, M.Q., *Thermal radiative transfer and properties*. 1992, New York: Wiley. xxii, 543 p.
6. Barron, R.F., *Cryogenic systems*. 2nd ed. Monographs on cryogenics. 1985, New York; Clarendon Press. xvii, 507 p.
7. Gilmore, D.G., *Satellite thermal control handbook*. 1994, El Segundo, Calif.: Aerospace Corporation Press.
8. Marquardt, E.D., J.P. Le, and R. Radebaugh, *Cryogenic material properties database*. Cryocoolers 11, 2001: p. 681-687.
9. C. K. Liu, L.G.N., A. F. Manikowski, Jr. *Sizing an emergency venting system for a cryogenic dewar*. in *Advances in Cryogenic Engineering*. 1986. Cambridge, MA: Plenum Press.

(62a) Optimization of a Louisiana Native Algal Co-Culture for Biofuel Production

Rong Bai[1], Maria Teresa Gutierrez-Wing[2], **Michael G. Benton**[1] and Kelly A. Rusch[2], (1)Cain Department of Chemical Engineering, Louisiana State University, Baton Rouge, LA, (2)Civil and Environmental Engineering, Louisiana State University, Baton Rouge, LA

Microalgae have been reported to possess significantly higher lipid productivity per unit area than traditional oil crops, a benefit for biofuel production. Microalgal strains such as *Nannochloropsis sp., Botryococcus braunii,* and *Chlorella vulgaris* have been reported to contain high lipid/oil content. However, these results are mostly at the lab scale. In large-scale, commercial culture systems, tolerance to local environmental conditions and invasive algal species significantly affect lipid/oil production. Therefore, native microalgal strains with high lipid productivity are likely better options for large-scale biofuel production.

A native fresh water microalgae strain from Louisiana was investigated. This work measures the impacts of irradiance and nitrate levels on total lipid yield and the fatty acid profile of this strain. The culture was harvested in the late exponential-stationary phase to maximize the lipid content, and Soxhlet extraction with chloroform/ methanol (2:1, v/v) was used for total lipid content evaluation. Microwaves and nanoparticle assisted Folch's extraction methods have been used to optimize lipid extraction. The fatty acid profiles of neutral lipids were analyzed by gas chromatography combined with mass spectrometry.

The results indicated the lipid yield and content at 400 and 800 mmol $m^{-2}s^{-1}$ was the highest and no significant differences (32.13 and 32.76 mg $L^{-1}d^{-1}$, 31% and 30%) between the two treatment levels were observed. Results also indicate that although the lower nitrate level increased lipid content, the increase did not compensate for a decrease in growth rate. The fatty acid profiles indicate the major components are 16- and 18-carbon fatty acids, and ~35% are saturated fatty acids. The fatty acid profiles do not vary significantly with the irradiance and nitrate levels.

See more of this Session: Advances In Algal Biorefineries I
See more of this Group/Topical: Sustainable Engineering Forum

(206g) Pyrolysis Vapors Upgrading Using Metal Oxides

Ayman Karim, Donghai Mei, Vanessa Lebarbier, Changjun Liu and Yong Wang, Pacific Northwest National Lab, Richland, WA

Pyrolysis VaporsUpgrading using Metal Oxides

Ayman M. Karim[1],Donghai Mei[1], Vanessa Lebarbier[1], Changjun Liu[1,2] and Yong Wang[1,2]

[1]PacificNorthwest National Laboratory

[2]WashingtonState University

The vapors generated during fast pyrolysis of biomasscontain a significant fraction of light oxygenates (acetic acid, hydroxyacetaldehyde, hydroxyacetone,..) which are produced through undesired fragmentationreactions during pyrolysis. Hydrodeoxygenation of the light oxygenates wouldresult in light alkanes which are not useful fortransportation fuels. Therefore. thelight compounds need to be converted to larger molecules before the hydrodeoxygenation/hydrotreating ofthe pyrolysis vapors (or bio-oil).

In this contribution, deoxygenation strategies for theupgrading of light oxygenates using metal oxides will be presented. The workcombines experimental and density functional theory (DFT) calculations toidentify metal oxide(s) for the

1- Upgrading of light oxygenates (building up) tofuel range compounds (C5+)

2- Deoxygenation ofbiomass depolymerization products (C5+ molecules)

We will show that ketonization and condensation on CeO_2based mixed oxides is an effective way of re-building the light oxygenates intofuel range molecules. CeO_2 based mixed oxides show a 90+% selectivity to acetone during the ketonization of acetic acid. The ketonizationreaction pathway on CeO_2 was investigated using density functionaltheory calculations and the role of the second metal on selectivity andtolerance to H_2O and CO_2 will be presented.

The deoxygenation of the pyrolysis vapors using partiallyreduced metal oxides will also be presented. We show that oxygen vacancies on MoO_3(010) can selectively cleave the C-O bond over C-C bonds.

Finally we will discussthe strategies and challenges to combine both approaches in a single upgradingstep.

See more of this Session: Catalytic Biofuels Refining I
See more of this Group/Topical: Fuels and Petrochemicals Division

(588a) Flexible Energetically Self-Sufficient Co-Production of Hydrogen and Formic Acid

Jorge Pena Lopez, Chemical Engineering Department, University of California, Los Angeles, Los Angeles, CA and Vasilios Manousiouthakis, Chemical Engineering Department, University of California at Los Angeles, Los Angeles, CA

The use of fossil fuels (gasoline, Diesel, etc.) for light vehicle based transportation has resulted in two major environmental quandaries: poor air quality in largely populated cities and increased carbon dioxide emissions into the Earth's atmosphere. The use of hydrogen fueled vehicles has been proposed as an avenue to address the aforementioned dilemmas. However, hydrogen production technologies have yet to achieve a generation system competitive with, for example, the production of gasoline. Several options exist for hydrogen production; one of the most explored has been the reforming of natural gas. However, conventional natural gas reformers release carbon dioxide into the atmosphere. A solution to this issue is redirecting carbon dioxide into formic acid, thus preventing its emission into the atmosphere.

In this work we propose a novel method to synthesize energetically self-sufficient and environmentally friendly process flowsheets, and we show hydrogen production as a case study. To this end we make use of commercially available technologies for the production of formic acid [1]. The resulting process flowsheet goes through heat and power integration techniques that allow the maximum utilization of energy in favor of hydrogen production, maintaining a self-powered plant and the carbon dioxide emissions at zero. Using the simulation tool UniSim Design, we show that there exists an optimal themodynamical and operational point for the generated process flowsheet.

[1] Reutemann W, Kieczka H. Ullmann's Encyclopedia of Industrial Chemistry, "Formic Acid". John Wiley & Sons, Ltd. 2000

See more of this Session: Fundamentals of Hydrogen Production and Utilization
See more of this Group/Topical: Environmental Division

(62b) Controls of Microalgal Biomass and Lipid Production In Municipal Wastewater-Fed Bioreactors

Belinda S.M. Sturm[1], Val H. Smith[2], Marie-Odile Fortier[1], Edward Peltier[1] and Frank Jerry deNoyelles[3], (1)Civil, Environmental & Architectural Engineering, University of Kansas, Lawrence, KS, (2)Ecology and Evolutionary Biology, University of Kansas, Lawrence, (3)Kansas Biological Survey, Lawrence

The use of wastewater as a nutrient source for algal growth as a biodiesel feedstock could increase the viability of this process and reduce wastewater discharge of N and P compounds. In this study, four algal bioreactors were fed continuously with municipal wastewater effluent over a six month period; these were monitored weekly to determine algal biomass and lipid production, nutrient removal, and carbon and nutrient availability. Algal biomass production ranged from 0.78 to 15.9 g dry weight m^{-2} d^{-1} over the experimental period. Algal reactors removed 19% of dissolved nitrogen and 43% of dissolved phosphorus from wastewater effluent. A stoichiometric analysis of particulates revealed that algal biomass had far lower C:P ratios (67:1) and N:P ratios (6:1), but higher C:N ratios (17:1), than Redfield values. These data suggest that the algae produced were rich in phosphorus, and that N was likely to be growth-limiting in these systems. Numerous studies also suggest that algal biomass can be enhanced by providing CO_2 in excess of normal atmospheric concentrations, and that algal biofuel production can contribute to CO_2 sequestration efforts. In order to optimize nutrient removal in wastewater-fed algal reactors, and to determine the potential for CO_2 sequestration, future reactor engineering research should continue to incorporate the expertise of ecological stoichiometry.

See more of this Session: Advances In Algal Biorefineries I
See more of this Group/Topical: Sustainable Engineering Forum

(173g) Binderless Cobalt Oxide Nanoparticle Films Through Electrophoretic Deposition, for Li-Ion Batteries

Don-Hyung Ha, Mohammad A. Islam and **Richard D. Robinson**, Materials Science and Engineering, Cornell University, Ithaca, NY

Nanoscalematerials for Li-ion battery electrodes have received great attention due totheir high surface-to-volume ratio and reduced Li-ion diffusion length whichresults in faster charge/discharge. Nanomaterialsalso show the ability to accommodate strain through volume expansion without pulverizationor capacity fading. In this work we have developed a hollow cobalt oxide NP filmfor Li-ion battery electrodes that require no additional support materials suchas carbon black or polymers to establish good electrical pathways. Thecolloidal cobalt NPs are prepared by decomposition of organometallicprecursors in non-aqueous conditions. The NP film is formed by using electrophoretic deposition (EPD). During EPD, a film isformed on both the cathode and anode but the film on the anode is slightlythicker than that of the cathode, which indicates that more particles arenegatively charged. The EPD deposited films are calcinedto form hollow cobalt oxide (Co_3O_4) NPs that act as the batteryanode.

Battery cycling of these anodes show lowdegradation and are near the theoretical capacity limits for this material. Theseresults are surprising since control experiments made from dropcastfilms form poor battery contacts under the same calcination and battery-test conditions.

See more of this Session: Nanomaterials for Energy Storage III
See more of this Group/Topical: Topical 5: Nanomaterials for Energy Applications

(429g) Algal Genotyping for Biofuel Production

Alice C. Jernigan, Chemical Engineering, University of Arkansas, Fayetteville, AR and **Christa N. Hestekin**, Ralph E. Martin Department of Chemical Engineering, University of Arkansas, Fayetteville, AR

Algae is a high growth source for biomass with a great potential to be a significantly eco-friendly bio-fuel source. Since different species of algae have different oil and lipid concentration it is important to genotype the algae and determine the best species for biofuel production. For genotyping, we examine the 16S and the 18S genes, which are highly conserved among prokaryotes and eukaryotes (respectively), but both have unique small variable regions that can be used to differentiate between species. We analyze the DNA using capillary electrophoresis single strand conformation polymorphism (CE-SSCP). In SSCP, double stranded DNA is heat denatured and snap cooled, which allows the single strands to form unique conformation depending on their sequence. These conformations can then be separated using electrophoresis. Using a reference database of pure culture species, we are working to identify algae from environmental samples being grown for biofuel production. By monitoring the environmental conditions and the corresponding changes in algal species, we can improve biofuel production potential.

See more of this Session: Alternative Fuels and Enabling Technologies I
See more of this Group/Topical: Fuels and Petrochemicals Division

(496g) Conversion of Waste Vegetable Oil to Biodiesel

Ashley Allen and **Farrah Steeley**, Chemical Engineering, Howard University, Washington, DC

Production of Biodiesel from Waste Vegetable Oil

Ashley Allen, Farrah Steeley, Dr. Jason Ganley, Dr. Ramesh Chawla

Howard University, Department of Chemical Engineering Washington, District of Columbia, U.S.A

A stable and reliable energy supply is essential for sustainable economic growth. In recent years, the sharp increase in energy prices has highlighted the need for alternative energy source to alleviate the dependence on fossil fuels as the main energy source for the transportation sector. Biodiesel has received a great deal of attention worldwide because it is a renewable fuel and of its environmental benefits in reducing pollutants emission from diesel engines.

Biodiesel is comprised of fatty acid methyl esters (FAME) that can be acquired from vegetable oil by transesterification with alcohol such as methanol in the presence of an alkaline catalyst. One of the advantages of the biodiesel utilization is that the feedstock can come from a variety of sources. Waste vegetable oil is an attractive option as a biodiesel feedstock. However, the main concerns when using waste vegetable oil are the presence of food solids and other undesirable chemicals in the mixture

This study focuses on the design, operation, and analysis of a cost effective filtration process that utilizes chemical and material engineering principles that incorporates sedimentation synthetic fiber, and aluminum wire mesh to eliminate undesirable particulates, which reduce the yield of biodiesel production. Waste vegetable oil was collected from personal use and local restaurants in the DC area. Samples of 100mL were then allocated and their physical properties were statistically analyzed before and after the filtration process. The produced biodiesel had a viscosity of 165 centistokes, a pH reading of 10.4, a 17 brix percentage refractive index, and a temperature of 20ºC. These results as well as future observations will be presented.

See more of this Session: Alternative Fuels and Enabling Technologies II
See more of this Group/Topical: Fuels and Petrochemicals Division

(288g) Speaking Truth to Power

Deborah Grubbe, BP Products North America, Inc., Chadds Ford, PA

Speaking truth to power

See more of this Session: Developing Leadership Skills
See more of this Group/Topical: Management Division

(429f) The Production of Fuel Grade Butanol From Algal Carbohydrates Utilizing a Hybrid Biological and Chemical Process

Thomas M. Potts, Chemical Engineering, University of Arkansas, Fayettville, AR, Michael Ackerson, Ralph E. Martin Department of Chemical Engineering, University of Arkansas, Fayettville, AR and Jamie A. Hestekin, Ralph E. Martin Department of Chemical Engineering, University of Arkansas, Fayetteville, AR

Butanol has the potential to become a superior renewable fuel, either as a mixture with petroleum derived gasoline or as a stand-alone, neat fuel. The promise of butanol is currently marred by the relatively high cost of production. This paper explores a new and novel approach to the production of fuel grade butanol by combining a bacterial fermentation front end with a catalytic hydrogenation back end. Thus, the fermentation of algal carbohydrates to butyric acid includes a dialysis step to concentrate the butyric acid and remove the inhibitory organic acid from the fermentation broth. The concentrated and de-watered butyric acid is then converted to butanol in a liquid phase catalytic hydrotreater. The kinetics of the hydrogenation is discussed, and a discussion of the economics of the hybrid processing system is presented.

See more of this Session: Alternative Fuels and Enabling Technologies I
See more of this Group/Topical: Fuels and Petrochemicals Division

(206c) High Pressure Catalytic Hydrodeoxygenation of Hydro-Pyrolysis Vapor Lignin Model Compounds

Sara L. Yohe, W. Nicholas Delgass, Fabio H. Ribeiro and Rakesh Agrawal, Chemical Engineering, Purdue University, West Lafayette, IN

Solving our nation's energy problems is of highest priority as developing a source of sustainable domestic energy would improve the energy security and independence of the U.S. and also positively impact the environment and the economy. The transportation sector poses a large challenge to sustainable energy development due to the high energy density fuel required, making liquid fuels favorable over fuel cells or electricity. Development of liquid transportation fuels from biomass provides a solution to creating a sustainable alternative to replace liquid fossil fuels. The H2Bioil (1) process, involving high pressure catalytic fast-hydropyrolysis of biomass followed by a second stage fixed-bed hydrodeoxygenation (HDO) reactor, has been proposed to generate transportation fuels from lignocellulosic biomass. During fast-pyrolysis, hundreds of vapor compounds are generated and, if condensed, would produce a bio-oil with high oxygen content (~35-40 wt %) and an energy content this is half that of gasoline (~17 MJ/kg). Therefore, the vapors must be upgraded, via catalytic hydrodeoxygenation, to increase the energy content, reduce oxygen content to within 3-5%, and increase pH and oil stability.

This presentation will focus on the catalytic hydrodeoxygenation aspect of the proposed process. Due to the large number of vapor compounds produced during fast-pyrolysis and the variety of oxygen functional groups they contain, model compound studies present a method for gaining fundamental understanding of the link between catalyst functionality and oxygen functional group. The lignin portion of biomass poses the greatest challenge in upgrading and conversion to transportation fuels, and therefore lignin derived model compounds were the focus of this study. Supported metal catalysts, such as Ru, have shown desirable hydrodeoxygenation activity of lignin model compounds containing phenol and methoxy oxygen functional groups at 350-800psi in hydrogen and 300-450oC without hydrogenating the aromatic ring.

References:

1) R. Agrawal, N. R. Singh, Synergistic Routes to Liquid Fuel for a Petroleum Deprived Future, AIChE Journal,55,7,1898-1905, 2009.

See more of this Session: Catalytic Biofuels Refining I
See more of this Group/Topical: Fuels and Petrochemicals Division

(206e) Upgrading of Short Oxygenates From Pyrolysis Oil Via Condensation Reactions for the Refining of Biofuels

Anirudhan Gangadharan, Chemical, Biological and Materials Engineering, University of Oklahoma, Norman, OK, Daniel E. Resasco, School of Chemical, Biological and Materials Engineering, University of Oklahoma, Norman, OK and Richard Mallinson, Chemical, Biological, and Materials Engineering, University of Oklahoma, Norman, OK

Torrefaction of biomass between 270-290 ^{o}C produces significant amounts of short oxygenates including acetol, acetic acid and acetone together with large amounts of water. The carbon number and energy content of these oxygenates are too low to be directly utilized for the production of transportation fuels and there is a need to build the carbon chain of these oxygenates. Condensation reactions are an important route to build the carbon chain to produce molecules in the gasoline and diesel range without the external addition of hydrogen and it is of interest to perform these reactions *in-situ* after the pyrolysis of biomass to prevent degradation reactions within the liquid bio-oil product. To do this, a suitable catalyst has to be designed that will selectively perform condensation reactions in the presence of a large excess of water, while resisting deactivation by furfural and phenolic compounds. In this work, a model study of pure and mixed feeds of acetol, acetic acid and water were carried out over metal oxides, with the aim of maximizing the yield to acetone. Acetone is a useful intermediate that can be further upgraded via self aldol condensation reactions and cross aldol condensation reactions with furfural to form fuel range carbon chains. It can also be hydrogenated to iso-propanol, which is a good alkylating agent for phenolic compounds. From the reaction of the mixed feed, acetone was obtained as the major product, with small amounts of cyclopentanones and other oxygenates. From the reaction of pure acetol, the yields of acetone, cyclopentanones and other oxygenates were comparable. It is believed that the pathway to acetone from acetol involves several possible steps. The dehydrogenation of acetol produces 2-oxopropanal that undergoes decarbonylation to give acetaldehyde. The ketonization of acetaldehyde gives acetone. Acetol also apparently undergoes cyclization through a series of steps to give various cyclic ketones. These are believed to form a surface pool that undergoes cracking to yield acetone.

See more of this Session: Catalytic Biofuels Refining I
See more of this Group/Topical: Fuels and Petrochemicals Division

(436b) Direct Production of Molecules In the Fuel Range by Selective Tailoring of Biomass Fast Pyrolysis

Piotr Gawecki[1], Fabio Ribeiro[1], W. Nicholas Delgass[1], Rakesh Agrawal[1], Andrew Smeltz[1], Matthew R. Hurt[2], David J. Borton II[2], Nelson R. Vinueza[2], Hilkka I. Kenttamaa[2], Nicholas J. Nugent[3] and William E. Anderson[3], (1)Chemical Engineering, Purdue University, West Lafayette, IN, (2)Chemistry, Purdue University, West Lafayette, IN, (3)Aeronautics and Astronautics, Purdue University, West Lafayette, IN

Pyrolysis of biomass is being evaluated as a promising technique for production of fuels and value-added chemicals in place of petroleum-based approaches. We have developed a technique that allows for direct, real time detection and analysis of pyrolysis products. Using sub-milligram quantities of biomass heated at rates up to 20,000 K s^{-1}, pyrolyzed directly in front of a linear quadrupole ion trap mass spectrometer, we are able to determine the true primary products, which are small fragments of the biomass building blocks. Furthermore, we have shown that it is possible to systematically tailor secondary reactions of these primary products by varying the temperature and residence time they spend in a flow tube. Specifically, we are able to restrict the product distribution to the number of carbons useful for making liquid fuels (C_{12}-C_{18} range). The findings from the sub-milligram reactor have been verified in a 10 g min^{-1} reactor.

See more of this Session: Biomass Pyrolysis II
See more of this Group/Topical: Fuels and Petrochemicals Division

(690e) Energy Storage and Generation From Thermopower Waves: Covalently Functionalized Thermal and Electrical Conduits

Joel T. Abrahamson[1], Changsik Song[2], Jennifer H. Hu[1], Jared M. Forman[1], Sayalee G. Mahajan[1], Nitish Nair[1], Wonjoon Choi[3], Eun-Ji Lee[2] and Michael S. Strano[1], (1)Department of Chemical Engineering, Massachusetts Institute of Technology, Cambridge, MA, (2)Department of Chemistry, Sungkyunkwan University, Suwon, South Korea, (3)Mechanical Engineering, Massachusetts Institute of Technology, Cambridge, MA

We have demonstrated through simulation and experiment that the nonlinear coupling between an exothermic chemical reaction and a nanowire or nanotube with large axial heat conduction accelerates the thermal reaction wave along the nano-conduit [1-3]. The thermal conduit rapidly transports energy to unreacted fuel regions, and the reaction wave induces a concomitant thermopower wave of high power density, producing electrical current in the same direction. At up to 7 W/g, this can be substantially larger than the power density offered by current micro-scale power sources (e.g. fuel cells, batteries) and even about seven times greater than that of commercial Li-ion batteries. MEMS devices and wireless sensor networks would benefit from such high power density sources to enable functions such as communications and acceleration hampered by present power sources.

Recently, we have covalently attached energetic molecules of mono-, di-, and trinitrobenzenes to single-walled carbon nanotubes via diazonium chemistry. Although covalent functionalization introduces defects that can scatter electrons and phonons, thermopower waves can still rapidly propagate and produce electricity on these decorated nanotubes. Differential scanning calorimetry confirms that the energetic molecules release heat, but thermopower waves require additional fuel adsorbed on the nanotubes to propagate.

[1] W. Choi, S. Hong, J. T. Abrahamson, *et al. Nature Materials*, 9, (2010), 423.

[2] W. Choi, J. T. Abrahamson, J. M. Strano, M. S. Strano. *Materials Today,* 13, (2010), 22.

[3] J. T. Abrahamson, *et al. ACS Nano* 5, (2011), 367.

See more of this Session: Energy Storage and Electrochemical Processes I
See more of this Group/Topical: International Congress on Energy 2011

Quantitative Comparison of Chemical and Biological Cardboard's Lignin Degradation

E. Isaza
Department of Chemical Engineering, Universidad de Los Andes, Bogotá, Colombia
R. Sierra R.
Advisor, Universidad de Los Andes, Bogotá, Colombia

ABSTRACT: An efficient way to increase biomass digestibility, thereby, make it usable in the production of second generation biofuels, is pretreatment. However, it demands important energy inputs which make it one of the most expensive stages in biofuels processing. Comparisons between chemical (oxidative with lime) and biological (with commercial *Laccase* extract) pretreatments were performed both by experimental measurements and by simulations. All experiments were performed following a full factorial experimental design, with lignin and sugar yields as response variables. The influence of the type of pretreatment (i.e., pretreatment yields) on the process that leads to biogasoline and jet fuel production (MixAlco®) was simulated in AspenPlus®. This simulation showed higher productivity and energy efficiency (15 and 60% respectively) for the chemical pretreatment over the biological pretreatment.

1 INTRODUCTION

First generation biofuels have emerged as an alternative to fossil fuels, whose price and environmental impacts have been subject of intense debate over the last years. From all biofuels, biodiesel and bioethanol are the most popular. They are currently produced using high energy crops such as soy, sugar cane, oil palm and corn starch as feedstocks (U.S. State Department, 2010). These resources, however, are characterized by such high commercial and alimentary value that they become restrictive for biofuels and generate issues of alimentary sustainability (Lee, Biological conversion of lignocellulisic biomasss to ethanol, 1997).

The utilization of more inexpensive feedstock in fermentative process such as lignocellulosic residues is a good baseline for the optimization of this technology (Lee, Biological conversion of lignocellulisic biomasss to ethanol, 1997). Good feedstock resources are agricultural, industrial, and municipal wastes among others. The *MixAlco®* process, created by Dr. Mark Holtzapple in Texas A&M University and patented by Terrabon L.L.C, uses this kind of waste to produce ethanol and higher energy alcohols.

In this process, four main stages may be differentiated: pretreatment, fermentation, chemical dewatering, gasification and cogeneration, thermal conversion involving hydrogen, and final conversion into biofuels. Among these, pretreatment is the most expensive (Gousamme, 2011).

In biological and some chemical pretreatments (such as lime pretreatment), biomass compositional changes are sought. These changes, include but are not limited to delignification. As a result, an increased biomass digestibility is observed. Other effects of pretreatment are: disruption of the crystalline structure of cellulose, increased pore size to enhance enzyme access to carbohydrates, and some degree of separation of the main lignocellulose constituent polymers.

Lignin is a highly heterogeneous tridimensional aromatic polymer, formed by three different units of phenylpropylene: coumaric alcohol, coniferyl alcohol and sinaphyl alcohol; bounded randomly through C-C and C-O-C type bounds (Sylvia, Fuhrman, Hartel, & Zuberer, 1998), (Lee, Biological conversion of lignocellulisic biomasss to ethanol, 1997).

Widely extended delignification processes such as Kraft pulping, (Lachental, Mortha, Sevillano, & Zaroubine, 2004) are controversial due to the use of environmental harmful chemicals. Delignification in alkali with possible oxygen addition is preferred because comparable delignification levels are obtained with less toxic waste products. Pretreatments such as oxidative and non-oxidative lime (i.e., calcium hydroxide ($Ca(OH)_2$), have been reported to promote lignin degradations close to 70% of the compound (Chang, Nagwani M, Kim, & Holtapple, 1999).

On the other hand wild occurring delignification processes are promoted by extracellular enzymes released by fungi, bacteria (Zhang, Kjonas, & Flurkey, 1999) or even plant cells; however, white rot fungi are the degraders by excellence (Lee, Biological conversion of lignocellulisic biomasss to ethanol, 1997). They generate extracellular enzymes such as *Laccase*, Manganese peroxidase and lignin peroxidase. Laccase (EC. 1.10, 32), benzendioloxigenoxidoreductase, is a copper containing enzyme capable of lignin degradation in the presence of oxygen. The only product of this transformation is water (Rodriguez Cuoto & Toca-Herrera, 2006). Because of this reason, this enzyme has been widely studied and used in industrial applications.

Since there are many advantages and drawbacks for both chemical and biological delignifications, it is desirable to formally compare these processes. However, a systematic comparison on the same basis (same feedstock, analytical procedures and general methods) was not found in the literature. This study aimed to make such comparison. To accomplish this goal, a full factorial design was implemented to both: oxidative lime delignification of cardboard (high in lignin content) and delignification of the same material using purified lacasse (provided by Sigma Aldrich)

The experimental data obtained from the best global conversion of carbohydrates was loaded to an existing preliminary simulation of the *MixAlco®* process (Gousamme, 2011) to determine which of the delignification processes is best on the basis of highest productivity and energy efficiency.

2 METHODOLOGY

2.1 *Chemicals and reagents*

2.1.1 Feedstock for pretreatments and substrate for enzymatic hydrolysis

Recycled cardboard is commonly used in Colombia to transport eggs. This biomass is characterized by high lignin content; a desirable feature for a delignification study. Untreated cardboard was the feedstock for pretreatment. The samples were obtained from *Kikes eggs*, Leader Company in the egg market at Colombia. Pretreated cardboard was used as substrate for enzymatic hydrolysis.

2.1.2 *Reagents for enzymatic delignification*

The selected enzyme was *Laccase Suberase* brand *Novartis*. It was activated in a phosphate buffer at pH 3, 4.5, 5, 5.5, 6, 6.5 and 7.

2.1.3 *Chemical pretreatment*

Hydrogen peroxide was obtained from *Panreac* at (%w/w) 0.5, 1, and 2. Calcium hydroxide $Ca(OH)_2$ also was obtained from Panreac.

2.1.4 *Enzymatic saccharification*

It was performed using cellulase 15 FPU obtained from *Sigma Aldrich*.

2.2 *Characterization of the selected biomass*

Quantification of biomass components (i.e., lignin, structural carbohydrates, and ash), was obtained following the laboratory analytical procedures (LAP) produced by the *National Renewable Energy Laboratory (NREL)* for lignocellulosic biomass (Sluiter A. , Hames, Ruiz, Scarlata, Sluiter, & Templeton, 2004). A summary of experimental protocols follows.

2.2.1 *Biomass conditioning*

In accordance to NREL Lap entitled "Preparation of Samples for Compositional Analysis" (Hames, Ruiz, Scarlata, Sluiter, Sluiter, & Templeton, 2008), all biomass was conditioned in a step previous to experimentation, by reducing: particle size (the fraction taken for experimentation was +40/-80 Thyler mesh) and the moisture content was reduced to less than 10%, by drying in a convection oven at 45°C for >24 h.

2.2.2 *Ash content determination*

The ash content of raw biomass was determined by exposing a known dry weight of biomass (7 replicates) to calcination during four hours in a muffle at 575°C. The weights of the resulting ashes were measured and related to the initial dry weight of each sample. (Sluiter A. , Hames, Ruiz, Scarlata, Sluiter, & Templeton, 2008).

2.2.3 *Structural sugars and lignin*

These components were measured in accordance to NREL laboratory analytical procedure entitled "Determination of Structural Carbohydrates and Lignin in Biomass" which involves two acid hydrolysis steps with H_2SO_4. The first stage is concentrated at 72% and the second diluted with 4%. This procedure results in solubilization of structural carbohydrates while lignin is precipitated (Scarlata, Slutier, Termpleton, Crocker,

Hames, & Ruiz, 2008). Sugar concentration is determined by the colorimetric method dinitrosalicilic acid or DNS, and lignin is separated and gravimetrically measured.

2.3 *Chemical pretreatment*

Cardboard (1 g), lime (0,5 g) and 45 mL of hydrogen peroxide were well mixed into a stainless steel reactor (Parr 500). Afterwards, the reactor was tightly closed, and submitted to a pretreatment at the temperature, hydrogen peroxide concentration and time of pretreatment that corresponds to the experiment according to the factorial design 2^3 summarized in Table 1.

Table 1. Factorial design of chemical pretreatment.

Factor	Leve l1	Lev el 2
Hydrogen peroxide concentration%	0.5	2
Operating time (min)	30	120
Operating temperature (°C)	80	150

Pretreated samples were neutralized using 5 N HCl, washed with abundant distilled water, filtered through Whatman #1 filter paper and weighted. Afterwards, the harvested biomass was divided into three parts. The first part was kept frozen to be submitted to enzymatic hydrolysis later. Another part was put to dry in a 45°C convection oven, to be used in lignin and sugar quantification. The last part was used to determine moisture content.

A parameter was calculated to determine total solids yield after pretreatment (Y_T) as follows:

$$Y_T = \frac{\text{recovered biomass (g)}}{\text{biomass before pretreatment (g)}} \qquad (1)$$

With this parameter, it is possible to determine structural carbohydrates and lignin yields (Y_s) (Y_L) as defined in equations (2) and (3),

$$Y_s = \frac{\text{recovered carbohydrates (g)}}{\text{carbohydrates before pretreatment (g)}} = Y_T \frac{C_0}{C_t} \qquad (2)$$

$$Y_L = \frac{\text{recovered lignin (g)}}{\text{lignin before pretreatment(g)}} = Y_T \frac{L_0}{L_t} \qquad (3)$$

Where, (L)=Lignin; (C)= Sugars; (0)=initial; and (t)=at the end of the pretreatment.

For lime pretreatment, kinetics of lignin and sugar degradation were modeled by standard power law (Eq. 4), accounting for two reacting moieties (fast and slow) as proposed by (Sierra, et al., 2011). Model parameters are obtained by minimization of the sum of squared residuals.

$$-r_i = -\frac{dY_i}{dt} = k_{if} M_{O_2}{}^{\beta if} Y_{if} + k_{is} M_{O_2}{}^{\beta is} Y_{is} \qquad (4)$$

Where, (i)=Lignin or sugars; (j)=slow or fast reaction; k_{ij}= rate constant for substance; R=universal constant law for ideal gases (8.314×10^{-3} kJ/(mol K); M_{O_2}=oxygen loading (kg initial oxygen/kg dry biomass); and β_{ij}= reaction order for hydrogen peroxide.

2.4 *Enzymatic pretreatment*

2.4.1 *Activity determination*

Activity determination of the enzyme, *Laccase Suberase* brand *Novartis*, was made according to the *Sigma Aldrich* protocol (Aldrich, 1997). This protocol recommends the use of 0.216 mM Syringaldazine solution, water and *Laccase Suberase*, to determine the activity by a continuous spectrophotometric rate determination (Aldrich, 1997). Because experiments were run at several different pH (Table 2), one result of this experimentation was the recommended pH.

2.4.2 *Selectivity*

Enzymatic delignification is supposed to be highly (if not completely) selective. Nevertheless, a selectivity experiment was performed in order to ratify that the enzyme was free of cellulases. For this purpose, the enzyme (20 μL of *Laccase*) was put in contact with 0.1 g of cellulose at pH of 6, 6.5 y 7, and room temperature and 55°C for a day.

2.4.3 *Experimental Design*

A full factorial design 3^3 was made in order to determine enzyme loading, substrate concentration, residence time and temperature that would maximize enzyme performance at optimum pH of 6.5 and optimum temperature of 55°C. (Baldrian, Fungal Laccasses-ocurrance and properties, 2006). It is summarized in Table 2.

Table 2. Factorial design for enzymatic pretreatment

Factor	Level 1	Level 2	Level 3
Enzyme loading (mL/g substrate)	0.0 1	0.1	0.5
Substrate concentration (w/w)	1	5	10
Residence time (min)	60	360	1440

Immediately after pretreatment, enzyme pretreated samples were submitted to thermal shock to denature the enzyme and therefore stop the reactions. Subsequently, they were profusely washed with distilled water.

Based on the obtained results a decrease on the number levels of the factorial design is proposed. Reducing: residence time(0-1 h), enzymatic loading (20,150 μL) and substrate concentration (2 g). For a better compression of the degradation of lignin it is modeled by Michaelis Menten:

$$-r_s = \frac{V_{max} \times (S)}{K_m + (S)} \qquad (5)$$

Where (S)=substrate concentration; K_m=affinity constant; and V_{max}=maximum degradation rate of substrate.

2.5 *Enzymatic hydrolysis*

To determine the effect of pretreatment in biomass digestibility, an enzymatic hydrolysis is made quantifying the cellulose fraction that is hydrolyzed by cellulase at optimum conditions.

The procedure is the *NREL Lap* for *Enzymatic Saccharification of Lignocellulosic Biomass* (Brown & Torget, 1996). It consists on exposing pretreated biomass with an excess of cellulase (15 FPU/g biomass) at 50°C and pH 4.8 for 3 days. The protocol is modified adding 1g of biomass instead of 1 g cellulose equivalent in biomass.

The monomers of glucose and xylose are determined later by DNS, and digestibility is calculated with relation (5) and (6):

$$Y_{saccharification} = \frac{\text{digeted celulose (g)}}{\text{added celulose (g)}} \qquad (5)$$

$$Y_{saccharification} = \frac{\text{sugars}_e \times V_T}{\text{biomass(g)} \times \text{SST} \times \text{sugars}_p} \qquad (6)$$

Where, $sugars_e$=sugar content of saccharification extract (mg sugars/ml); V_T=total volume of saccharification (10.6 ml), biomass =pretrated biomass= 0.1 g; SST= total solid content in pretreated biomass; and $sugars_p$ =sugar content of pretreated biomass (g sugars/g pretreated biomass).

2.6 *Global conversion*

The combined effect of pretreatment and enzymatic hydrolysis a global recuperation factor is found:

(6)

Mentioned relation allows choosing the conditions of chemical and enzymatic pretreatment to load in the *MixAlco®* process simulation, where higher values of correspond to superior efficiency in pretreatment, therefore, preservation of sugars, breakage of crystalline structure of celluloses and hemicelluloses and higher lignin degradation.

2.7 *MixAlco® process simulation*

The simulation of the *MixAlco®* process, is modified from a existing one made by (Goussame,2011) where experimental data of both pretreatments are loaded. Accomplishing finally, the comparison of the two pretreatments, at a preliminary stage, because the simulation is an ideal modeling of the process.

The energy index was defined by (Gousamme, 2011) as,

(7)

3 RESULTS AND DISCUSSION

3.1 *Biomass characterization*

The obtained results of biomass characterization protocols are, that non pretreated biomass is composed in dry weight: 43% by lignin, 3% by ash, and a 44% of sugars, Illustration 1.

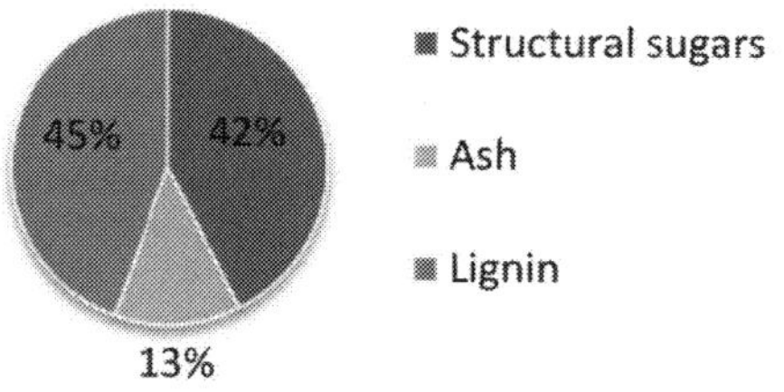

Ilustration1. Composition of cardboard in dry weight.

3.2 *Chemical pretreatment*

3.2.1 *Lignin degradation*

The average of the experimental results obtained for factorial design 2^3, for chemical pretreatment are presented in Illustration 2 with respective standard deviation.

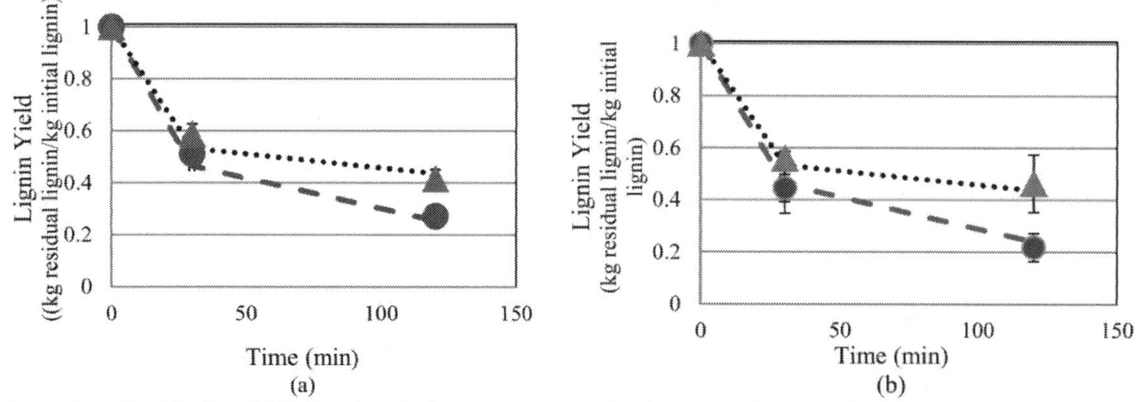

Illustration 2. Lignin yield for chemical pretreatment with lime Ca(OH)2 and respective kinetic model. (a) 80°C(b)150°C. (··▲··) 2% hydrogen peroxide, (━●━)0.5% hydrogen peroxide.

According to Illustration 2 at higher peroxide concentration, temperature and time, greater lignin degradation. This can be explained because higher hydrogen peroxide concentration generates more quantity of superoxide ions responsible of breaking C-C and C-O-C bounds.

As shown in Ilustration2, the time that superoxide ions take in breaking the lignin chain ends, where the degradation begins, is half hour. With higher residence time an increase of the lignin degradation is measurable, because the numerical relation between superoxide ions and attack sites become more efficient.

The highest delignification is at two hours, with lime ($Ca(OH)_2$) at 150°C and hydrogen peroxide at 2%.

Calculated lignin yields are presented as cutter lines in the operating rank of each pretreatment. Data is obtained by least squares minimization of the model presented by Sierra ET all, 2009, whose solution generates the parameters show in Table 3.

Table 3. Achieved values for parameters of lignin degradation with lime $Ca(OH)_2$.

Paramet er	Value	Units
Y_{if0}	0.545	g residual lignin/g initial lignin
a_{iif}	0.0597	min^{-1}
B_f	0.0257	
E_{iif}	0.0825	kJ/mol
a_{ifs}	4.96E-07	min^{-1}
B_s	2.10	
E_{iis}	0.878	kJ/mol

3.2.2 *Sugar Yields*

Average of experimental data obtained for sugar yield in chemical pretreatment are presented as marks in Illustration 3.

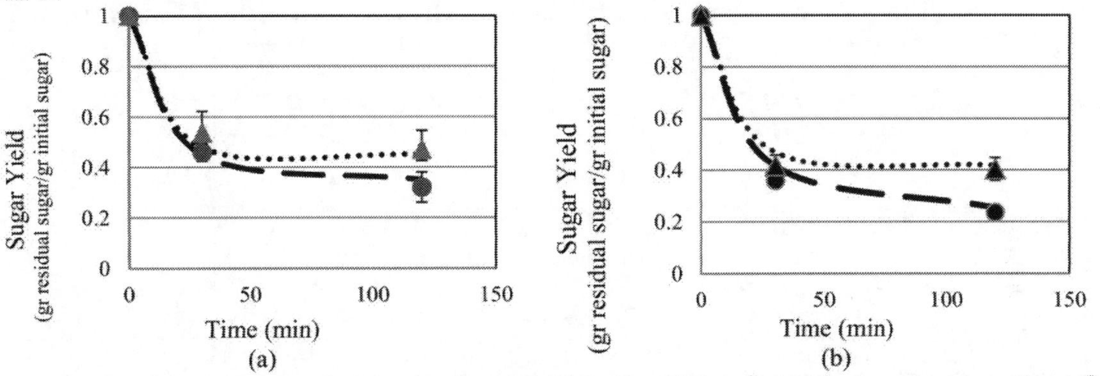

Illustration 3. Sugar yield for chemical pretreatment (a) 80°C, (b) 150°C. (··▲··) hydrogen peroxide at 2% , (–●–)hydrogen peroxide.0.5%

The behavior of sugar yield is similar to lignin yield, because at higher operation conditions lower sugar yield, this according to the presence of superoxide ions and harder thermal conditions.

Comparing part (a) and (b) of the Illustration 3, can be observed that sugars are sensitive to temperature degradation because at every pretreatment with equal time and peroxide concentration and higher temperature lower sugar yield is obtained.

The effect of time in sugar degradation is least at lower hydroxide concentration.

The kinetic modeling is made same as the lignin degradation, the obtained parameters are:

Table 4. Obtained values of kinetic parameters for sugar degradation with lime Ca(OH)2.

Parameter	**Value**	**Units**
Y_{if0}	0,489	g residual sugar/g initial sugar
a_{iif}	0,00156	min^{-1}
B_f	1,02	
a_{ifs}	911×102	min^{-1}
B_s	4,315	
E_{iif}	11,8	kJ/mol
E_{iis}	0,869	kJ/mol

3.3 *Enzymatic pretreatment with Laccase*

3.3.1 *Activity determination and enzyme selectivity*

The higher activity is 864 U ml^{-1} at pH 6.5, Illustration 4.

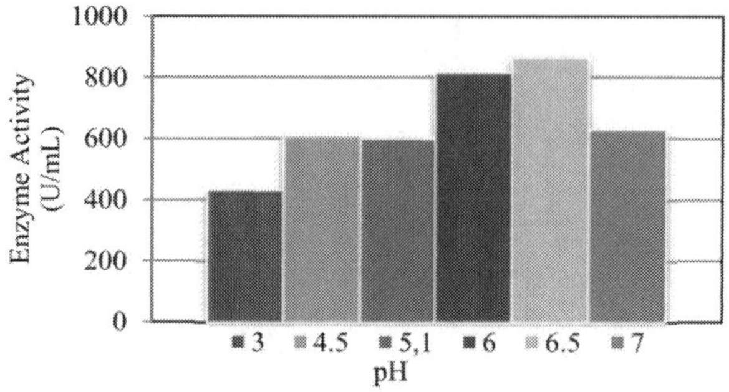

Illustration 4. Activity of the enzyme.

Obtained results for enzyme selectivity showed changes of 10^{-3} g, attributed to moisture content in the samples.

3.3.2 *Enzymatic lignin degradation*

Enzymatic experimental average results of the first substrate level (2 g) of the factorial design 3^3 are presented in Illustration 5. Only this level is presented because the behavior obtained for the other two levels of substrate, 1.5 and 1.0 g are very similar; (where at every enzymatic loading there is an increase of the lignin

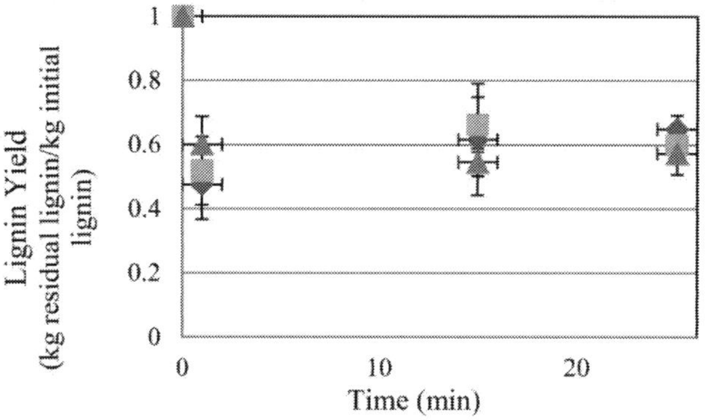

Illustration 5. Enzymatic delignification (◆) 20 μL, (■) 150 μL, (▲) 1000 μL.

For this reason the experimental conditions were cut to: a maximum of an hour, for residence time, and two levels of enzymatic loading, 20 and 150 μL, with a single substrate concentration, 2 g.

In Illustration 6, the results of the last Laccase experimentation are presented where the higher lignin degradation is 0.45(g residual lignin/g initial lignin), accomplish by 150 μL of enzyme at one hour. While the minimum yield is 0.75(g residual lignin/g initial lignin) obtain with 20 μL of Laccase at 30 min.

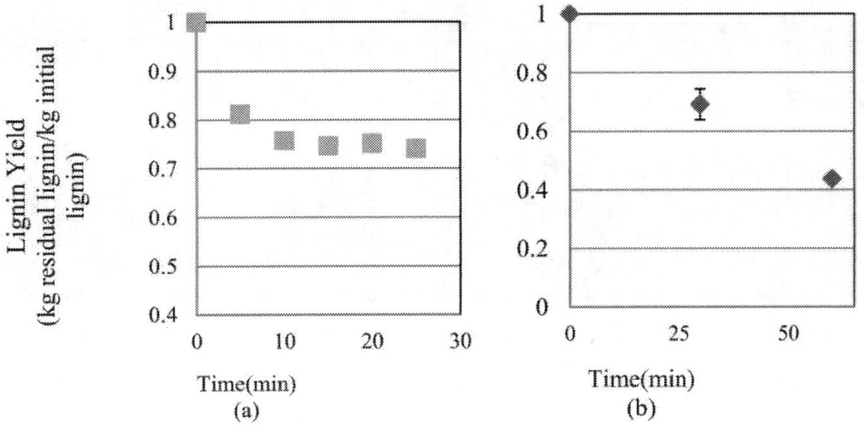

Illustration 6. Enzymatic delignification of 2 g of substrate (a) 20µL enzyme, (b) 150 µL enzyme.

The kinetic modeling for the enzymatic lignin degradation was made by the equation of Michaelis Menten as a first approximation.

The Lineweaver-Burk linearization was used to model the kinetic parameters, for 20 y 150 µl of enzyme data.

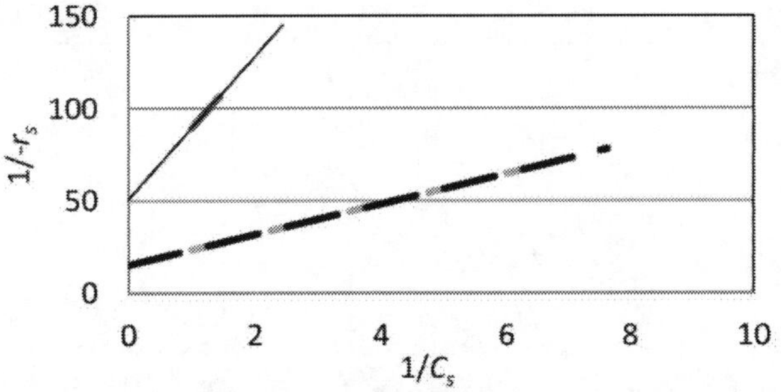

Illustration 7. Linewever-Burk for 2 g of biomass. (- -) 20 µL, ( ) 150 µL.

The values obtained for the kinetic parameters are presented at the Table 5.

Table 5.Obtained values for the enzymatic degradation of lignin.

Parameter	Level	Value	Units
V_{max}	20µl	0,0646	U mg^{-1}
K_m	20µl	0,529	g cm^{-3}
V_{max}	150µl	0,0195	U mg^{-1}
K_m	150µl	0,752	g cm^{-3}

3.3.3 *Sugar Yield*

As shown at Illustration 8, it can be identified that the higher sugar yield is obtained with enzymatic pretreatment at ten minutes and 20 µL of enzyme, while the lowest is also obtained with 20µL at 20 minutes.Thereby time is an important factor at this enzymatic loading level.

At 150 µL of enzyme, sugar yield keeps near 0.6 (g recover sugar/g initial sugar) in the selected time range.

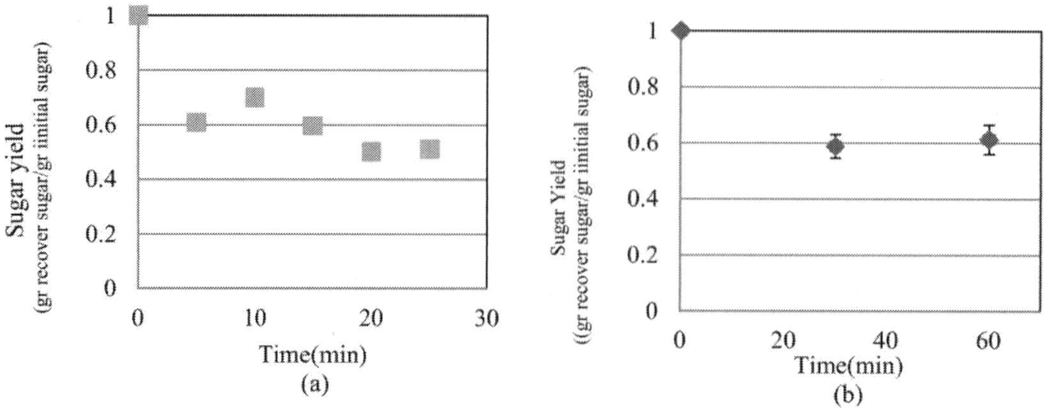

Ilustration8. Sugar yield for 2 g of substrate (a) 20 µL of enzyme (b) 150 µL of enzyme.

3.4 *Enzymatic saccarification*

Higher digestibility values were achieved for chemical pretreatment with the higher level of: time (2 h) and hydrogen peroxide concentration(2%), but since results are similar with different temperatures, Illustration 9, it is necessary to calculate global sugar conversion to select the best chemical pretreatment conditions.

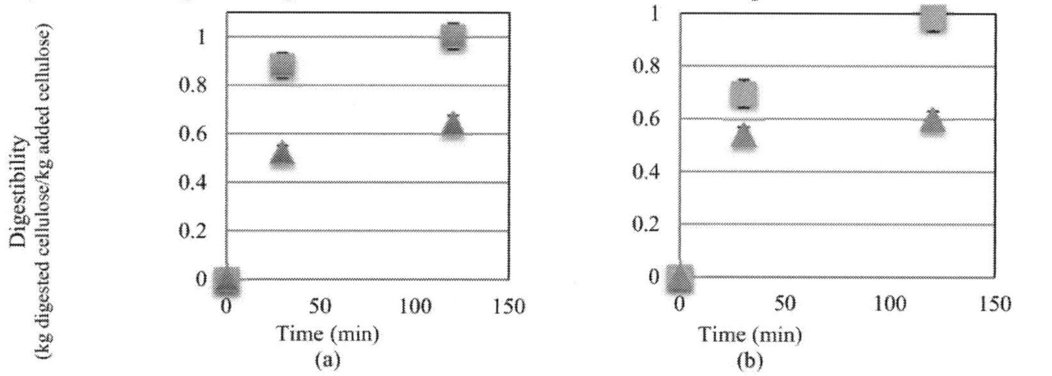

Illustration 9. Digestibility of chemical pretreated biomass.(a) 80ºC, (b) 150ºC. (▲) 0.5% peroxide,(■)2% peroxide.

Highest digestibility for enzymatic pretreatment is at one hour of pretreatment with *Laccase,* with 150 µL of enzyme loading, Illustration 10.

The biomass, pretreated with 20 µL of *Laccase* obtained the highest digestibility at 10 minutes of pretreatment, which can be justified with the fact that low enzyme concentration triggers quicker enzyme action because there is lower enzyme competition, but also is less efficient compared to 150 µl of enzyme where are more competition but higher lignin degradation.

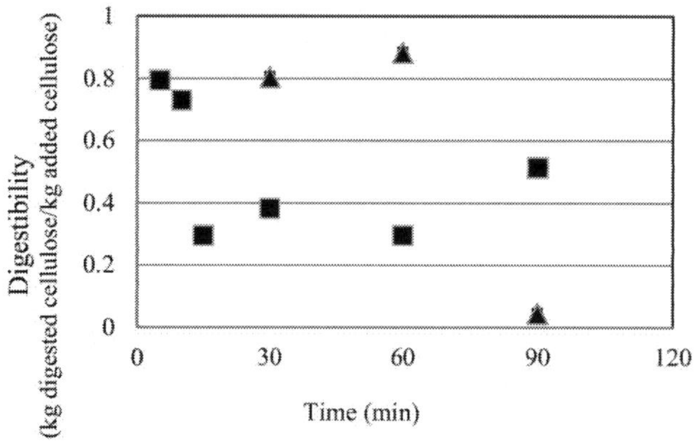

Illustration 10. Digestibility enzymatic pretreated biomass (■) 20µL and (▲)150 µL de enzyme.

The calculation of global conversion is necessary to determine which chemical and enzymatic pretreatment is the proper one to implement at Aspen® simulation of the *MixAlco®* process.

The global conversion results are presented in a contour graph, made using *MiniTab Statistical Software©.*

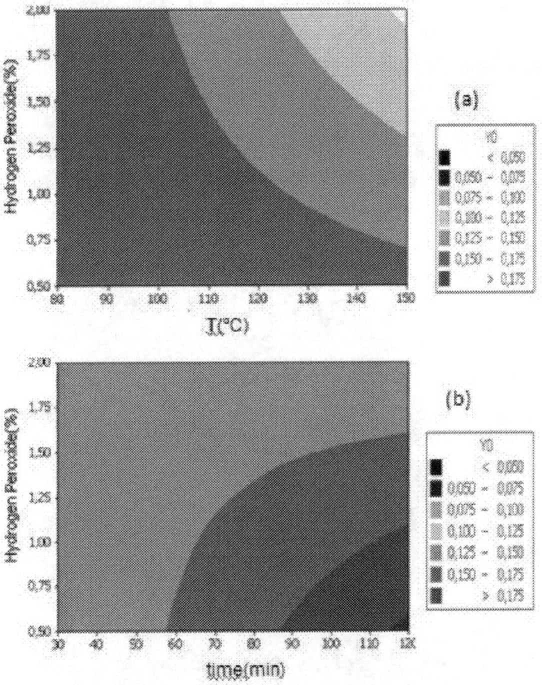

Illustration 11.Contour graph of global conversion for chemical pretreated biomass. (a) %Peroxide vs. temperature(°C),(b) %peroxide vs. time(min).

The global sugar conversion, Y_0, of chemical pretreatment samples are shown in Illustration 11, where at less peroxide concentration, lower temperature y higher time results in the best sugar conversion which can be explained by sensibility of sugar degradation to temperature and lower superoxide ions concentration.

The global sugar conversion for enzymatic pretreated biomass is presented in Illustration 12, where the zone with higher sugar conversion correspond to one hour, 150 μL of enzyme loading which can be explained due to enzyme selectivity to lignin degradation and the inverse relation between enzyme competivity and residence time.

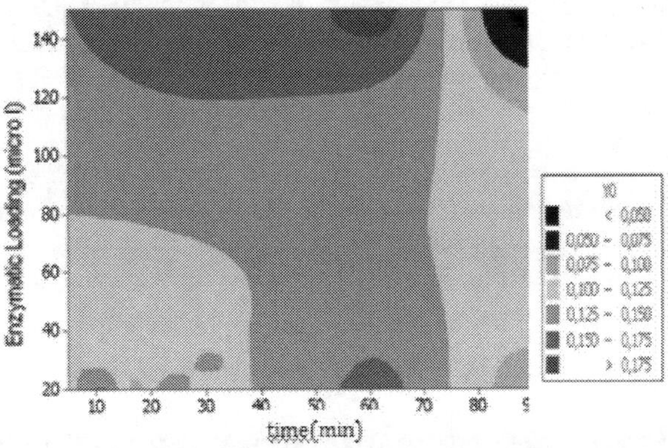

Illustration 12. Contour graph for global sugar conversion for enzyme pretreated biomass.

The best global conversion value was achieved with enzyme pretreated biomass at one hour of pretreatment and 150 μL of enzyme because the recuperation factor Y_t was close to one and the sugar degradation was low compare to chemical degradation recuperation factor close to ,5 and higher sugar degradation.

3.5 *Aspen Plus® Modeling*

The pretreatment conditions chosen to be loaded at the *MixAlco®* process simulation, design by (Gousamme, 2011) are the corresponding to chemical pretreatment 0.5% peroxide, 80°C y 120 minutes and the corresponding to enzymatic pretreatment with 150 μL and one hour of residence time.

The higher biofuel production was achieved with chemical pretreatment with an energy index of 4,477 vs. 4,189 (kJ out/kJ in) of the enzymatic pretreatment.

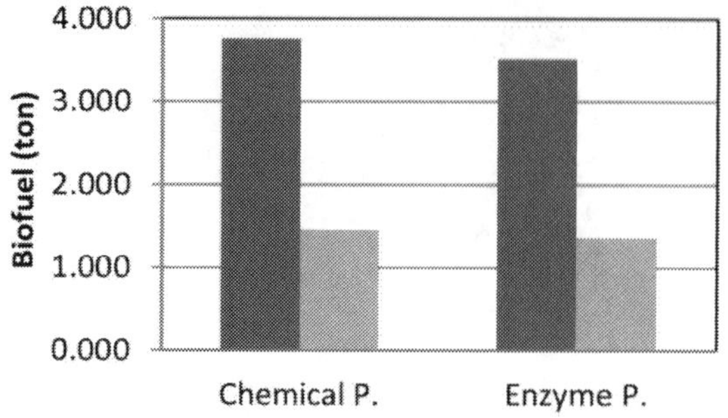

Illustration 13. Bar graph for biofuel production with chemical and enzyme pretreatment. ( ) Jet biofuel, ( ) gasoline.

4 CONCLUSIONS

Comparing chemical with enzymatic pretreatment, the enzymatic is more sensitive to time and pH, which complicates the implementation of this kind of pretreatment in the industry.

Enzymatic pretreatment also achieved the best values for digestibility and global sugar conversion.

The simulation, final step of the comparison between chemical and biological pretreatment, gives better results for the chemical pretreatment in biofuel generation and energy index.

5 BIBLIOGRAPHY

1. Aldrich, S. (1997 11-03). *Enzymatic Assay of LACCASE.* Retrieved 2011 12-03 from http://www.sigmaaldrich.com/etc/medialib/docs/Sigma/General_Information/2/laccase.Par.0001.File.dat/laccase.pdf
2. Baldrian, P. (2006). Fungal Laccasses-ocurrance and properties. FEMS Microbial Review 30 , 215-242.
3. Brown, L., & Torget, R. (1996). *Enzymatic Saccharification of lignocellulosic Biomass.* NREL.
4. Carvajal Montoya, S., & Giraldo Garcia, D. (2010). Simulación del proceso MixAlco en Aspen Plus y análisis de sensibilidad de los parámetros involucrados en las principales etapas del proceso. Bogota D.C: Proyecto de grado, Universidad de los Andes.
5. Chang, V., Nagwani M, M., Kim, C., & Holtapple, M. (1999). *Oxidative Lime Pretreatment of High-Lignin Biomass.* College Station: Texas A&M.
6. David, C., Salvador, S., Dirion, J., & Quintard, M. (2003). Determination of reaction scheme for cardboard thermal degradation using thermal gravimetricanalysis. *Journal of Analitycal an d Applied Pyroysis* , 307-232.
7. Fogler, S. H. (2008). Elementos de ingeniería de las reacciónes químicas. Naucalpan de Juarez, mexico: Pearson Prentice Hall.
8. Gandi, J., Holtzapple, M. T., Ferrer, A., Byers, M. F., Turner, N. D., Nagwani, M., et al. (1997). Lime treatment of agricultural residues to improve rumen digestibility. *Animal Feed and Science Technology* , 195.
9. Gousame, P. (2011). Simulation of the MixAlco Process. *Special Project* .
10. Granda, C. B., & Holtzapple, M. T. (2008). Part 1: Comparison of Three Biomass Conversion Platforms. In *Carboxylate Platform: The MixAlco Process.* Texas: Department of Chemical Engineering Texas A&M University.
11. Hames, B., Ruiz, R., Scarlata, C., Sluiter, A., Sluiter, J., & Templeton, D. (2008). *Preparation of Samples for Compositional Analysis.* Cole Boulevard, Golden, Colorado : National Renewable Energy Laboratory.
12. Hayatsu, R., Winans, R., McBeth, R., Scott, R., Moore, L., & Stuider, M. (1979). Lignin-like polymers in coals. *Nature* , 278.

13. Holtzapple, M., Davison, R. R., Ross, M. K., Aldrett-Lee, S., Nagwani, M., Lee, C.-M., et al. (1999). Biomass Conersion to Mixed Alcohol Fuels Using the MixAlco Process. *Applied Biochementry and Biotechnology* , 609-630.
14. Kaar, W. E., & Holtzapple, M. T. (2000). Using lime pretreatment to facilitate the enzymic hydrolysis. *Biomass and Bioenergy* , 189.
15. Lachental, D., Mortha, G., Sevillano, R.-M., & Zaroubine, M. (2004). Isolation of residual lignin from softwood kraft pulp. Advantages of the acetic acid acidolysis method. *Compites Rendus Biologies* , 911-916.
16. Lee, J. (1997). Biological conversion of lignocellulisic biomasss to ethanol. *Journal of Biotechnoly* , 3-4.
17. Lee, J. (1997). Biological conversion of lignocellulosic biomass to ethanol. *Journal of Biotechnology* , 4.
18. Mandels, M., Hontz, L., & Nystrom, J. (1974). Enzymatic Hydrolysis of Waste Cellulose. *Biotechnology and Bioengineering* , 1471-1493.
19. Morozova, O., Shumakovich, G., Gorbacheva, M., Shleev, S., & Yaropolov, A. (2007). "Blue" Laccases. *Biochemestry (Moscow) , 73* (10), 1396-1412.
20. Mosier, N., Wyman, C., Dale, B., Elander, R., Lee, Y., Holtappple, M., et al. (2005). Features o promising technologies for pretreatment of lignocellulosic biomass. *Bioresource technology* , 673.
21. Pham, V., Holtazpple, M., & El-Halwagi, M. (2010). Techno-economic analysis of biomass to fuel ocnversion. *Journal of Microbiology and Biotechnology* .
22. Phillips, T. (2006 йил June). Structural Formula of Ethanoic Acid reacting with Propanol to form Propyl-Ethanoate.
23. Robert J. Wooley, V. P. (1996 йил Abril). Development of an ASPEN PLUS Physical Property Database for Biofuels Components. Golden, Colorado, Estados Unidos.
24. Rodriguez Cuoto, S., & Toca-Herrera, J. L. (2006). Lacasses in the textile industry. *Biotechnology and Molecular Biology review* , 115-120.
25. Scarlata, C., Slutier, J., Termpleton, a. D., Crocker, D. A., Hames, B., & Ruiz, R. (2008). *Determination of Structural Carbohydrates and Lignin in Biomass LAP.* Cole Boulevard, Golden, Colorado: National Renewable Energy Laboratory.
26. SE HOON, K. (2004). *LIME PRETREATMENT AND ENZYMATIC HYDROLYSIS.* Texas, USA: Texas A&M University Dissertation.
27. Sierra, R., & Holtzapple, M. T. (2009). Selectivity and Delignification Kinetics for Oxidative and Non-Oxidative Lime Pretreatment of Poplar Wood - Part III: Long-therm. *(Under Review)* .
28. Sierra, R., & Holtzapple, M. T. (2009). Selectivity and Delignification Kinetics for Oxidative Short-therm Lime Pretreatment of Poplar Wood - Part I: Constanr Pressure. *(Under Review)* .
29. Sierra, R., Smith, A., Granda, C., & Holtzapple, M. (2008). Producing fuels and chemicals from lignocellulosic biomass. *Chemical Engineering Progress* , 510-518.
30. Sluiter, A., Hames, B., Ruiz, R., Scarlata, C., Sluiter, J., & Templeton, D. (2008). *Determination of Ash in Biomass (LAP).* Cole Boulevard, Golden, Colorado: National Renewable Energy Laboratory.
31. Sluiter, A., Hames, B., Ruiz, R., Scarlata, C., Sluiter, J., & Templeton, D. (2004). *Determination of Total Solids in Biomass LAP.* Cole Boulevard, Golden, Colorado: National Renewable Energy Laboratory.
32. Smufit Kappa. (2011). *Preguntas frecuentes.* Retrieved 2011 6-Mayo from http://www.smurfitkappa.com.co/DropdownMenu/Contact/FAQ/Recycling/
33. Sylvia, D., Fuhrman, J., Hartel, P., & Zuberer, D. (1998). Decomposition of Naturraly Occuring Organic Materials. In *Principles and applications of soil microbiology.* (pp. 228-234). USA: Prentice Hall.
34. Tchobanoglous, G., Theissen, H., & Eliassen, R. (1982). Producción de desechos sólidos. In *DESECHOS SÓLIDOS PRINCIPIOS DE INGENIERÍA Y ADMINISTRACIÓN* (pp. 67-82). Ambiente y los Recursos Naturales Renovables.
35. Terrabon, L. (2008). *Terrabon Renewable Energy Solutions.* Retrieved 2009 30-Octubre from MixAlco Process: http://www.terrabon.com/mixalco_overview.html
36. Tisma, M., Znidarsic-Plazl, P., Plazl, I., Vasic-Racki, D., & Zelic, B. (2009). *Oxidation of Coniferyl Alcohol Catalized by catalized from Trametes versicolor.* Croacia, Eslovenia.
37. Tuomela, M., Vikman, M., Hatakka, A., & Itavaara, M. (2000). Biodegradation of lignin in a compst enviroment: a review. *Bioresource Technology* , 169-183.
38. U.S. State Department. (2010 йил 15-March). *Raw materials of biofuels.* Retrieved 2011 йил 11-March from http://www.america.gov/st/develop-spanish/2010/March/20100409143620CMretroP3.654116e-02.html
39. UNL. (2004). *Capítulo 2. Aspectos Generales.* From bibliotecavirtual.unl.edu.ar:8180/tesis/.../1/.../07-Capitulo%202.pdf
40. Van Soest, P. J. (1973). Lignin. In *Nutritional ecology of the ruminant* (p. 193). Cornell, USA: Cornell University Press.
41. Yamauchi, K., & Fukushima, K. (2004). The regulation from guaicyl to syringyl lingnin in the differenciating xylem of Robinia pseudoacacia. *Comptes Rendus Biologies* , 791-797.
42. Zhang, J., Kjonas, R., & Flurkey, W. (1999). Does N-hydroxyglycine inhibit plant and fungal laccases? Phytochemestry , 775-783.

(436d) The Influence of Non-Condesible Gas Recycling On the Yield of Fractional Catalytic Pyrolysis Products

Foster Agblevor, Biological Engineering, Utah State University, Logan, UT, **Ofei Mante**, Biological Systsems Engineering, Virginia Polytechnic Institute and State University, Blacksburg, VA, S. Ted Oyama, Chemical Engineering, Virginia Polytechnic Institute and State University, Blacksburg, VA and Ron McClung, BASF Catalysts LLC, Iselin, NJ

The fractional catalytic pyrolysis of biomass generates a considerable amount of non-condensible gases stemming from the cracking of the primary biomass degradation products. In some gases, the gas fraction can exceed 50 wt% of the biomass. This gas can conceivably be either recycled to fluidize the reactor or combusted to fuel the pyrolysis reaction. The important question to ask is whether the recycling of this gas influences the product yield and composition. We investigated both the recycled non-condensible gases (NCG) , carbon monoxide, carbon dioxide and synthetic mixture of gases for their effect on the catalytic pyrolysis of hybrid poplar wood. The NCG in all cases appeared to increase the yield of the liquid products and as well as increase the yield of the hydrogen fraction. However, the viscosity of the oils appeared to have increased with the recycling of the gases. Carbon dioxide alone also appeared to affect the viscosity of the liquid products and underwent the water gas shift reaction producing more hydrogen. The carbon dioxide appeared to have very little effect on the pyrolysis process. In general, the carbon dioxide appeared to be inert. The hydrocarbons in the NCG also appeared to undergo some reactions and therefore increasing the yield of the products. Thus, the recycling of the NCG in the fractional catalytic pyrolysis process has a beneficial effect on the entire process.

See more of this Session: Biomass Pyrolysis II
See more of this Group/Topical: Fuels and Petrochemicals Division

(496a) Production of Biodiesel Fuel From Tall Oil Fatty Acids

Jamie A. Hestekin[1], Robert Babcock[2], Thomas M. Potts[3] and Kris White[2], (1)Ralph E. Martin Department of Chemical Engineering, University of Arkansas, Fayetteville, AR, (2)University of Arkansas, Fayetteville, AR, (3)Chemical Engineering, University of Arkansas, Fayettville, AR

Tall oil fatty acids are a bi-product of the paper industry consisting of almost 100% free-fatty acids. Although this feedstock is ideal for biodiesel production, there has been relatively little study on its conversion to biodiesel. Thus, the purpose of this study was to investigate the high temperature reaction of methanol with tall oil at subcritical and supercritical pressures to produce fatty acid methyl esters. This work successfully reacted tall oil fatty acids with supercritical and subcritical methanol in a continuous tubular reactor, resulting in a temperature controlled reaction. Conversions at subcritical pressures of 4.2 MPa and 6.6 MPa were 87 percent and 89 percent, respectively. Pressure seemed to have very little correlation to conversion in both regimes, and conversions were comparable between the two. Additionally, it was also found that tall oil fatty acids react well with methanol to give comparable conversions at the relatively low molecular flow ratio of less than 10 in methanol. Both of these observations suggest that hydrolyzed triglycerides or free fatty acid feedstocks would make the primary biodiesel reaction and the subsequent separation and purification significantly less expensive to operate.

See more of this Session: Alternative Fuels and Enabling Technologies II
See more of this Group/Topical: Fuels and Petrochemicals Division

(522b) Improving Corn Stover Reactivity In Low Severity Pretreatment Through Deacetylation and Disc-Refining

Xiaowen Chen[1], Joseph Shekiro[1], mary Ann Franden[1], Min Zhang[1], Bonwook Koo[2], Sunkyu Park[2] and Melvin Tucker[1], (1)National Bioenergy Center, National Renewable Energy Laboratory, Golden, CO, (2) Department of Forest Biomaterials, North Carolina state university, Raleigh, NC

The current "state-of-art" dilute acid pretreatment to produce bio-ethanol from corn stover requires severe conditions to overcome biomass recalcitrance, which, in turn, offers 70~80% xylose monomer yield and 5~7% furfural yield. However, high acid loadings and high reaction temperatures harm the economic feasibility of the process. In addition, the amount of acetate and furfural produced during pretreatment is in a range that strongly inhibits cell growth and impedes ethanol fermentation. To decrease the impact of these issues, deacetylation, pretreatment with lower acid concentrations, lower temperatures, and integration of disc-refining was carried out in this work. The results show that deacetylation with dilute NaOH (0.1M) combined with low severitydilute-acid pretreatment increased corn stover reactivity by increasing the xylose monomer yield by 20% in pretreatment while degradation of xylan to furfural remained under 2%. Glucose yield increased by 10% and xylose yield by 20% during dilute solids enzymatic hydrolysis as a result of deacetylation of the feedstock prior to pretreatment. Particle size reduction by mechanical refining after pretreatment is shown to assist achieving> 90% cellulose yield in high solids (20 wt% total solids) enzymatic hydrolysis. Bioscreen C growth assays and mini fermentation at high solids (20wt%) showed less toxicity of deacetylated pretreated corn stover and enhanced xylose utilization by 20%.

See more of this Session: Developments In Biobased Alternative Fuels I
See more of this Group/Topical: Sustainable Engineering Forum

(51a) Multi-Physics Computational Simulation of Drilling Ultra-Deep Wells

Jeremiah N. Mpagazehe and C. Fred Higgs III, Mechanical Engineering, Carnegie Mellon University, Pittsburgh, PA

Increases in the demand for energy have generated the need to search for oil at greater depths beneath the earth's surface. In particular, offshore ultra-deep drilling has now become an important component in energy production. During ultra-deep drilling, temperatures and pressures can rise to over 300° F and 10,000 psi, respectively. For the economically viable and efficient recovery of oil in these extreme conditions, a better understanding of the interaction among drilling components such as the drill bit, the drilling fluid and the rock formation is needed. In this work a multi-physics drilling model is presented which can help to elucidate such interactions. The model uses computational fluid dynamics (CFD) to model the drilling fluid and the discrete element method (DEM) to model the rock formation. Simulations are performed demonstrating the interaction between the drill bit, the drilling fluid, and the rock formation.

See more of this Session: Recent Advances In Drilling Under Extreme Conditions
See more of this Group/Topical: Fuels and Petrochemicals Division

(106e) Thermopower Wave Dynamics: Nanoscale Alternating Current Sources and Geometry Effects

Joel T. Abrahamson[1], Wonjoon Choi[2], Sayalee G. Mahajan[1], Nicole S. Schonenbach[1], Jungsik Park[1], Michael P. Walsh[1], Jae-Hee Han[1], Kourosh Kalantar-Zadeh[3] and Michael S. Strano[1], (1)Department of Chemical Engineering, Massachusetts Institute of Technology, Cambridge, MA, (2)Mechanical Engineering, Massachusetts Institute of Technology, Cambridge, MA, (3)School of Electrical and Computer Engineering, RMIT University, Melbourne, Australia

The nonlinear coupling between exothermic chemical reactions and a nanostructure with large heat conduction results in a self-propagating thermal wave guided along the nano-conduit in the direction of its greatest thermal diffusivity [1,2]. The resulting reaction wave pushes charge carriers along, creating a concomitant thermopower wave and electrical current in the same direction. For carbon nanotubes, examples of one-dimensional conduits, we have measured high power density (> 7 kW/kg) from such waves. However, two-dimensional conduits such as graphene should also provide new avenues for control of wave velocity and temperature due to additional degrees of control from geometry and boundaries. We have developed the theory and mathematics of such waves for one- and two-dimensional conduits.

For one-dimensional conduits, our models predict that certain values of the chemical reaction kinetics and thermal parameters produce oscillating reaction wave velocities with well-defined frequency and amplitude [3]. We demonstrate such oscillations experimentally using a cyclotrimethylene-trinitramine/multiwalled carbon nanotube system, which produces frequencies in the range of 400 to 5000 Hz. The propagation velocity oscillations and the frequency dispersion are well described by our theoretical calculations and are linked to oscillations in the voltage generated by the reaction. These thermopower oscillations may enable new types of nanoscale power and signal processing sources.

[1] W. Choi, S. Hong, J. T. Abrahamson, *et al. Nature Materials*, 9, (2010), 423.

[2] W. Choi, J. T. Abrahamson, J. M. Strano, M. S. Strano. *Materials Today,* 13, (2010), 22.

[3] J. T. Abrahamson, *et al. ACS Nano* 5, (2011), 367.

See more of this Session: Nanomaterials for Energy Storage II
See more of this Group/Topical: Topical 5: Nanomaterials for Energy Applications

(349c) The Influence of Nickel Phosphide On the Fractional Catalytic Pyrolysis of Hybrid Poplar Wood

Foster Agblevor, Biological Engineering, Utah State University, Logan, UT, S. Ted Oyama, Chemical Engineering, Virginia Polytechnic Institute and State University, Blacksburg, VA, Ofei Mante, Biological Systsems Engineering, Virginia Polytechnic Institute and State University, Blacksburg, VA and Ron McClung, BASF Catalysts LLC, Iselin, NJ

The pyrolytic conversion of biomass to into liquid fuels has the potential to become drop-in transportation fuels if some of its negative attributes such as high total acid number (TAN), high viscosity, and poor storage stability are overcome. We investigated nickel phosphide supported on fluid catalytic cracking (FCC) catalysts. The pyrolysis was conducted in a bubbling fluidized bed reactor using multiple pyrolysis media comprising supported nickel phosphide and kaolin. The reaction was conducted at 450 C and the run lasted for about one hour. The total liquid yield was 40 wt% and the water yield was 15 wt%. the char/coke yield was 22.5 wt% and gas yield was 22 wt%. The C-nmr analysis of the fractional catalytic pyrolysis oils showed high level ketone derived products compared to the control sample. There was also significant cracking of the sugar derivatives. Thus, the nickel phospide catalyst has a potential of improving the properties of the fractional catalytic pyrolysis oils.

See more of this Session: Biomass Pyrolysis I
See more of this Group/Topical: Fuels and Petrochemicals Division

(453e) New Modeling Framework of Enzymatic Hydrolysis of Cellulose

Yang Zhang, Deparment of Chemical Engineering, Michigan Technological University, Houghton, MI and Wen Zhou, Chemical Engineering, Michigan Technological University, Houghton, MI

Cellulose has long been recognized as a renewable carbohydrate source for human energy use. It is available in large quantities in plant biomass. In recent years, for both economic and environmental reasons, intense research efforts have been directed at the utilization of cellulose to produce short-chain soluble sugar oligomers, such as glucose or cellobiose, which are subsequently or concurrently metabolized by microorganisms to produce biofuels such as bioethanol. The hydrolytic conversion of cellulose into soluble, fermentable sugars is achieved by means of cellulase systems which contain three major types of enzymatic activities: (1) endoglucanases which cut beta(1,4) glucosidic bonds randomly at all internal bond sites of insoluble cellulose chains in solid substrates; (2) exoglucanases which cut bonds only at the ends of insoluble cellulose chains; and (3) b-glucosidases which hydrolyze soluble oligomer sugars into glucose. The endoglucanases and exoglucanases thus depolymerize long solid cellulose chains into short soluble sugars and this is believed to be the rate-limiting step of the cellulose hydrolysis process.

In this work, we develop a general framework for a realistic rate equation modeling of cellulose hydrolysis using noncomplexed cellulase. Our proposed formalism, for the first time, takes into account explicitly the time evolution of the random substrate morphology resulting from the hydrolytic cellulose chain fragmentation and solubilization. This is achieved by integrating novel geometrical concepts to quantitatively capture the time-dependent random morphology, together with the enzymatic chain fragmentation, into a coupled morphology-plus-kinetics rate equation approach. In addition, an innovative site number representation, based on tracking available numbers of beta(1,4) glucosidic bonds, of different site types, exposed to attacks by different enzyme types, is presented. This site number representation results in an ordinary differential equation (ODE) system, with a substantially reduced ODE system size, compared to earlier chain fragmentation kinetics approaches. This formalism enables us to quantitatively simulate both the hydrolytically evolving random substrate morphology and the profound, and heretofore neglected, morphology effects on the hydrolysis kinetics. By incorporating the evolving morphology on an equal footing with the hydrolytic chain fragmentation, our formalism provides a framework for the realistic modeling of the entire solubilization process, beyond the short-time limit and through near-complete hydrolytic conversion.

See more of this Session: Fundamentals of Biomass Utilization
See more of this Group/Topical: Environmental Division

(453f) Theoretical Analysis of Biodiesel Production Under High Temperature and Pressure Conditions

Fernando Olmos, Chemical and Biomolecular Engineering, UCLA, Los Angeles, CA and Vasilios Manousiouthakis, Chemical Engineering Department, University of California at Los Angeles, Los Angeles, CA

Typically, biodiesel is synthesized by the transesterification of vegetable oil in heterogeneous fashion, i. e. in multiple liquid phases since the reactants, methanol and vegetable oil, are immiscible. In this work, we propose that the synthesis of fatty acid methyl esters (FAME), which after purification become biodiesel, take place in a solution where methanol, vegetable oil, and FAME are all homogeneously dissolved. Carrying out the reaction scheme in this manner eliminates any mass transfer effects between the oil and methanol phases. In addition, the overall reaction rate is further enhanced by the gravity based removal of the glycerol final product from the reaction mixture. The proposed process is modeled as a multiphase reactor with three phases (vapor, FAME, glycerol) in equilibrium with each other. Reactions are only considered to be carried out in the FAME phase. The mathematical model constitutes a Differential-Algebraic-Equation (DAE) problem that is solved through the implementation of a hybrid Runge-Kutta/Newton-Raphson solver. Several temperature and pressure conditions are considered to explore the operational characteristics of this process. Comparisons with the original heterogeneous biodiesel production process are carried out.

See more of this Session: Fundamentals of Biomass Utilization
See more of this Group/Topical: Environmental Division

(560a) Evaluation of Modified Feedstocks for Deconstrctionability

Seema Singh, Joint BioEnergy Institute, Emeryville, CA

Today, carbon-rich fossil fuels, primarily oil, coal and natural gas, provide 85 percent of the energy consumed in the United States. Fossil fuel use increases CO2 emissions, increasing the concentration of greenhouse gases and raising the risk of global warming. The high energy content of liquid hydrocarbon fuels makes them the preferred energy source for all modes of transportation. In the US alone, transportation consumes around 13.8 million barrels of oil per day and generates over 0.5 gigatons of carbon per year. This has spurred research into alternative, non-fossil energy sources. Among the options (nuclear, concentrated solar thermal, geothermal, hydroelectric, wind, solar and biomass), only biomass has the potential to provide a high-energy-content transportation fuel. Biomass is renewable resource that is carbon-neutral.

Currently, biofuels such as ethanol are produced largely from grains, but there is a large, untapped resource (estimated at more than a billion tons per year) of plant biomass that could be utilized as a renewable, domestic source of liquid fuels. Well-established processes convert the starch content of the grain into sugars that can be fermented to ethanol. Plant-derived biomass contains cellulose, which is more difficult to convert to sugars. The development of cost-effective and energy-efficient processes to transform cellulose in biomass into fuels is hampered by significant roadblocks, including the lack of specifically developed energy crops, the difficulty in separating biomass components, low activity of enzymes used to deconstruct biomass, and the inhibitory effect of fuels and processing byproducts on organisms responsible for producing fuels from biomass monomers.

The Joint BioEnergy Institute (JBEI) is one of three US Department of Energy Bioenergy Research Centers that will address these roadblocks in biofuels production. JBEI draws on the expertise and capabilities of three national laboratories (Lawrence Berkeley National Laboratory (LBNL), Sandia National Laboratories (SNL), and Lawrence Livermore National Laboratory (LLNL)), two leading US universities (University of California campuses at Berkeley (UCB) and Davis (UCD)), and a foundation (Carnegie Institute of Washington at Stanford University) to develop the scientific and technological base needed to convert the energy stored in cellulose into transportation fuels and commodity chemicals. Established scientists from the participating organizations are leading teams of researchers to solve the key scientific problems and develop the tools and infrastructure that will enable other researchers and companies to rapidly develop new biofuels and scale production to meet US transportation needs, and to develop and rapidly transition new technologies to the commercial sector. This talk will present a summary of the organization and efforts at JBEI, highlight the efforts on the evaluation of modified feedstock for deconstructionability and development of novel biomass pretreatment methods that enable the efficient conversion of biomass into next-generation biofuels.

(39c) Band Gap Tuning with CZTS-Based Nanocrystal-Ink Solar Cells

Grayson Ford[1], Rakesh Agrawal[1] and **Hugh W. Hillhouse**[2], (1)School of Chemical Engineering, Purdue University, West Lafayette, IN, (2)Chemical Engineering, University of Washington, Seattle, WA

$Cu_2ZnSn(S,Se)_4$ is of particular interest due to the achievement of 9.6% efficient laboratory scale devices using a non-vacuum, hydrazine-based deposition process. While hydrazine poses safety challenges from a manufacturing perspective, an alternative non-vacuum deposition approach has achieved 7.2% efficiency via the selenization of Cu_2ZnSnS_4 (CZTS) nanocrystals deposited by roll coating. The resulting $Cu_2ZnSn(S,Se)_4$ (CZTSSe) film has a band gap of approximately 1.1 eV, similar to $CuInSe_2$. $CuInSe_2$ device performance is improved by incorporation of Ga, which widens the band gap. We hypothesized that similar bandgap tuning and improvements to CZTS devices may be possible by substituting a lower atomic number group IV element for some of the Sn. Substitution with germanium is an attractive route, as germanium's crustal abudance of 1.5 mg/kg compares favorably with the 2.3 mg/kg crustal abundance of Sn, the rarest element in CZTS. Here we report a synthesis of $Cu_2Zn(Sn_{1-x}Ge_x)S_4$ (CZTGS) nanocrystals and the performance of $Cu_2Zn(Sn_{1-x}Ge_x)(S_ySe_{1-y})_4$ (CZTGSSe) solar cells fabricated by sintering the nanocrystal films with elemental selenium vapor. We find that the band gap of the CZTGS nanocrystals and the CZTGSSe solar cell can be rationally controlled by adjusting the Ge/(Sn+Ge) ratio. Solar cells fabricated from Cu_2ZnGeS_4 nanocrystal films yielded a power conversion efficiency of 0.51%. However, $Cu_2Zn(Sn_xGe_{1-x})S_4$ nanocrystals with a Ge/(Ge+Sn) ratio 0.7 yielded devices with an efficiency of 6.8% when synthesized to be Cu-poor and Zn-rich. This result opens the possibility of forming Ge gradients to direct minority carriers away from high recombination interfaces and significantly improve the device efficiency of CZTSSe-based solar cells.

See more of this Session: Nanomaterials for Photovoltaics I
See more of this Group/Topical: Topical 5: Nanomaterials for Energy Applications

(588c) Solar Thermochemical Hydrolysis of Metal Nitrides for H2 Production and Integrated Storage In Form of Ammonia

Ronald Michalsky and Peter Pfromm, Department of Chemical Engineering, Kansas State University, Manhattan, KS

Solar thermochemical hydrolysis of metal nitrides forH_2 production and integrated storage in form of ammonia

RonaldMichalsky and Peter H. Pfromm

Department of Chemical Engineering, Kansas State University, Manhattan, Kansas, USA

Conversion ofabundant but intermittent solar energy at high temperatures to high valuechemical energy stored in the products of a solar thermochemical reaction cycle,such as H_2 or syngas has been studied widely. Ammonia (NH_3)has been proposed more recently as hydrogen carrier for sustainabletransportation fuel. Easily liquefied and transported, NH_3 exceedsthe H_2 storage requirements of the Department of Energy. Modifieddiesel engines can combust NH_3 releasing mainly H_2O and N_2. H_2 can also be recovered catalytically on board a vehicle from $NH_3$for subsequent combustion.

Solarthermochemical NH_3 synthesis from steam and nitrogen produces asustainable solar fuel and avoids the storage problem for H_2.Production of metallic nitrides by reduction of their metal oxides is a high-temperatureand energy-intensive process that may take advantage of concentrated solarradiation as inexpensive and sustainable source for process heat. This workpresents a solar thermochemical NH_3 synthesis process sequence ofnitride hydrolysis splitting H_2O and absorbing protons released inthe formation of NH_3 at ambient pressure, and endothermic metaloxide reduction and nitridation driven by concentrated solar energy.

A thermodynamicrationale is presented. Various characteristic metals are selected andexperimentally explored studying yield and kinetics of breaking the N_2triple bond during nitridation and liberating NH_3 when hydrolyzingthe nitride. Experimental data will focus transition metal interactions on N_2uptake from the gas phase and nitrogen distribution in the solid bulk materialstudied using various solid-state analytical techniques. Protonation of thenitrogen liberated during steam hydrolysis of various ionic, covalent,intermediate and interstitial nitrides is examined establishing a mass balanceon the nitrogen atom. The yield of NH_3 will be correlated with thenitride ionicity and with the activation energy of nitride hydrolysis to thusdiscuss the trade-off between desirable high yields of metal nitridation andnitride hydrolysis on one side and undesirable strong bonds between the metalliccomponent and oxygen formed during hydrolysis.

See more of this Session: Fundamentals of Hydrogen Production and Utilization
See more of this Group/Topical: Environmental Division

(523b) Isoconversional Methods Applied to Kinetic Modeling Oxidative Alkaline Pretreatment of Sugarcane Bagasse

Rocio Sierra, Cuellar Gerardo, Simbaqueba Andres, Auza Nicolas and Copete Diego, Chemical Engineering, Universidad de los Andes, Bogota, Colombia

Delignification (decrease of lignin content) has shown to be a valid method to significantly increase lignocellulosic feedstocks digestibility. This operation (widely known as pretreatment) makes these abundantly available and inexpensive feedstocks usable in fermentation processes whose main goal is to produce biofuels and other chemicals. Because pretreatment is one of the most expensive processes in biofuels production plants, it has been extensively studied; however, a throughout literature search shows that development of kinetic models to better understand, optimize and control pretreatment, are very scarce if available at all.

This study uses sugarcane bagasse which is a residue from the sugar industry abundantly produced particularly near and in the Equator zone. This lignocellulosic resource is traditionally used in combustion processes to supply energy to the production plant; however life cycle analysis have shown that if lignocellulose is used as substrate in fermentation processes, the energy yield would be much closer to the theoretically available energy.

In this work, sugarcane bagasse is submitted to an alkaline pretreatment that uses an oxidative agent, systematically applied at widely varied ranges of temperature and time using a full factorial experimental design

Model fitting methods do not achieve a clean separation between the temperature dependent Arrhenius constant and the reaction model. On the contrary, the isoconversional methods evaluate the activation energy values at progressive degree of conversion without modelistic assumptions. This work applies this methodology to calculate activation energy for alkaline pretreatment of sugarcane bagasse and compares with previous work where the traditional curve fitting is employed.

See more of this Session: Developments In the Pretreatment of Lignocellulosics for Bioconversion II
See more of this Group/Topical: Sustainable Engineering Forum

(522e) Poly(oxymethylene) Dimethyl Ethers As Components of Tailored Diesel Fuel - Properties, Synthesis and Purification Concepts

Jakob Burger, Laboratory of Engineering Thermodynamics, Technische Universitt Kaiserslautern, Kaiserslautern, Germany

Poly(oxymethylene) Dimethyl Ethers as components of tailored diesel fuel - properties, synthesis and purification concepts

Jakob Burger [1]*, Markus Siegert [2], Eckhard Ströfer [2], Michael Nilles [2], Hans Hasse [1]

[1] Laboratory of Engineering Thermodynamics, University of Kaiserslautern, Germany;

[2] BASF SE Company, Ludwigshafen, Germany
*Corresponding author: Jakob Burger, Laboratory of Engineering Thermodynamics, University of Kaiserslautern, Germany. Email: jakob.burger@mv.uni-kl.de

Keywords: Fuel additive, Soot reduction, Process development

Poly(oxymethylene) Dimethyl Ethers (POMDMEs), oligomers of the general structure CH_3-O-(CH_2-O)$_n$-CH_3, open a new route for tailoring diesel fuels. POMDMEs belong to the group of oxygenates which reduce soot formation in the combustion when added to diesel fuels. They can be produced on a large scale based on Methanol which can be produced from renewable resources. This contribution presents the process chain, possible synthesis routes and some fuel related properties of POMDMEs. Comparisons to other oxygenates, e.g. Dimethyl Ether, are given.

A particularly favourable route for the large scale production of POMDMEs in which they are formed from Methylal and Trioxane is proposed. The chemistry behind this route is given in detail. Measurements of the reaction kinetics and the reaction equilibrium were carried out in a batch reactor using acidic ion exchange resin as heterogeneous catalyst. The parameters of a pseudo-homogeneous model were fitted to the results of these experiments. The formation of side-products was studied experimentally and concepts to minimize them are discussed. Based on the reaction model a process concept was developed. It consists of a reactor and a downstream distillation sequence to yield pure POMDMEs of the most favourable chain length n=3-5.

The use of distillation technology allows an easy scale-up to huge single train plants which can potentially fulfil the need for millions of tonnes of the innovative oxygen carrier POMDME needed for the supply of clean diesel fuel in the future.

See more of this Session: Developments In Biobased Alternative Fuels I
See more of this Group/Topical: Sustainable Engineering Forum

(378f) Can a Liquid Hydrogen Storage Tank BLEVE?

Subramanya Nayak, Sergio Garcia, Victor Carreto and Sam Mannan, Chemical Engineering, Texas A&M University, College Station, TX

Abstract

A BLEVE (boiling-liquid expanding-vapor explosion) occurs when a tank containing liquefied gas ruptures, resulting in the instantaneous release of the vapor, boiling liquid, missiles (tank fragments), and blast wave[1]. The most common type of the BLEVE is caused when the external wall of the tank is exposed to a fire[1]. The continuous heat transfer from the fire to the tank wall increases the temperature and pressure inside the tank, until the tank fails. This sudden failure of the tank results in an explosive vaporization of its contents.

In passenger cars the LH_2 is stored as a cryogenic liquid at -240° C and at moderate pressures (20 to 30 bars)[2]. It is believed that the LH_2 will not BLEVE, since it is stored at such a low temperature[3]. However, LNG, which is also stored as a cryogenic liquid presents a potential for BLEVE[4]. Consequently, in this study we have theoretically investigated if the LH_2 storage tanks will BLEVE.

A theoretical model considering transient mass and energy transfer is developed and solved numerically[5]. It is assumed that the LH_2 tank is engulfed in a fire and the fire is caused due to spilled gasoline or hydrogen. Different tank materials and insulations are tested. The transient model has a capability to predict the time required for the LH_2 tank to BLEVE. This prediction will be useful for the fire fighters and emergency responders when dealing with LH_2 tank engulfed in a fire. Finally the consequences due to the BLEVE of LH_2 tank are also studied.

References

1. Abbasi, T. and S. A. Abbasi (2007). "The boiling liquid expanding vapour explosion (BLEVE): Mechanism, consequence assessment, management." Journal of Hazardous Materials 141(3): 489-519.
2. Rigas, F. and S. Sklavounos (2005). "Evaluation of hazards associated with hydrogen storage facilities." International Journal of Hydrogen Energy **30**(13-14): 1501-1510.
3. C. J. H. van den Bosch, R. A. P. M. W. (2005). Methods for the calculation of physical effects - due to the releases of hazardous materials (liquids and gases) - 'Yellow Book'. The Hague,

Committee for the Prevention of Disasters.
4. Pitblado, R. (2007). "Potential for BLEVE associated with marine LNG vessel fires." Journal of Hazardous Materials 140(3): 527-534.
5. Runnels, D. S. R. (2008). AFFTAC: Analysis of Fire Effects on Tank Cars.

See more of this Session: Hydrogen Storage Systems: Designs, Analysis, and Safety Aspects
See more of this Group/Topical: International Congress on Energy 2011

Examination of Tropical Cyclogenesis using the High Temporal and Spatial Resolution JRA-25 Dataset

Masato Sugi
Forecast Research Department, Meteorological Research Institute, Japan
Correspondence: msugi@mri-jma.go.jp

INTRODUCTION

Imperfect models alone cannot draw a reliable picture of the real world, while temporally and spatially insufficient observation data alone cannot draw a very detailed picture of the real world. Reanalysis may be regarded as an optimum combination of an imperfect model and insufficient observation data, which can produce the most reliable and detailed dataset to describe the real world.

Regarding the future tropical cyclone (TC) activity, a recent very high resolution GCM experiment indicates a reduction in the TC frequency as a whole, with an increase in the frequency of very intense TCs (Oouchi et al. 2006). On the other hand, in the recent IPCC report, it is summarized as "Based on a range of models, it is likely that future tropical cyclones (typhoons and hurricanes) will become more intense, with larger peak wind speeds and more heavy precipitation associated with ongoing increases of tropical SSTs. There is less confidence in projections of a global decrease in numbers of tropical cyclones." (IPCC, 2007). There is no doubt that for a more reliable projection of the future tropical cyclone activity, it is necessary to better understand the physical processes associated with the genesis and development of tropical cyclones, and to assess the capability of GCMs to simulate these processes.

In the present study, in order to understand the physical processes of tropical cyclogenesis, we examine the formation stage of typhoons using the high temporal and spatial resolution JRA-25 dataset. Because of the high quality of the tropical precipitation in the JRA-25 dataset (Onogi et al. 2006), we can expect tropical disturbances and their environmental circulations are analyzed reasonably well in JRA-25.

DATA

The main data used in the present study is the 6 hourly, 1.25 degree latitude-longitude grid, standard pressure level JRA-25 data. For the JRA-25 precipitation data, 1.125 degree latitude-longitude grid, 6 hour forecast data is used. The CDA file is used to check if the TCR data (tropical cyclone wind profile retrieval data) is used or not in each time. To examine the convective activity around the tropical cyclones, 1 hourly, 0.05 degree latitude-longitude grid GMS satellite black body temperature (TBB) data is used. In addition, data from a very high resolution (horizontal grid size 20km) GCM experiment is used for the purpose of comparison. The period of analysis is the three months from August to October in 2004. During this period 14 typhoons are formed, 8 in August, 3 in September and 3 in October. Formation stages of 12 typhoons out of the 14 typhoons are examined, and three of them, T0416 (CHABA), T0418 (SONGDA) and T0422 (MA-ON), are analyzed in some detail.

DEFINITON OF TC GENESIS

In the present study, we define TC genesis time as the time when TCR data is first used for the TC in JRA-25. TCR data is included in JRA-25 when maximum wind $\geq$ 30 kts in the Neumann's best track data (Hatsushika et al. 2006). On the other hand, the JMA TC genesis time is defined as the time when maximum wind $\geq$ 34 kts, when tropical depression (TD) becomes named typhoon (TS). Because of the difference in the threshold value for TC genesis, the TCR genesis time is usually 6 to 18 hours earlier than the JMA genesis time. In addition, we should note that the maximum winds estimated from satellite images are considerably different in different best track dataset. The maximum wind in JTWC best track data is usually stronger than that in JMA best track data, particularly in the mature stage of TCs. For example, the maximum wind speed of T0416 (CHABA) at mature stage is 110kts in JMA best track data, while it is 155kts in JTWC best track data.

INTENSIFICATION OF VORTEX

We made maps of vorticity at 925hPa every 6 hours from August to October in 2004 over the tropical western North Pacific (Figures not shown). In these maps we can see many weak vortices, some of which develop into typhoons. The central question regarding TC genesis is why some of these vortices develop into TCs but many others do not. We first examine the development processes of the vortices that eventually developed into typhoons. Intensification of vortices during the TC genesis period, from 2 days before the genesis time to 2 days after the genesis time, is shown in Figure 1 for the 12 typhoons formed in August to October in 2004. There was a concern that the inclusion of TCR data might cause a discontinuous intensification of vortex, but the Figure 1 shows that there is no such gaps in the intensity before and after the genesis time. In some case there is a rapid intensification around the genesis time. For example, a rapid intensification can be seen around genesis time of T0419 or T0422, but the intensification continues sometime after the genesis time, and it may be real intensification rather than the artificial one caused by the inclusion of TCR data at genesis time. All the intensification curves of vortices of the 12 TCs are plotted together in Figure 2, with their average. The average curve shows slow intensification before the genesis time and more rapid intensification after the genesis time. This suggests that a self intensification process starts to take place around the genesis time.

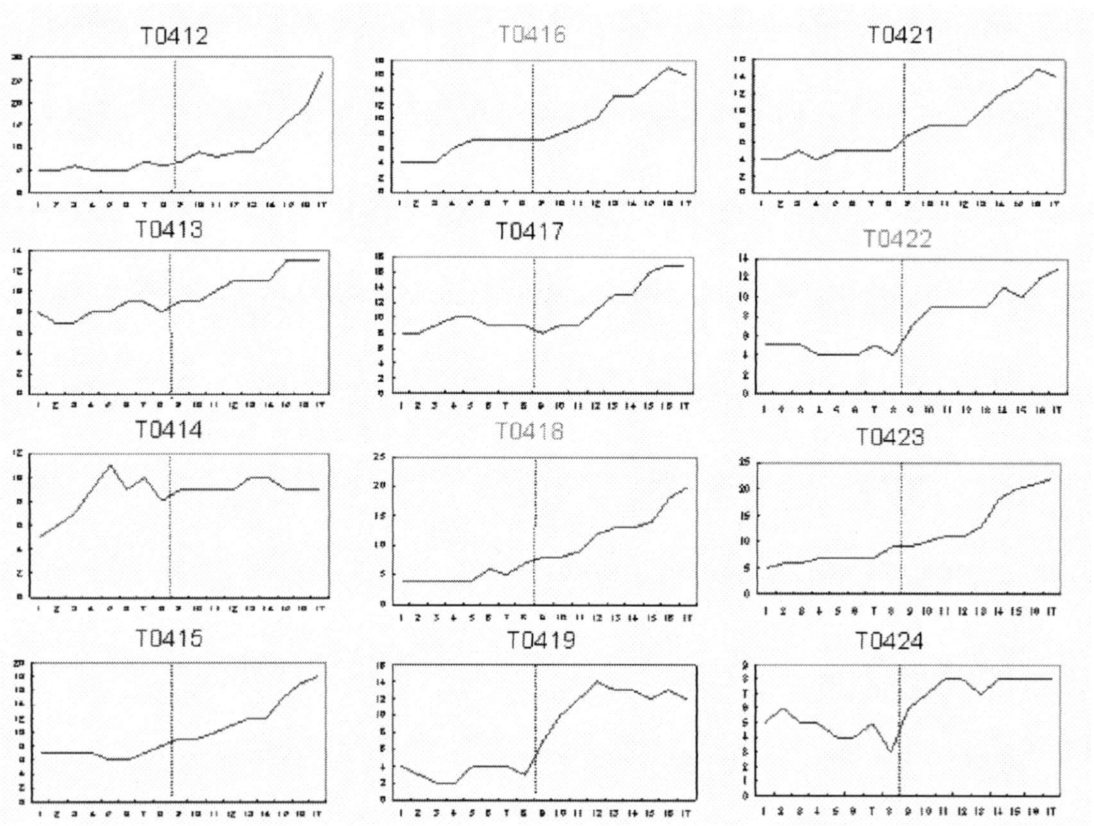

Figure 1 Variation of the intensity of vortex at 925hPa during the 4 day period centered at the genesis time for 12 typhoons formed in August to October of 2004.

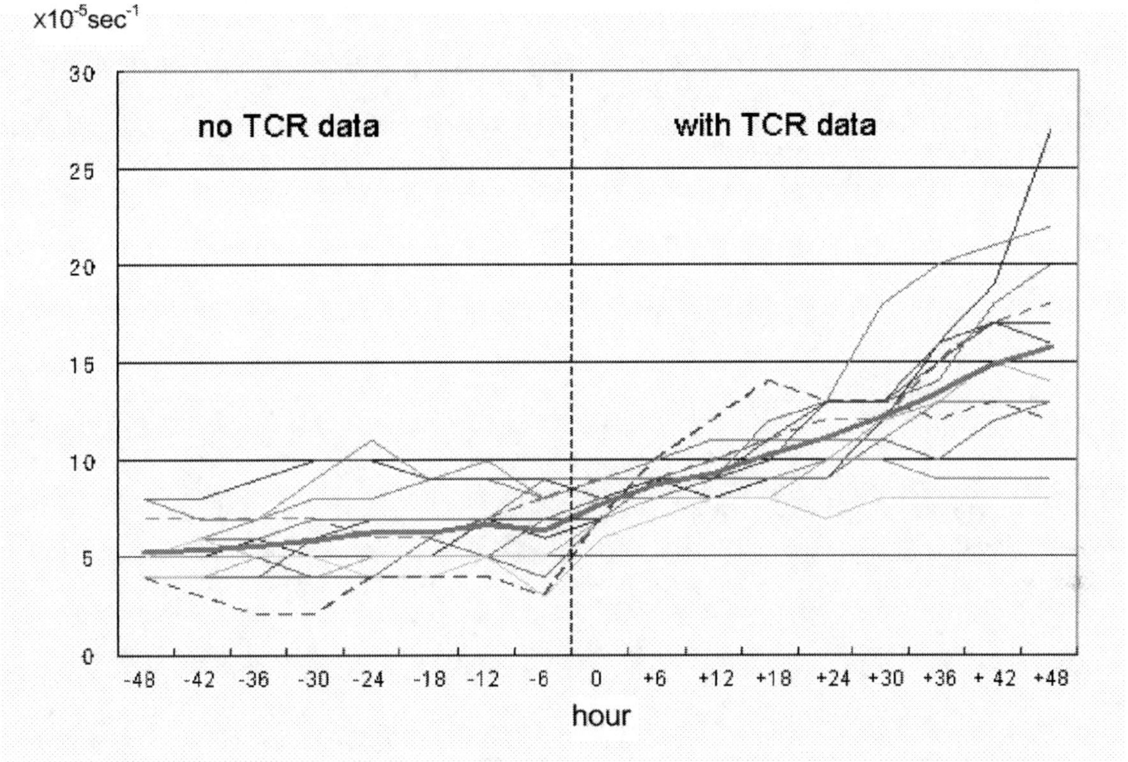

Figure 2 The same as Figure 1 but all curves for 12 typhoons are plotted together, with their average (thick red curve). 0 hour is the genesis time.

CASE STUDIES

We examine genesis processes of three typhoons in some detail. Figure 3 shows the evolution of vorticity, sea level pressure, convective clouds (GMS TBB), precipitation and vorticity generation term $-(\zeta+f)D$ for the TC0416 (CHABA) during the 4 days period centered at the genesis time (18UTC of 18 August 2004). The vertical cross sections of vorticity show the upward development of the vortex from bottom to top, suggesting that the vorticity is generated in the boundary layer and advected upward to free atmosphere by convections. Before the genesis time when the TCR data is first included, a development of vortex together with the surface low pressure area can be seen, indicating that the tropical cyclogenesis process is well captured by the JRA-25 data. Concentrated active convective clouds can be seen near the vortex center after the genesis time, while some scattered active convective clouds can be seen before the genesis time. The role of these scattered convective clouds during the genesis stage seems to be producing a large scale convergence in the boundary layer, leading to an intensification of the vortex through the vorticity generation term. Although the 6-hour forecast precipitations in the JRA-25 are not well concentrated near the vortex center, we can see vorticity generation in the boundary layer near the vortex center. This indicates that the large scale convergence associated with the ensemble of scattered convections around the vortex play an important role in the development of vortex.

The evolutions of the other two typhoons (T0418 and T0422) are almost similar to T0416 (Figures not shown), although there are some minor differences. For both typhoons, we can see upward development of vortex similar to T0416 during the TC genesis period. A development of surface low pressure area and an intensification of vorticity

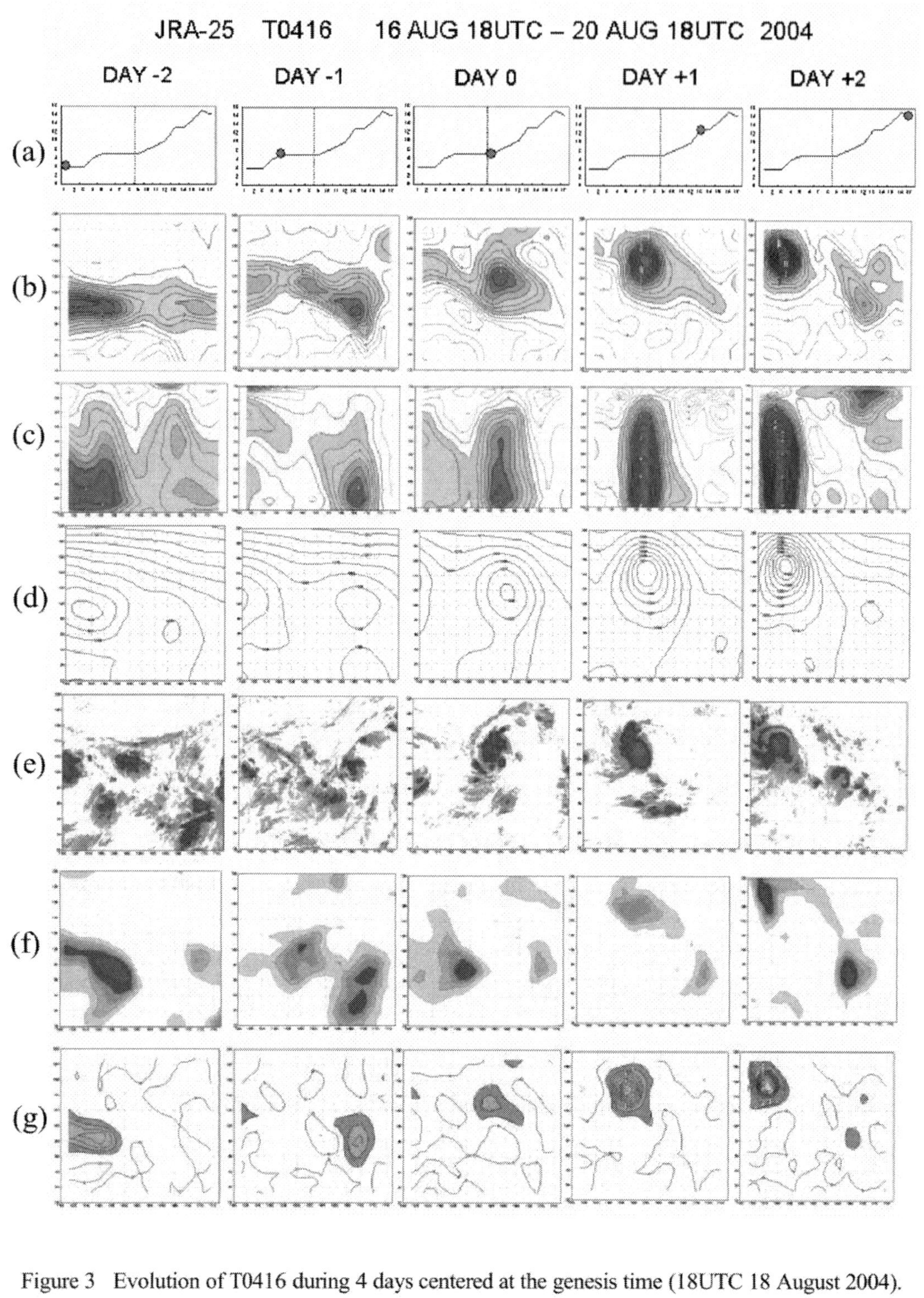

Figure 3 Evolution of T0416 during 4 days centered at the genesis time (18UTC 18 August 2004). (a) Variation of intensity of vortex. (b) Vorticity at 925 hPa. (c) Vertical cross section of vorticity . (d) Sea level pressure. (e) Convective clouds. (f) 6hr forecast precipitation. (g) vorticity generation term ($-(\zeta+f)D$)

generation term around the vortex center can be seen in both cases. We can also see some scattered convective clouds around the vortex center before the genesis time, and concentrated convective clouds near the vortex center as in the T0416 case. Furthermore, a similar evolution of vortex during the TC genesis period is also seen in the very high resolution GCM simulation.2.

CONCLUSIONS

The main conclusions of this study are:

1) High resolution JRA-25 dataset is very useful for TC genesis study. TC genesis process is reasonably well captured by the JRA-25. TCR data makes no significant gaps in the intensity of TC before and after TC genesis.

2) There are many weak vortices ($\leq 5 \times 10^{-5}$ sec^{-1}) in the boundary layer over the tropical western North Pacific. Some of them develop into typhoons but many of them do not develop.

3) It seems that the vorticity generation in the boundary layer through convergence of absolute vorticity ($-(\zeta+f)D$) and its upward advection to free atmosphere by convection are the most important processes for the development of the vortex.. Further quantitative analysis of these processes is an important future work.

4) During the genesis stage of TC, scattered active convective clouds can be seen around the vortex center. The role of these scattered convective clouds during the genesis stage seems to be producing a large scale convergence in the boundary layer, leading to an intensification of the vortex through the vorticity generation term.

5) To assess the capability of GCMs to simulate TC genesis processes, a comparison of GCM simulation with cloud resolving model (CRM) simulation is also an important future work.

ACKNOWLEDGMENT

The author would like to express sincere thanks to the members of JRA-25 project, who helped the author to use the high resolution JRA-25 dataset and CDA files. Thanks are extended to Dr. Mike Fiorino who prepared the high quality TCR data for the JRA-25.

REFERENCES

Hatsushika, H., J. Tsutsui, M. Fiorino, and K. Onogi, 2006: Impact of Wind Profile Retrievals on the Analysis of Tropical Cyclones in the JRA-25 Reanalysis. *J. Meteor. Soc. Japan*, **84**, 891-905.

IPCC 2007: Climate Change 2007: The Physical Science Basis. Cambridge University Press, 996pp

Onogi, K., J. Tsutsui, H. Koide, M. Sakamoto, S. Kobayashi, H. Hatsushika, T. Matsumoto, N. Yamazaki, H. Kamahori, K. Takahashi, S. Kadokura, K. Wada, K. Kato, R. Oyama, T. Ose, N. Mannoji and R. Taira 2007: The JRA-25 Reanalysis. *J. Meteor. Soc. Japan*, **85**, 369-432.

Oouchi, K., J. Yoshimura, H. Yoshimura, R. Mizuta, S. Kusunoki, A. Noda, 2006: Tropical Cyclone Climatology in a Global Warming Climate as Simulated in a 20km-Mesh Global Atmospheric Model: Frequency and Wind Intensity Analysis. *J. Meteor. Soc. Japan*, **84**, 259-276.

(373g) Indirect Carbon Air Fuel Cells

Klaus Hellgardt[1], G. Kelsall[2], P. Kaewpet[2] and O. Agbede[2], (1)Chemical Engineering, Imperial College London, London, United Kingdom, (2)Imperial College London, London, United Kingdom

The chemical energy in fossil fuels can be converted directly and efficiently into electrical energy by a direct carbon fuel cell with a calculated equilibrium potential of ca. 1.02 V, which is almost temperature independent for carbon dioxide (CO_2) formation from a carbonaceous solid fuel. Separate evaluation and operation of a liquid metal based reformer (or partial oxidation reactor) in conjunction with a solid oxide electrolyte based electrochemical cell demonstrates our novel approach of an 'indirect' carbon-air fuel cell, incorporating a liquid metal anode as a solvent for the anodically-evolved oxygen, coupled to an external reformer supplied with oxygen from the liquid metal. The concurrent production of electricity, heat as well as fuels such as synthesis gas ($CO + H_2$) as functions of temperature and other experimental variables will be presented.

See more of this Session: Fuel Cell Technology II
See more of this Group/Topical: Fuels and Petrochemicals Division

(641f) Secondary Thermochemical Xylo-Oligomer Hydrolysis of High-Solids Dilute-Acid Pretreatment Slurries

Erik M. Kuhn[1], Joseph Shekiro[1], Nick J. Nagle[2] and Richard T. Elander[1], (1)National Renewable Energy Laboratory, Golden, CO, (2)National Bioenergy Center, National Renewable Energy Lab, Golden, CO

Dilute-acid pretreatment is a critical step in the biochemical conversion of biomass to fuels and chemicals, releasing hemicellulose as oligomeric and monomeric xylose and rendering the cellulose fraction readily digestible by enzymes. It has been difficult to achieve very high yields of monomeric xylose during pretreatment, since acid hydrolysis also converts xylan to various oligomers and non-fermentable sugar degradation products. An alternative strategy for the production of high yields of monomeric xylose is to perform secondary thermochemical hydrolysis on the dilute-acid pretreated material. This secondary hydrolysis, typically performed at lower temperatures than the primary dilute-acid pretreatment step, liberates additional monomeric xylose while minimizing the yield of sugar degradation products. This work presents the results of laboratory and pilot scale primary and secondary thermochemical hydrolysis experiments over a wide range of conditions. Our results suggest that reduced severity conditions for primary dilute-acid pretreatment provide the highest overall yields of monomeric xylose after secondary thermochemical hydrolysis for the conditions tested. The results of our LC-MS investigations suggest that higher-molecular weight xylo-oligomers are converted to xylose during secondary thermochemical hydrolysis, but that a small fraction of the xylan in corn stover is converted to stable but non-fermentable compounds, such as xylo-oligomers, during both primary and secondary hydrolysis.

See more of this Session: Chemical and Catalytic Conversions and Processes for Renewable Feedstocks
See more of this Group/Topical: Sustainable Engineering Forum

Considerations for Research, Development and Commercialization/Scale-up of Biomass Pretreatment Processes

Dale Monceaux[1], *Richard C. Agar*[1] *and Charles E. Wyman*[2], *(1)AdvanceBio, LLC, Milford, OH, (2)College of Engineering-Center for Environmental Research and Technology, University of California, Riverside, Riverside, CA*

Global efforts to cost effectively convert biomass-based polysaccharides to sugars via enzymatic hydrolysis and subsequently to fuels and chemicals have been ongoing since the early discoveries by the US Army Natick Research & Development Command. These, and other early developments related to enzymatic hydrolysis of cellulose were followed by advancements in microbial genetic engineering. Genetically modified microorganisms capable of utilizing xylose-rich hemicellulose fractions of biomass feedstocks in addition to the cellulose derived glucose, promised improved ethanol yields and process economics. The past fifteen years of research has led to other major advancements in cellulase protein chemistry as well as an improved understanding of the microstructure of lignocellulose, moving the world ever closer to commercially viable biomass-based fuels and chemicals.

Early research ultimately revealed the sensitivity of the feedstock, pretreatment, enzyme and organism regime. This, in turn has exposed the need for consistency in methodology as research proceeds from concept to commercial-scale operations.

Much current activities focusing on the commercialization of biomass based fuels and chemicals hinge on the development of cost-effective biomass pretreatment processes. This "first step" in the process of converting biomass to end-products is critical to successful commercialization efforts for a number of reasons. First, it provides reactive cellulose that can be efficiently enzymatically hydrolyzed to oligomeric or monomeric glucose. Second, it efficiently solubilizes hemicellulose fractions of the biomass, producing oligomeric and or monomeric xylose, mannose, galactose, arabinose, etc. Third, it produces minimal reaction by-products that inhibit enzymatic hydrolysis and fermentation processes. Last, it is mechanically and economically scalable.

Many pretreatment technologies employ thermo-chemical reactions that reduce the recalcitrance of ligno-cellulosic matrices, solubilize fractions of the biomass and expose reactive sites to subsequent reactions. Various pretreatment technologies are under development, each with a range of design basis conditions. **Table 1** illustrates general conditions for six common pretreatment chemistries.

Pretreatment	Temp (°C)	Residence Time (min)	Water Loading*	Catalyst	Catalyst Loading**	Catalyst Recyclable
AFEX	150	30	200	Anhydrous NH_3	152	Yes (95-95% of loaded)
Dilute Acid	140	40	891	1% H_2SO_4	9	No
Lime	120	240	1500	CaOH/O_2	100/100 psi	Maybe/Ne
Liquid Hot Water	200	10	663	Water	N/A	N/A
SAA	160	60	765	15% Aqueous NH_3	135	Yes, ecess
SO_2	180	10	900	5% SO_2/Steam	5/?	Yes, excess

* Water Loading = L/100 kg of dry biomass input to pretreatment
** Catalyist Loading = kg/100 kg of dry biomass input to pretreatment

Table 1

Typical Thermochemical Pretreatment Chemistries

Reaction kinetics varies significantly with feedstock type, mechanical pretreatment, catalyst type, catalyst loading, water loading, residence time and temperature. Tight control of key parameters is required to maximize conversion efficiency. Elevated catalyst concentration and residence times increase reaction rate but also xylose destruction as illustrated in **Table 2**. Consequently, optimum reaction conditions require mechanical designs that deliver controlled residence times under near-plug flow conditions.

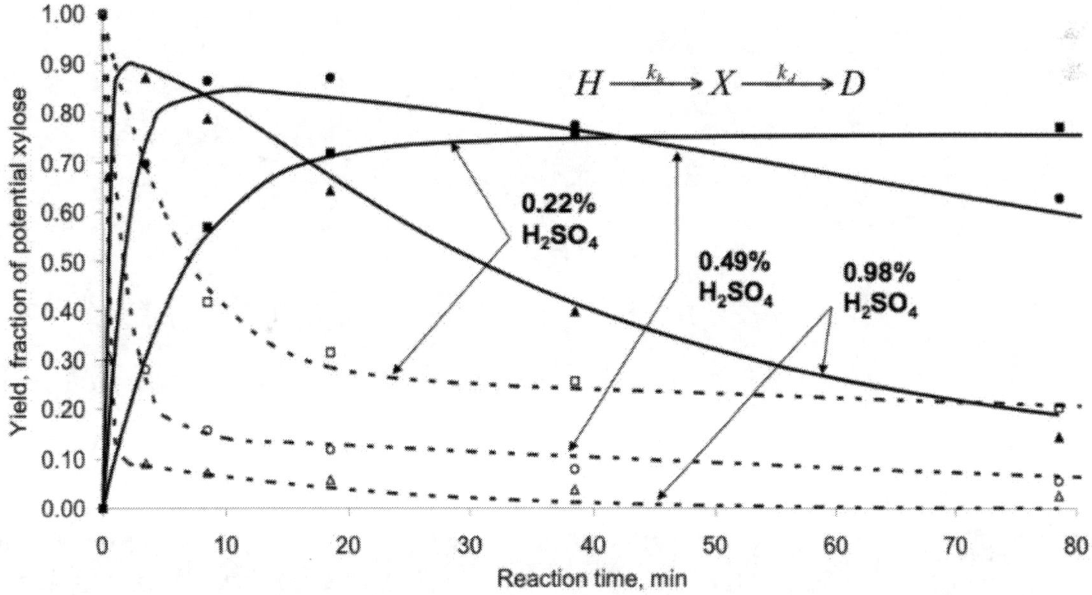

Table 2

Xylose Yield in Pretreatment vs Time at 160°C

Resource limitations, including feedstock supply, workspace, staffing and funding constrain the research and development efforts at each stage. Real-world feedstock properties such as foreign matter content, chemical composition variability, recalcitrance, bulk density, etc. introduce constraints that further complicate research and commercialization efforts. Feedstock selection dramatically impacts capital investment and production cost. Feedstock recalcitrance in conjunction with chemical composition effectively determines yield, and therefore the quantity of feedstock required per unit of fuel or chemical produced. Feedstock chemical composition and bulk density, in conjunction with reactor residence time establishes the size of the major components of the capital and energy intensive pretreatment reactor unit operation, as illustrated in **Table 3**.

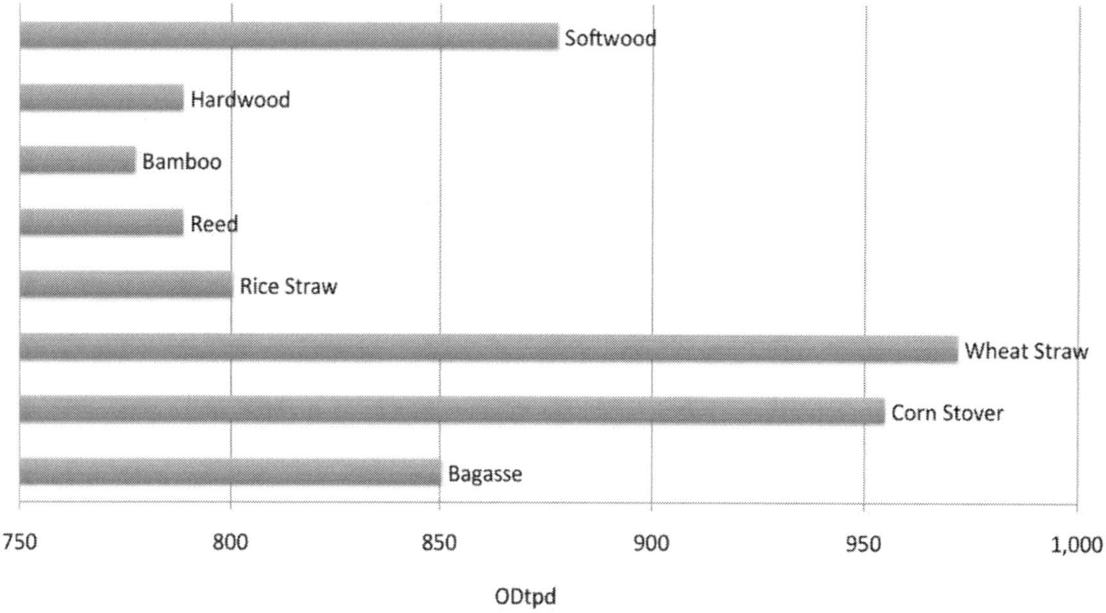

Table 3

Estimated Feedstock Tonnage per 100 Megaliters of Ethanol

High yields are critical to distribute both operating and capital costs. Low operating costs, including chemicals, energy and labor, are essential to providing adequate margins for return on capital. General operating cost categories must be lower for a new process than for established technologies making conventional products since low capital costs are essential to minimize financial exposure. The driving forces to contain capital investment include general technologies and processes that minimize equipment sizes, operate at lower pressures, operate at neutral or higher pHs as well as thermal processes that operate at lower temperatures and biochemical processes that operate at elevated temperatures.

Technical compromises frequently lead to approaches that ultimately expand the research and development program effort, schedule and budget. The impact of decisions made early in the technology development phase of a project may not fully materialize until late in the project after significant design and engineering expenditures have been made. The time period between research decision-making and engineering-based capital cost estimating delays the commercialization or leads to technology dead-ends. Visualization of the expected commercial scale process should occur very early in the technology research and development effort. A process "scale-down" approach to optimize research and technology development, scale-up and commercialization efforts has proven to be and effective strategy to accelerate project development and finance.

(508c) Heterogeneous Catalysts in the Dehydration of Fructose to HMF

Sylvia Reiche, Xiao Chen Zhao, Klaus Friedel, Matthew Aronson, Raoul Blume, Edward Kunkes, Jean-Philippe Tessonnier, Malte Behrens, Dangsheng Su, Robert Davis, Robert Schlögl

Several crucial steps in the upgrading of biomass-derived raw materials involve acid catalyzed reactions, such as the hydrolysis of cellulose or the dehydration of sugars. Although heterogeneous catalysts are desired in large scale industrial processes, their application is limited by several challenges concerning the conversion of these highly functionalized feedstocks.

In our work, we tested a wide range of heterogeneous acids for the dehydration of fructose into 5-hydroxymethyl furfural (HMF) in 2-butanol solvent. The main focus was on carbon based catalysts, due to their structural diversity and hydrothermal stability. The amount of acidic functional groups determined by titration varied with the carbon support between 0.1 mmol/g for functionalized multi-walled carbon nanotubes (MWCNTs) and 1.7 mmol/g for amorphous carbon prepared by glucose pyrolysis (AGP). However, tests involving exposure of the catalyst to heated reaction solvent showed that part of the acidic groups is released in the solution (i. e. leaching) and the reaction is subsequently catalyzed at least partly homogeneously. The described leaching phenomenon was observed for all materials tested, including polymer resins, such as Nafion and Amberlyst. In case of functionalized AGP, the system retained 0.42 mmol/g of acid functional groups after four consecutive pretreatments in reaction solvent (2-butanol) at reaction temperature (130°C).

A second aspect of catalyst deactivation is the surface poisoning by secondary polymerization products, i. e. humins. This phenomenon was further investigated by comparing the activity towards dehydration of fructose with activities in a reference reaction, i. e. the esterification of acetic acid with ethanol. The esterification reaction was chosen because this reaction does not show deactivation due to secondary products. Although some systems, such as the pretreated AGP catalyst, were stable and recyclable in the esterification reaction, they significantly deactivated already after the first reaction run of fructose dehydration. The accumulation of secondary polymerization products on the catalyst surface was observed by inter alia scanning electron microscopy (SEM).

In conclusion, different deactivation mechanisms were identified for the use of heterogeneous catalysts in the dehydration of fructose. The leaching of functional groups can be suppressed by suitable pretreatments of the catalysts. Adsorption studies suggest that solvent choice can influence the extent of humin accumulation. However, a thorough recycling requires calcinable, hydrothermally stable catalysts.

See more of this Session: Catalytic Biomass Conversion to Chemicals I

(35d) Evaluation of Ni-Mo Oxide (Ni-MoOx) Electrocatalyst for Li-Air Battery

Jamie Gomez[1], Egwu E. Kalu[2], Ruben Nelson[3], Mark H. Weatherspoon[3] and Jim P. Zheng[3], (1)Chemical & Biomedical Engineering, Florida A&M University, Tallahassee, FL, (2)Department of Chemical & Biomedical Engineering, Florida A& M Univ. and Florida State Univ., Tallahassee, FL, (3)Electrical & Computer Engineering, Florida A&M University, Tallahassee, FL

Evaluation of Ni-Mo Oxide (Ni-MoO_x) Electrocatalystfor Li-air Battery

Jamie Gomez[1], Egwu E. Kalu[1],Ruben Nelson[2],

Mark H. Weatherspoon[2],Jim P. Zheng [2]

[1]Department of Chemical and Biomedical Engineering

[2]Department of Electrical & Computer Engineering

FAMU-FSU College ofEngineering

Tallahassee, FL,32310

The direct electro-reductionof oxygen as the active cathode process in the Li-Air battery has beenidentified as a major advantage of the Li-Air battery system. Unfortunately,the challenges on the air cathode system include the identification ofappropriate catalyst for oxygen reduction in the non-aqueous media. The presenttalk will focus on the synthesis of composite Ni-Mo oxide electrocatalyston woven and non-woven carbon fiber. The approach differs from the currentpractice that utilizes carbon particles as the catalyst support and usesclassical impregnation method for the synthesis of catalyst metal nanoparticles.

In our approach, electroless Ni-Mo composite metal was deposited onwoven/non-woven carbon fiber. The oxidation of the Nickel/molybdenum metalcoated on woven/non-woven carbon support was performed electrolyticallyto form Nickel/molybdenum oxide. A three-electrode system was used at constantpotential to activate and form active metal catalyst oxides.

Preliminaryresults show that the metal oxide catalysts formed on the woven/non-woven-carbon cathode can enhance the discharge capacity of the lithium air cell. Thespecific discharge capacity of the nickel/molybdenum oxide (Ni-MoO_x) catalyzed woven-carbon cathode wasdetermined to be 1210 mAh g^{-1} whereas cobalt oxide (CoO_x)catalyzed carbon cathode yielded a discharge capacity of 687 mAh g^{-1} at charging voltages of 4.2 V. Theseresults indicate not only the superior performance of the Ni-MoO_x catalyzed woven-carbon cathode but also thesignificant role the oxide electrocatalyst plays inthe electrochemistry of the air electrode. Figure 2 compares the preliminarydischarge times of different electrocatalysts. Resultson the kinetics of oxygen reduction in non-aqueous media using Ni-MoOxelectrocatalyst will be presented including thetexture and micro-texture characterization of the composite metal oxidecatalyzed carbons. Correlation s of electrocatalystsynthesis conditions with their electrochemical performance will be presented.

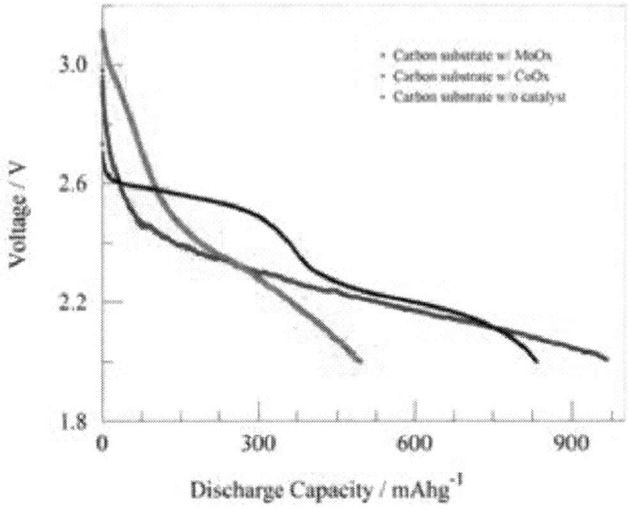

Figure 1: Discharge capapcity ofLi-air battery using Ni-MoO$_x$ catalyzedwoven- carbon cathode at 0.314mA/ cm^2

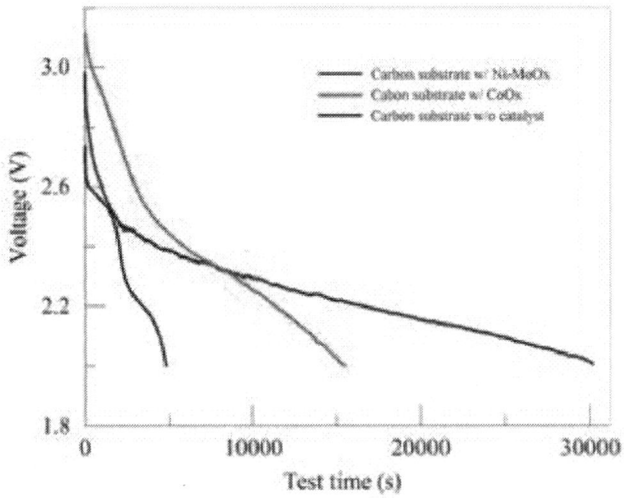

Figure 1: Comparisonof discharge profiles of Li-air battery using different catalyzed woven- carboncathodes at 0.314 mA/cm^2

(38e) Mechanism of Li Adsorption On Carbon Nanotube-Fullerene Hybrid System: A First-Principles Study

Wonsang Koh[1], Ji Il Choi[2], Kevin Donaher[3], Seung Geol Lee[2] and **Seung Soon Jang**[2], (1)School of Physics, Georgia Institute of Technology, Atlanta, GA, (2)School of Materials Science and Engineering, Georgia Institute of Technology, Atlanta, GA, (3)Department of Applied Physics and Applied Mathematics, Columbia University, New York, NY

The lithium (Li) adsorption mechanism on the metallic (5,5) single wall carbon nanotube (SWCNT)-fullerene (C_{60}) hybrid material system is investigated using first-principles method. It is found that the Li adsorption energy (-2.649 eV) on the CNT-C_{60} hybrid system is lower than that on the peapod system (-1.837 eV) and the bare CNT (-1.720 eV), indicating that the Li adsorption on the CNT-C_{60} hybrid system is more stable than on the peapod or bare CNT system. This is due to the C_{60} of high electron affinity and the charge redistribution after mixing CNT with C_{60}. In order to estimate how efficiently Li can utilize the vast surface area of the hybrid system for increasing energy density, the Li adsorption energy is calculated as a function of the adsorption positions around the CNT-C_{60} hybrid system. It turns out that Li preferably occupies the mid-space between C_{60} and CNT and then wraps up the C_{60} side and subsequently the CNT side. It is also found that the electronic properties of the CNT-C_{60} system, such as band structure, molecular orbital and charge distribution, are influenced by the Li adsorption as a function of the number of Li atoms. From the results, it is expected that the CNT-C_{60} hybrid system has enhanced the charge transport properties in addition to the Li adsorption, compared to both CNT and C_{60}.

See more of this Session: Nanomaterials for Energy Storage I
See more of this Group/Topical: Topical 5: Nanomaterials for Energy Applications

(278g) Catalytic Conversion of Biomass Into Carboxylic Acids Via APPO Process

Hongfei Lin, Chemical and Materials Engineering, University of Nevada, Reno, NV

Converting raw solid biomass into liquid or gas phase intermediates which can be readily upgraded to hydrocarbon fuels is the most challenging or energy intensive step during the biofuel production process due to the complex composition and structure of biomass. We are developing cost competitive catalytic biomass deconstruction strategies using aqueous phase partial oxidation (APPO) for releasing platform intermediates from solid biomass with a focus on the production of organic acid streams that can be readily upgraded and converted to commodity chemicals or infrastructure compatible liquid hydrocarbon fuels. The APPO process breaks down solid biomass into water soluble and stable intermediate organic acid molecules over heterogeneous catalysts under mild operation conditions. Oxidation catalysts have been developed and tested for short durations with promising results using cellulose and lignin as the model feeds. A broad range of biomass species including woody biomass and corn stover have been evaluated to be converted into the water soluble organic acids with the APPO process. The oxidation catalysts have been proven not to be feedstock specific allowing full utilization of the various biomass species. We are working on optimizing the catalytic system to realize high yield of the organic acids at minimal dilution for enhanced process efficiency and to control the composition of the APPO product stream.

See more of this Session: Catalytic Biofuels Refining II
See more of this Group/Topical: Fuels and Petrochemicals Division

(288f) Effective Communications

Gautam Pillay, Department of Chemical Engineering, Rowan University, Glassboro, NJ

Effective communication between management, employees, external stakeholders, and partner organizations (often geographically separated).

See more of this Session: Developing Leadership Skills
See more of this Group/Topical: Management Division

(288h) Dhamma Tools for Managers

Gavin Towler, UOP, Des Plaines, IL

This paper will describe methods for maintaining your composure, motivating your team, dealing with obstacles, getting results and becoming a better leader, while maintaining a harmonious working environment.

See more of this Session: Developing Leadership Skills
See more of this Group/Topical: Management Division

(35a) Transport Properties for Ionic Liquids Used In Next Generation High Capacity Batteries

Vyran George, Chemical and Biomolecular Engineering, Georgia Institute of Technology, Atla nta, GA

Using room temperature ionic liquids (RTILS) as electrolytes in Li-based batteries provide important safety and performance benefits over the more volatile, combustible organic solvents currently used. However, in order to gain a more profound understanding of transport in these electrolytes, properties such as conductivity, transference numbers, and salt activity must be measured as a function of salt concentration. Experimental Methods are proposed for calculating a complete set of transport properties for a room temperature ionic liquid (RTIL) system. These data will be used in conjunction with concentrated solution theory equations to develop a robust performance model for advanced batteries.

See more of this Session: Lithium Battery Technology and Materials
See more of this Group/Topical: Fuels and Petrochemicals Division

(288e) Dealing with Uncertainty In Decision Making

Diana Matonis, Flucomp Enterprises, Hickory Hills, IL

What is the difference between risk and uncertainty?

What are some quantitative methods that aid decision making under uncertainty?

What are some qualitative methods that aid decision making under uncertainty?

How can decision makers effectively combine qualitative and quantitative methods?

See more of this Session: Developing Leadership Skills
See more of this Group/Topical: Management Division

(453b) Reaction Mechanism of Catalytic Oxidative Depolymerization of Cellulosic Biomass

Hongfei Lin, Chemical and Materials Engineering, University of Nevada, Reno, NV

While biomass recalcitrance is one of the biggest challenges in biofuel industry, Nature has the key to unlock this puzzle. When a plant grows, the cell wall undergoes re-structuring. Recent biological studies found that hydroxyl radical (•OH), a very short-lived and strongly oxidative radical, is harnessed by the growing plant cells to loosen their cell walls by nonenzymatically cleaving the wall polysaccharides, although the physiological significance of this loosening mechanism is unclear. This nonenzymatic process sheds light on developing chemical oxidation methods for biomass deconstruction under mild reaction conditions. We are developing cost competitive catalytic biomass deconstruction strategies using aqueous phase partial oxidation (APPO) for releasing platform intermediates from solid biomass with a focus on the production of organic acid streams that can be readily upgraded and converted to commodity chemicals or infrastructure compatible liquid hydrocarbon fuels. The reaction mechanism of the APPO process has been investigated. A free radical route is presented to interpret the pathway of cleaving cellulosic biomass via the APPO process and the functionalities of oxidation catalysts are correlated to the reaction mechanism. By understanding the reaction mechanism, we are able to optimizing the catalytic system to realize high yield of the organic acids and to manipulate the selectivity of the APPO process.

See more of this Session: Fundamentals of Biomass Utilization
See more of this Group/Topical: Environmental Division

(491b) Progress In Biomass Trait Screening Using a High Throughput Digestibility Platform

Nick Santoro[1], Shane Cantu[2] and Jonathan Walton[2], (1)Great Lakes Bioenergy Research Center, Michigan State University, East Lansing, MI, (2)Michigan State University, East Lansing, MI

A key step in the development of a viable lignocellulosic ethanol industry is the improvement of the plants that are to be used as biomass feedstocks. We have previously described a high throughput cell wall digestibility platform (HTDP) to facilitate the identification and evaluation of feedstock plants. Cell wall digestibility is assayed by the amount of free Glc and Xyl released after chemical pretreatment and incubation with a commercial cellulase mixture (Accellerase). One key component of the HTDP is a custom-designed workstation that grinds and dispenses individual plant biomass to within 0.1 mg accuracy. Combining this workstation with a second key component, a laboratory information management system (LIMS), provides the basis for generating a quantitative measure of cell wall digestibility on a vast number of samples.

The results presented in this talk summarize cell wall digestibility screening projects carried out in the two years since the HTDP opened. A collection of approximately 7,000 *Brachypodium* EMS mutant M2 lines have been screened. Classes of outliers are identified and studies are underway to further characterize the mutant lines. A collection of approximately 1,200 *Arabidopsis* T-DNA lines with insertions in cell wall biosynthesis-related genes have been screened for differences in cell wall digestibility. The chemical and genetic characterization of these lines is presented. Corn lines comprising a University of Wisconsin Diversity Set have been screened recently using corn stalk core samples. An initial survey of outliers is presented. Results for transgenic poplars containing knock-down insertions for decreasing the expression of the CCR gene are also presented. The poplar screening studies compare the cell wall digestibility of the transgenic and wild type poplar in response to a variety of pretreatment conditions as well as to geographical field location. Substantial differences in cell wall digestibility of the transgenic and wild type poplar are shown as the above parameters are varied.

In Summary, the HTDP is an extremely useful tool, demonstrating utility as a screening tool to identify improved plants from large sample populations (Brachypodium). In addition, the HTDP is flexible to accommodate novel sampling methods (corn stalk cores). Furthermore, once interesting plants are identified, they can be further tested in the HTDP using a number of varied pretreatment conditions (poplar).

See more of this Session: Advances In Biofuels: DOE Bioenergy Research Centers I
See more of this Group/Topical: Sustainable Engineering Forum

(288c) Strategies for Motivating Employees

Charlene A. Wall-Warren, BASF, Florham Park, NM

Strategies for motivating employees

See more of this Session: Developing Leadership Skills
See more of this Group/Topical: Management Division

(110a) Investigating Fundamental Electrochemical Processes In a Li-O2 Battery

Bryan McCloskey, Science and Technology, IBM Almaden Research Center, San Jose, CA

Li-air batteries have received significant attention as a potential high specific energy alternative to current state-of-the-art rechargeable Li-ion batteries. Nevertheless, published Li-air embodiments have only achieved a small fraction of their enormous theoretical specific energy (11,700 Wh/kg of Li) with limited rechargeability, and many operational aspects, including lithium-oxygen electrochemistry, appear to be poorly understood. This presentation will outline our initial efforts to understand some fundamental properties of non-aqueous Li-O_2 electrochemistry.

Among the many important challenges facing the development of Li-air batteries, our research has focused on processes occurring at the porous carbon cathode. To better understand the electrolyte solvent's role in producing the desired rechargeable cathode reaction (*i.e.*, $Li^+ + O_2 + 2e^- \rightarrow Li_2O_2$), quantitative Differential Electrochemical Mass Spectrometry (DEMS), coupled with *ex-situ* chemical analysis of cell cathodes, was used to study the applicability of carbonates and dimethoxyethane in Li-O_2 cells.

Our studies have revealed that carbonate-based electrolyte solvents, similar to those used in Li-ion batteries, irreversibly decompose upon Li-O_2 cell discharge to form undesired electrodeposits composed mainly of lithium carbonate and lithium alkyl carbonates. Employing dimethoxyethane (DME) as a solvent, however, mainly produces lithium peroxide on discharge, but both Li_2O_2 decomposition (the desired electrochemistry) and DME oxidation (a parasitic side reaction) occur during cell charging. These results clearly indicate that oxidative and reductive electrolyte stability in the presence of Li_2O_2 is critically important to produce a truly rechargeable Li-air battery.

See more of this Session: Novel Battery Chemistry and Technology
See more of this Group/Topical: Fuels and Petrochemicals Division

(35e) Carbon Impact On Lithium-Sulfur Battery Performance

Galen Suppes, Michael J. Gordon, Ramsey Hilton and Benjamin John Roweburrough, Chemical Engineering, University of Missouri Columbia, Columbia, MO

Several carbons were evaluated for supporting sulfur in a lithium sulfur battery. The carbon properties had a significant impact on performance.

See more of this Session: Lithium Battery Technology and Materials
See more of this Group/Topical: Fuels and Petrochemicals Division

(738f) Design of a Self-Supporting Natural Gas Liquefaction Process for Accessing Stranded off-Shore Resources

Achim Wechsung[1], Audun Aspelund[2], Truls Gundersen[2] and Paul I. Barton[3], (1)Department of Chemical Engineering, Massachusetts Institute of Technology, Cambridge, MA, (2)Energy and Process Engineering, Norwegian University of Science and Technology, Trondheim, Norway, (3)Chemical Engineering, Massachusetts Institute of Technology, Cambridge, MA

Natural gas accounts for 24% of world energy consumption today, consumption is projected to grow by 44% until 2035 and natural gas is expected to become increasingly important since it combusts more cleanly and produces power more efficiently than other fossil fuels. On the one hand, close to 60% of the world's natural gas reserves are found in three countries: Russia, Iran, and Qatar. On the other hand, most of the future production growth is predicted to take place in non-OECD countries which may lead to increasingly strained and unreliable supply chains to consumers in developed nations. Stranded reserves account for the majority of the resource base and LNG will play a key role in exploiting these. Lastly, it is anticipated that public policy requiring capture and sequestration of carbon dioxide will become more widespread.

Stranded off-shore gas yields will require new technologies to make production economically viable. Traditional natural gas infrastructure for offshore production necessitates large capital expenditures to connect yields with onshore processing facilities. Ifnatural gas could be liquefied onboard dedicated platforms and then transported with LNG tankers to customers, most of the inflexible infrastructure becomes superfluous. Additionally, since LNG can be traded much more easily, long-term contracts couldbe replaced by marketing the gas on a global gas market.

Currently, liquefying natural gas uses large amounts of energy to provide refrigeration. At the receiving terminal, LNG is re-gasified using seawater or air as a heat source. However, the "cold" of the LNG could be recycled thus improving exergy efficiency of the supply chain. For example, nitrogen and captured carbon dioxide can be liquefied providing the required heat to gasify LNG. The liquefied N2 and CO2 can be loaded aboard the empty LNG tanker which returns to the off-shore site. There, the liquefied gases will provide the required cooling to liquefy natural gas and are vented to the atmosphere (N2) or sequestered in geological formation such as depleted gas and oil fields (CO2).

While N2 and CO2 vaporize at constant temperature, natural gas condenses over a temperature interval. It is possible to modify the composite curves by adding compressors and expanders to track the condensation curve of natural gas effectively. The design goal is to find a flowsheet that is self-contained in the sense that no additional power or cooling is necessary for its operation and uses the least nitrogen.

(737e) Novel O2/H2O-Selective Membrane for Metal/Air Batteries

Wei Liu, Pacific Northwest National Laboratory, Richland, WA

This presentation will describe new membrane materials and structures that can be used to encapsulate the cathode of a metal/air battery for its operation under ambient air conditions. The membrane barrier is needed to allow transport of oxygen from air into the cathode for the electrochemical reaction and to block permeation of water vapor that can degrade the battery. In this work, two kinds of membranes are found effective for selective permeation of oxygen over water vapor: immobilized oil membrane, supported dense polymer thin-film membrane. The immobilized oil membrane is prepared by loading silicone oil into the pores of a thin (~50μm), porous support sheet of suitable pore sizes and surface property for retaining of the oil by capillary force. The supported thin-film membrane was prepared by coating a dense layer of PTFE (polytetrafluoroethylene) material (<10mm thick) on a thin porous metal sheet support (~50μm). Both membranes enable stable operation of a lithium/air battery in ambient air (~20% RH) in a much longer time than the conventional porous diffusion membrane. With attachment of a quality membrane on the cathode, the battery operated in ambient air even shows better performance than the same battery housed in a dry box. Systematic membrane preparation, characterization, and battery testing results will be presented. The membrane has large potential for other metal/air battery systems.

See more of this Session: Energy Storage and Electrochemical Processes II
See more of this Group/Topical: International Congress on Energy 2011

(411d) Reclamation of Spent Drilling Fluids by Cross-Flow Microfiltration

Yanling Wu, Environmental Solutions Applied Research, M-I SWACO, A Schlumberger Company, Houston, TX

Oil- and synthetic-based drilling fluids can become contaminated with significant quantities of low-gravity solids, which cannot be removed by conventional solids control equipment and methods. The accumulation of low-gravity solids degrades the performance of costly oil- and synthetic-based drilling fluids in several ways and often generates large quantities of spent drilling fluids not suitable for drilling activities. As a result of rigorous environmental regulations moving toward zero discharge, drilling wastes are the focus of attention in the oil and gas exploration industry.

In this research, pressure driven liquid-phase microfiltration (MF) was investigated for the first time as a potential solution to the removal of low-gravity solids from spent oil- and synthetic-based drilling fluids, where fine solids are retained by the membrane and clean oil passes through the membrane and is recovered as permeate. The results demonstrated that cross-flow MF is capable of breaking the strong emulsions of drilling fluids and recovering clean base oil in the permeate, independent of drilling fluid formulation and composition while minimizing chemical handling and the variability in chemical-based processes. Understanding the impact of each operating parameter on permeate flux will allow us to realize the full potential of cross-flow MF as a spent drilling fluids treatment technology. In this study several operating parameters such as trans-membrane pressure, back-pulse, cross flow velocity and temperature were investigated to optimize permeate flux, minimize membrane fouling and maximize oil recovery using a membrane with a nominal pore size of 0.1 micron. A membrane-cleaning protocol for the fouled membranes was also developed.

See more of this Session: Solids-Liquid Separation Processes for the Petroleum Industry
See more of this Group/Topical: Separations Division

(51b) MODELING the MECHANICS of Drilling Bit

Jeen-Shang Lin, National Energy Technology Laboratory, U.S. Department of Energy, And Department of Civil and Environmental Engineering, University of Pittsburgh, Pittsburgh, PA, **Yaneng Zhou**, Department of Civil and Environmental Engineering, University of Pittsburgh, Pittsburgh, PA, Isaac K. Gamwo, National Energy Technology Laboratory, US Department of Energy, Pittsburgh, PA, Aslam M. Kabir, U.S. Department of Energy, National Energy Technology Laboratory, Pittsburgh, PA and Jamie L. Brown, U.S Department of Energy, National Energy Technology Laboratory, Morgantown, WV

The mechanics of rock drilling is complex. This complexity hinders the development of modeling tool that can serve as a reliable predicative tool. By reliable tool we mean that the cutting forces and fragmental process can be reasonably estimated. From a modeling perspective, advancing a drilling bit constitutes a sequence of challenging problems: A contact problem first arises as a drilling bit interacts with a target rock. This is followed by the task of determining when and if the rock would fail. If the rock does fail, a modeler is further faced with the problem of how to initiate the fragmentation process. Subsequently, this sequence of problems repeats itself as the drilling progresses. This study adopted a ground up modeling approach by starting first with the development of a single cutter model. Both the finite element method and the discrete element methods were employed. After extensive work on both linear and circular cutting modeling, a preliminary full drilling bit was built. These models are still being tested, and some of the results will be presented.

See more of this Session: Recent Advances In Drilling Under Extreme Conditions
See more of this Group/Topical: Fuels and Petrochemicals Division

(498a) State-of-the-Art Techniques and Challenges in Characterization of New Hydrogen Storage Materials

Andre Levchenko, Setaram Inc, Pleasanton, CA

The new hydrogen storage materials discovered in the last few years present a number of challenges to the research community from a viewpoint of materials characterization. These include both evaluation of the performance in real-world applications of hydrogen storage materials and understanding the underlying fundamental mechanisms controlling their properties. This presentation gives an overview of state-of-the-art characterization techniques for hydrogen storage materials. It covers simultaneous TPD-gas sorption/mass-spectrometry measurements, volumetric PCT isotherm measurements on thin-film and extremely small metal-hydride samples, as well as direct van't Hoff and in-situ calorimetric measurements of enthalpy of formation of hydrides. Challenges in performing measurements on particular types of materials or under demanding conditions are discussed.

See more of this Session: Aspects of Hydrogen Storage Materials
See more of this Group/Topical: International Congress on Energy 2011

AUTHOR INDEX

AUTHOR INDEX

AUTHOR INDEX

AUTHOR INDEX

AUTHOR INDEX

AUTHOR INDEX

AUTHOR INDEX

AUTHOR INDEX